Factor Analysis and Dimension Reduction in R

Factor Analysis and Dimension Reduction in R provides coverage, with worked examples, of a large number of dimension reduction procedures along with model performance metrics to compare them. Factor analysis in the form of principal components analysis (PCA) or principal factor analysis (PFA) is familiar to most social scientists. However, what is less familiar is understanding that factor analysis is a subset of the more general statistical family of dimension reduction methods.

The social scientist's toolkit for factor analysis problems can be expanded to include the range of solutions this book presents. In addition to covering FA and PCA with orthogonal and oblique rotation, this book's coverage includes higher-order factor models, bifactor models, models based on binary and ordinal data, models based on mixed data, generalized low-rank models, cluster analysis with GLRM, models involving supplemental variables or observations, Bayesian factor analysis, regularized factor analysis, testing for unidimensionality, and prediction with factor scores. The second half of the book deals with other procedures for dimension reduction. These include coverage of kernel PCA, factor analysis with multidimensional scaling, locally linear embedding models, Laplacian eigenmaps, diffusion maps, force directed methods, t-distributed stochastic neighbor embedding, independent component analysis (ICA), dimensionality reduction via regression (DRR), non-negative matrix factorization (NNMF), Isomap, Autoencoder, uniform manifold approximation and projection (UMAP) models, neural network models, and longitudinal factor analysis models. In addition, a special chapter covers metrics for comparing model performance.

Features of this book include:

- Numerous worked examples with replicable R code
- Explicit comprehensive coverage of data assumptions
- Adaptation of factor methods to binary, ordinal, and categorical data
- Residual and outlier analysis
- Visualization of factor results
- Final chapters that treat integration of factor analysis with neural network and time series methods

Presented in color with R code and introduction to R and RStudio, this book will be suitable for graduate-level and optional module courses for social scientists, and on quantitative methods and multivariate statistics courses.

G. David Garson is Professor Emeritus in the School of Public and International Affairs, NCSU, specializing in advanced research methodology. His most recent works are *Data Analytics for the Social Sciences: Applications in R* (Routledge, 2022) and *Multilevel Modeling: Applications in STATA, IBM, SPSS, SAS, R, & HLM* (Sage, 2020).

Professor Garson may be contacted at gdavidgarson@gmail.com.

Factor Analysis and Dimension Reduction in R

A Social Scientist's Toolkit

G. David Garson

Routledge
Taylor & Francis Group
LONDON AND NEW YORK

Designed cover image: © Getty

First published 2023
by Routledge
4 Park Square, Milton Park, Abingdon, Oxon OX14 4RN

and by Routledge
605 Third Avenue, New York, NY 10158

Routledge is an imprint of the Taylor & Francis Group, an informa business

British Library Cataloguing-in-Publication Data
A catalogue record for this book is available from the British Library

ISBN: 978-1-032-24668-0 (hbk)
ISBN: 978-1-032-24669-7 (pbk)
ISBN: 978-1-003-27969-3 (ebk)

DOI: 10.4324/9781003279693

Typeset in Bembo
by codeMantra

Access the Support Material: https://www.routledge.com/9781032246697

“This book is dedicated to my beloved wife, Kathryn Kallio, who has supported me so well in this project as in everything in my life.” – Dave Garson, May 2022

Contents

Figures

Preface

Factor analysis in the form of principal components analysis (PCA) or principal factor analysis (PFA, aka common factor analysis) is familiar to most social scientists. However, what is less familiar is the understanding that factor analysis is a subset of the more general statistical family of dimension reduction methods. This book provides coverage, with worked examples, of a large number of dimension reduction procedures. In addition, model performance metrics to compare procedures are detailed, an essential aspect covered little if at all in most texts on factor analysis. Moreover, by taking advantage of a wide variety of R language statistical packages, this book highlights underutilized capabilities of even traditional PCA and PFA procedures. Features of this book include explicit comprehensive coverage of the data assumptions of factor procedures; adaptation of factor methods to binary, ordinal, and categorical data; residual and outlier analysis; and visualization of factor results. Final chapters treat integration of factor analysis with neural network and time series methods.

As indicated in the subtitle of this book, its purpose is to expand the number of tools in the toolkit of the social scientist challenged with dealing with factor-related research questions. These questions, which are outlined in Chapter 1, often go to the heart of what social science is about. Methods in this book are needed whenever the research question involves any of the following four crucial issues:

1 How many dimensions does the outcome of interest have, and does this number of dimensions conform to existing theory?
2 How may a large number of measurements (e.g., 80 survey questions) be reduced to a much smaller number of underlying dimensions (e.g., six constructs) for purposes of more informative modeling? How can the convergent and discriminant validity of scales for these constructs be established?
3 How may subjects, participants, or clients be segmented into meaningful groups?
4 How may groups be compared cross-sectionally or over time based on differences in their factor structure?

In addition, factor methods may be used to predict a dependent variable, impute missing values, identify outliers, analyze residuals, and uncover common method bias.

For the social scientist, the teaching and using of research methods has been a constant process of learning new tools and procedures. As social scientists we need to ride the wave of the paradigm shift associated with data analytics and the R language, not fear the learning curve all new things bring with them. I hope this book can be a small contribution to understanding of one of the most important domains of statistical analysis for social science topics, that of factor analysis and dimension reduction.

G. David Garson
School of Public and International Affairs
North Carolina State University
May, 2022
Email: gdavidgarson@gmail.com

Acknowledgments

I would like to thank the dozens of reviewers, anonymous and otherwise, who provided valuable feedback on the proposal for this work, and for the work itself, though all errors are my own, of course. The extensive R community is something all authors using R, myself included, must acknowledge and praise. The authors of R packages are too numerous to enumerate here but are listed in the "Authors" section of the "Description" file of their respective online documentation (in RStudio, select the "Packages" tab, then click on the package name).

Part I

Multivariate analysis of factors and components

1 Factor analysis

Research questions it addresses

Introduction

Factor analysis is used to uncover the dimensions or latent structure of a set of variables. It reduces the high-dimensional attribute space associated with a larger number of observed variables in the raw data to a low-dimensional space characterized by a smaller number of factors or components. As such it is typically a "non-dependent" procedure which does not assume any particular measured variable is the outcome or dependent variable (DV). However, factor scores may be saved for each observation and may be used as predictors or DVs in a model. In a survey instrument with 80 items, for instance, factor analysis might reduce attribute space to seven factors, one of which is an outcome variable of interest. Whereas a model with 80 nodes might be near-uninterpretable, a model with seven factors may provide what is needed to throw light on some phenomenon of research interest.

The hypothetical 80 survey items are examples of "indicator" variables. Synonyms are observed variables, measured variables, or manifest variables. Conventionally, the factor or component is seen as the "reality" which causes the values of indicators. For instance, "job satisfaction" may be a factor and various survey items pertaining to job satisfaction may reflect the abstract but real construct of job satisfaction. For this reason, the observed variables may be called "reflective indicators" and conceptually the arrows in the graphical representation of the factor model go from job satisfaction to appropriate indicators.

In contrast, in the structural equation modeling (SEM) approach to confirmatory factor analysis (CFA), the model may also be reflective but it is also possible to be "formative", with the arrows going from the indicators to the factor. Formative models imply the indicators constitute the factor. For instance, if the factor is "philanthropy", the indicators might be "Dollars given to religion", "Dollars given to education", "Dollars given to health causes", ..., "Dollars given to other". In formative models, it is important that the indicators be comprehensive (hence the inclusion of an "other" item). Leaving one out changes the meaning of "philanthropy". In comparison, in the usual reflective model, the indicators only have to be representative of the construct, not comprehensive, and in principle leaving one out does not change the meaning of the construct. In modeling terms, a principal component analysis (PCA) model's arrows also go from the observed measures to the component and but they are not viewed as formative. Rather than seeing a PCA component as being constituted by its indicators, components are seen as being determined by a linear variate of the observed measures, which may be representative or comprehensive.

DOI: 10.4324/9781003279693-2

The origins of factor analysis are often attributed to Charles Spearman (1904) but many methodologists have contributed to its evolution. Notable among these was Louis Leon Thurstone (1931, 1935, 1937, 1940, 1947), a pioneer in the field of psychometrics. An important advocate of factor analysis in social science and inventor of the screen test was the British-American psychologist Raymond Cattell (1952, 1978). Also in psychology, the study of factor analysis has been closely tied to the study of human cognitive abilities. Critical contributions were made in this area by John Carroll (1989, 1993).

Factor analysis is typically used for exploratory purposes. Exploratory factor analysis (EFA) comes in various flavors:

1 PCA is used primarily for dimension reduction with dimensions called "components". There is a version used with categorical data called categorical PCA (CATPCA) as well as versions for binary or mixed-level data.
2 Principal or common factor analysis (PFA) is used primarily for causal interpretation, with dimensions called "factors". Sometimes PFA is labeled simply "factor analysis" but in this volume, we use "factor analysis" as an umbrella term for all types. SEM uses a type of PFA and in the context of SEM, factors are called "latent variables".
3 More broadly, any procedure which maps data from observed high-dimensional space to low-dimensional factor or component space is a form of dimension reduction. This mapping process is called "embedding", a term which shows up in the name of some procedures (e.g., "locally linear embedding" or "stochastic neighbor embedding").

In contrast to EFA, which does not require the researcher to posit the number or labeling of factors beforehand, CFA does. CFA is most commonly a subset of SEM and, while highlights are outlined in this volume, CFA is treated much more fully in Garson (2018), *Structural Equation Modeling*, among other SEM texts.

The R language has a very broad array of tools for factor analysis and dimension reduction, many of which are explained in this book. A basic introduction to R and RStudio is presented in Appendix 2 of this book. In addition, the applications in later chapters of this book provide step-by-step hands-on explanations for the R code underlying each example and even each figure in the book.

Purposes of factor analysis

A search of peer-reviewed articles using Summon (searches multiple bibliographic databases) uncovered 553,597 hits for "factor analysis" since 1/1/2000 through 1/1/2022; 496,923 for "PCA"; 178,369 for "PFA or PAF or common factor analysis"; and 46,872 for "dimension reduction". While not all of these search hits are relevant, nonetheless it may be concluded that topics covered in this book are very widely used in academic research.

Factor analysis could be used for any of the following purposes:

- To reduce a large number of variables to a smaller number of dimensions for modeling purposes. A large number of observed variables may make modeling unduly complex. Moreover, by subsuming several observed variables (indicators) into a factor score, greater stability may be achieved than by using any single measure individually.

- Reducing the dimensionality of the data may decrease redundancy and noise in the data.
- To validate a scale or index by demonstrating that its constituent items load on the same dimension, and to drop proposed scale items which cross-load on more than one dimension. A good scale is not only reliable (e.g., by Cronbach's alpha; Cronbach, 1951) but its unidimensionality may also be established by factor analysis.
- To select a subset of variables from a larger set, based on which original variables have the highest correlations with a factor or component.
- To establish that multiple tests measure the same dimension, thereby giving justification for administering fewer tests. Factor analysis originated a century ago with Charles Spearman's attempts (Spearman, 1904) to show that a wide variety of mental tests could be explained by a single underlying intelligence or IQ factor (a notion now rejected, by the way).
- To create a set of factors to be treated as uncorrelated variables as one approach to handling multicollinearity in such procedures as multiple regression. Components and factors are uncorrelated when orthogonal rotation is used. Orthogonal rotation is typically the default in many factor analysis packages.
- To identify clusters of cases and/or to identify outliers.
- To determine network groups by determining which sets of people cluster together (factor analysis can cluster variables, observations, or both)

Limitations of factor analysis

A non-technical analogy: A mother sees various bumps and shapes under a blanket at the bottom of a bed. When one shape moves toward the top of the bed, all the other bumps and shapes move toward the top also, so the mother concludes that what is under the blanket is a single thing – her child. Similarly, factor analysis takes as input a number of measures and tests, analogous to the bumps and shapes. Those that move together are considered a single thing, which it labels a factor. That is, in factor analysis the researcher is assuming that there is a "child" out there in the form of an underlying factor, and he or she takes simultaneous movement (correlation) as evidence of its existence. If correlation is spurious for some reason, this inference will be mistaken, of course. Therefore it is important when conducting factor analysis that possible variables which might introduce spuriousness, such as common anteceding causes or common method bias, be taken into account.

Like all statistical procedures, factor analysis is not a panacea. In particular, the interpretability and usefulness of factor analysis results is diminished if sample size is small, factor loadings are low, factors are highly correlated, and/or the number of factors is large (Lim & Jahng, 2019).

Most PCA or PFA implementations fall under the general linear model (GLM) family of procedures. Therefore they make many of the same assumptions as multiple regression: linear relationships, interval or near-interval data, untruncated variables, proper specification (relevant variables included, extraneous ones excluded), lack of high multicollinearity, and multivariate normality for purposes of significance testing. However, other dimension reduction procedures do not require continuous-normal data.

CATPCA, for instance, is designed for categorical data. As another example, functional PCA (fPCA) is a nonparametric method (normality is not assumed).

When used to cluster variables, factor analysis generates a table in which the rows are the observed indicator variables and the columns are the factors which explain as much of the variance in these variables as possible. The cells in this table are factor loadings. The meaning of the factors must be induced by seeing which variables are most heavily loaded on which factors. This inferential labeling process can be fraught with subjectivity as diverse researchers may impute different labels for the same factor. Sometimes researchers forego labeling factors altogether for this reason. Chapter 2 expands on the assumptions and limitations of factor analysis.

Common research questions associated with factor analysis

Question A: May I use factor analysis on sub-interval data?

While factor analysis as traditionally practiced falls within the GLM family of statistics and therefore assumes continuous measurement, variants of factor analysis can handle sub-interval data and nonlinear relationships. For instance, variants exist for binary, ordinal, and mixed data. Kernel PCA (kPCA) is an example of a nonlinear analog to ordinary PCA, which is linear. In this book, factor analysis with sub-interval data is treated in Chapters 8, 11, and 13.

Question B: How many dimensions are there in my data?

Factor analysis and related dimension reduction procedures address important theoretical questions, such as the question of how many underlying dimensions there are in the data. For instance, one scholar may have written that each nation of the world falls into one of five types of financial administration regimes. Using a collection of measured variables related to financial administration, factor analysis can determine if this five-dimensional theory is supported by the data. If it is not, factor analysis suggests how many dimensions are compatible with the data. In some cases, even a one-dimensional solution may suffice to account for the majority of variation in the data.

Question C: What are the best measures for my construct and how should I weigh them?

Imagine that a researcher wished to measure "favorability" toward a presidential candidate. A survey has been taken with many items, such as questions about favorability toward the candidate's stand on abortion, foreign policy, social security, or other issues; as well as personality items like agreement that the candidate projects a good presidential image or is of "good moral character"; or other items such as "I identify as a member of the candidate's party" or "If the candidate becomes president, I expect to be better off." One approach would be to create a favorability score based on all items weighted equally. However, the researcher instead could consider "favorability" to be best represented by the items which load most heavily on the first factor, which always explains the greatest share of the variance in the entire set of factors thought by the researcher a priori to represent favorability. Items could be weighted by the relative size of their loadings on this first factor. Or factor scores on Factor 1 might be used as a measure of candidate favorability.

Question D: How do people in my sample cluster?

"Biplots", discussed later in this book, automatically visualize the clustering of observations (subjects) on the factors formed by the variables in the researcher's data. Variables are also clustered by factor. However, the research design may be focused on the cluster structure of the subjects specifically. Consider the 435 members of the U. S. House of Representatives to be the subjects and let 200 roll-call votes on environmental issues be the variables. In conventional factor analysis (observations as rows, variables as columns), factor analysis would reveal if being "pro-environment" is a single dimension or if there are multiple types of (dimensions of) being pro-environment. Representatives would have a factor score on each dimension. These scores could be used in cluster analysis to establish groupings of Representatives for each type of being pro-environment. Alternatively, rows can be variables (votes) and columns can be Representatives. Factor analysis would then address the research question, "For 200 environmental votes as a set, what is the factor (voting bloc) structure of the U. S. House of Representatives?"

Question E: How do I use factor analysis in R to compare groups?

The researcher may wish to compare groups, such as men versus women subjects, using factor analysis. Using survey data on 773 respondents from the United States, the steps below show one among several ways to compare male and female groups in R. Multigroup analysis using CFA (discussed in Chapter 5 and in Garson, 2018) provides a more sophisticated but more complex approach to using factor analysis in R to compare groups.[1]

A note on the use of package prefixes below and throughout this book: A package prefix is simply the name of the package from which a command comes followed by two colons, then followed by the command. Below, "base::" is the prefix and setwd() is the command. For instructional reasons, we have included package prefixes throughout. However, if the package in question is 'base' or one of the other preloaded packages in the R System Library such as base, datasets, graphics, grDevices, stats, or utils, it is not necessary to include the package prefix. For that matter, the package prefix may be omitted if the command in question is part or one and only one loaded package. However, those learning R may find Including package prefixes useful in understanding where each command comes from.

```
# Read in the data
base::setwd("C:/Data")
music <- utils::read.csv("musicvars.csv", header=TRUE, sep = ",",
stringsAsFactors=TRUE)
# Reduce music to complete cases version, dropping cases with miss-
ing values.
music <- music[stats::complete.cases(music),]
# Create vector with 10 music genres (use names(music2)to get column
numbers)
# Column 18 is "sex", coded 1=male and 2=female.
# After limiting" music" to these variables, sex is column 11.
musicColumns <- base::c(2,3,5,6,8,10,11,13,14,15,18)
music <- music[,musicColumns]
```

```
# Verify that music has 773 cases.
base::nrow(music)
[Output:]
[1] 773

# Create male and female subset data frames
malemusic <- base::subset(music, sex == 1)
# Verify that malemusic has 384 observations
base::nrow(malemusic)
[Output:]
 [1] 384
femalemusic <- base::subset(music, sex == 2)
# Verify that femalemusic has 389 observations
base::nrow(femalemusic)
[Output:]
[1] 389

# Obtain eigenvalues for the male set, based on the correlation ma-
trix excluding sex (col.  11)
maleEigen <- base::eigen(stats::cor(malemusic[,-11]))
maleEigen$values
[Output:]
[1] 3.0570210 1.5452578 1.4870934 0.9212345 0.6917717
[6] 0.5831385 0.5447392 0.4292571 0.3965653 0.3439216

# Run a common factor analysis on the males. Ask for three factors
based on eigenvalues >= 1.
maleResults <- psych::fa(malemusic[,-11], nfactor=3,
rotate="varimax")
maleResults
[Partial output:]
Standardized loadings (pattern matrix) based upon correlation matrix
           MR1   MR3   MR2  h2   u2 com
bigband   0.58  0.13  0.32 0.46 0.54 1.7
blues     0.26  0.61  0.25 0.50 0.50 1.7
blugrass  0.13  0.06  0.63 0.41 0.59 1.1
classicl  0.78  0.07 -0.14 0.63 0.37 1.1
country  -0.06 -0.06  0.69 0.48 0.52 1.0
hvymetal -0.11  0.39 -0.13 0.18 0.82 1.4
jazz      0.41  0.55  0.02 0.48 0.52 1.8
musicals  0.72  0.04  0.08 0.53 0.47 1.0
opera     0.71  0.08  0.04 0.51 0.49 1.0
rap       0.05  0.49  0.02 0.24 0.76 1.0
…
Cumulative Var        0.22 0.33 0.44

# Obtain eigenvalues for the female set.
femaleEigen <- base::eigen(stats::cor(femalemusic[,-11]))
femaleEigen$values
```

```
[Output:]
[1] 2.9746369 1.4788794 1.1803504 1.1247983 0.8097635
[6] 0.6427335 0.5496842 0.4608008 0.4079954 0.3703576

# Run a common factor analysis on the females. Ask for four factors
based on eigenvalues >= 1.
femaleResults <- psych::fa(femalemusic[,-11], nfactor=4,
rotate="varimax")

femaleResults
[Partial output:]
Standardized loadings (pattern matrix) based upon correlation matrix
           MR1   MR2   MR3   MR4   h2   u2  com
bigband   0.56  0.20  0.33 -0.26 0.54 0.46 2.4
blues     0.14  0.91  0.12  0.02 0.87 0.13 1.1
blugrass  0.29  0.15  0.68  0.01 0.57 0.43 1.5
classicl  0.70  0.17 -0.01  0.01 0.52 0.48 1.1
country  -0.03 -0.08  0.55 -0.08 0.31 0.69 1.1
hvymetal -0.02  0.02 -0.02  0.44 0.19 0.81 1.0
jazz      0.23  0.57 -0.07  0.08 0.39 0.61 1.4
musicals  0.67  0.15  0.11 -0.18 0.51 0.49 1.3
opera     0.70  0.08  0.06  0.15 0.52 0.48 1.1
rap      -0.01  0.04 -0.06  0.48 0.23 0.77 1.0
...
Cumulative Var        0.19 0.32 0.41 0.47
```

For the sample "music" data, two of the primary comparisons between the male and female subsets are:

1 The two genders do not have the same factor structure on the music items. Four factors are required to explain 47% of the variance in the data for females, while three factors are required to explain 44% of the variance for males.
2 For females heavy metal and rap are most associated with their own separate factor, while for males both of these music genres are most associated with a factor with which jazz and blues are also most associated.

While neither the female nor the male factor models may be considered strong (neither explains 50% or more of the variance in the 12 measured music indicator variables), there are clear differences between groups which may be of interest to the researcher.

Question F: How do I know if my factors are really subfactors of a more comprehensive construct?

When predictor constructs are based on factor scores, the assumption in orthogonal models is that each factor is independent of all the others and not a subfactor or any of them. In oblique models, the assumption is that the factors are correlated and the constructs they represent overlap according to the degree of inter-factor correlation. But there is a third possibility, which is that one or more factors may be a subfactor of a hierarchically higher factor. To explore a hierarchical factor model is the role of second-order (or higher order) factor modeling, described in Chapter 7 of this book.

To simplify, raw measures are subjected to an orthogonal factor analysis, generating factor scores for some optimal number of factors. These factor scores may be used as the input for a second-level oblique factor analysis to determine if first-level factors are nested within one or more second-order factors. Depending on the data, it may even be possible to generate one or more third-order factors, with level 1 factors nested within level 2 factors which are nested within level 3 factors.

Question G: How may I use factor analysis to predict a dependent variable?

Factor analysis is often labeled a "non-dependent" procedure. This means that none of the input variables are treated as a DV. From this, the naïve user might wrongly infer that if predicting a DV is central to his or her research question, then some procedure other than factor analysis must be used. However, factor analysis may be performed on a set of predictors, associating these predictor items with the extracted factors, with factor scores computed for each observation/subject. This leads to three possible prediction strategies, all of which assume that the dataset also contains a DV of interest which here we will hypothetically call "TestOutcome":

1. Factors and factor scores may be computed without TestOutcome, which is treated as a supplemental variable not used in creating the factors. After factor analysis, TestOutcome may be predicted from factor scores a cross-validation design. The predictive model may have greater stability because causes reflect constellations of causes (the factors) and do not rely on noise-prone single raw items. This is discussed in Chapter 10.
2. The "psych" package supports the `principal()` function for PCA and the `fa()` function for PFA (common factor analysis). `Its predict.psych()` function "finds predicted factor/component scores from a factor analysis or components analysis of data set A predicted to data set B." The usage is:

   ```
   base::library(psych)
   stats::predict(object, data, old.data, ...)
   ```

   For information, type:

   ```
   base::library(psych)
   utils::help(predict.psych)
   ```

   This help page contains examples.

3. It is possible to enter TestOutcome as an active rather than supplemental variable when factors are calculated. The researcher may look at the factor on which TestOutcome is most highly loaded to see what other variables also load highly on that factor. In this way, the TestOutcome is located within a cluster of related variables. This knowledge may help the researcher formulate causal theory related to TestOutcome.

Question H: Can factor analysis help me understand the effect of outliers on my results?

Outlying observations may be defined at various levels of stringency (e.g., beyond two versus beyond three standard deviations from a variable's mean). Defining outlierness in this way, in terms of an observation's value on a variable in relation to the mean of that variable creates univariate measures of outlierness. That is, outlierness is computed one variable at a time and a given observation may be an outlier on one variable but not on another, though outlierness on the DV may be of particular interest. In R, as illustrated in Chapters 2 and 6 for example, boxplots may be used to identify and visualize univariate outliers.

Often, however, the researcher may well prefer a multivariate measure of outlierness, based on all the predictor variables in the model. Biplots provide a multivariate approach to observation-level outliers. Each observation may be plotted in factor space, as in Chapter 6, and an ellipse (e.g., a 95% ellipse as in Figure 6.7) may be drawn such that observations outside the ellipse are defined as multivariate outliers. Variables may also be outliers in factor analysis. Each variable has an observed correlation with each extracted factor. Using factor scores it is possible to create an expected (predicted) correlation matrix, This expected correlation matrix may be subtracted from the observed matrix, yielding variable-level factor residuals. This is simply the traditional O minus E definition of residuals, here applied at the variable level. This is discussed in Chapter 7 and elsewhere in this book.

Question I: How may I represent my factors spatially?

Some R factor analysis packages support "supplemental variables" (cf. Chapter 10) which are not used to compute factors but which nonetheless remain in the dataset. When the data are saved along with factor scores for observations, if there are latitude-longitude or address supplemental variables, these may be saved also. When the dataset, including both factor and supplemental spatial information, is imported into a geographic information system (GIS) package such as ArcGIS, the researcher may create choropleth, dot-density, contour, or other maps utilizing the factor scores. There are GIS packages for R[2] as well as a few experimental spatial factor analysis packages in R, such as "SpatialFA" from James-Thorson-NOAA.[3]

Question J: How can factor analysis be used to tell if I have common method bias?

A research design has common method bias if variations in responses reflect the instrument or methods of measurement rather than the true construct that the instrument is intended to uncover. Harman's single-factor test attempts to address this issue using factor analysis. In this test, all measures are entered into an EFA (e.g., see Greene & Organ, 1973; Aulakh & Gencturk, 2000). The researcher then examines the unrotated factor output to determine the number of factors needed to account for the variance in the data (usually this is the number of factors with eigenvalues of 1.0 or higher). In a single factor accounts for most of the variance, this flags a possible common method bias problem. Some researchers (e.g., Iverson & Maguire, 2000) have sought to create a more sophisticated version of the Harman technique by using confirmatory rather than EFA (CFA is covered in Chapter 5).

A somewhat similar logic applies to the general methods factor test. If there is common method bias, this line of reasoning holds, then it should be present in all extracted factors in a factor analysis. Therefore one could construct a two-level factor model (see Chapter 7) and see if there is a second-order factor which explains a substantial amount of the variance in the data (e.g., over 25%). If there is, this is evidence of possible common method bias. However, the second-order factor might represent something other than method bias (e.g., a second-order factor might represent general intelligence, not common method bias, in a study in which the first-order factors were domain-specific (e.g., verbal or math intelligence).

The Harman and related tests, however, have been sharply criticized by Podsakof et al. (2003: 889), who noted, "common method variance would have to completely account for the covariances among the items for it to be regarded as a problem [under the Harman test] in a particular study. Clearly, this assumption is unwarranted." Controlling for common method variance is a complex topic on which researchers have not reached consensus, though Podsakof et al. (2003: 898) present a series of recommendations. The Harman and general factor tests may flag common methods bias, but they may also generate false positives.

The reader should be aware that because of the complexities in identifying and controlling for common methods bias, research journals often disparage the use of post-hoc statistical diagnosis of common methods bias. Instead, the researcher is thought to be better served by seeking to reduce common methods bias at the measurement and instrument design stage, prior to data collection. This topic goes beyond the scope of this book, but efforts to mitigate possible methods bias include use of multi-method, multi-trait (MTMM) designs, using alternative forms with different response formats, alternating positive and negative wording of items, physically separating in the instrument items for the same construct, and many other measures. A web search for "reducing common method bias in survey research" or the like will reveal many lists of such tips. See also Jordan and Troth (2019).

Notes

1 Running factor analysis on two groups means each analysis is based on a fraction of the total cases in the data, which may make findings of non-significance more common. In contrast, multigroup CFA is based on the total of cases across groups. There are many other differences between the two approaches. The approach illustrated in this section should be regarded as exploratory.

2 Zev Ross has posted a list of GIS packages for R: https://www.zevross.com/blog/2019/05/01/unscientific-list-of-popular-r-packages-for-spatial-analysis/.

3
```
install.packages("remotes")
remotes::install_github("James-Thorson/spatial_factor_analysis")
```

2 Assumptions and limitations of factor analysis

Introduction

All statistical methods have assumptions and limitations, and factor analysis and dimension reduction procedures are no exception. In this chapter, we seek to treat the most important considerations. Additional discussion of assumptions and limitations also occurs in later chapters with regard to specific procedures (e.g., kernel PCA discussed in Chapter 11 is a nonlinear approach which does not have the same assumptions about linearity as does ordinary principal components analysis). Attending to the assumptions and limitations of factor analysis is not "fine print" which may be safely ignored. Rather, the content in this chapter is critical to the proper use of factor analysis and avoiding common types of misuse.

Existence of underlying dimensions

Factor analysis assumes that underlying dimensions are shared by clusters of variables. If this assumption is not met, the "garbage in, garbage out" (GIGO) principle applies. Factor analysis cannot create valid dimensions (factors) if none exist in the input data. It can, however, generate invalid factors even from random data. In such cases, factors generated by the factor analysis algorithm will not be meaningful. Likewise, the inclusion of multiple definitionally similar variables representing essentially the same thing will lead to results which do not advance theory.

The usual way of testing for factorability is to use the Kaiser-Meyers-Olkin (KMO) test, which provides a measure of sampling adequacy (MSA) for any data frame. MSA is shown below for the data frame "USArrests", which is preloaded in R's "datasets" library. The KMO() function is found in the "psych" package.

```
base::library(psych)
psych::KMO(USArrests)
[Output:]
Kaiser-Meyer-Olkin factor adequacy
Call: psych::KMO(r = USArrests)
Overall MSA =  0.65
MSA for each item =
  Murder  Assault UrbanPop     Rape
    0.62     0.64     0.50     0.78
```

DOI: 10.4324/9781003279693-3

The usual cutoff for judging data to be suitable for factor/component analysis is that the overall MSA should be 0.60 or higher. For the USArrests data, it is 0.65 which meets the KMO criterion. If MSA were lower, one strategy would have been to drop the variable with the lowest MSA (here, UrbanPop), then compute KMO again, and continue one variable at a time until overall MSA rises to 0.60 or higher. Another rule of thumb is to drop variables with MSA < 0.60.

Another screening method for exploring whether underlying dimensions may be present is to see if the correlation matrix has several coefficients of 0.30 or higher (Hair et al., 2019; Tabachnick & Fidell, 2019). This can be done by creating a correlogram with the `corrplot()` command from the "corrplot" package or by creating a correlation network plot using the `network_plot()` command from the "corrr" package.

In R, there are several alternative ways to plot or graph correlations, including `corPlot()` from the "psych" package. We use the "USArrests" data as an example since, with only four variables (fewer than recommended for actual factor analytic research), output is easier to understand. With a large data frame, the methods below become even more useful. "USArrests" is part of R's built-in "datasets" package.

Setup for the correlogram table in Figure 2.1:

First, we create the correlation matrix. This does not filter out correlations < 0.30 but correlations near 0 are color-coded white and do not show.

```
# Step 1: Create a correlation matrix object from USArrests.
corrmatrix <- stats::cor(USArrests)

# Step 2: Run corrplot() on the correlation matrix just created.
base::library(corrplot)
corrplot::corrplot(corrmatrix,
method="number",
diag=FALSE,
order="AOE",
title="Correlogram for USArrests",
mar=base::c(0,0,1,0))
```

Alternatively, we may create a correlation network diagram with the same information but in a more visual format by using the `network_plot()` command from the "corrr" package, resulting in Figure 2.2.

```
# Step 1:  Load the required library, "corrr", which must be in-
stalled first.
base::library(corrr)

# Step 2: Create a correlation dataframe from corrmatrix.
corrmatrix <- stats::cor(USArrests)
cordf <- corrr::as_cordf(corrmatrix, diagonal=1)

# Step 3: Create a network correlation plot for correlations >=
0.30.
g <- corrr::network_plot(cordf, min_cor = 0.30,
colors=base::c("darkgreen", "white", "royalblue2"), curved=TRUE)
```

```
# Step 4: Display the graph.
# NOTE: if arrows don't show in the RStudio Viewer, go to Plots >
Zoom
base::plot(g)
```

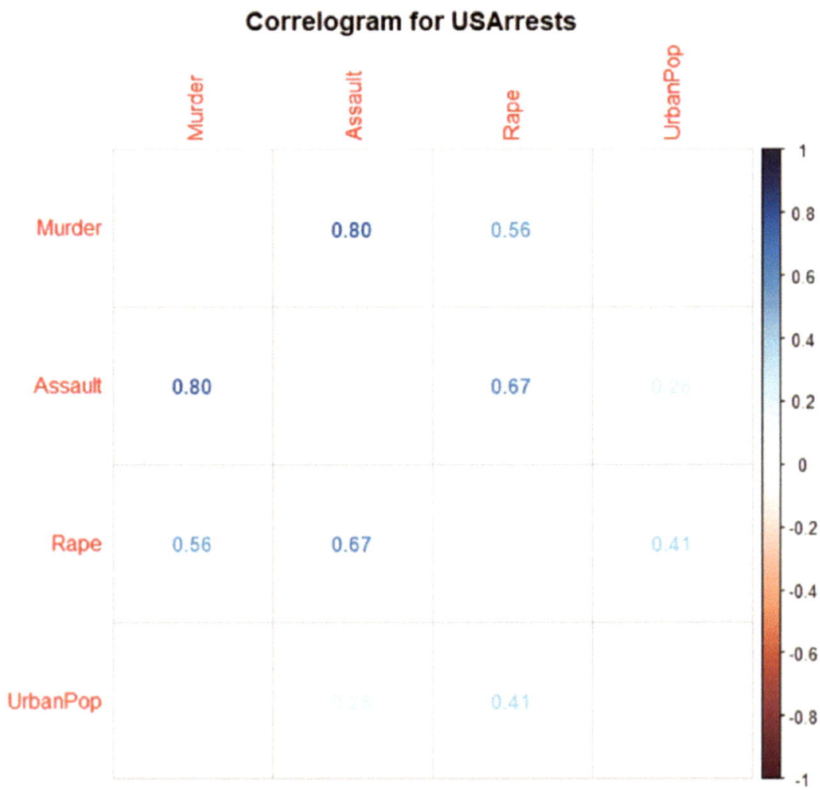

Figure 2.1 Correlogram table for USArrests

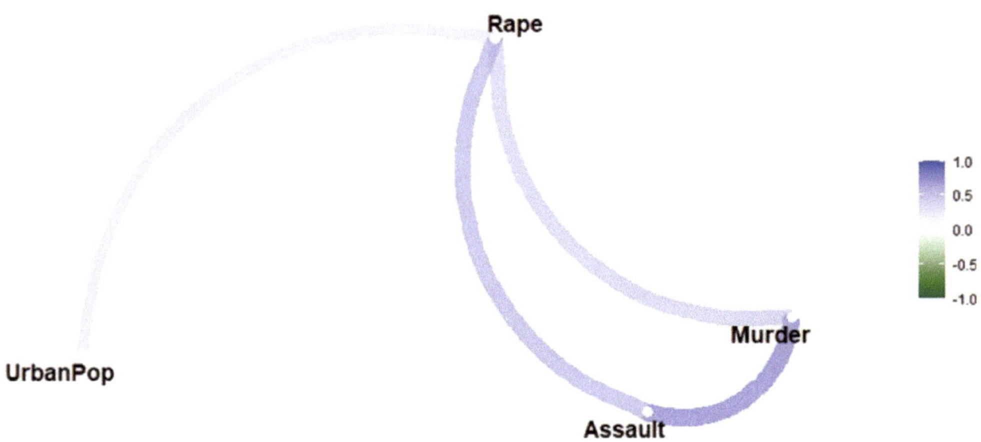

Figure 2.2 Correlation network diagram for USArrests

Figure 2.2 reflects filtering to eliminate correlations below 3.0. The darker blue the connecting correlation path, the stronger the correlation. Based on bivariate correlations, we see that Murder-Assault-Rape is likely to form a single factor.

Proper specification/no selection bias

The exclusion of relevant variables and the inclusion of irrelevant variables in the correlation matrix being factored will affect, often substantially, the factors which are uncovered. Although social scientists may be attracted to factor analysis as a way of exploring data whose structure is unknown, knowing the factorial structure in advance helps select the variables to be included and yields the best analysis of factors. Also, if one deletes variables arbitrarily in order to have a "cleaner" factorial solution, erroneous conclusions about the factor structure will result. See Kim and Mueller (1978a: 67–68).

For exploratory factory analysis, Thurstone (1947) recommended at least three measured variables per factor (Kim & Mueller, 1978b: 77). A greater number may give more stability to the factor. However, "the more, the better" may not be true when there is a possibility of suboptimal factor solutions ("bloated factors"). Cattell (1978) called these "bloated specifics." Too many too similar items will mask true underlying factors, leading to suboptimal solutions. For instance, items like "I like my office", "My office is nice", "I like working in my office", etc., may create an "office" factor when the researcher is trying to investigate the broader construct of "job satisfaction". To avoid suboptimization, the researcher should start with a small set of the most defensible (highest face validity) items which represent the range of the factor (e.g., ones dealing with work environment, coworkers, and remuneration in a study of job satisfaction). Assuming these load on the same job satisfaction factor, the researcher then should add one additional variable at a time, adding only items which continue to load on the job satisfaction factor, and noting when the factor begins to break down. This stepwise strategy results in the most defensible final factors.

For confirmatory factor analysis, there is no specific limit on the number of variables to input. Using confirmatory factor analysis in structural equation modeling, having several or even a score of indicator variables for each factor will tend to yield a model with more reliability, greater validity, higher generalizability, and stronger tests of competing models, than will CFA with two or three indicators per factor, all other things equal. However, the researcher must take account of the statistical artifact that models with fewer variables will yield apparent better fit as measured by SEM goodness of fit coefficients, all other things equal.

Proper specification of the number of factors

Proper specification of the number of factors to be extracted is crucial to analysis. As discussed in greater detail in the section on extraction methods in Chapter 3, having too many factors suboptimizes on overly narrow factors, conflating common factors and obscuring the "big picture", while having too many factors may create factors on which only one or two indicators load, increasing the chances that the model will not replicate well. Tabachnick and Fidell (2013: 617) write, "Failure to measure some important factor may distort the apparent relationships among measured factors". For this reason, the researcher may explore a range of models with varying numbers of extracted factors rather than just set that number solely on a data-driven basis such as the Kaiser criterion of eigenvalues of 1.0 or greater.

While selecting the number of factors to extract is best done based on the theoretical utility of a solution with the chosen number of factors, often after exploration of a range of models, there are a number of numerical and graphical methods, including:

- Kaiser criterion based on eigenvalues of 1.0 or greater, or variants such as percent explained, mean eigenvalue, or the Joliffe criterion.
- Scree plots, also using eigenvalues in a graphic format.
- Empirical scree tests, based on standard error, slope acceleration factor, optimal coordinates, or the Kaiser rule.
- Parallel analysis, based on eigenvalues greater than random.
- Minimizing reproduced correlation residuals.

To avoid redundancy, these methods of selecting an appropriate number of factors to extract are discussed and illustrated in later chapters of this book rather than here.

An outlier variable (as opposed to outlying observations, discussed in a later subsection) may be defined as one which displays a low correlation with all factors important enough to be extracted and which has a low squared multiple correlation (SMC) with all other measured variables. Such variables often load on one of the factors extracted later in the factor analysis process. Often they are the only variable to load on one of these weak factors ("singlets") though sometimes a second variable also loads ("doublets"). This type of factor is usually unreliable. To explore further, the researcher may add additional measures also thought to represent the weak factor ("friends") in an effort to make it stronger. Usually, however, weak factors based on outlier variables are not extracted and do not enter into the analysis.

Data homogenous on factor structure

If more than one sample is involved, these should not be pooled unless previously the researcher first has shown that each sample has a similar factor structure. "Similar" means that each sample optimizes on the same number of factors and that indicator variables load on the same factors in a similar manner. Merging heterogeneous samples obscures relationships in the data and is rightly disparaged in factor analysis as it is for other statistical procedures.

Valid imputation of factor labels

Factor analysis is notorious for the subjectivity involved in imputing factor labels from factor loadings, so much so that some researchers eschew labeling factors at all. For the same set of factor loadings, one researcher may label a factor "work satisfaction" and another may label the same factor "personal efficacy", for instance. Factor interpretations and labels must have face validity and/or be rooted in theory. One recommended practice is to have a panel not otherwise part of the research project assign items to factor labels. A rule of thumb is that at least 80% of the assignments should agree. Alternatively, the researcher may wish to involve a panel of subject matter experts in the imputation process through an iterative Delphi consensus-building process (see Garson, 2014). Ultimately, however, there is no "correct" solution to the labeling problem.

Data level

Continuous-normal data are assumed for standard PCA and PFA. If the data consist of ordinal items, then categorical principal component analysis (CATPCA) may be used. There are also PCA variants for binary and mixed-level data. However, Kim and Mueller (1978b: 74–75) note that ordinal data may be used in ordinary factor analysis if it is thought that the assignment of ordinal categories to the data does not seriously distort the underlying metric scaling. Likewise, these authors allow use of dichotomous data if the underlying metric correlations between the variables are thought to be no more than moderate (0.7). Nonetheless, using ordinal data for an underlying continuous variable is a form of measurement error which attenuates correlation, making resulting factor loadings harder to interpret. Put another way, factor loadings might well be higher and thus more interpretable were there no measurement error.

Attenuation is low for ten- and seven-point Likert items. Based on simulation studies, Lozano et al. (2008) found use of seven-point scales or higher was optimal for factor validity. Use of five-point items is widespread, however, and is supported by some methodologists (DiStefano, 2002; Mueller & Hancock, 2019). Use of four-point items is regarded as minimal (Lozano et al., 2008). Use of three-point Likert items is disparaged because attenuation is greatest. For dichotomies, Rummel (1970) suggested discarding variables with a 9:1 split or worse. Note that if several ordinal items are combined into a scale, it is widely accepted practice to treat the scale variable as continuous.

It also must be considered that exploratory factor analysis is used for exploratory purposes, not usually for confirmatory purposes. In exploratory research, the value of including ordinal variables may outweigh the cost of some attenuation of correlation. To illustrate the effects of attenuation, we use the "USJudgeRatingss" data frame, which is in R's built-in "datasets" package. We show below how the correlation of INTG (Judicial Integrity rating) and RTEN (Worth Retaining rating) is increasingly attenuated as binning becomes more severe. Only in unusual circumstances will binning increase correlation. When that happens it may be because binning has pulled in outliers, which also diminishes correlation.

```
utils::data(USJudgeRatings)
dat <- USJudgeRatings        # Copy to shorter name for convenience
base::options(digits = 3)
base::library(sjmisc)        # Used for binning (must be installed
first)
x <- dat$INTG                # Copy INTG to shorter name for convenience
y <- dat$RTEN                # Copy RTEN to shorter name for convenience

# Initial correlation with continuous, unbinned variables.
stats::cor(x,y)
[Output:]
   [1] 0.937

# Correlation with  x (INTG)  in 8 bins.
# split_var() bins a variable into a specified number of groups of
equal size (here, 8).
INTG8  <- sjmisc::split_var(x, n=8, as.num=TRUE)
```

```
# Correlation of x (INTG)  binned into 8 levels with y (RTEN):
stats::cor(INTG8,y)
[Output:]
   [1] 0.879

# Correlation for 6, 4, and 2 bins (code not shown, but same as
above).
# Also shows run when both variables are dichotomized.
Correlation with 6 bins: [1] 0.873
Correlation with 4 bins: [1] 0.852
Correlation with 2 bins: [1] 0.716
Correlation when both variables are dichotomized: [1] 0.672
```

Anything that undermines correlation, which assumes continuous-normal data, also undermines usual forms of factor analysis (Carroll, 1985; Warner, 2007). Texts on factor analysis routinely stipulate that variables be continuous at data level (Walsh, 1996; Puth, Neuhäuser, & Ruxton, 2015; Bandalos, 2018). Correlation is attenuated when data are binned, restricting range. Outliers and missing data may also undermine correlation, on which factor analysis is based. Chapter 8 discusses forms of factor analysis adapted for ordinal, binary, and mixed data levels.

Categorical variables with similar splits will necessarily tend to correlate with each other, regardless of their content (see Gorsuch, 1983). This is particularly apt to occur when dichotomies are used. The correlation will reflect similarity of "difficulty" for items in a testing context; therefore such correlated variables are called *difficulty factors*. The researcher should examine the factor loadings of categorical variables with care to assess whether common loading reflects a difficulty factor or substantive correlation. Improper use of dichotomies can result in too many factors.

Dichotomous data may pose problems. Shapiro, Lasarev and McCauley (2002) used simulation methods to study biases of factor analysis in a dataset of dichotomous variables. "Our work has shown", they concluded,

> that application of standard rules to 19 randomly generated and independently created dichotomous variables could result in models containing five factors, which explained approximately 30 percent of the total variance. Even more troubling is the realization that rotated loadings in excess of 0.40, the traditional cutoff used by investigators, occurred more than 95 percent of the time in our randomly generated data set. If, as our simulation demonstrated, similar results can be obtained using randomly generated data, we are forced to reconsider the existence of syndromes found in earlier studies, especially those discovered through factor analysis of dichotomous variables.

That is, dichotomous data tend to yield many factors (by the usual Kaiser criterion), and many variables loaded on these factors (by the usual 0.40 cutoff), even for randomly generated data. Chapter 8 of this book addresses factor analysis adapted for binary items.

Problems arise even when the number of categories is greater than two. Spurious factors may be created not because items are similar in meaning but because they are similar in difficulty (Gorsuch, 1974; Lawrence et al., 2004). Treating ordinal variables as interval is a form of measurement error and hence involves attenuation of correlation. This is why basing exploratory factor analysis (EFA) on a matrix of polychoric correlations, which are designed for ordinal data, results in higher factor loadings and higher eigenvalues as a rule. Monte Carlo studies by Joreskog and Sorbom (1986) uphold the desirability of basing EFA on polychoric matrixes when using ordinal data, as does research by Muthen and Kaplan (1985) and by Gilley and Uhlig (1993). This is treated in a later section of this book on "Polychoric and heterogeneous PCA".

Polychoric correlation

For ordinal and mixed data, some researchers prefer to input a polychoric correlation matrix rather than use a Pearsonian matrix. Polychoric correlation matrices can be created as illustrated below. We use the "music" data object created below to illustrate, computing a polychoric correlation matrix using the `corCi()` function of the "psych" package. Note that there is also a `polychoric()` function in the "psych" package as well, but it will not work if a variable has more than 8 levels. This topic is explored more fully in Chapter 8 in its section on "PCA for mixed data levels".

```
# Step 1: Load the psych package, from William Revelle (2019, 2021).
base::library(psych)

# Step 2: Set the working directory and read in the "music" data.
base::setwd("C:/Data")
music <- utils::read.csv("musicvars.csv", header=TRUE, sep = ",",
stringsAsFactors=TRUE)

# Step 3: Create music2 as a complete cases version of music
music2 <- music[stats::complete.cases(music),]

# Step 4: Create a vector with the columns in a subset of music2 to
analyze.
# Column index numbers below are obtained from the names(music2)
command.
musicColumns <- base::c(2,3,5,6,8,10,11,13,14,15)

# Step 5: Use corCi(), with poly=TRUE, sending results to the object
"polyoutput".
# The corCi() function also produces a correlation plot if
plot=TRUE. This is Figure 2.3.
# Go to Plot > Zoom to see all labels.
# NOTE: This command may take a minute or two before output appears.

polyoutput <- psych::corCi(music2[,musicColumns],
    keys = NULL,
    n.iter = 100,
    p = 0.05,
    cex=3,
    diag=FALSE,
    overlap = FALSE,
    poly = TRUE,
    plot=TRUE)
```

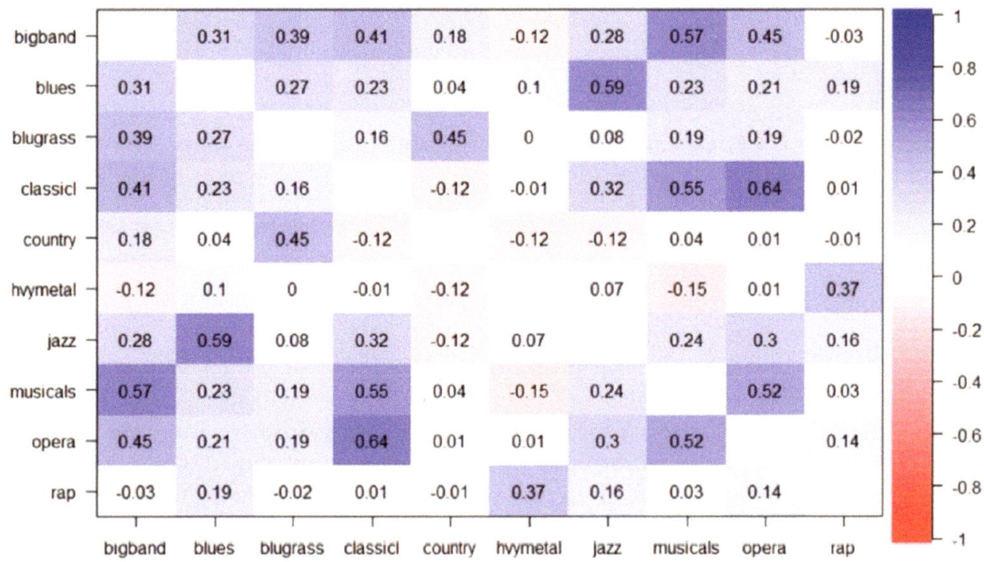

Figure 2.3 Polychoric correlation plot.

```
# Step 7 (optional). Observe that polyoutput is not yet a matrix (it
is a list).
base::class(polyoutput)
# Output:
   [1] "psych"  "cor.ci"

#  Step 8. Polychoric correlations are in the rho element of
polymatrix.
# Below we convert this to an object of class "matrix".
polymatrix <- base::as.matrix(polyoutput$rho)

# Step 9. Optionally, view the polychoric matrix in the Viewer.
View(polymatrix)

#Step 10. Print out one of the polychoric correlations
polymatrix["bigband","blues"]
# Output:
   [1] 0.3063139

# Step 11 (optional). Compare Pearson, Kendall, and Spearman cor-
relations for the same data.
stats::cor(music2$bigband,music2$blues, use="complete.obs",
method="pearson")
# Output:
   [1] 0.2919103
```

```
stats::cor(music2$bigband,music2$blues, use="complete.obs",
method="kendall")
# Output:
   [1] 0.2362031

stats::cor(music2$bigband,music2$blues,use="complete.obs",
method="spearman")
# Output:
   [1] 0.275061
```

Linearity

Factor analysis is based on correlation or covariance, both of which assume linearity. Standard PCA therefore is a linear procedure though, as we will see, there are nonlinear versions of PCA. In its usual form, PCA assumes (1) a linear relationship between measured variables and the factor/component they represent, and (2) linearity is invariant across any groupings of observations (Fabrigar & Wegener, 2012; Hair et al., 2019). Scatterplot should not reveal any marked departure from linearity. Note that the smaller the sample size, the more important it is to screen data for linearity. Scatterplots showing linearity or lack thereof may be created with the `pairs.panels()` command from the "psych" package, among other ways. This is illustrated later in this book. The linearity assumption notwithstanding, even using standard PCA, as with multiple linear regression, nonlinear transformation of selected variables may be a pre-processing step.

Multivariate normality

Multivariate normal distribution of data is required for significance tests related to standard PCA and PFA. If significance testing is not needed by the researcher, as when data are a complete enumeration of the universe of interest, factor analysis does not require distributional assumptions except variables should have the same distribution because variables from different distributions cannot correlate at a 1.0 level even when both are perfectly ordered from low to high values. Note also that a less-used variant of factor analysis, maximum likelihood factor analysis, does assume multivariate normality. The smaller the sample size, the more important it is to screen data for normality. Nonetheless, normality is not considered one of the critical assumptions of factor analysis. Tabachnick and Fidell (618) write, "To the extent that normality fails, the [factor analysis] solution is degraded but may still be worthwhile."

The normality assumption particularly affects significance testing of coefficients. If one is just interested in clustering factors or developing factor scores, significance testing may not be needed, provided variables are not heavily skewed or kurtotic or from distributions with markedly different shapes. In correlation, significance testing is used with random samples to determine which coefficients cannot be assumed to be different from zero. However, in factor analysis, the issue of which variables to drop is assessed by identifying those with low communalities since these are the ones for which the factor model is "not working." Communalities are a form of effect size coefficient, whereas significance also depends on sample size. As noted below, it is still true that factor analysis also requires adequate sample size, in the absence of which the factor scores and communalities may be unreliable. If variables come from markedly different underlying distributions, correlation and factor loadings will be attenuated as they will be for other violation of factor analysis assumptions since violations represent measurement error.

Normality is often assessed using a Q-Q (quantile-quantile) plot. The Q-Q plot tests if two data series (e.g., observed and estimated data) have the same or a similar distribution. An ideally normal distribution has all points on or very close to the 45-degree line in the Q-Q plot. Below we produce a Q-Q plot for the variable "Murder" using the `mardia()` command from the "psych" package. The data are in "USArrests", which is a data frame in R's built-in datasets library. The Q-Q plot is shown in Figure 2.4.

```
base::library(psych)
utils::data(USArrests)
psych::mardia(USArrests$Murder, plot=TRUE)
[Output other than the plot not shown].
```

The `mardia()` command also outputs numeric tests of multivariate skewness and kurtosis, both forms of non-normality. These are discussed in the following subsection. While it can be seen in Figure 2.4 that there is a departure from normality on the low end of the range of values for "Murder", most points roughly follow the 45-degree line. The numeric tests of skewness and kurtosis suggest both are within acceptable bounds.

Skew and kurtosis

When the distributional shape of two variables is markedly different, correlation is attenuated, biasing the results of factor analysis. For instance, if one variable is normally distributed and another variable is heavily skewed, their correlation will be less than 1.0 even when the cases in each variable are both arranged from low to high values. In contrast, with two variables with the same (usually normal) distribution, correlation is 1.0 when high values in the first are perfectly paired with high values of the second and

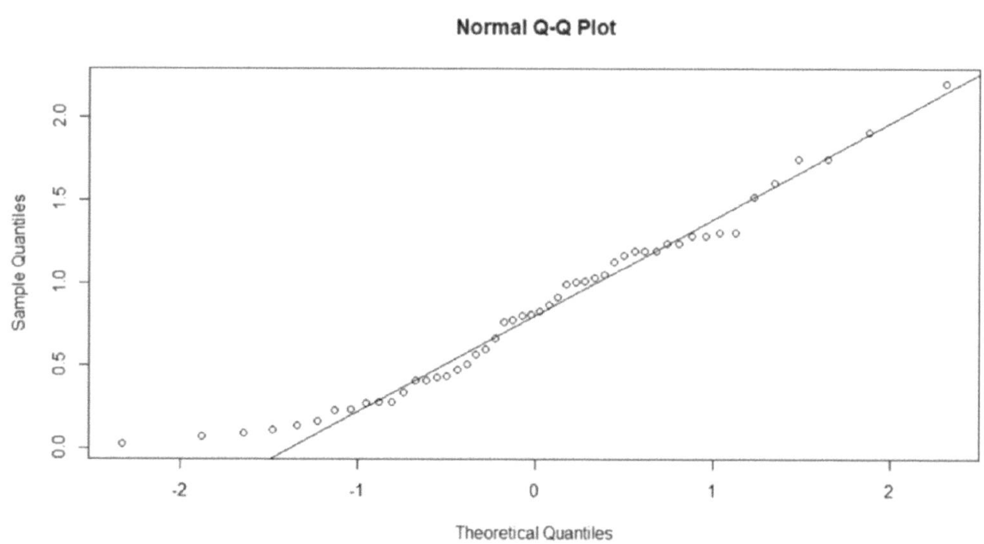

Figure 2.4 Q-Q plot of "Murder" from the "USArrests" data frame.

low values in the first are perfectly paired with low values in the second. Also, if one variable has positive skew and a second variable has negative skew, their correlation is attenuated and analysis is weakened. Note that though high skew is more often checked in articles which appear in the literature, high kurtosis can affect correlation and therefore factor analysis even more.

So what is the cutoff for too much skew or kurtosis? A widely accepted rule of thumb is that skew should not be outside the range of plus and minus 2 (some say 3), and that univariate kurtosis should not be greater than 7.0 (Curran, West, & Finch, 1996; Bandalos, 2018). Some say univariate kurtosis need only be between plus and minus 10. For multivariate kurtosis, kurtosis should not be greater than 3.0 to 5.0 (Bentler, 2005; Mueller & Hancock, 2019).

For a perfectly normal distribution, skew is 0 and univariate kurtosis is 3. Univariate skew and kurtosis may be checked in R with the `describe()` command from the "psych" package. This is illustrated below for the variable "Murder" from the "USArrests" dataset in R's built-in "datasets" package. The USArrests data have no missing values.

```
base::library(psych)
utils::data(USArrests)
psych::describe(USArrests$Murder)
[Output:]
        vars  n mean   sd median trimmed  mad
    X1     1 50 7.79 4.36   7.25    7.53 5.41
        min  max range skew kurtosis   se
    X1 0.8 17.4  16.6 0.37     -0.95 0.62
```

With skew = 0.37 and kurtosis = −0.95, the variable "Murder" is not considered severely skewed or kurtotic.

The `describe()` function gives univariate tests of skew and kurtosis. A stricter multivariate criterion for skew and kurtosis is Mardia's test, also implemented by the "psych" package (see Mardia, 1970). Note that the `mardia()` command would accept a data frame and all its variables as input, such as USArrests, giving overall skew and kurtosis tests for all four variables in USArrests.

```
psych::mardia(USArrests$Murder, plot=FALSE)
[Output:]
    Mardia tests of multivariate skew and kurtosis
    Use describe(x) the to get univariate tests
    n.obs = 50   num.vars =  1
    b1p =  0.14   skew =  1.14  with probability  <=  0.57
      small sample skew =  1.29  with probability <=  0.51
    b2p =  2.05   kurtosis =  -1.37  with probability <=  0.1
```

Mardia's skew statistic is 1.14, indicating absence of severe departure from normality. Mardia's kurtosis = −1.37, also indicating a lack of severe departure from normality. If sample size is small (e.g., <20) then the small sample corrected skew value would be used, but in this case, the inference would be the same. Both p-values of skewness and kurtosis statistics should be greater than 0.05, which they are here, to conclude the input data are multivariate normal.

Homoskedasticity

In regression, homoskedasticity means that error variance should be the same across the range of the outcome variable. If data are heteroskedastic the regression line fits some ranges better than others and the estimate of the DV is a "bad average" of fit. Since factors are regression-like linear functions of measured variables, homoskedasticity of the relationship is also assumed. However, homoskedasticity is not considered a critical assumption of factor analysis if heteroskedasticity is not extreme. Lewin-Koh and Amemiya (2003; see also Jones, 2001) introduced model-fitting procedures for heteroskedastic factor analysis but these are not yet widely available in statistical software, including R, at least at this writing.

No influential outliers

Outlying observations can impact correlations heavily and thus distort factor analysis. There are four common strategies for handling observation-level outliers:

1 Delete them. However, as the outliers are usually real data rather than coding errors, deletion is usually disparaged.
2 Run with and without outliers. This have-it-both-ways strategy highlights the effect of outliers but also shows sample results with outliers removed.
3 Run without outliers but analyze outliers as a separate group. This shows how the model applied to most cases differs from the model applied to outliers. The danger is that the outlier model may be based on a few cases.
4 Run the whole sample, outliers included, but discuss the bias effects of the outliers.

The first step in checking for outliers is to run frequencies on one's variables to check for out-of-range values due to data entry errors. Typically such variables are set to missing (NA in R), not removed from the dataset. To flag outliers not due to data entry error, one may use Mahalanobis distance, leverage, or other influence measures to identify cases which are multivariate outliers, then possibly remove them from the analysis prior to factor analysis. As removal is controversial (after all, outliers are real data), alternatively one may create a dummy variable set to 1 for cases with high Mahalanobis distance and set to 0 for other cases, then regress this dummy on all other variables. If this regression is non-significant (or simply has a low R-squared for large samples) then the outliers are judged to be close to random and there is less danger in retaining them. The ratio of the beta weights in this regression indicates which variables are most associated with the outlier cases.

One method of identifying outliers is to use boxplots such as that in Figure 2.5. The `boxplot()` function is part of R's built-in "graphics" package. This shows that there are two outliers for the variable "Rape" in the data frame "USArrests", which is one of the datasets in R's built-in "datasets" package. The R code below not only produces the boxplot but also lists the outlier values and the rows in USArrests which contain

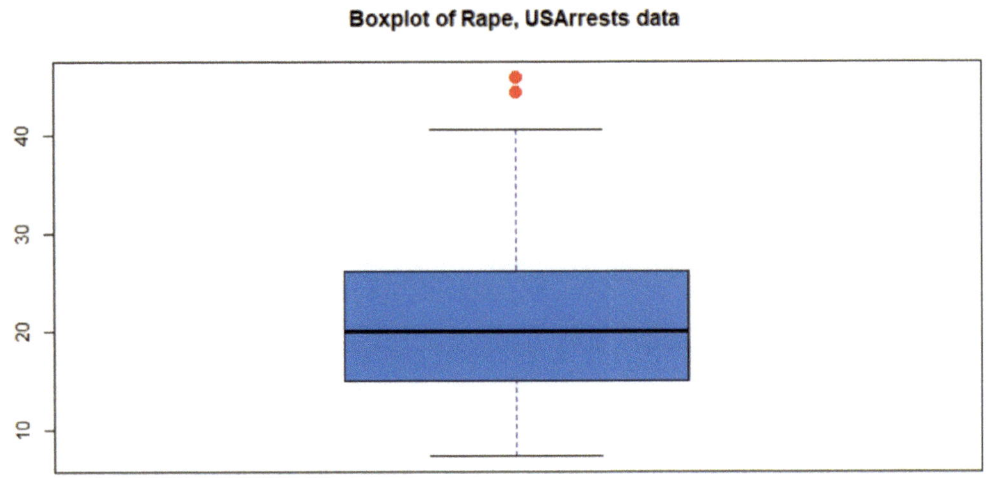

Figure 2.5 Boxplot of rape showing outliers.

them. We find that the two outliers are rows 2 and 28 in the USArrests data frame, corresponding to Alaska and Nevada.

```
utils::data(USArrests)
boxvalues <- graphics:: boxplot(USArrests$Rape,
    plot=TRUE,
    pch=16,
    cex=1.5,
    whiskcol = "blue",
    outcol="red",
    boxfill="dodgerblue",
    main="Boxplot of Rape, USArrests data")
boxvalues$out
[Output:]
    [1] 44.5 46.0

outliers <- base::which(USArrests$Rape %in% boxvalues$out)
outliers
[Output:]
     [1]  2 28
```

In Figure 2.5, the blue box shows the distance from the first to third quartile. The horizontal black line in the middle is the median. The horizontal whiskers at the top and bottom show the range of the data when outliers are excluded. The red dots are the outliers. Outliers are defined by the interquartile range (IQR) criterion. The IQR is the distance between the first (25th percent) quartile and the third (75th percent) quartile. Outliers are points which are more than 1.5 * IQR below the first quartile or 1.5 * IQR above the third quartile. Some scholars (e.g., Streiner, 2018) prefer the more conservative definition which identifies outliers as points outside 2.2 ⋆ IQR.

The IQR criterion is a univariate method of identifying outliers. A multivariate method of spotting outliers is to use Mahalanobis distance (D^2). The `mahalanobis()` function calculates D^2 and is part of R's built-in "stats" package. The general format for this command is `mahalanobis(x, center, cov)`, where x is the data in data.frame or matrix format, center is a vector of means of the variables, and cov is the covariance matrix. For the same example USArrests data, the command thus is:

```
utils::data(USArrests)
D2values <- stats::mahalanobis(USArrests, base::colMeans(USArrests),
stats::cov(USArrests))
D2values
[Partial output:]
    Alabama          Alaska         Arizona        Arkansas
     2.3361665     15.1680638      5.7240749       1.4744001
   ...
       Vermont         Virginia     Washington  West Virginia
     7.1116744      0.2960040      2.2992245       3.8954199
     Wisconsin          Wyoming
     2.3234779      0.5746895
```

As an outlier criterion, we can create p-values to flag cases which are significant based on the size of their D^2 values. The p-value is calculated with the `pchisq()` function from R's "" package, based on the chi-square statistic of the Mahalanobis distance with k − 1 degrees of freedom, where k is the number of variables (4 for USArrests).

```
# Get p-values and add them as a variable/column to USArrests
USArrests$pvalue <- stats::pchisq(D2values, df=3, lower.tail=FALSE)
utils::head(USArrests)
[Output:]
              Murder Assault UrbanPop Rape      pvalue
   Alabama      13.2     236       58 21.2 0.505627788
   Alaska       10.0     263       48 44.5 0.001678516
   Arizona       8.1     294       80 31.0 0.125834075
   Arkansas      8.8     190       50 19.5 0.688191245
   California    9.0     276       91 40.6 0.088881640
   Colorado      7.9     204       78 38.7 0.159931953
```

Finally, list the outliers, which are the states with significant p-values. Typically for this purpose, a conservative alpha value is used, such as the 0.001 level. For these data, there are no 0.001 outliers. Below we show the 0.01 and 0.05 outliers. We use the `subset()` function, which is part of R's "base" package.

```
outliers <- base::subset(USArrests, pvalue <= 0.01)
outliers
[Output:]
```

```
                  Murder Assault UrbanPop Rape      pvalue
    Alaska            10     263       48 44.5 0.001678516
    North Carolina    13     337       45 16.1 0.005559972
```

```
outliers <- base::subset(USArrests, pvalue <= 0.05)
outliers
[Output:]
               Murder Assault UrbanPop Rape      pvalue
Alaska           10.0     263       48 44.5 0.001678516
Georgia          17.4     211       60 25.8 0.022747021
Mississippi      16.1     259       44 17.1 0.049506261
Nevada           12.2     252       81 46.0 0.043223413
North Carolina   13.0     337       45 16.1 0.005559972
Rhode Island      3.4     174       87  8.3 0.020491348
```

As a further consideration, extreme outliers can distort the D2 statistic itself, leading some researchers to prefer using "robust Mahalanobis distances" (DeSimone, Harms, & DeSimone, 2015). Robust Mahalanobis distances are calculated by the "faoutlier" package (Chalmers & Flora, 2015). The `method="mcd"` option below specifies the minimum covariance determinant method.

The faoutlier print method lists in descending order the row numbers of the states with the most significant robust Mahalanobis distances. For the example data, the first eight are below p = 0.001 and may be considered outliers by this criterion. The most outlying case is row 2 in USArrests, which is Alaska. The p <= 0.001 threshold is recommended by Hair et al. (2019) and Tabachnick and Fidell (2019).

```
base::library(faoutlier)  # "faoutlier" must be installed first.
robustD2 <- faoutlier::robustMD(USArrests, method="mcd")
base::print(robustD2, digits=3)
[Output:]
         mah     p  sig
  2   46.291 0.000 ****
  10  39.489 0.000 ****
  28  37.294 0.000 ****
  24  34.545 0.000 ****
  33  31.360 0.000 ****
  11  25.851 0.000 ****
  6   25.496 0.000  ***
  5   22.306 0.000  ***
  18  18.203 0.003   **
  40  17.129 0.004   **
```

The `outlier()` command for the "psych" package will also plot and label cases (here, states) which are multivariate outliers based on Mahalanobis distance (D-squared). The `bad=5` option specifies that the five most outlying states be labeled (for these data, this is 10% of the sample). The resulting plot is Figure 2.6.

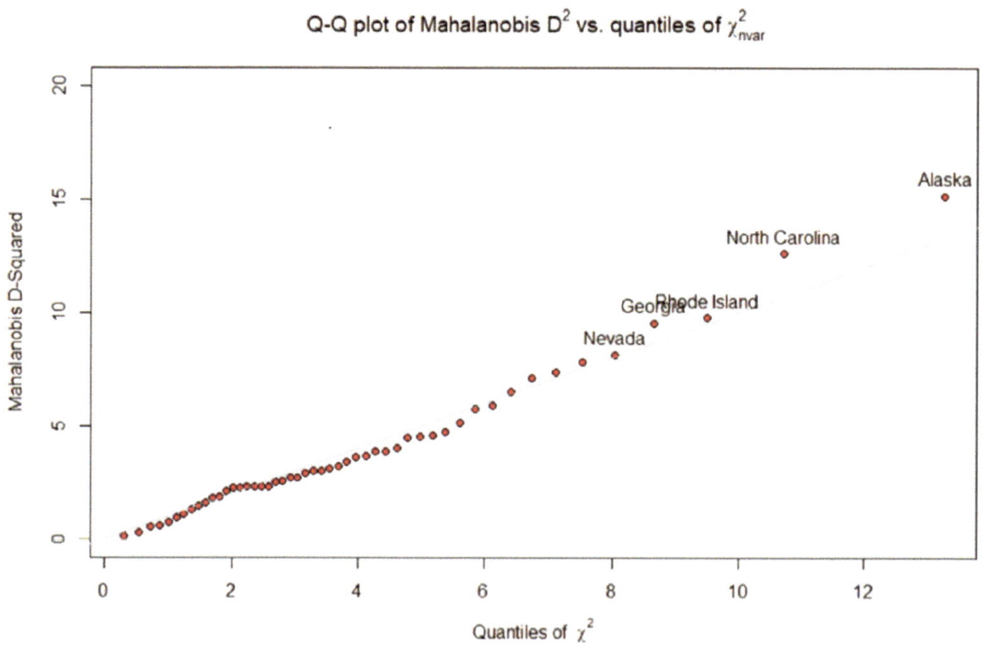

Figure 2.6 Outlier states by Mahalanobis D-Squared.

```
psych::outlier(USArrests, bad=5, plot=TRUE, na.rm=TRUE,
bg=base::c("red"), pch=21, ylab="Mahalanobis D-Squared",
ylim=base::c(0,20))
```

The `outlier()` command simultaneously lists all observations (states) by D-square:

```
       Alabama        Alaska          Arizona        Arkansas  California
         2.336        15.168            5.724           1.474       6.520
      Colorado   Connecticut         Delaware         Florida     Georgia
         5.168         3.120            5.921           4.556       9.556
        Hawaii         Idaho         Illinois         Indiana        Iowa
         7.384         2.700            2.557           1.286       2.094
        Kansas      Kentucky        Louisiana           Maine    Maryland
         0.566         3.675            4.574           3.039       3.244
 Massachusetts      Michigan        Minnesota     Mississippi    Missouri
         3.449         2.247            1.618           7.837       0.940
       Montana      Nebraska           Nevada   New Hampshire  New Jersey
         1.095         0.756            8.139           2.256       4.034
    New Mexico      New York   North Carolina    North Dakota        Ohio
         2.313         2.927           12.610           4.507       1.837
      Oklahoma        Oregon     Pennsylvania    Rhode Island       South
                                                                 Carolina
         0.122         3.039            1.804           9.784       4.736
  South Dakota     Tennessee            Texas            Utah     Vermont
         2.717         3.633            3.897           2.539       7.112
      Virginia    Washington    West Virginia       Wisconsin     Wyoming
         0.296         2.299            3.895           2.323       0.575
```

Not shown here, the command plot(robustD2,ylab="Mahalanobis D2", xlab="State ID") will also generate a bar graph of robust Mahalanobis distances for all states in the USArrests data frame.

While dropping outliers is controversial and often not recommended, discriminant analysis may provide one means of doing so. Tabachnick and Fidell (2013: 658) outline a procedure in which an influence measure such as Mahalanobis distance is used to create a dummy variable coded 0 = not an outlier, 1 = is an outlier. Then, stepwise discriminant analysis is used to identify the variables which predict the dummy variable with a significance level of 0.001. The observations predicted to be dummy=1 are removed, creating a reduced dataset which is then subject to factor analysis.

No influential missing data

Ideally, of course, the researcher's dataset has no missing data. Second best is that missing values are missing completely at random (MCAR), something which is unlikely but which if true means missing values may be ignored. Third best is that there are not too many missing values and those that are missing are imputed based on other variables in the dataset (see Garson, 2019). In a large dataset with 5% missing or less, researchers often allow listwise deletion of cases with missing values (Tabachnick & Fidell, 2019). With 10% or more missing, data imputation is typically recommended (Newman, 2014).

To determine if a given dataset has missing values, such as the USArrests dataset used in previous sections, one may use this command, which shows zero missing in this instance:

```
base::sum(base::is.na(USArrests))
[Output:]
    [1] 0
```

Many statistical commands support options which allow the researcher to choose between listwise deletion or pairwise deletion when missing cases are encountered. The latter, however, is sometimes disparaged as its use leads to different coefficients having different sample sizes. The fa() factor analysis command, defaults to pairwise deletion but this may be overridden with its use = option. The fa() command has simple data imputation available in its impute = "mean" or impute = "median" options. Type help(<name of command>) and examine its syntax to see what options are available for handling missing values or doing imputation.
Examples of different options are below:

```
# Obtain mean using listwise deletion.
base::mean(USArrests$Murder, na.rm = TRUE)
[Output:]
    [1] 7.79

# Obtain correlation using listwise deletion.
stats::cor(USArrests$Murder, USArrests$Rape, use = "complete.obs")
[Output:]
    [1] 0.564
```

```
# Obtain correlation using pairwise deletion.
stats::cor(USArrests$Murder, USArrests$Rape, use = "pairwise.com-
plete.obs")
[Output:]
   [1] 0.564
```

There are many packages in R to examine patterns of missingness and to impute missing values. A common imputation package is "mice", which stands for multiple imputation with chained equations. Package "mimi" implements imputation using maximum likelihood multiple imputation. The "impute_mle" package also uses maximum likelihood. The popular "VIM" package supports four methods of imputation: hot-deck imputation, k-nearest neighbor imputation, regression imputation, and iterative robust model-based imputation (Kowarik & Templ, 2016). The "naniar" package contains extensive functions for analysis of missing values. The CRAN page at https://cran.r-project.org/web/views/MissingData.html discusses many more data imputation packages available in R.

Moderate to moderate-high intercorrelations without multicollinearity

Applying factor analysis to a correlation matrix with only low intercorrelations may generate a solution with nearly as many factors as there are original variables, thereby defeating the dimension reduction purpose of factor analysis. Some researchers require correlations > 0.30 to conduct factor analysis. Alternatively, an overall MSA (measure of sampling adequacy, from KMO) of 0.60 or higher indicates data will factor well.

As a corollary, a measure with restricted or low variance, such as when all students score A's on a test, will have low correlations with other measures. Such a variable may not load well on any extracted factor.

Because factor analysis is sensitive to correlations, the researcher must pay careful attention to problems which can undermine correlation. These include the presence of outliers, sub-continuous levels of measurement, poorly distributed categorical variables (e.g., worse than 9:1 split for binary variables; categorical levels with count less than 5), and missing data.

Note that high bivariate correlations are not sufficient evidence that the data lend themselves to factor analysis. For instance, such correlations could be limited to just two variables and therefore would not be evidence for a more wide-ranging factor structure in the data. As a rule of thumb, there should be multiple high correlations linking indicator variables, even though other correlations may not be high.

An alternative to Kaiser's measure of sampling adequacy (MSA) is examination of the anti-image correlation matrix. This matrix contains the negatives of the partial correlations of each pair of observed measures, controlling for other variables in the model. If off-diagonal coefficients in the anti-image correlation matrix are mostly low, this is a sign that correlations are adequate to infer that the researcher's data are suitable for factor analysis.

Absence of high multicollinearity

Some researchers employ factor analysis precisely because measured variables (indicators) display multicollinearity. Factors based on multicollinear items will be orthogonal, meaning the constructs they represent are independent. On the other hand, multicollinearity increases the standard error of factor loadings, making them less reliable and thereby making more difficult the process of inferring labels for factors. To avoid this, some researchers advocate the combining of collinear terms or elimination of collinear terms prior to factor analysis. Other researchers go so far as to advocate foregoing factor analysis in the face of multicollinearity, instead simply picking the most meaningful of the collinear indicators, admitting multicollinearity, and proceeding with analysis with the comfort of knowing just what the selected indicator represents rather than having to rely on factor scores whose exact meaning is difficult to assess.

KMO statistics may be used to detect multicollinearity in a factor analysis, or data may first be screened using VIF or tolerance in regression. Multicollinearity also may be tested by computing the SMC of each indicator variable when it is treated as a dependent variable predicted by all the other variables in the model. While there is no agreed-upon threshold, SMC values should not approach 1.0. A value of 0.80, for instance, does not flag multicollinearity.

No perfect multicollinearity

Singularity in the input matrix, also called having an ill-conditioned matrix, arises when two or more variables are perfectly redundant. Singularity prevents the matrix from being inverted and prevents a solution.

Sphericity

Bartlett's test of sphericity should be significant. Bartlett's test tests the null hypothesis that the covariance matrix is proportional to an identity matrix (an identity matrix would indicate all dependent variables were uncorrelated). A finding of significance indicates that the null hypothesis is rejected, meaning that at least one of the inter-correlations is non-zero. Put another way, significance means there is at least minimal adequate correlation among variables to justify proceeding with factor analysis. However, Bartlett's test is extremely sensitive to sample size and returns significance for most social science data. Therefore its use is recommended only for small samples, as when the sample size is five times the number of variables or less (Tabachnick & Fidell, 2013: 619).

Adequate sample size

Samples must be large enough that correlations are reliably estimated. Recommendations for appropriate sample size for factor analysis vary widely in the literature. A sample size of 200 with at least 10 times the number of variables is an arbitrary "in the ballpark" recommendation. However, recommendations in the literature vary from 100 to 1,000 observations and 3 to 20 times the number of variables.

Larger sample sizes are required when …

- ariables are binary rather than continuous
- Variables are ordinal rather than continuous
- There are missing data
- Communality is low
- here are few indicators per factor (e.g., 3)
- he number of factors or components is large

In general, smaller sample sizes may be tolerated when correlations of indicator variables are strong and the number of factors is small.

There is no universal rule for minimum sample size applicable to all research situations (Reise, Waller, & Comrey, 2000: 290). Alternative arbitrary "rules of thumb" include those below. These are not mutually exclusive: Bryant and Yarnold (1995), for instance, endorse both the STV ratio and the Rule of 200 (see below). There is near universal agreement, however, that factor analysis is inappropriate when sample size is below 50.

- Rule of 10. There should be at least ten cases for each item in the instrument being used.
- TV ratio. The subjects-to-variables ratio should be no lower than 5 (Bryant and Yarnold, 1995)
- Rule of 100: The number of subjects should be the larger of five times the number of variables, or 100 (Gorsuch, 1983; Child, 2006). Even more subjects are needed when communalities are low and/or few variables load on each factor (Hatcher, 1994).
- Rule of 150: Hutcheson and Sofroniou (1999) recommended at least 150–300 cases, more toward the 150 end when there are a few highly correlated variables, as would be the case when collapsing highly multicollinear variables.
- Rule of 300. There should be at least 300 cases (Norušis, 2011). Comrey and Lee (1992) stated that 100 participants is poor, 200 is fair, 300 is good, 500 is very good, and 1,000 is excellent. At least 300 cases are needed when communalities are low, there are few factors, and only three or four indicators per factor (Tabachnick & Fidell, 2013: 618).
- Rule of 500. There should be at least 500 cases when communalities are low and there are a large number of weakly determined factors (Tabachnick & Fidell, 2013: 618).
- There should be at least a 10:1 ratio of cases to indicator variables and preferably 20:1 (Benson & Nasser, 1998; Osborne & Banjanovic, 2016; Hair et al., 2019).
- Binary data require three to ten times as many cases compared to continuous data (Pearson & Mundfrom, 2010; Rouquette & Falissard, 2011).
- Ordinal data require 350–400 cases for ten indicators/factor; 400–450 cases for seven to nine indicators/factor; and 500–600 cases for five to six indicators/factor (simulation study by Rouquette & Falissard, 2011).
- Significance rule. There should be 51 more cases than the number of variables, to support chi-square testing (Lawley and Maxwell, 1971).
- Mundfrom, Shaw and Ke (2009: 166–167), based on simulation studies, found

  > with a variables-to-factors ratio of at least 7, and even with low communality, the minimum necessary sample size for excellent agreement is never greater than 180 and, in most cases, less than 150. On the other hand, with a variables-to-factors ratio of 3, the number of factors between 3 and 6, and low communalities, the minimum necessary sample size is at least 1,200.

  They also summarized (p. 159),

  > in general, minimum sample sizes appear to be smaller for higher levels of communality; minimum sample sizes appear to be smaller for higher ratios of the number of variables to the number of factors; and when the variables-to-factors ratio exceeds 6, the minimum sample size begins to stabilize regardless of the number of factors or the level of communality.

3 Fundamental concepts in factor analysis

Research modes: EFA vs. CFA

Exploratory factor analysis (EFA) and confirmatory factor analysis (CFA) are sometimes defined in terms of the programs used. In one definition of terms, EFA is associated with stand-alone factor analysis programs while CFA is associated with a structural equation modeling (SEM) approach. In a different vocabulary, some scholars equate EFA with principal factor analysis (PFA, also called common factor analysis), as through the `fa()` command in R. In this vocabulary, EFA is counterposed to principal components analysis (PCA), which is not considered to be factor analysis at all. However, in this volume, we prefer to define EFA and CFA by characteristics associated with their research modes, whether data extraction is by PFA or PCA. Often a given statistical program may be used in an EFA or CFA research mode, depending on the researcher's study design.

As defined in this book, the exploratory mode of factor analysis typically has these traits:

- The researcher is seeking to uncover the underlying structure of a set of observed variables. It is not known in advance how many factors suffice to describe the data, nor which measured variables are associated with which factor. It is not even known if there is a well-defined factor structure at all.
- No prior theory is assumed. Rather, the purpose of EFA is to explore the data for purposes of theory construction.
- EFA may be used to explore the number of factors appropriate for a later CFA model and to screen out weak indicators of these items.

The confirmatory mode of factor analysis typically has these traits:

- The researcher specifies the factor model in advance, including how many factors there are, what their labels are, and what indicator variables go with each.
- Specification of the model is typically based on prior theory, either from the literature or posited by the researcher.
- The purpose of CFA is to confirm that the model assumptions are consistent with the data. That is, CFA seeks to determine if the number of factors and the loadings of measured (indicator) variables on them conform to what is expected on the basis of pre-existing theory.
- Indicator variables are selected on the basis of prior theory and factor analysis is used to see if they load as predicted on the expected factors (or components). The researcher's assumption is that each factor is associated with a specified subset of

DOI: 10.4324/9781003279693-4

indicator variables (Kim & Mueller, 1978b: 55). The researcher seeks to determine, for instance, if measures created to represent a factor/latent variable belong to the same underlying dimension and thereby establish that they really belong together.
- While CFA may be accomplished using stand-alone factor analysis software, more often the CFA research mode is implemented through SEM software (e.g., lavaan in R). SEM programs provide goodness of fit measures for the factor model. In the context of SEM, the factor model is called the measurement model. It is that portion of the model with just the latent variables (a type of factor) and their indicators (observed measures).[1] SEM software uses a form of common or PFA rather than PCA, which is usually the default in EFA.

The distinction between "exploratory" and "confirmatory" can be exaggerated. It is not uncommon to use a primarily EFA procedure such as PCA to confirm theory-based expectations, such as demonstrating that items in a psychological scale all sort as expected on the same factor. Likewise, initial CFA results are often revised in the SEM model development process as the researcher examines the data in an exploratory fashion.

Estimation methods

There are different ways of estimating coefficients in component or factor analysis. Some packages assume a particular method (e.g., `factanal()` assumes maximum likelihood estimation), not giving the user any choice. Other packages offer a range of estimation methods from among which to choose (e.g., `fa()` gives ten choices. The `fm=` option in the `fa()` command sets the factor estimation method. The default for `fa()` is "minres". The choices are:

- "minres": minimizes the residual matrix using an OLS procedure
- "uls": also the minimum residual method
- "pa": the principal axis factoring solution
- "ols": similar to "minres" but uses the empirical first derivative
- "wls": uses a weighted least squares (WLS) solution
- "gls": uses generalized weighted least squares (GLS)
- "ml": does maximum likelihood factor analysis
- "minchi": minimizes the sample size weighted chi square when treating pairwise correlations with different number of subjects per pair
- "minrank": does a minimum rank factor analysis
- "alpha": does alpha factor analysis

The default estimation method for any given program may not be what the researcher wants. Moreover, different estimation methods may lead to different coefficient values and different interpretations of the results. There is no one universal "best" estimation method. The researcher may wish to try different estimation methods of the same data to see which provides the results most useful for theory construction or modeling.

Data extraction: PCA vs. PFA

Though there are different methods for dimension reduction, PCA and PFA are the two leading approaches to extracting underlying dimensions from a set of variables.[2] These are the forms of factor analysis most likely to be found in the literature. In standard form, both assume that the data being factored are continuous-normal.

PCA is typically used in exploratory research in which the researcher does not have a causal model but rather simply wants to reduce a larger number of items (e.g., an 80-item survey) to a smaller number of underlying latent dimensions (e.g., seven components) for purposes of modeling. Dimensions are called "components". PCA components reflect a mixture of common, unique, and error variance of indicator variables. In the data matrix analyzed by PCA, 1's are put on the diagonal, reflecting the fact that the total variance is being analyzed. When used with varimax or other orthogonal rotation, PCA components are independent, each representing the total variance of the indicator variables.

Synonyms for PFA are principal axis factoring (PAF), common factor analysis, or simply "factor analysis" (FA). PFA is often the choice for use in confirmatory research when seeking to demonstrate that an existing model fits the data. As such it is used in conjunction with causal modeling techniques such as path analysis, partial least squares modeling, and SEM. In PFA, dimensions are called "factors". The PFA method seeks to reflect common variance in the variables, using communalities to eliminate unique and error variance from the indicator variables. In the data matrix analyzed by PFA, communalities are put on the diagonal, reflecting the fact that only common (shared) variance is being analyzed. When used with varimax or other orthogonal rotation, PFA factors are independent, each representing the shared variance of the indicator variables. Because PFA is based on common rather than total variance of the indicator variables, PFA may not reproduce the observed correlation matrix as well as PCA.

PCA and PFA compared

For most datasets, PCA and PFA will lead to similar substantive conclusions (Velicer & Jackson, 1990; Wilkinson, Blank, & Gruber, 1996; Hoelzle & Meyer, 2013). However, the two differ when the following conditions apply:

- The greater the variance among communality estimates for different variables (communality, discussed below, is the percent of variance in an indicator variable explained jointly by all the factors)
- The smaller the sample size
- The fewer the input variables

Other comparisons of PCA and PFA are shown in the table below.

PCA vs. PFA: part 1

Principal components analysis (PCA), a.k.a. components analysis or factor analysis	*Principal factor analysis (PFA), a.k.a. principal axis factoring (PAF), common factor analysis, or factor analysis*
Analyzes a correlation matrix in which the diagonal contains 1's. (This is not equivalent to analyzing the covariance matrix.)	Analyzes a correlation matrix in which the diagonal contains the commonalities. (This is equivalent to analyzing the covariance matrix, which is also what structural equation modeling does.)

PCA accounts for the total variance of variables. Factors, properly called components, reflect the common variance of variables plus the unique variance. That is, manifest variables may be conceptualized as reflecting a combination of total variance and unique variance explained by the components, plus error variance not explained by the components.	Common factor analysis accounts for the covariation among variables. Factors reflect the common variance of the variables, excluding unique (variable-specific) variance. That is, manifest variables maybe conceptualized as reflecting a combination of common variance explained by the factors, plus unique and error variance not explained by the factors.
Components seek to reproduce the total variable variance as well as the correlations. That is, PCA accounts for the total variance of the variables.	Factors seek to reproduce the correlations of the variables. That is, PIA accounts for the covariation among the variables.
PCA is thus a variance-focused approach.	PFA is thus a correlation-focused approach.

PCA vs. PFA: part 2

Principal components analysis (PCA), a.k.a. components analysis or factor analysis	*Principal factor analysis (PFA), a.k.a. principal axis factoring (PAF), common factor analysis, or factor analysis*
For the first component, PCA creates a linear equation which extracts the maximum total variance from the variables; for the second component, PCA removes the variance explained by the first component and creates a second linear equation which extracts the maximum remaining variance, etc., continuing until the components can explain all the common and unique variance in a set of variables.	PFA seeks the least number of factors which can account for the covariance shared by a set of variables. For the first factor, PFA creates a linear equation which extracts the maximum covariance from the variables; for the second component PFA removes the covariance explained by the first component and creates a second linear equation which extracts the maximum remaining covariance, etc., continuing until the factors can explain all the covariance in a set of variables.
Normally, components are orthogonal to (uncorrelated with) one another, though an oblique (correlated) option is available.	Normally, components are orthogonal to (uncorrelated with) one another, though an oblique (correlated) option is available.
Adding variables to the model will change the factor loadings.	In principle, it is possible to add variables to the model without affecting the factor loadings.
PCA is used when the research purpose is data reduction or exploration. PCA is not used in causal modeling (e.g., not used with structural equation modeling).	PFA is used when the research purpose is theory confirmation and causal modeling. A type of PFA is built into structural equation modeling programs, for instance.
In SPSS: Analyze, Data Reduction, Factor; click the Extraction button; from the "Method" drop-down, select "Principal components".	In SPSS: Analyze, Data Reduction, Factor; click the Extraction button; from the "Method" drop-down, select "Principal axis factoring".

PCA determines the factors which can account for the total (unique and common) variance in a set of variables. This is appropriate for reducing attribute space. PCA is appropriate for most social science research purposes and is the most often used form of factor analysis. Conceptually, components are seen as being determined by a linear variate of the observed indicator measures.

Common factor analysis (PFA) is appropriate for testing whether a single factor can account for the bulk of the common variance in the set, though PCA also may be used to test dimensionality. Common factor analysis has the disadvantage that it can generate negative eigenvalues, which are meaningless. Conceptually, the PFA model's arrows go from the factors (latent variables) to the observed measures, which are seen as being caused by the latent variables. This is consistent with SEM. Synonyms for SEM are covariance structure analysis, covariance structure modeling, and analysis of covariance structures.

PFA determines the least number of factors which can account for the common variance in a set of variables. Common factor analysis is appropriate for causal modeling, which focuses on covariances. Widaman thus wrote (2018: 829), "PCA should never be used if the goal is to understand and represent the latent structure of a domain; only FA techniques should be used for this purpose". The view that PCA should not be used for causal analysis of a domain is a widely shared one in spite of the fact that substantive results are often the same for PFA and PCA (Carroll, 1978, 1985; Fabrigar et al., 1999; Finch, 2013; Bandalos, 2018; Haig, 2018) and in spite of the fact that the literature contains many instances of the use of PCA for confirmatory purposes.

Other data extraction methods

Not counting tools for dimension reduction treated in Part II of this book, there are other traditional but less-used methods of extraction in addition to PCA and PFA, including those listed below.

- *Maximum likelihood factoring (ML)*: Developed by Derrick Norman Lawley in the 1940s, maximum likelihood factoring is based on a linear combination of variables to form factors, where the parameter estimates are those most likely to have resulted in the observed correlation matrix, using ML methods. Put another way, ML estimation seeks to reproduce the population factor loading matrix that generated the sample correlation matrix. Put a third way, the ML algorithm seeks factor loadings which maximize the likelihood of sampling the observed correlation matrix. See Lawley and Maxwell (1971).

  ML methods assume multivariate normality and a random sample. When data are non-normal, ML estimates may be biased. Correlations are weighted by each variable's uniqueness. Uniqueness is 1 minus the communality of a variable. Maximum likelihood factoring generates a model chi-square goodness-of-fit test. Factors may be tested for significance, as for purposes of confirmatory research. The researcher can increase the number of factors one at a time until a satisfactory goodness of fit is obtained. ML extraction is most appropriate when data are normal-continuous, the sample is large, and factors are relatively strong. Warning: for large samples, even very small improvements in explaining variance can be significant by the goodness-of-fit test and thus may lead the researcher to select too many factors.
- *Iterated principal axis factoring (IPAF)*: This variant of PFA uses an iterated algorithm to obtain estimates which are more robust against non-normality than those from ordinary PFA or ML. When non-normality is severe, IPAF or ULS are recommended (Nunnally & Bernstein, 1994; Fabrigar et al., 1999). The usual `fa()` command for PFA in R defaults to a single iteration, but the researcher may specify any number in the `n.iter=` option. The number of iterations until convergence is limited by the `max.iter=` option, which defaults to 50, but the researcher may change this. Also changeable is the `min.err=` option, which defaults to 0.0001 as

the criterion for stopping iteration when the change in communalities is less than this amount.

IPAF in R is also implemented by the `fapa()` command from the "fungible" package, usually with the option `communality = "SMC"`. In the SMC option, initial communalities are estimated using the squared multiple correlations of each indicator after regressing the indicator on the remaining variables. See Widaman and Herringer (1985).

- *Unweighted least squares (ULS) factoring*: ULS is based on minimizing the sum of squared differences between observed and estimated correlation matrices, not counting the diagonal. That is, ULS minimizes squared residual correlations. ULS is the same extraction as the ordinary least squares (OLS) and minimum residual (MINRES, an earlier label for ULS) extraction methods. Iterated principal axis (IPA) extraction also converges on an OLS solution (Briggs & MacCallum, 2003). Least squares extraction methods attempt to reproduce the sample (as opposed to population) correlation matrix. Unlike ML, these methods do not assume a normal distribution or a random sample. Least square extraction methods have been found in simulation studies to perform better than ML methods when the model is misspecified with too many factors, sample size is small, and/or factors were weak (Briggs & MacCallum, 2003).
- *Image factoring* is based on the correlation matrix of predicted variables rather than raw measurements of variables. Each variable is predicted from the others using multiple regression. The "image" in the name refers to the fact that each variable is reflected in the predicted scores. Regression is used to create empirical factors reflecting common variance, with unique and error variance of the indicator variables eliminated.
- *Minimum residual factoring* also extracts factors from the correlation matrix, but this method ignores the diagonal of the correlation matrix.
- *Generalized least squares (GLS) factoring*: based on adjusting ULS by weighting the correlations inversely according to their uniqueness (more unique variables are weighted less). Put another way, GLS seeks to minimize squared residual correlations like ULS, but first weights variables by common variance. Uniqueness is 1 − h2, where h2 is the communality. Like ML, GLS also generates a chi-square goodness-of-fit test. The researcher can increase the number of factors one at a time until a satisfactory goodness of fit is obtained.
- *Canonical factor analysis*, also called *Rao's canonical factoring*, is a different method of computing the same model as PCA, which uses the principal axis method. CFA seeks factors which have the highest canonical correlation with the observed variables. CFA is unaffected by arbitrary rescaling of the data.
- *Alpha factoring* is based on maximizing the reliability of factors, assuming variables are randomly sampled from a universe of variables. All other methods assume cases to be sampled and variables fixed. The goal of alpha factoring is to maximize the generalizability (reliability) of the orthogonal factors which are created. Alpha factoring is most relevant to research designs involving the sampling of indicator variables from a population of indicator variables.

Nothing prevents a researcher from seeking a solution under two or more extraction methods rather than just one. A test of the stability of a factor structure is that the structure is similar across multiple extraction and rotation methods (Tabachnick & Fidell, 2013: 638).

Number of dimensions to extract

Trade-off considerations

The more factors are extracted, the greater the variance explained in the indicator variables by the set of factors. The maximum number of factors is the number of measured variables. This number of factors will explain 100% of the variance in the variables. However, there is a trade-off between the percent of variance explained and model parsimony.

The researcher's factor model is misspecified if too few or too many factors/components are retained. Having too many factors suboptimizes on overly small or narrow factors, obscuring the "big picture". Too few factors mean that common factors are conflated, also obscuring the true factor structure of the domain. Having too few factors is generally regarded as even more harmful than having too many. When there are too few factors, indicators loading on a given factor may not appear to represent the same dimension. On the other hand, when there are too many factors, factors may be "singlets" or "doublets", meaning only one or two indicators load on them. Such factors have been found to have low communality and not to replicate well (Fabrigar & Wegener, 2012). Factors with three or more indicators are preferred. Hair et al. (2019: 144) likens the tradeoff decision to selecting the focus on a microscope: "Too high or too low an adjustment will obscure a structure that is obvious when adjustment is just right".

Therefore selecting the number of factors to extract is a critical research decision. While there are a variety of empirical criteria discussed in this volume, such as the Kaiser criterion or parallel analysis, there is no universal agreement on the best method, nor should there be. The best number of factors is the one which yields the most interpretable solution consistent with theory. The researcher may wish to examine a range of solutions from the highest to the lowest number of factors, selecting the number which meets this criterion (Fabrigar et al., 1999; Flora, 2018).

Eigenvalues and percent of variance explained

Eigenvalues may provide a criterion for selecting how many factors to extract. As discussed in greater detail below, the eigenvalue for a given factor measures the variance in all the variables which is accounted for by that factor. Eigenvalues are also called "characteristic roots". The ratio of eigenvalues is the ratio of explanatory importance of the factors with respect to the variables. The first factor extracted always is the largest and the last is the smallest. If a factor has a low eigenvalue, then it is contributing little to the explanation of variances in the variables and might be ignored because it is adding little information compared to more important factors.

Note that the eigenvalue is not itself the percent of variance explained but rather a measure of amount of variance in relation to total variance. Since variables are standardized to have means of 0 and variances of 1, total variance is equal to the number of variables. A factor's eigenvalue may be computed as the sum of its squared factor loadings for all the variables.

For orthogonal solutions, *trace* is the sum of variances for all factors, which is equal to the number of variables since the variance of a standardized variable is 1.0. A factor's eigenvalue divided by the trace is the percent of variance it explains in all the variables, often labeled *percent of trace* in computer output. Computer output usually lists the factors in descending order of eigenvalue, along with a cumulative percent of trace for as many factors as are extracted. It is common for the first two factors to explain a majority of the variance in the data.

Setup for this section

This section lists the steps in creating variables used in the subsequent subsections on the Kaiser criterion and the VSS criterion.

```
# Loading the needed library and data setup
base::library(psych)
base::setwd("C:/Data")
music <- utils::read.csv("musicvars.csv", header=TRUE, sep =
",", stringsAsFactors=TRUE)

# Create music2 as a complete cases version of music, dropping cases
with missing values.
music2 <- music[stats::complete.cases(music),]

# Create a subset of music2 to analyze and copy it into the object
"dat".
# Column index numbers below refer to bigband, blues, blugrass,
classicl, country, hvymetal,
# jazz, musicals, opera, and rap. Use names(music2) to see these
column index numbers.
musicColumns <- base::c(2,3,5,6,8,10,11,13,14,15)
musicsubset <- music2[,musicColumns]
dat <- musicsubset

# musicsubset has 10 variables, so this is the maximum number of
factors
maxfactors <- base::length(musicsubset)
```

The Kaiser criterion

Eigenvalues provide one way to determine how many factors to extract. By the widely-used Kaiser criterion, extraction stops when the eigenvalue for the next factor to extract falls below 1.0 (Kaiser, 1958). The Kaiser criterion is also called the Kaiser-Harris criterion or the K1 rule (Harris, 1962). Though originated earlier by Guttman in 1954, the criterion is usually referenced in relation to Kaiser's 1960 work which relied upon it. Be aware that the Kaiser rule may overestimate or underestimate the true number of factors. The preponderance of simulation study evidence suggests it usually overestimates the true number of factors, sometimes severely so (Lance, Butts, & Michels, 2006). The Kaiser criterion is the default in most computer programs but it is advisable to look at more than one criterion when determining the number of factors to extract.

Below we run PFA on music2 for the columns in musicColumns. PFA using the `fa()` command from the "psych" package is explained in a later section. Here we are only interested in applying the Kaiser criterion to select the number of factors to extract. Note that we will get the same eigenvalues even if the number of factors (set by `nfactors=`) is changed. Model results are placed in the object "PFAresult".

```
PFAresult <- psych::fa(r=music2[,musicColumns],nfactors = 2,
fm="pa", rotate ="varimax")

# Eigenvalues are stored in the $e.values element of the result.
PFAresult$e.values
[Output:]
 [1] 2.9590359 1.5174835 1.3416894 1.0199456 0.7703126
 [6] 0.6109687 0.5346705 0.4621152 0.4108412 0.3729376
```

The output above shows that the first four factors have eigenvalues above 1.0. Therefore, by the Kaiser criterion, we would extract four factors. We would do this in the `fa()` command by changing `nfactors=` to 4 and in output, we would have four-factor columns labeled PA1, PA2, PA3, and PA4. Note that prior to extraction, `fa()` computes ten factors, which is the maximum possible since there are ten music preference variables, but with `nfactors=4` as an option, `fa()` only extracts four factors.

Very simple structure (VSS) and minimum average partial (MAP) criteria

Two other criteria for selecting the number of factors to extract are VSS and MAP. The latter is computed as part of VSS output. Both are associated with the concept of "simple factor structure". There is no scientific rule for what constitutes a simple factor structure, but by common rule of thumb, all main loadings should be 0.70 and all others below 0.3. Cross-loadings occur when a variable is loaded in the 0.3 to 0.7 range on each of two or more factors. If a variable cross-loads, it is a "complex variable" in factor analysis terms. Researchers usually wish to avoid complex variables as they may interpretation of factors more ambiguous and more difficult. Simple factor structure also may be assessed by looking at the item complexity and mean item complexity coefficients, as discussed in a later section.

Criteria for simple factor structure are arbitrary. By common social science practice, a variable is considered to be important to the factor if it is 0.30 or higher. Some statistical packages produce a simplified factor loading table which omits loadings less than 0.3. There are other proposed cut-offs. Norman and Streiner (1994: 139) give this alternative formula for minimum loadings when the sample size, N, is 100 or more: Min FL = 5.152/[SQRT(N − 2)]. Another arbitrary rule-of-thumb terms loadings as "weak" if less than 0.4, "strong" if more than 0.6, and otherwise as "moderate". The meaning of the factor loading magnitudes varies by research context. For instance, loadings of 0.45 might be considered "high" for dichotomous items but for Likert scales, a 0.6 might be required to be considered "high". In any event, these criteria are arbitrary.

VSS seeks to select the number which gives the simplest factor structure. Simple factor structure is usually the most interpretable factor structure since each variable loads mainly on one factor, without appreciable cross-loadings. "VSS" stands for "very simple structure". The `vss()` function from the "psych" package, authored by William Revelle (2019, 2021), computes a factor analysis and provides VSS coefficients. The highest VSS coefficient flags the number of factors giving the best very simple structure. See also Revelle and Rocklin (1979).

In the R syntax below, the first argument must be a data frame or correlation matrix object. Other VSS options are these:

`n=`	Specifies the number of factors to examine. It is recommended that n be higher than the number one expects to wind up with.
`rotate=`	Specifies the rotation method. May be one of "none", "varimax", "oblimin", or "promax". The default is "varimax".
`fm=`	Specifies the factor method. May be one of "pa" (Principal Axis Factor Analysis), "minres" (minimum residual [OLS] factoring), "mle" (Maximum Likelihood), or "pc" (Principal Components). The default is "minres".
`plot=`	If TRUE (the default), a VSS fit plot is produced, with number of factors on the x axis and VSS fit on the y axis.
`nobs=`	Note that if using a correlation matrix rather than raw data for input, the number of observations must be input using the `nobs=` option. For instance, `VSS(my.matrix,n.obs=500)`. Otherwise the default is 1,000 observations.

To start, we run VSS on the "dat" data created above, using the PFA method and varimax rotation.

```
VSSresult <- psych::VSS(dat, n = maxfactors-1, nobs = base::nrow(-
dat), SMC=FALSE, rotate ="varimax", fm = "pa", plot=TRUE)

# Display results. A plot of VSS fit by number of factors is also
produced. See Figure 3.1.

VSSresult

[Output:]
    Very Simple Structure
    Call: vss(x = x, n = n, rotate = rotate, diagonal = diagonal, fm
    = fm, n.obs = n.obs, plot = plot, title = title, use = use, cor
    = cor,nobs = ..1, SMC = FALSE)

    Although the VSS complexity 1 shows  7  factors, it is probably
    more reasonable to think about  1  factors
    VSS complexity 2 achieves a maximimum of 0.84  with  7  factors

    The Velicer MAP achieves a minimum of 0.04  with  1  factors
    BIC achieves a minimum of  -6.03  with  4  factors
    Sample Size adjusted BIC achieves a minimum of  28.23  with  5
    factors

Statistics by number of factors
  vss1 vss2   map dof   chisq     prob sqresid  fit
1 0.53 0.00 0.038  35 6.3e+02 2.7e-110    7.34 0.53
2 0.52 0.64 0.051  26 4.2e+02  3.5e-72    5.62 0.64
3 0.60 0.72 0.063  18 1.5e+02  4.8e-22    3.92 0.75
4 0.66 0.77 0.084  11 6.7e+01  4.3e-10    2.98 0.81
5 0.71 0.83 0.143   5 4.6e+01  1.1e-08    2.00 0.87
6 0.63 0.82 0.196   0 1.4e+01       NA    1.71 0.89
```

```
7 0.72 0.84 0.273   -4 4.4e+00         NA     1.10 0.93
8 0.64 0.82 0.483   -7 3.8e-01         NA     0.69 0.96
9 0.63 0.82 1.000   -9 1.9e-04         NA     0.31 0.98

  RMSEA BIC SABIC complex  eChisq     SRMR eCRMS eBIC
1 0.148 398   509      1.0 8.6e+02 1.1e-01 0.126  631
2 0.140 244   327      1.3 5.0e+02 8.5e-02 0.111  324
3 0.096  26    83      1.3 1.4e+02 4.5e-02 0.071   22
4 0.081  -6    29      1.3 3.9e+01 2.4e-02 0.048  -34
5 0.102  12    28      1.2 2.2e+01 1.8e-02 0.053  -12
6    NA  NA    NA      1.4 5.6e+00 9.0e-03    NA   NA
7    NA  NA    NA      1.4 1.6e+00 4.8e-03    NA   NA
8    NA  NA    NA      1.3 1.3e-01 1.3e-03    NA   NA
9    NA  NA    NA      1.3 9.3e-05 3.7e-05    NA   NA
```

Discussion of VSS output above

VSS output produces a variety of measures which could be used to select the best number of factors to extract. Not all factor methods (set by the `fm=` option) produce all measures.

Value

- vss1: Select the highest VSS for dimension 1. The output above suggests seven dimensions (since vss1 is highest for row 7 = 0.72). Note: VSS does not work well if the data are very factorially complex. Simulations suggest that item-level complexities should be no greater than 2. Additional VSS statistics are stored in an element of the results object (VSSresult). Type `VSSresult$vss.stats` to view them.
- vss2: The same for dimension 2. At two decimals, this criterion suggests extracting 7 factors (in row 7, vss2 = 0.84).
- map: This is Wayne Velicer's (1976) Minimum Average Partial (MAP) criterion. By this criterion, the best solution minimizes MAP. Here that is the 1-dimension solution. MAP and VSS will not always agree on the optimal number of factors to extract. Simulation studies have found MAP to be generally accurate (Zwick & Velicer, 1986) though Gorsuch (1983) found MAP did not work well when only a few variables loaded on some components. The MAP algorithm computes a partial correlation matrix after extracting each principal component, then MAP computes the average squared off-diagonal partial correlation. This average is at a minimum when the best number of components has been extracted. This is because as common variance is partialled out, the MAP criterion diminishes. After all common variance has been partialled out and only unique variance remains, the MAP criterion will rise when additional components are extracted. Put another way, the MAP algorithm retains only components with common variance. O'Connor (2000) also presents programs for MAP.
- dof (degrees of freedom), chisq (chi square), and prob (probability that the residual matrix > 0) all appear if using methods other than `fm="pc"`. The rule of thumb is to keep extracting factors until the chi square of the residual matrix is not significant. Here this suggests a five-dimensional solution but these coefficients could not be computed for dimensions 6 through 9).

- sqresid and fit: The sqresid statistic is the square of the residuals (residual * residual). Then totalresid is the sum of squared residuals from the main diagonal (that is, it is sum(sqresid) – diagonal * sum(diag(sqresid)). The fit is 1 – (totalresid/totaloriginal) = 1 – (sumsquared residuals/sumsquared original) of off-diagonal elements. For these data fit reaches a maximum(0.98) for the nine-dimensional solution.
- RMSEA: RMSEA is the root mean square error or approximation. Lower RMSEA is a better fit. By common rule of thumb, RMSE should be 0.05 or less to be "good" and 0.08 or less to be "adequate". Here, the lowest RMSEA is achieved for the four-dimensional solution (0.081).
- BIC and SABIC: This is the Bayesian Information Criterion for each number of factors. Lower is better (less error). SABIC is sample-adjusted BIC. Here, BIC suggests a four-dimensional solution while SABIC suggests a five-dimensional solution.
- complex: This is the overall complexity coefficient. Closer to 1.0 corresponds to less complexity and a simpler factor structure. Here the one-dimensional solution has a complexity of 1.0, but this is true by definition of one-factor solutions and therefore is dismissed. The lowest complexity for the remaining solutions is that for the five-dimensional solution (1.2).
- Not a statistic, but sound advice repeated in VSS help documentation, is to keep "Extracting factors as long as they are interpretable".
- Other measures:
  - eChiSq: the empirically found chi square
  - eRMS: Empirically found mean residual
  - eCRMS: Empirically found mean residual corrected for df
  - eBIC: The empirically found BIC based upon the eChiSq
  - cfit.1 to cfit.n: : VSS fit of complexity 1 through n
  - cresidual.1 to cresidual.n: sum squared residual correlations for complexity 1 through

The VSS plot

The `VSS()` command computes fit statistics for each possible solution (here the one to nine dimensions solutions), storing them in the $vss.stats element.

```
VSSresult$vss.stats
[Partial output:]
     cfit.1    cfit.2    cfit.3
1 0.5316611 0.0000000 0.0000000
2 0.5210758 0.6412972 0.0000000
3 0.6010659 0.7210020 0.7497947
4 0.6556170 0.7701104 0.8021091
5 0.7096894 0.8280566 0.8642952
6 0.6331239 0.8174162 0.8670417
7 0.7211239 0.8423398 0.8961195
8 0.6421124 0.8244443 0.8981985
9 0.6320853 0.8176615 0.8993640

     cfit.4    cfit.5    cfit.6
1 0.0000000 0.0000000 0.0000000
2 0.0000000 0.0000000 0.0000000
```

```
3 0.0000000 0.0000000 0.0000000
4 0.8097385 0.0000000 0.0000000
5 0.8705598 0.8726966 0.0000000
6 0.8856965 0.8895647 0.8908852
7 0.9212602 0.9269759 0.9288671
8 0.9321099 0.9483612 0.9539302
9 0.9385290 0.9657114 0.9740463

     cfit.7    cfit.8    cfit.9
1 0.0000000 0.0000000 0.0000000
2 0.0000000 0.0000000 0.0000000
3 0.0000000 0.0000000 0.0000000
4 0.0000000 0.0000000 0.0000000
5 0.0000000 0.0000000 0.0000000
6 0.0000000 0.0000000 0.0000000
7 0.9296283 0.0000000 0.0000000
8 0.9553066 0.9557994 0.0000000
9 0.9786715 0.9797402 0.9800478
```

These fit coefficients are those used for the y axis in the VSS plot in Figure 3.1. Line 1 corresponds to the cfit.1 column (which is fit for dimension 1), line 2 to the cfit.2 column (fit for dimension 2), and so on. These are the VSS fit of complexity 1, the VSS fit of complexity 2, and so on.

If our prime concern is fit for the most important dimension, which is dimension 1, then we would focus on the line marked "1". For this line, fit increases linearly through

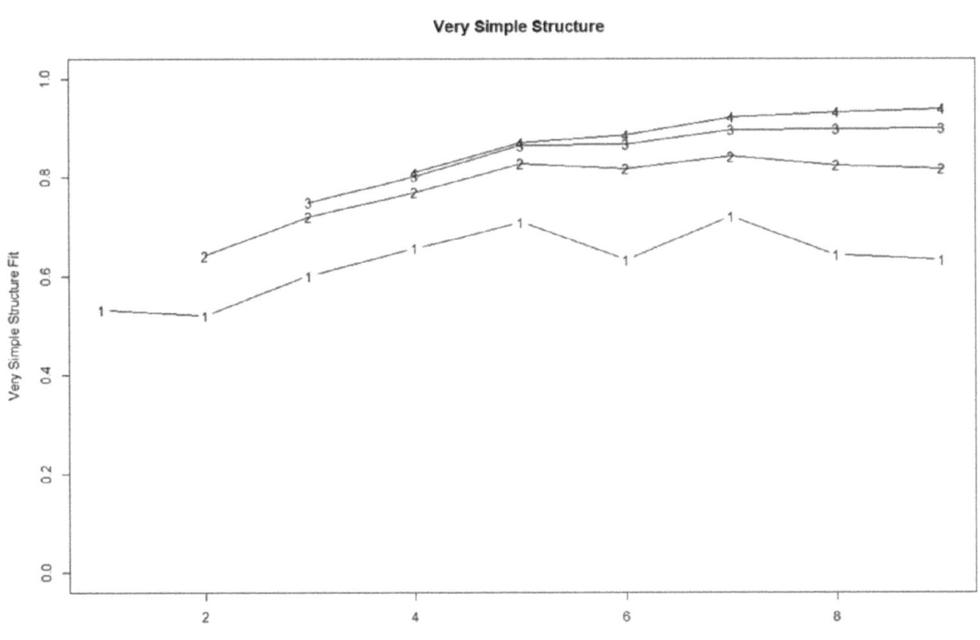

Figure 3.1 VSS plot for VSSresult.

a number of factors = 4, then declines. The same is true for dimension 2. On this basis, we might select the four-factor solution. However, the interpretability of the coefficients should be determining.

Other criteria

In addition to the Kaiser criterion, VSS, and MAP, there are several other possible criteria for determining how many factors to extract. The researcher may consider multiple criteria when making a choice of the number of factors to extract.

- *Comprehensibility.* Though not a strictly mathematical criterion, we put this first because this criterion should be paramount. The researcher should consider limiting the number of factors to those which provide readily comprehensible factors. Often this is the first two or three. Using the Kaiser criterion, VSS, MAP, or other methods below, the researcher may determine an appropriate range of solutions to investigate. For instance, the Kaiser criterion may suggest three factors and the scree test below may suggest 5, so the researcher may request 3-, 4-, and 5-factor solutions and select the solution which generates the most comprehensible factor structure.
- Scree plot: The Cattell (1966) scree test plots the components as the X axis and the corresponding eigenvalues as the Y axis. As one moves to the right, toward later components or factors, the eigenvalues drop. When the drop ceases and the curve makes an elbow toward a less-steep decline, Cattell's scree test says to drop all further components after the one starting the elbow. The theory is that the eigenvalues of common factors will roughly form a descending straight line, whereas the eigenvalues of error factors will form a different line with a smaller slope. The plot may be seen as the shape of a cliff with rubble (scree) at the bottom. This rule is sometimes criticized for being amenable to researcher-controlled "fudging." That is, as picking the "elbow" can be subjective because the curve has multiple elbows or is a smooth curve, the researcher may be tempted to set the cut-off at the number of factors desired by his or her research agenda. Researcher bias may be introduced due to the subjectivity involved in selecting the elbow. The scree criterion may result in fewer or more factors than the Kaiser criterion. Gorsuch (1983) suggests that scree plots work best when each factor has variables with high loadings, communalities are high, and sample size is large. Scree plots are illustrated in the later section on the `prcomp()` command.
- Empirical scree tests, discussed in Chapter 6, seek to provide a quantitative criterion for use in conjunction with scree plots, making judgment less subjective.
- Parallel analysis (PA): Parallel analysis, also known as Humphrey-Ilgen parallel analysis, is now often recommended to assess the true number of factors (Velicer, Eaton, & Fava, 2000: 67; Lance, Butts, & Michels, 2006). It has been found to perform well (Zwick & Velicer, 1986; Auerswald & Moshegan, 2019) even for non-normal data and even if non-Pearsonian correlations are used (Garrido, Abad, & Ponsoda, 2013) or if ordinal data are used (Cho, Li, & Bandalos, 2009; Garrido, Abad, & Ponsoda, 2013). PA selects the factors whose eigenvalues are greater than random. The actual data are factor analyzed, and separately a factor analysis is conducted repeatedly and eigenvalues are averaged for a matrix of random numbers representing the same number of cases and variables. For both actual and random solutions, the number of factors on the x axis and cumulative eigenvalues on the y axis is plotted. Where

the two lines intersect determines the number of factors to be extracted. As the PA method may recommend slightly too many factors and the MAP method discussed earlier may recommend too few, some authors recommend using both (cf., Velicer, Eaton, & Fava, 2000; Bandalos, 2018). Parallel analysis is treated in Chapter 4, which is on PFA.

- Variance explained criteria. Some researchers simply use the rule of keeping enough factors to account for 90% (sometimes 80%) of the variation. Where the researcher's goal emphasizes parsimony (explaining variance with as few factors as possible), the criterion could be as low as 50%.
- The Joliffe criterion is a less used, more liberal rule of thumb which may result in twice as many factors as the Kaiser criterion. The Joliffe rule is to crop all components with eigenvalues under 0.7 and extract those which remain.
- The mean eigenvalue rule uses only the factors whose eigenvalues are at or above the mean eigenvalue. This strict rule may result in too few factors.
- Using reproduced correlation residuals to validate the choice of number of factors. Reproduced correlations are those in the correlation matrix of original items which would result based on the assumption that the computed factors were the true and only factors. For any given pair of variables, the reproduced correlation is the product of their factor loadings on the first factor plus the product on the second factor, etc., for all factors. The diagonal values are the extracted communalities. Reproduced correlation residuals or the residual correlation matrix is the matrix of differences between the reproduced and actual correlations. The closer the residuals are to zero (i.e., low or non-significant), the more confidence the researcher has in his or her selection of the number of factors in the solution. The reproduced correlation residuals matrix may help the researcher to identify particular correlations which are ill reproduced by the factor model with the current number of factors. By experimenting with different models with different numbers of factors, the researcher may assess which model best reproduces the correlations which are most critical to his or her research purpose.
- Having the appropriate number of factors improves fit, which in turn means residuals are lower. Ideally with good fit, most residual correlation matrix values are very low, approaching zero. Observing that such values are low is another criterion for judging that the number of factors is appropriate.
- Before dropping a factor below the chosen eigenvalue cut-off, the researcher should check its correlation with the dependent variable. A seemingly minor factor can have a large correlation with the dependent variable, in which case it should not be dropped. Also, as a rule of thumb, factors should have at least three high, interpretable loadings. Fewer may suggest that the researcher has asked for too many factors.

Item complexity and simple factor structure

Hoffman's index of complexity measures the extent to which a variable contributes to "simple factor structure", discussed above in the section on VSS. In perfect simple factor structure, any given variable loads on one and only one factor or component. If complexity = 1.0 for a variable, then it loads on only one factor, though not necessarily strongly. If complexity = 2 for a variable, then it loads on two variables. And so on.

Below is output for the complexity coefficient using the "ability.cov" data. These data are results of six ability tests (e.g., "reading") and are provided in R's built-in "datasets" package. While many factor analytic programs produce the complexity coefficient, as an example we use output from the `principal()` command from the "psych" package. The bare-bones R code below is also treated in the discussion of oblique rotation in Chapter 7, but the complexity coefficient may be computed for orthogonal rotations also.

```
utils::data(ability.cov)                    # data on six ability tests/
base::library(psych)                        # Has the principal() command
covmatrix <- ability.cov$cov                # Create the covariance matrix
corrmatrix <- stats::cov2cor(covmatrix)     # Convert to a correlation
                                              matrix

# Create a factor analysis results object with the principal()
command, promax rotation
principalTestsPromax  <- psych::principal(corrmatrix,
    nfactors = 2,
    n.obs = 112 ,
    rotate = "promax",
    scores=TRUE,
    oblique.score=FALSE)

# List the complexity coefficients for the six variables, contains
in the "complexity" element
# of the results object, which here is principalTestsPromax.
principalTestsPromax$complexity
[Output:]
 general  picture   blocks     maze  reading    vocab
1.979076 1.016459 1.000156 1.072308 1.002460 1.003237

# Compute mean complexity.
base::mean(principalTestsPromax$complexity)
[Output:]
[1] 1.178949
```

Most of the time a given variable loads to some degree, however minimal, on all factors. This gives rise to decimal values of the complexity coefficient shown above. The complexity value represents the degree to which the variable contributes to simple factor structure (1.0) or deviates from it (higher than 1.0). The complexity value informs the researcher how much a given variable reflects a single factor or component and thus contributes to a simple factor structure. For instance, in the output above the variable "general" ability test has a complexity coefficient of 1.98, meaning it loads on more than one factor, detracting from simple factor structure.

Mean complexity is simply the average of complexity values for all variables in the analysis. Complexity is the number of factors measured by a variable and mean complexity reflects how many columns (factors) of the loadings matrix the set of variables appears in on average. Mean complexity helps the researcher assess the number of factors or components needed to account for the observed variables. That is, mean complexity can be used as a test of the hypothesis that the number of factors or components in the researcher's current model is sufficient to yield a result close enough to a simple structure that it may be interpreted.

There is no established cut-off for what is a high enough mean complexity to call for re-specifying the model with a larger number of factors. However, a search of the literature finds values even as high as 1.5 being interpreted as supporting the hypothesis that the current number of factors is sufficient. This author suggests that a mean complexity of 1.5 or higher flags the need to assess the number of factors by running a model with more factors and comparing fit measures. See Hofman (1978) and Pettersson and Turkheimer (2010). With mean complexity here being 1.18, the two-factor model requested in the `principal()` command is judged sufficient.

Rotation of axes

Rotation serves to make factor analysis output more understandable. Rotation is usually necessary to facilitate the interpretation of factors. Put another way, for solutions with two or more factors, prior to rotation the first axis will lie in between the clusters of variables and in general the variables will not sort well on the factors. Rotation of the axes brings variables closer to the axes, increasing the factor loadings of each variable. Ordinarily, correlated variables will load on the same rotated factor. After rotation, variables are more clearly differentiated by factor. The meaning of a factor is imputed from the variables which load most heavily upon it.

Note that this does not mean fit is improved, only interpretation. In factor analysis, eigenvalues give the percent of variance in all input variables explained by each factor. The sum of eigenvalues is not affected by rotation but rotation will alter the eigenvalues (and percent of variance explained) of individual factors and will change the factor loadings. Since alternative rotations may explain the same variance (have the same total of eigenvalues) but have different factor loadings, and since factor loadings are used to intuit the meaning of factors, the consequence is that different meanings may be ascribed to the factors depending on the rotation – a problem often cited as a drawback to factor analysis. If factor analysis is used, the researcher may wish to experiment with alternative rotation methods to see which leads to the most interpretable factor structure.

For orthogonal rotations, the factor loading is the correlation of the row variable with the column factor. Thus, for PCA, components represent total variance of the indicator variables and the loading correlation of a given indicator variable likewise reflects both common and unique variance. Only loadings above some pre-determined threshold (e.g., 0.32) are used in imputing factor labels. When indicator variables have a restricted range (are more homogenous), as in the case of 0–1 binary coding, correlations of variables with factors are attenuated and a lower threshold may be used.

For PFA, factors represent common variance of the indicators, stripped of unique and error variance, In an orthogonal rotation, the loading correlation of an indicator variable still reflects both common and unique variance, as it does for PCA.

For oblique rotations, loadings in the pattern matrix are not correlations in the usual sense but rather reflect the unique relationship (stripped of common variance) between a variable and its component or factor. Loadings in the structure matrix are correlations reflecting total (both common and unique) relationships between a variable and its component or factor. However, in oblique analysis, the structure correlations are inflated due to the correlation of factors or components (that is, the correlation reflects direct effects plus indirect effects by way of common variance shared with other indicator variables). Loadings in the pattern matrix tend to differentiate variables by factor more clearly than do loadings in the structure matrix, which is why pattern matrix loadings are used more frequently. However, when the pattern matrix is used, then imputation of labels is based on unique relationships of the indicators with the factors, not on total (unique plus common) relationships as in correlations for orthogonal factor loadings.

For a different purpose, some researchers will use two rotation methods of the same type (e.g., oblimin and promax, both of the oblique type) to demonstrate the stability of the model across different rotation solutions. More broadly, Gorsuch (1983: 330) urged researchers to seek to assure robustness by factoring "the data by several different analytic procedures and hold sacred only those factors that appear across all the procedures used" (that is, by various combinations of estimation, extraction, and rotation methods).

There are multiple types of rotation, some described below. There is no single "correct" rotation. Rather, a large number of rotations are possible, some more informative to theory-building or theory confirmation than others. Because of this "rotational indeterminacy" (Fabrigar & Wegener, 2012), the selection of rotation ultimately rests with the researcher's judgment about what is most useful for investigation of theory.

No rotation

No rotation is the default in some packages (e.g., SPSS) but often rotation is needed in order to obtain interpretable factors. However, if the aim is simple dimension reduction without interpretation of factors, then no rotation may suffice. The original, unrotated principal components solution maximizes the sum of squared factor loadings, efficiently creating a set of factors which explain as much of the variance in the original variables as possible. That is, the amount of variation explained in the data is reflected in the sum of the eigenvalues of all factors.

However, unrotated solutions are hard to interpret because the initial orientation of axes is arbitrary and because variables tend to load on multiple factors. Because factor interpretation is usually important to the analyst, the "no rotation" option is often rejected and instead a rotation method is selected. The most common rotation choice is varimax, discussed further below.

The next few figures show various rotations for a set of variables dealing with views on music genres and political issues. Figure 3.2 shows the relation of these variables to components 1 and 2 when there is no rotation.

Varimax and orthogonal rotations

Varimax is an orthogonal rotation of the factor axes (Kaiser, 1958). "Orthogonal" means that the resulting factors or components are independent (uncorrelated). Orthogonal rotations include varimax, quartimax, equamax, parsimax, and others. Put another way, "orthogonal" means that the factor axes are made to be at right angles in n-dimensional space. Ideally, after rotation variables cluster along one of the rotated axes. If variables could cluster along axes, but not at right angles (right angles are orthogonal), then oblique rotation may be necessary. Sometimes it is not feasible for variables to cluster along either orthogonal or oblique axes and there is no simple factor structure possible.

Varimax maximizes the variance of the squared loadings of a factor (column) on all the variables (rows) in a factor matrix. The effect is to simplify the loadings matrix by differentiating the original variables by factor. Loadings which are high after extraction

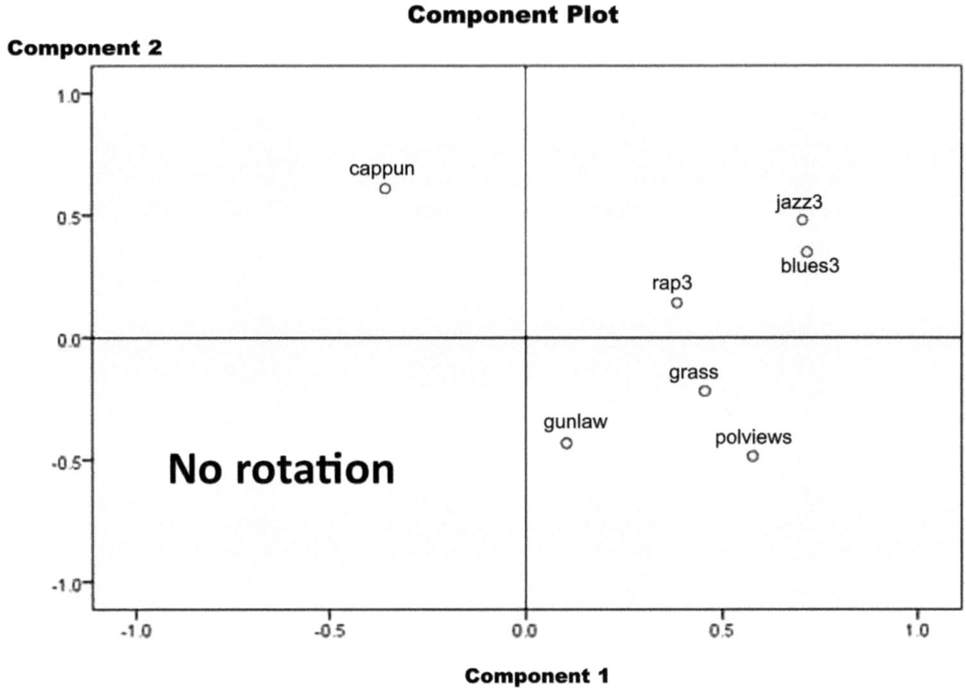

Figure 3.2 PCA solution, no rotation.

will be higher still after varimax rotation, and those low after extraction will be lower still. Each rotated factor will tend to have either large or small loadings for any particular variable. Factors will be uncorrelated. A varimax solution yields results which make it as easy as possible to identify each variable with a single factor. Tabachnick and Fidell (2013: 644) note, "Varimax also tends to reapportion variance among factors so that they become relatively equal in importance". This is the most common rotation option and is the default for orthogonal rotation in most statistical packages (Figure 3.3).

In the varimax rotation above, component 1 differentiates the music variables while component 2 differentiates the variables pertaining to political views.

Other orthogonal rotation methods are listed below.

- *Quartimax* is an orthogonal rotation method alternative to varimax. Quartimax minimizes the number of factors needed to explain each variable. The consequence is that the first factor tends to be a single general factor on which most variables are loaded to a high or medium degree. Where varimax maximizes the variance of loadings on each factor, quartimax maximizes the loadings on each variable. Though both are orthogonal methods, quartimax may not yield the same solution as varimax. Varimax is much more popular than quartimax because the researcher is usually much more interested in simple factors, reflected in clearly differentiated variable loadings on the factors, as when each variable loads strongly on just one column (factor). Quartimax would be appropriate if the researcher's interest is in simple variables, reflected in clearly differentiated factor loadings on the variables, as when each factor has a high loading on just one row (variable).

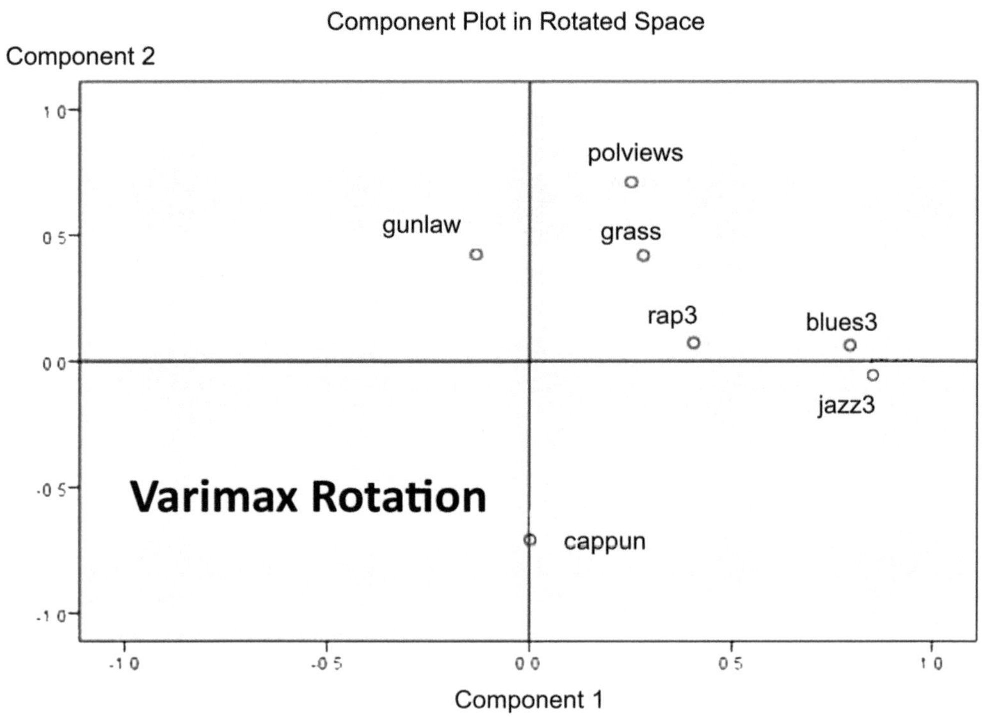

Figure 3.3 PCA solution, varimax rotation.

- *Equamax* rotation is a compromise between varimax and quartimax criteria by simplifying both factors (columns) and variables (rows). Mulaik (1972) found equamax results to be erratic unless the researcher can posit the true number of factors with confidence.
- *Parsimax* rotation is an orthogonal rotation which also simplifies both factors and columns. Parsimax is supported by SAS software.

Oblimin and oblique rotations

Sometimes just called "oblique rotation", oblimin rotation (Jennrich & Sampson, 1966) is the most common choice when one wishes for a non-orthogonal (oblique) solution – that is, one in which the factors are allowed to be correlated. Other oblique forms of rotation include promax, quartimin, and procrustes. The orthoblique rotation may be either oblique or orthogonal.

Oblique rotation will result in a more accurate solution in terms of obtaining higher eigenvalues than for orthogonal methods like varimax. Also, oblique rotation is consistent with the fact that correlation among social science variables is the rule, not the exception. Some researchers strongly recommend it for these reasons (e.g., Schmitt, 2011). However, the researcher runs the risk of diminished interpretability of the factors. Sometimes researchers run oblique rotation first, then proceed to an orthogonal solution if factors are weakly or not correlated in the oblique solution, thus approaching orthogonality. For instance, Finch (2020) recommends orthogonal rotation if the correlation between factors is 0.20 or less (factors explain 4% of each other) in an oblique solution. Roberson et al. (2014) and Tabachnick and Fidell (2019) say 0.32 or less (factors explain 10% of each other; there is 10% or more overlap in variance among the factors). Other researchers advocate performing both orthogonal and oblique rotation on the same data in order to better explore the dynamics of factors being studied.

Since oblique rotations allow the factors to be correlated, in statistical output a "factor correlation matrix" is generated whenever oblique rotation is requested. In contrast, when an orthogonal method such as varimax is selected, no factor correlation matrix is produced. This is because, in an orthogonal solution, the correlation of any factor with another is constrained to be zero. In some packages, a "component transformation matrix" is output, showing the correlation of the factors before and after rotation.

Oblique rotation is often defended on the grounds that (1) it reflects the real world more accurately, since real-world constructs are often correlated; and (2) the resulting factor structure is simpler and more interpretable. However, oblique rotation is less common than varimax rotation precisely because the correlation of factors makes much more problematic the greatest difficulty of factor analysis, which is imputing the meaning of the factors from the factor loadings for purposes of labeling the factors. In general, oblique solutions will result in more cross-loaded variables than will standard orthogonal rotations.

However, occasionally an oblique rotation will still result in a set of factors whose intercorrelations approach zero. This, indeed, is the test of whether the underlying factor structure of a set of variables is orthogonal. Also, oblique rotation is necessary as part of hierarchical or higher-order factor analysis (HFA). HFA, which is discussed in Chapter 7, seeks to identify higher-order factors on the basis of correlated lower-level ones.

When should oblique rotation be used? In CFA, if theory suggests two factors are correlated, then this measurement model calls for oblique rotation. In EFA, the researcher does not have a theoretical basis for knowing how many factors there are or what they are, much less whether they are correlated. Researchers conducting EFA usually assume the measured variables are indicators of two or more fully *different* factors, a measurement model which implies orthogonal rotation. EFA is more widely used than CFA in social science than is oblique rotation, though both oblique and orthogonal rotations are common.

When choosing a factor analysis strategy, oblique rotation may be used as a filter. Data may be first analyzed by oblique rotation and the factor correlation matrix is examined. If the factor correlations are small (e.g., <0.32, corresponding to 10% explained), then the researcher may feel warranted in assuming orthogonality in the model. If the correlations are larger, then covariance among factors should be assumed (e.g., in SEM, one adds double-headed arrows connecting latent variables). For purposes other than modeling, such as seeing if test items sort themselves out on factors as predicted, orthogonal rotation is the rule.

A different and more common strategy is to use PCA with varimax rotation as a baseline, then experiment by varying the number of factors, the extraction method, and both orthogonal and oblique rotations until the researcher produces the "solution with the greatest scientific utility, consistency, and meaning: this is the solution which is interpreted" (Tabachnick & Fidell, 2013: 647).

Common oblique rotation methods are listed below.

- *Oblimin*, also called direct oblimin, seeks to minimize the cross-products of the loadings of indicator variables on the factors. A delta parameter is available to constrain factor correlations. The "delta" factor can be specified to constrain the degree of correlation allowed in the solution, with 0 allowing high correlation and a large negative delta (e.g., −4) constraining the solution to be nearly orthogonal (uncorrelated factors). Oblimin is quite sensitive to the delta setting. Howard (2016) recommended that delta be set to 0 (usually the default), which gives maximally oblique factors (45 degrees rather than 90 as in orthogonal rotations). Values greater than 0, approaching 0.8, allow high intercorrelations, so much so that factors may not be distinguishable.

- *Promax* is a popular oblique alternative to oblimin. Under promax, orthogonally rotated factor axes are rotated a second time to oblique orientations. Promax is computationally faster than the oblimin method and therefore is sometimes used for very large datasets (Hendrickson & White, 1964). Thompson (2004: 43) opined, "promax rotation is almost always a good choice" when oblique rotation is wanted. The kappa (k) setting in promax determines the maximum amount of inter-factor correlation but the solution in promax is less sensitive to kappa than oblimin is to the analogous delta setting. Though both are oblique methods, promax does not necessarily yield the same solution as oblimin.
- *Parsimax* rotation is an orthogonal rotation which also simplifies both factors and columns. Parsimax is supported by SAS software.
- *Orthoblique* rotation gives an orthogonal rotation when factor loadings are rescaled and gives an oblique solution if loadings are not rescaled.

Eigenvalues

Eigenvalues are variances. Specifically, eigenvalues summarize in compact format the variance in a correlation matrix. The first eigenvalue reflects variances along the most important dimension of the matrix. The second eigenvalue reflects variances along the second most important dimension, after the first is controlled. And so on for remaining eigenvalues. The factor associated with the first eigenvalue will account for the largest percentage of variance explained in the matrix, with other percentages explained being in descending order. The factor associated with the last eigenvalue explains the smallest percent of variance explained.

While eigenvalues may be output from a variety of factor analytic programs in R, a simple method is to use the `eigen()` command from R's built-in "base" package. Below is output for the eigenvalues using the "ability.cov" data. These data are results of six ability tests (e.g., "reading") and are provided in R's built-in "datasets" package.

```
utils::data(ability.cov)                # data on six ability tests/
base::library(psych)                    # Has the principal()
command
covmatrix <- ability.cov$cov            # Create the covariance
matrix
corrmatrix <- stats::cov2cor(covmatrix) # Convert to a correlation
                                          matrix
```

Based on the ability data, we now compute eigenvalues for the "corrmatrix" object. There are six eigenvalues because that is the maximum number of factors for the ability data, which has six variables. We see that the first factor is the largest and the last factor is the smallest. Only the first two factors have eigenvalues above 1.0, which is the Kaiser criterion for choosing how many factors to extract.

```
eign <- base::eigen(corrmatrix)
eignvals <- eign$values
eignvals
[Output:]
   [1] 3.0768236 1.1396875 0.8171869
   [4] 0.4113127 0.3550744 0.1999149
```

The same eigenvalues will be returned by other factor analysis programs that use the same parameters. Below, for instance, is the eigenvalue output for the principal() command from the "psych" package.

```
p_out<- psych::principal(corrmatrix)
p_out$values
[Output:]
   [1] 3.0768236 1.1396875 0.8171869
   [4] 0.4113127 0.3550744 0.1999149
```

Note, however, that there are different methods of decomposing a correlation matrix. The `eigen()` command and `principal()` use spectral decomposition, also called eigendecomposition. However, other decomposition methods exist, which may give different eigenvalue coefficients. These include singular value decomposition (SVD, used by the `prcomp()` program and by `FactorMineR::PCA()`, with a slightly different algorithm), Cholesky decomposition, and QR decomposition. For these, use the `svd()`, `chol()`, and `qr()` commands instead of the `eigen()` command. All are also from R's "base" package.

A factor's eigenvalue divided by the sum of all eigenvalues is the percent that factor explains of all the variance in the data. Thus, for the ability data above, the percent of variance explained by the first factor is 51.3%.

```
pctexplained <- eignvals/base::sum(eignvals)
base::round(pctexplained,3)
[Output:]
   [1] 0.513 0.190 0.136 0.069
   [5] 0.059 0.033
```

Eigenvectors and factor loadings

Eigenvectors are the coordinates used to plot variables in factor space if the researcher is factoring variables. If the researcher is factoring observations, then eigenvectors are the coordinates used to plot observations in factor space. Factor or variable coordinates are synonyms for the eigenvectors.

Factor loadings, also called variable loadings, are usually defined as the eigenvectors times the square root of the eigenvalues. There is one loading per variable per factor. Loadings vary from 0 to 1. The higher the loading, the more that variable contributes to the meaning of the factor. Technically, factor loadings pertain to PFA, while component loadings pertain to PCA. In this section, however, we will use "factor" generically to refer to both PCA and PFA.

Factor contributions, also called variable contributions, are usually defined as the square of the eigenvectors. The sum of squared eigenvectors across all variables for a given component is always 1. Put another way, the contribution of a given variable to the importance of each component is the square of its eigenvectors. Factor loadings and factor contributions are different. Both show the same relative order of importance of variables to each component and thus both may be used to ascribe a label to a factor. However, the metrics are different.

The prevailing social science definition for a loading is expressed in the formula: loadings = eigenvectors times the square root of the eigenvalues. Factor loadings represent the correlations of the variables with the factors. While the order of importance of variables by factor loadings is the same as for factor contributions, the metric determining the actual coefficients differs. Factor loadings are more appropriate for judgments social scientists often make about which variables are "heavily loaded" or "weakly loaded" on any given factor. Since factor loadings are correlations of variables with factors, this is a familiar way for social scientists to assess strength.

The factor loading table is central to factor analysis. In this table, the columns are the factors and the rows are the variables. Cell entries are the loadings of the row variable on the column factor. The higher the loading, the more important that variable is to the meaning of the factor. Factors are most interpretable when there is a simple factor structure (Thurstone, 1947). By arbitrary but common rule of thumb, loadings less than 0.30 are "weak", more than 0.70 are "strong" (though some use the less stringent cutoffs of 0.40 and 0.60). Those in between are "moderate". Since loadings are correlations of the row variable with the column factor, 0.30 corresponds to less than 10% explained ($0.3^2 = 0.09$) while 0.70 corresponds to almost half explained.

Simple factor structure improves interpretability of the factor solution. Not only does it support the scientific principle of parsimony (Harman, 1976) but, as Le et al. (2010: 112) note, simple factor structure upholds the principle that "the purpose of science is to uncover the relatively simple deep structure principles or causes that underlie the apparent complexity observed at the surface structure level". A factor structure is simple to the extent that each variable loads heavily on one and only one factor. Usually, rotation is necessary to achieve simple structure, if it can be achieved at all. Rotation changes eigenvectors and factor loadings but not the sum of eigenvalues. Oblique rotation does lead to simpler structures in most cases, but it is more important to note that oblique rotations result in correlated factors which may be difficult to interpret or differentiate. Simple structure is only one of several sometimes conflicting goals in factor analysis.

Factor loadings may be used to impute labels to the different factors. The researcher examines the most heavily loaded indicators in each column and assigns a factor label. Even better, labels may be assigned by consensus among a panel of subject-matter experts. By one rule of thumb, loadings should be 0.70 or higher to confirm that independent variables identified a priori are represented by a particular factor, on the rationale that the 0.7 level corresponds to about half of the variance in the indicator being explained by the factor (Comrey & Lee, 1992). However, the 0.7 standard is a high one and real-life data may well not meet this criterion. Hair et al. (1998) call loadings above 0.6 "high" and those below 0.4 "low". Morin et al. (2020: 1052) called loadings of 0.50 or higher "fully satisfactory". Other researchers, particularly for exploratory purposes, will use a lower level such as 0.4 for the central factor and 0.25 for other factors (Raubenheimer, 2004). Loadings may be above a particular cutoff but still may be non-significant, leading to not interpreting them as different from zero.

A variety of cutoffs have been proposed for factor loadings in orthogonal models or pattern coefficients in oblique models. All cutoffs, of course, are arbitrary. It is more important to understand that factor loadings must be interpreted in the light of theory, not by arbitrary cutoff levels. A corollary is that the researcher should establish in advance of "crunching the numbers" what minimum level is to be considered to be of interest (e.g., 0.3 for exploratory research, 0.4 for confirmatory research). Establishing cutoff standards a priori eliminates the bias which may come from a posteriori judgments about what minimum level is "enough" to report.

Unfortunately, different R programs use such terms as "factor coordinates", "factor contributions", and "factor loadings" differently. Later chapters on specific factor analytic packages in R discuss the meaning of these terms in relation to particular packages. A worked example of orthogonal factor loadings is found in the later subsection on "PCA: Standard deviations, variances, eigenvectors, eigenvalues, contributions, and loadings" in the "recipes" section in Chapter 13.

Oblique rotation returns both a "structure matrix" as in ordinary (orthogonal) factor or component analysis, and it returns a "pattern matrix". These represent two sets of factor or component loadings. Oblique rotation is discussed in Chapter 7.

Communality and uniqueness

Communalities are the sum of the squared factor loadings for all factors for a given variable and is the variance in that variable accounted for by all the factors. In a complete PCA, with all factors extracted, this will be 1.0, or 100% of the variance. The ratio of the squared factor loadings for a given variable (row in the factor matrix) shows the relative importance of the different factors in explaining the total variance of the given variable. Note, however, that rotation redistributes variance among the factors and among variables within factors. Therefore, the importance of a given factor or variable will differ in the rotated solution compared to the rotated solution. However, the sum of squared factor loadings, which is the communality of a variable, will not change between the unrotated and rotated solution.

Uniqueness of a variable is simply 1.0 minus the communality. The "1.0" represents 100% of the variability of the variable and uniqueness represents total variability minus common variance, which is represented by the communality.

Communality, h^2 (sometimes written as "h2" in computer output), may also be defined as the squared multiple correlation (SMC) for an indicator variable as a dependent variable using all the factors as predictors. The communality varies from 0 to 1. Values greater than 1 indicate that the solution is problematic (e.g., there may be too little data or the model may be extracting the wrong number of factors). Communality measures the percent of variance in a given variable explained by all the factors jointly and may be interpreted as the "reliability" of the indicator. As a rule of thumb, the researcher may consider dropping from the factor model any indicators with a reliability (communality) below 0.50 since for those indicators more than half of their variance is due to measurement error (Kline, 2013: 173).

Communalities may be low for a given variable for three common reasons:

1 The item is not well related to the domain which the factor or component represents as inferred from the loadings of other indicators;
2 The item's reliability as a measure is low; and
3 One or more additional factors may be required to represent the item.

When an indicator variable has a low communality (e.g., <0.50), the factor model is not working well for that indicator and possibly it should be removed from the model. However, there is considerable disagreement on the lower threshold for communality. Osborne, Costello and Kellow (2008), for example, have suggested that a communality above 0.4 is acceptable. Child (2006) suggested removing indicators if their communality is below 0.2. Tabachnick and Fidell (2013: 618) considered communalities to be "high" if all communalities were 0.60 or greater or if their mean was 0.70 or higher. It is common in the literature to regard a communality (SMC) of 0.70 or greater to be "high", implying stable factors. Low communalities imply that the indicator variables poorly define the factors.

A communality of 0.75 seems high but is meaningless unless the factor on which the variable is loaded is interpretable, though it usually will be. A communality of 0.25 seems low but may be meaningful if the item is contributing to a well-defined factor. That is, what is critical is not the magnitude of the communality coefficient per se, but rather the extent to which the item plays a role in the interpretation of the factor, though as a rule, this role is greater when communality is high.

The variance of a variable may be divided between common and unique variance. Common variance is variance which can be explained by the variable's correlation with other variables in the model and by the factors, which are based on these correlations. A communality of 1.0 means that the entire variance of the given variable is common variance. If the communality exceeds 1.0, there is a spurious solution, which may reflect too small a sample or the researcher has too many or too few factors. If communality is 0, then the entire variance of the variable is unique variance, which is made up of variable-specific variance plus error variance, neither of which can be explained by correlations with other variables in the model or by the factors. For further discussion see Field (2013).

Communality for a variable is computed as the sum of squared factor loadings for that variable. Recall that r-squared is the percent of variance explained, and since factors are uncorrelated if orthogonal rotation is used, the squared loadings may be added to get the total percent explained, which is what communality is. Communality does not change when rotation is carried out.

Communalities are also used to differentiate PCA from PFA (also called common factor analysis or FA). In PCA, the diagonal contains 1's and components reflect the total variance of the indicator variables. In PFA, the diagonal contains the estimated communalities and the factors reflect the common variance of the indicator variables. Only if communalities approach being the same for all variables will the PFA and PCA solutions be similar (Velicer & Jackson, 1990; Fava & Velicer, 1992).

Below we illustrate communalities using the `fa()` factor analysis program from the "psych" package. We put model results in an object labeled "FAresult" for the unrotated solution and "FAresult_r" for the varimax rotated solution. Communalities are found in the "communality" element of the result object. The `fa()` command defaults to extracting just one factor, which we make explicitly with the `nfactors=` option. Output below demonstrates that communalities are the same before and after rotation.

```
FAresult <- psych::fa(corrmatrix,rotate ="none", nfactors=1)
FAresult$communality
[Output:]
```

```
    general   picture    blocks      maze   reading     vocab
  0.6436742 0.2771739 0.4668680 0.1622683 0.5204772 0.4846112

FAresult_r <- psych::fa(corrmatrix,rotate ="varimax", nfactors=1)
FAresult_r$communality
[Output:]
    general   picture    blocks      maze   reading     vocab
  0.6436742 0.2771739 0.4668680 0.1622683 0.5204772 0.4846112
```

Uniquenesses are found in the "uniquenesses" element of the result:

```
FAresult$uniquenesses
[Output:]
    general   picture    blocks      maze   reading     vocab
  0.3563258 0.7228261 0.5331320 0.8377317 0.4795228 0.5153888
```

Using 0.50 as the cutoff, the output above shows that the ability tests general and reading are adequately explained by the factor model. Using 0.40 as the cutoff, the variables blocks and vocab are adequately represented for exploratory purposes. The ability tests picture and maze are not well-explained by the model.

By default, the `fa()` command extracted only one factor. However, the ability data might be hypothesized to need three factors: verbal (reading, vocab), spatial (blocks, maze), and reasoning (general, picture). We therefore re-run the model, specifying that three factors be extracted. When we print out the communalities, we see that they are changed from the one-factor solution. In the three factor solution, all communalities except maze are above the 0.50 level and may be considered adequately represented by the factor model.

```
FAresult3 <- psych::fa(corrmatrix,rotate ="varimax", nfactors=3)
FAresult3$communality
[Output:]
    general   picture    blocks      maze   reading     vocab
  0.5590111 0.7830058 0.6715222 0.4201014 0.9602979 0.6639396
```

When we print out the results of the three-factor solution, we see that communalities are in the h2 column and uniquenesses are in the u2 column. The factor loadings in the MR1, MR2, and MR3 columns show that reading and vocab load most heavily on MR1, picture loads most heavily on MR2, and blocks and maze load most heavily on MR3. The general ability test is cross-loaded on all three factors. Also, the blocks test is cross-loaded on MR2 and MR3.

Note that loadings, communalities, and uniqueness would differ for a solution with a different number of requested factors.

```
FAresult3
[Partial output:]
  Factor Analysis using method =  minres
```

```
Call: psych::fa(r = corrmatrix, nfactors = 3, rotate = "varimax")
Standardized loadings (pattern matrix) based upon correlation
matrix

         MR1  MR2  MR3   h2   u2 com
general 0.50 0.39 0.39 0.56 0.44 2.8
picture 0.13 0.86 0.15 0.78 0.22 1.1
blocks  0.23 0.53 0.59 0.67 0.33 2.3
maze    0.11 0.09 0.63 0.42 0.58 1.1
reading 0.96 0.14 0.10 0.96 0.04 1.1
vocab   0.78 0.12 0.19 0.66 0.34 1.2

                       MR1  MR2  MR3
SS loadings           1.87 1.22 0.97
Proportion Var        0.31 0.20 0.16
Cumulative Var        0.31 0.52 0.68
Proportion Explained  0.46 0.30 0.24
Cumulative Proportion 0.46 0.76 1.00
...
```

Factor and component scores

If a factor is ascribed the label "job satisfaction", then the factor score for that factor for a given observation is the job satisfaction score for that observation. In PCA, factor scores may be termed "component scores". Factor scores are quite different from factor loadings. Factor scores are coefficients of *observations* (cases) on the factors, whereas factor loadings are coefficients of *variables* on the factors. Unless the data matrix has been transposed to factor observations, a factor score is a score for a given individual, case, or observation on a given factor or component. It represents "how much" of each component an individual observation is associated with.

"Factor scores" usually refer to the default type, which is "regression scores", based on OLS estimates. Regression factor scores have a mean of 0. Factor scores can be correlated even when an orthogonal factor extraction method is employed. Computing factor scores allows one to look for factor outliers. Also, factor scores may be used as variables in subsequent modeling.

While factor scores may be computed for oblique rotations, this is problematic since oblique rotations have overlapping (correlated) factors and because the importance of a variable in an oblique rotation does not depend on a single matrix such as the loadings matrix in orthogonal rotations. Rather, for oblique rotations, the pattern matrix, the structure matrix, and the factor correlation matrix must be taken into account. Tabachnick and Fidell (2013: 651) thus write, "if factor scores...are to be used as IVs or DVs in other analyses...then orthogonal rotation has distinct advantages".

Some researchers like to use factor scores as variables in a model based on orthogonal rotation because then the dimensions represented by the factors are uncorrelated. That is, modeling with factor scores can be used for orthogonalization of predictors in multiple regression. That is, for orthogonal rotations, factor scores for one factor will be close to independent (uncorrelated) with respect to scores for another factor. In a case where there is multicollinearity, one may use the component scores in place of the X scores, thereby assuring there is no multicollinearity of predictors. Note, however, that this orthogonalization comes at a price. After orthogonalizing in this manner, instead of explicit variables, one is modeling in terms of factors, the labels for which are difficult to impute. Statistically, multicollinearity is eliminated by this procedure, but in reality, it is hidden in the fact that all variables have some loading on all factors, muddying the purity of meaning of the factors, especially when the solution falls short of a simple factor structure. If two multicollinear variables load heavily on the same factor, then both contribute to the meaning (label) of that factor as imputed by the researcher. However, when imputing a label, the standard errors of the two multicollinear variables are inflated, diminishing the researcher's ability to distinguish between them.

It is important for the researcher to realize that use of factor scores is controversial. Grice (2001: 431) noted that "an infinite number of ways for scoring the individuals on the factors could be derived that would be consistent with the same factor loadings". Factor scores may be computed in a variety of ways and there is no one best way accepted as applicable to all circumstances. However computed, factor scores should have a high multiple correlation with their indicator variables (Mulaik, 2018, recommended 0.95 or higher; Gorsuch, 1988, recommended at least 0.80). The `fa()` program for PFA from the "psych" package, for instance, generates measures of factor score adequacy. Without explaining the substance of this bare-bones one-factor PFA model, the output listing these measures looks like this for the three-factor model created with the `fa()` command using varimax rotation on the same "corrmatrix" example data based on six ability test measures. This model was the FAresult3 object created above.

```
FAresult3
[Partial output:]
...
Measures of factor score adequacy
                                                  MR1  MR2  MR3
Correlation of (regression) scores with factors  0.98 0.88 0.77
Multiple R square of scores with factors         0.95 0.77 0.59
Minimum correlation of possible factor scores    0.91 0.53 0.18
```

These measures of "factor score adequacy" are also called "indeterminacy indices" because they reveal whether the factor score estimates seem to be representative of the factors. The measures are indeterminant because there are an infinite number of ways factor scores might be computed and still be consistent with the given loadings (Revelle, 1921; Grice, 2001). As evident from partial output above, there are three lines of measures:

- *Correlation of regression scores with factors*: This is the square root of multiple R squared, reflecting the correlation of the observed indicators with the factors.
- *Multiple R square of scores with factors*: This is the percent of variance that the observed indicators predict in the factor score, or vice versa.
- *Minimum correlation of possible factor scores*: This index ranges from −1 to +1 and is always less than the first two measures but still should be high. If multiple R-squared is lower than 0.5, this statistic will be negative, indicating poor factor score adequacy.

There are a number of alternatives to the usual regression-based factor scores. These include Bartlett scores, Anderson-Rubin scores, and summation scores. For regression, Bartlett, and Anderson-Rubin scores, factor scores are generated for each observation. Every observation is associated with a factor score on every factor. Those measured variables which load more strongly on a given factor will be weighted more when factor scores are computed. Those variables which load more weakly on a given factor will be weighted less. Thus the factor scores for a given factor will be constructed based on all the input variables, but the variables more heavily loaded on that factor will count the most.

- *Bartlett scores* may be preferred over default regression scores on the argument that they better conform to the original factor structure. In a simulation study, Beauducel (2005: 157) concluded, "the extremely high congruency of the Bartlett component pattern with the original factor pattern makes them most compatible with the common practice in exploratory and confirmatory factor analysis". Bartlett scores are computed by minimizing the squared residuals as in regression scoring, but the residuals are weighted by the reciprocal of the unique standard deviation. Bartlett scores have a mean of 0 and may be correlated.
- *Anderson-Rubin scores* are a modification of Bartlett scores to ensure orthogonality. Thus Anderson-Rubin scores for one factor are uncorrelated with Anderson-Rubin scores for another factor. Anderson-Rubin scores have a mean of 0 and a standard deviation of 1.
- *Summation scores* are an entirely different approach to creating scores for observations. Prior to summation to create scale values, variables are transformed to be on the same scale (assuming equal weighting of items is desired). The summation method then uses factor analysis to identify which variable might compose a scale. The measured variables which load heavily on a given factor and which do not cross-load much on other factors (e.g., not more than 0.3) are candidates for inclusion in the scale. Having used factor analysis to identify potential scale items, the researcher should then check whether the items proposed for the scale are reliable. Some variables may be dropped if they cross-load in the factor analysis stage or if their inclusion diminishes reliability. If the final set of items is reliable as evidenced by having Cronbach's alpha more than or equal to at least 0.70. Having been shown to be unidimensional by factor analysis, then the items judged reliable may be combined in a scale by some summation method. Others have suggested that reliability in the 0.80 s are acceptable for non-critical work while below 0.70 is inadequate, even for exploratory research (Kline, 2013). DeVellis (2017) has suggested that Cronbach reliability should be 0.90 or higher for clinical decisions. Nonetheless, one sees an alpha >= 0.60 cutoff sometimes used in the literature. Variables not meeting the alpha criterion are dropped. The remaining items may be said to provide the basis for a unidimensional, reliable scale. If there is a sufficient number of variables, those with the highest loadings and highest alphas may be used in the scale, discarding weaker items To improve interpretability, after summation results may be rescaled to vary from 0 to 100.

Model fit

There is no single measure of goodness of fit for factor analysis, whether for PCA or PFA or for other dimension reduction procedures. Moreover, researchers do not agree about the usefulness of all measures, nor about cutoff levels for acceptable fit. The researcher should look at multiple measures. In Chapter 12 we discuss various metrics for assessing the results of dimension reduction procedures, including PCA. Some model fit measures available for many (but not all) factor analytic procedures are listed alphabetically below and treated in empirical sections later in this book.

- AIC and BIC: There is no cutoff level, but when comparing factor models for the same measured variables, the model with lower AIC (Akaike information criterion) or BIC (Bayesian information criterion) has less error. These measures, particularly BIC, penalize for lack of model parsimony. However, it has been found that BIC overestimates the number of factors as sample size goes up (Schmitt et al., 2018).
- *Communalities*: Ideally, no variable is poorly measured by the factors, indicated by all communalities being equal to or greater than 0.5.
- *Cophenetic_correlation*: Cophonetic correlation is discussed in Part II, in the section on "Performance quality metrics for dimRed". This is the correlation of the lower triangle of the distance matrix in high dimensional (component or factor) space with that in low dimensional (input) space.
- *Distance_correlation*: Distance correlation also is discussed in Part II, in the section on "Performance quality metrics for dimRed". This quality measure calculates the distance correlation between the distance matrices in high-dimensional (observed data) and low-dimensional (component) space.
- *Eigenvalues*: In a good model, eigenvalue should meet some established criterion for selecting the number of factors to extract. Usually, this is the Kaiser criterion of >= 1.0, but parallel analysis, scree plot, and other methods are also used.
- *Factor loadings*: Ideally, the strongest loading of each variable is strong (>= 0.6) and there are no or few cross-loadings in the 0.3–0.6 range.
- *Fit based upon off-diagonal values*: This is a measure of how well the off-diagonal elements in the correlation matrix are reproduced by the factor model. In the result object for the `fa()` command, this is the "fit.off" element. This can be seen as a type of pseudo-R-Squared measure which is approximately $(1 - \text{resid}^2/\text{cor}^2)$, where resid values are from the residual matrix and cor values are from the correlation matrix. The closer to 1.0, the better the "fit based upon off-diagonal values". There is also a "fit" element, which is how well the entire correlation matrix is reproduced.
- *Mean item complexity*: As discussed in the previous section, this is a measure of how each variable contributes to simple factor structure. There is also an overall mean item complexity measure for the model as a whole. Simple factor structure makes models more interpretable. A model with mean item complexity < 1.5 is considered to meet this fit criterion.
- *Mean_R_NX and AUC_lnK_R_NX*: These are discussed in Part II, in the section on "Performance quality metrics for dimRed". R_NX may be thought of as indicating the relative improvement over a random embedding of observed data in component space.
- *Multiple R-square of scores with factors*: This is the squared correlation of the regression scores with the factors. The higher, the more the factors predict the scores and vice versa.
- *Q-local and Q_global*: These are discussed in Part II, in the section on "Performance quality metrics for dimRed". Q_local is a measure of factoring quality with respect

to the local variables (the input variables) while Q_global score is the same thing but as applied to the global variables, which are the component dimensions.

- *Reconstruction_rmse*: This is discussed in Part II, in the section on "Performance quality metrics for dimRed". Reconstruction error is the difference between the input data matrix and the same data matrix as reconstructed based on the observations' scores on the components. Lower is better.
- *Root mean square error of approximation (RMSEA)*: By common rule of thumb, values <= 0.05 are good fit, 0.015–0.08 are adequate fit, and >0.08 are inadequate or poor fit. Some allow 0.06 as the cutoff for good fit. One strategy is to run the model with various numbers of factors and select the solution with the smallest number of factors where RMSEA is below the cutoff.
- *Root mean square of residuals (RMSR)*: This statistic tends to run lower than RMSEA but nonetheless the same cutoffs are used in the literature. One strategy is to run the model with various numbers of factors and select the solution with the smallest number of factors where RMSR is below the cutoff.
- *Total_correlation*: This is discussed in Part II, in the section on "Performance quality metrics for dimRed". This is the total correlation of the variables with the axes as a measure of performance quality of a dimensionality reduction.

Notes

1 This is in contrast to the structural model in SEM, which also includes interrelationships among the latent variables.
2 The term "principal factor analysis" (PFA) is used slightly more often than "principal axis factoring" (PAF), but both refer to the same thing. In this volume we use PFA.

4 Quick start

Principal factor analysis (PFA) in R

Introduction

Principal factor analysis (PFA) is also called principal axis factoring (PAF), common factor analysis, or just factor analysis. PFA is often what researchers mean when they refer to "factor analysis". PFA and principal component analysis (PCA) were compared and contrasted in Chapter 3. PCA decomposes the data matrix into a set of linear variates called "components", based on the correlations among variables, where correlations reflect both common and unique variance. In contrast, PFA decomposes the data matrix into a set of linear variates called "factors", based on the covariances among variables, where covariances reflect common but not unique variance.

That is, the variance of an observed variable in PFA reflects three parts:

1 the portion explained as common variance by the factors;
2 a unique variance portion; and
3 error variance.

Unique and error variances are not explained by the factors. Analysis of covariance, embedded in PFA, is seen by some researchers as more appropriate for causal analysis of real-world data because it does not conflate the common and unique variances of observed variables.

Elsewhere this author has written about a dataset on national judges' scores of athletic events (see Garson, 2015). An observation in this dataset was an "enthusiast" judge who also scored the same events. The enthusiast's rating scores were quite different from those of the national judges. If the data are standardized, however, every judge would have a mean score of 0 and a standard deviation of 1.0. With standardization, which is built into correlation and is the default in some PCA and PFA programs, the enthusiast judge is far less different from the others in absolute terms even though the enthusiast's scores go up and down in parallel with the unstandardized data, meaning that correlations remain the same. Covariances do not remain the same.

To understand real-world effects, it is necessary to look at covariances, which are the unstandardized correlations of each variable with each other controlling for other variables in the model. Causal interpretation is theoretically sound only by using a covariance-based approach, which partitions effects of variables between common variance and unique variance after controlling for other variables. That is, PFA is appropriate for causal analysis and PCA is not. However, in practice, it is common for

DOI: 10.4324/9781003279693-5

researchers using PFA to ask for centering and scaling (hence standardization) of data anyway, in order to make variables comparable and not give undue weight to variables whose scale is of large magnitude.

The differences between PFA and PCA can easily be exaggerated. For instance, a study by Guadagnoli and Velicer (1988) found that PFA and PCA solutions differed little in terms of substantive inferences. Stevens (2002) also found few differences, provided there were 10–20 measured variables with communalities of 0.70 or higher. In the presence of low communalities (e.g., <0.4); however, differences between PFA and PCA are likely to emerge.

PFA is usually implemented in R by using the `fa()` command from the "psych" package, which is R's system library and does not need explicit installation with the `install.packages()` command. Activation with the base::library(psych) command may still be needed, depending on the local setup. Other packages for PFA are `factanal()` from the "stats" package and `pfa()` from the "StatDA" package. The former is treated in Chapter 9.

Setup and example data

To illustrate PFA, we use the "ability.cov" data, which is one of the datasets in the "datasets" package in R's system library. The data are from Bartholomew (1987); Bartholomew and Knott (1990); and Bartholomew, Knott and Moustaki (2011). The data come from a psychology study concerning six ability and intelligence tests which were administered to 112 individuals. Type `help(ability.cov)` for more information. The variables in the study were these:

- *general*: a non-verbal measure of general intelligence using Cattell's culture fair test.
- *picture*: a picture-completion test
- *blocks*: block design
- *maze*: mazes
- *reading*: reading comprehension
- *vocab*: vocabulary

Typically, researchers factor analyze more variables than six. However, this small example keeps output and figures less complex and this may aid understanding of the PFA method.

To load the data it is enough to type its name. The ability.cov object is a list object. Typing its name lists the elements it contains. The elements are ability.cov$cov (the covariance matrix0, ability.cov$center (the means, all 0), and ability.cov$n.obs (the number of observations, which is 112). We will refer to this as the "ability" data.

```
base::setwd("c:/Data")        # Set the local directory
base::library("stats")        # Has the fa() command
base::library("psych")        # Has the factanal() command
utils::data(ability.cov)      # Invoke the data
base::class(ability.cov)      # Show that it is of object class "list"
[Output:]
   [1] "list"
```

```
ability.cov                # List the $cov, $center, and $n.obs ele-
ments of ability.cov
[Output:]
   $cov
            general picture  blocks    maze reading    vocab
   general   24.641   5.991  33.520   6.023  20.755   29.701
   picture    5.991   6.700  18.137   1.782   4.936    7.204
   blocks    33.520  18.137 149.831  19.424  31.430   50.753
   maze       6.023   1.782  19.424  12.711   4.757    9.075
   reading   20.755   4.936  31.430   4.757  52.604   66.762
   vocab     29.701   7.204  50.753   9.075  66.762  135.292

   $center
   [1] 0 0 0 0 0 0

   $n.obs
   [1] 112
```

Based on this list object, we can create the object "covmatrix" for the covariance matrix.

```
covmatrix <- ability.cov$cov
covmatrix

[Output is the same as above.]
```

Using the `cov2cor()` function of the "stats" package, we can create the object "corrmatrix", which puts the data in correlation matrix format. The correlation matrix is used as input for later procedures.

```
base::options(digits=3)
corrmatrix <- stats::cov2cor(covmatrix)
corrmatrix
[Output:]
            general picture blocks  maze reading vocab
   general    1.000   0.466  0.552 0.340   0.576 0.514
   picture    0.466   1.000  0.572 0.193   0.263 0.239
   blocks     0.552   0.572  1.000 0.445   0.354 0.356
   maze       0.340   0.193  0.445 1.000   0.184 0.219
   reading    0.576   0.263  0.354 0.184   1.000 0.791
   vocab      0.514   0.239  0.356 0.219   0.791 1.000
```

Parallel analysis

Parallel analysis is an alternative to scree plots or the Kaiser criterion for selecting the number of components or factors to extract. Though not limited to PFA, we present it here in order to illustrate the `fa.parallel()` command from the "psych" package. This command works for both PCA and PFA. Here we run the command for both but one may set `fa = "pc"` for just PCA or `fa = "fa"` for just PFA. Results are shown in Figure 4.1.

```
psych::fa.parallel(corrmatrix,
    fa="both",
    n.iter=1000,
    show.legend= TRUE)
graphics::abline(h=1, col="black")
```

Figure 4.1 contains two parallel analyses. The lower blue and red lines are for a PFA solution. The blue line is the scree plot based on eigenvalues extracted from the ability data. By the Kaiser criterion, which is to extract factors with eigenvalues of 1.0 or greater, we would ask for a one-factor PFA solution (note that the horizontal black line is the eigenvalue = 1.0 reference line). By the scree criterion, we would ask for a three-factor PFA solution since that is where the point of inflexion is for the lower blue line. Parallel analysis provides a third basis for selecting the number of factors to extract, suggesting selection of two factors (all those above the red line, indicating the factors which reflect better-than-random information).

The upper blue and red lines are for a PCA solution. For the PCA solution, by the Kaiser criterion, we would extract two components. By the parallel analysis criterion, which is to extract dimensions with better-than-random information (those above the red line), we would extract 1 component, which would be consistent with believing all six psychological tests measure a single latent variable, ability.

Parallel analysis thus compares the blue observed scree line with a red scree line for a simulated dataset of equal size but composed of random values. By this criterion, we would ask for a two-factor PFA solution since the second factor, while below the Kaiser eigenvalue = 1.0 line, is above the red scree line for random data. That is, the second factor will contain non-random information and is retained under the parallel analysis method. For this chapter, we follow parallel analysis for PFA and request a two-factor solution.

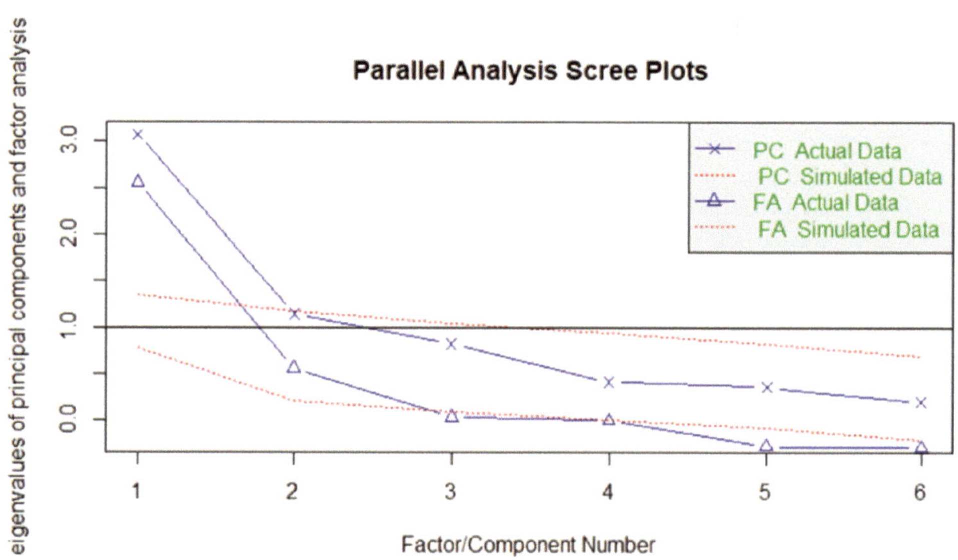

Figure 4.1 Parallel analysis for the ability data.

Orthogonal PFA with `fa()`

In this section, we continue to use the "ability" data as embedded in the covmatrix and corrmatrix data objects. We also choose to ask for an orthogonal solution. Oblique solutions with the `fa()` command are discussed in Chapter 7. As before, run PFA on the ability test variables, putting results in the object "faResultVarimax". We specify principal axis factoring (pa) as the method, with varimax rotation.

```
# Setup
base::setwd("C:/Data")
base::library(psych)

# Run the PFA model on the ability test data.
faResultVarimax <- psych::fa(r=corrmatrix,nfactors = 2, fm="pa",
rotate ="varimax", covar=FALSE, max.iter=1000)
```

Discussion of `fa()` options used above: Type `help(fa)` for more explanation and additional options.

- The `r=` option names a covariance matrix, a correlation matrix, or a raw data matrix. If raw data, the correlation matrix will be found using pairwise deletion. If covariances are supplied, they will be converted to correlations unless covar = TRUE.
- The `nfactors=` option is set to the number of factors to extract.
- Options for `rotate=` are "none", "varimax", "quartimax", "bentlerT", "equamax", "varimin", "geominT" and "bifactor" are orthogonal rotations. "Promax", "promax", "oblimin", "simplimax", "bentlerQ", "geominQ" and "biquartimin" and "cluster" are possible oblique rotations. The default is oblimin but we use varimax.
- If `covar=FALSE`, the correlation matrix is used, which is the default. If TRUE then the covariance matrix is used. Note that variable-level item complexities are only output when `covar=FALSE`.
- The fm= option sets the factor method. The default is "minres". Choices are:
  - "minres": minimizes the residual matrix using an OLS procedure.
  - "uls": also the minimum residual method.
  - "pa": the principal axis factoring solution, which we use here.
  - "ols": similar to "minres" but uses the empirical first derivative.
  - "wls": uses a weighted least squares (WLS) solution
  - "gls": uses generalized weighted least squares (GLS)
  - "ml": does maximum likelihood factor analysis
  - "minchi": minimizes the sample size weighted chi square when treating pairwise correlations with different numbers of subjects per pair.
  - "minrank": does a minimum rank factor analysis.
  - "alpha": does alpha factor analysis
- The `max.iter=` option sets the maximum number of iterations allowed to achieve convergence. The default is 50.

View the solution:

```
faResultVarimax
[Output:]
   Factor Analysis using method =  pa
   Call: psych::fa(r = corrmatrix, nfactors = 2, rotate = "varimax",
   covar = FALSE, max.iter = 1000, fm = "pa")
   Standardized loadings (pattern matrix) based upon correlation
   matrix

            PA1  PA2  h2   u2   com
   general 0.49 0.57 0.57 0.432 2.0
   picture 0.16 0.59 0.38 0.623 1.1
   blocks  0.18 0.89 0.83 0.166 1.1
   maze    0.13 0.43 0.20 0.798 1.2
   reading 0.93 0.20 0.91 0.089 1.1
   vocab   0.80 0.23 0.69 0.313 1.2

                          PA1  PA2
   SS loadings           1.83 1.75
   Proportion Var        0.30 0.29
   Cumulative Var        0.30 0.60
   Proportion Explained  0.51 0.49
   Cumulative Proportion 0.51 1.00

   Mean item complexity =  1.3
   Test of the hypothesis that 2 factors are sufficient.

   The degrees of freedom for the null model are  15  and the
   objective function was  2.48
   The degrees of freedom for the model are 4  and the objective
   function was  0.07

   The root mean square of the residuals (RMSR) is  0.03
   The df corrected root mean square of the residuals is  0.06

   Fit based upon off diagonal values = 0.99
   Measures of factor score adequacy
                                                    PA1  PA2
   Correlation of (regression) scores with factors  0.96 0.92
   Multiple R square of scores with factors         0.91 0.85
   Minimum correlation of possible factor scores    0.82 0.71
```

Discussion of **fa()** output above

1 That mean item complexity is only 1.3 supports the hypothesis that two factors are sufficient. That is, the loadings matrix is sufficiently close to simple factor structure that it should be interpretable. We also see, in the "com" column, that only the "general" test, with an item complexity of 2.0, diminishes simple factor structure since it cross-loads between the two requested dimensions.
2 Fit based upon off-diagonal values is relatively strong (0.99). A second fit measure is root mean square error of residuals (RMSR). In a well-fitting model, RMSR should be 0.05 or below for "good fit" and 0.08 or below for "adequate" fit. Here, at 0.03, fit is good by this criterion, though RMSR corrected for degrees of freedom is 0.06, which is adequate. A third fit measure is multiple R-squared of regression scores with factors, which is 0.91 for PA1 and 0.92 for PA2, which most researchers would consider acceptably high. When various fit measures point in different directions, this is a sign that the model is "marginal" rather than "strong" but here the model would be considered strong.
3 The variables contributing strongly (>0.6) to the PA1 factor are reading and vocab, making this the verbal factor. Only blocks contributed strongly to the PA2 factor, with picture and maze contributing moderately. The general test was crossloaded moderately on the two requested factors. It is not easy or necessary to infer labels for PA1 and PA2 based on the loadings, but we might call PA1 the "verbal ability" factor and call PA2 the "spatial ability" factor. Because labeling is subjective, we might bring in a panel of experts in a Delphi process to seek external consensus regarding the labels. Or we might forego labeling altogether.
4 Together, PA1 and PA2 cumulatively explain dome 60% of the variance in the ability items. Specifically, the PA1 and PA2 factors each contribute about 30% to variance explained. The lower the variance explained, the more the researcher may wish to specify a model with more factors and/or other variables.
5 Variables whose highest loading with any factor is less than 0.5 are by rule of thumb considered candidates for dropping from the two-factor model. For these data, the maze test is weak by this criterion and might be dropped from the factor model.

Factor scores in **fa()**

If the researcher wishes to obtain observation-level scores for each factor, these are easily obtained in **fa()** by adding the option `scores=`. This only works if the input is raw (individual-level) data, not a correlation matrix, as it was in this example.

Choices for the `scores=` option:

- `scores = "regression"`. This is the default, equivalent to `scores=TRUE`.
- `scores = "tenBerge"`: A correlation-preserving method "probably preferred for most solutions" unless the correlation matrix is improper.
- `scores = "Thurstone"` is based on simple regression.
- `scores= FALSE`. Do not compute factor scores.

To illustrate factor scores, we must turn to a different dataset based on individual-level observations, not correlation matrix input. We use the classic "iris" data, where observations were 150 plants. This data frame is part of R's built-in "datasets" package and does not need to be loaded. The command below asks for the usual regression-type scores.

```
# Run the fa() model for a 3-factor solution.
faResultScores <- psych::fa(r=iris[,1:4], nfactors = 3, fm="pa", ro-
tate ="varimax", scores="regression")

# List the first five observations.
utils::head(faResultScores$scores,5)
[Output:]
          PA1    PA2    PA3
    [1,] -1.04 -1.002 -0.123
    [2,] -1.40 -0.521  0.165
    [3,] -1.41 -0.348  0.121
    [4,] -1.39  0.214 -0.262
    [5,] -1.02 -0.828 -0.248
```

Save the factor scores to the object "fscores", which, in turn, is saved as a new column in the "iris" data frame. This is saved only to the iris data frame in memory.

```
fscores <- faResultScores$scores # Save scores to an object in
memory.

iris$fscores <- fscores          # Save scores to iris, but just in
memory.
```

To save to file, use the `save.csv()` command. The "iris" data in memory is saved to the "iris2.csv" file so as not to erase "iris.csv".

```
utils::write.csv(iris, file="iris2.csv")
```

Beta weights in `fa()`

Beta weights are used to find the principal components from the data and to find the factor score estimates for existing cases. Weights, of course, depend on the scoring method requested. Since beta weights are associated with scoring observations, the result object is usually based on observation-level rather than correlation matrix input. Beta weights for the variables will be in the "weights" element of the result object.

To get beta weights for the iris example, which was based on input of observation-level data, we used the iris results object labeled "faResultScores" above.

```
betaweights_iris <- faResultScores$weights
betaweights_iris
[Output]
                  PA1     PA2    PA3
    Sepal.Length  0.289 -1.5990  0.399
    Sepal.Width   0.249 -0.0828 -0.335
    Petal.Length  0.968  3.0895 -3.634
    Petal.Width  -0.195 -1.3527  3.263
```

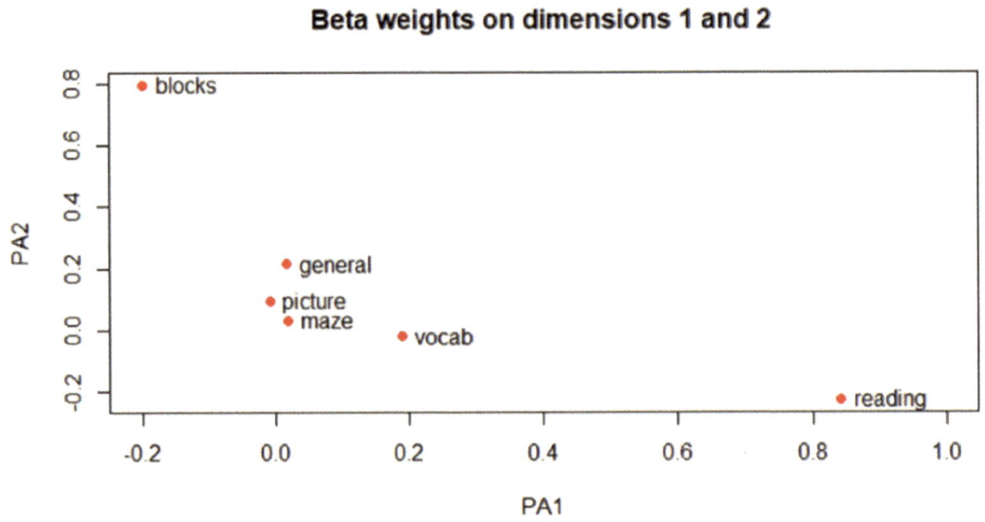

Figure 4.2 Plot of beta weights on dimensions 1 and 2.

To get beta weights for the ability test example, which was based on input of a correlation matrix, we used the ability tests results object labeled "faResultVarimax", also above.

```
betaweights_ability <- faResultVarimax$weights

betaweights_ability
[Output:]
                  PA1     PA2
   general  0.01770  0.2150
   picture -0.00799  0.0969
   blocks  -0.19831  0.7939
   maze     0.01916  0.0303
   reading  0.84178 -0.2240
   vocab    0.19059 -0.0204
```

Notice that whether input was individual-level data as with the iris data or involved correlation matrix input as with the ability tests data, the $weights element of the results object returns variable-level weights, which are the beta weights.

Beta weights may be plotted to show the importance of each variable in predicting the requested dimensions and scores for them. For the ability tests example, the visualization is Figure 4.2. Beta weights are not the same as factor loadings of variables and variables do not have the same rank order in both. Weights are standardized regression coefficients used to estimate (predict) the factors. Loadings are correlations of the variables with the factors. A higher beta weight means the variable is more important in estimating the factor. A higher loading means the variable contributes more to the meaning of the factor in the sense of explaining more of its variance and that the factor explains more of the variance in the variable. These are two contrasting meanings of "importance of the variable". Variable importance is usually interpreted in terms of loadings, not beta weights.

```
# Plot beta weights
base::plot(betaweights_ability, type="p", pch=16, col="red",
xlim=base::c(-.2,1.0), main="Beta weights on dimensions 1 and 2")

# Add labels
graphics::text(betaweights_ability[,1],betaweights_ability[,2], la-
bels=base::rownames(betaweights_ability), pos=4)
```

5 Quick start

Confirmatory factor analysis in R

Overview

As discussed at the start of Chapter 3, confirmatory factor analysis (CFA) in the sense of the researcher employing a confirmatory research design may be used with any of a variety of factor analytic procedures. However, in Chapter 5 we use the more specific meaning of CFA as using structural equation modeling (SEM) to confirm the researcher's factor model.

SEM has two steps: (1) confirming the measurement model (which is a type of factor model) and (2) testing the structural model (which is causal connections among the factors). CFA refers to the first step. In SEM, constructs based on indicator variables are called "latent variables" rather than "factors". The measurement model, which is the CFA step, includes specifying the number and labels of the latent variables and specifying which manifest variables (indicators) measure each construct. Constructs may also be called factors, components, or dimensions. The CFA model thus includes:

- How many constructs are modeled?
- Upon what measured variables (indicators) are the constructs based?
- Are the error terms for measured variables correlated?
- If error terms are correlated, are they correlated across constructs or only within?
- Is a higher-order factor model called for?

In exploratory factor analysis (EFA), every indicator loads on every factor, though some much more than others. In CFA, the model is almost always constrained so that only specified indicators load on the factor to which their content corresponds. If the measurement model (constructs and their specified indicators) is confirmed by CFA, the researcher may proceed to the second step, which is analysis of the structural model. Structural analysis deals with causal relationships among the constructs (dimensions, factors). Structural modeling is outside the scope of this book (instead see Kline, 1998: ch. 7; Garson, 2018).

Because in CFA the number of factors and the assignment of measured indicator variables to each is done a priori, simple factor structure is assured though goodness of fit (GOF) to the data is not assured. By the same token, there is no need for factor rotation in CFA because indicators are aligned with their factors by a priori constraint.

If EFA is for model generation, CFA is for model testing. In CFA, the model is tested based upon the researcher's a priori design. The researcher posits beforehand such things as the number and labels for the latent variables and the selection of appropriate

DOI: 10.4324/9781003279693-6

indicators. In PCA, PFA, and other forms of EFA, this need not be done. In fact, a major purpose of EFA is often to allow the algorithms to uncover how many constructs there are and what measures do or do not load on them. Thus, in EFA the model typically is revealed a posteriori.

DiStefano (2016: 167) concisely enumerated the key assumptions of CFA: "linearity, independent observations, sufficiently large sample size, multivariate normality of indicators, and a correctly specified model". Use of ordinal data further violates the continuous measurement assumptions of ordinary CFA. It is unlikely that the researcher fully meets all CFA assumptions. Therefore, the researcher is on much safer grounds if he or she examines at least two models and if the conclusion is that one model fits the data better than others. Examining a single model and concluding that it is the true model is an untenable position.

The distinction between confirmatory and exploratory or between a priori and a posteriori is not a clear-cut choice between CFA and EFA (Child, 2006; Hoelzle & Meyer, 2013). This is because one may use PCA, PFA, or other "exploratory" techniques to test an a priori model posited by the researcher. To simplify, under EFA the researcher may posit a model a priori, then see if the PCA or other computer algorithms generate a factor structure consistent with that model. In both EFA and CFA, GOF measures such as RMSEA and AIC exist to assess whether the a priori model is consistent with the data. What distinguishes CFA is that an a priori model is required, not optional. Second, CFA allows for modeling of error, as is discussed in the next section.

Testing error in the CFA measurement model

Using SEM, the researcher can explore CFA models with the assumption that error terms for the indicator variables are uncorrelated. Or the researcher may assume certain correlations among the error terms. Such measurement error terms represent causes of variance due to unmeasured variables as well as random measurement error. Depending on theory, it may well be that the researcher should assume unmeasured causal variables will be shared by indicators which will therefore correlate. SEM testing for correlated error may well be merited. Also, modification indexes (discussed further below) may suggest adding one or more correlated errors to the model.

Correlated error specified in the model tests the possibility that indicator variables correlate not just because of being caused by a common factor (the construct, which is not directly measurable but which determines the values of the indicators), but also due to common or correlated unmeasured variables. This possibility would be ruled out if the fit of the model specifying uncorrelated error terms was as good as the model specifying correlated error.

Other CFA tests

In CFA analysis, it is common to use "likelihood ratio tests", also called "redundancy tests", to compare one model with another. This might be to compare a full model with a model in which one indicator for a latent variable is omitted. Or the researcher might compare a model in which all indicators were constrained to be uncorrelated with an unconstrained one in which certain indicators are allowed to correlate. If the constrained model is not significantly worse than the unconstrained one, the researcher rejects the hypothesis that correlated error among indicators exists.

Another type of testing in CFA is the "measurement invariance test". This tests whether a set of indicators reflects a latent variable equally well across groups in the sample. For instance, are the indicators for "tolerance" weighted the same, say, between an American group and a Japanese group? If they are, then the meaning of tolerance measured by the instrument is valid across cultures. If they are not, then the meaning of tolerance differs between the two cultures. The constrained model is one in which factor loadings are specified to be equal between groups for each class of the grouping variable. If the unconstrained model is not significantly better than the constrained-equal model, then the researcher concludes that the indicators are valid across groups. This procedure is also called "multiple group CFA". If the model fails this test, then it is still necessary to examine each indicator for group invariance, since some indicators may still be invariant.

A third type of test is the "orthogonality test". In this test, correlations among latent variables (or among indicators for a given latent variable) are constrained to be 0. Absence of a causal arrow between two indicators or latent variables in CFA implies a zero constraint. If the unconstrained model is not significantly better than the constrained one, the factors in the model can be considered orthogonal (uncorrelated, independent).

Goodness-of-fit measures

There are a variety of GOF criteria in SEM and therefore also in CFA. A primary metric is the deviance, also called model chi-square, the likelihood ratio, or −2 log likelihood (−2LL). This derives from the fact that the primary estimation method, maximum likelihood (ML) generates a measure of error called the likelihood (L). When the log of L is multiplied by minus 2, this measure assumes a chi-square distribution which may be used to perform significance tests. The −2LL coefficient is model deviance or model chi-square and is used in likelihood ratio tests. Any two nested models may be compared by an LR test. Nested means all the effects in the smaller model must be present in the larger model.

When models are not nested, comparison is often done by using penalized versions of the deviance. Information theory measures such as Akaike's information criterion (AIC) and the Bayes information criterion (BIC) are common examples. The model with the lower AIC, for instance, may be judged the better-fitting model even if the models are not nested.

There are many other GOF measures for CFA. For this reason, it is commonly recommended that the researcher look at the deviance (model chi-square) and at least two others. Many of the other criteria use formulas which compare the researcher's model-implied covariance matrix with the observed covariance matrix. To the extent that the researcher's model can reproduce the observed covariance matrix, the model is judged well-fitting. A covariance, of course, is the unstandardized correlation between two variables. If anywhere in the causal model drawn by the researcher there is no connecting arrow, the researcher is implying zero covariance for those two variables. The researcher may also set a specific effect size for a given covariance. Baseline GOF measures assess how well the model-implied covariance matrix matches the observed one.

GOF is affected by saturation. If the researcher's model contains all possible causal arrows, there is complete saturation of the model. In a completely saturated model, the model-implied covariance matrix always matches the observed covariance matrix perfectly. The take-away is that if the researcher is getting very high baseline GOF measurements, this might be due to the researcher's model too closely approaching the saturated model. The objective of SEM is to find the least saturated (most parsimonious) model which still has good fit. As a check against being misled by saturation of the model, there are "parsimony-adjusted" GOF measures. These should be checked by the researcher whenever baseline measures seem implausibly high. CFA models are not saturated, however, because indicator variables are modeled as uncorrelated (there are no connecting arrows) and in a measurement model, there are no regression arrows connecting the latent variables.

Below is a brief synopsis of GOF measures for SEM. Not all measures are supported by all statistical packages, R packages included. See Garson (2018) and other texts for a fuller discussion.

- *Model chi-square* (deviance) was discussed above. Whether its value should be significant or not depends on the model to which the researcher's model is being compared. The usual test compares the researcher's model to the saturated model (a model with all possible connecting arrows), non-significance is desired since that means the researcher's more parsimonious model is as explanatory as the saturated model, which explains variance perfectly. Put another way, non-significance means that the variance-covariance matrix of the researcher's model is not significantly different from the observed variance-covariance matrix reflected in the saturated model.

  When the researcher's model is compared to the null model (a less common test), significance is expected since the researcher's model should have significantly less error than the null model.

  It is important to note that like other significance tests, even trivial differences may be reported as "significant" if the sample is large enough. For this reason, it is routine to discount LR tests for larger samples and instead examine other GOF measures.
- *Hoelter's critical N* is used to judge if sample size is too large to use model chi-square as a criterion to reject the researcher's model based on a significant model chi-square value when comparing it to the saturated model. If sample size is larger than Hoelter's critical N, then even trivial differences between the observed and model-implied covariance matrices may be significant by chi-square.
- *Information theory measures like AIC and BIC.* Information theory measures are based on absolute fit of model-implied covariances to observed covariances. As such, they represent modifications of model chi-square for varying purposes such as penalizing for lack of model parsimony and/or sample size. These measures are appropriate when comparing models which have been estimated using ML estimation. They are not interpreted for a single model but rather, when comparing models, lower is a better fit. Unlike model chi-square, there is no significance test for the difference between information theory measures because there is no known distribution of the difference in information theory measures between models. The lowest information theory measure may flag the best-fitting model of those considered but that may not include the correct model. Information theory measures are not to be used to compare models derived from different datasets. Rather, data and variables are assumed to be fixed across models. The researcher should not "mix and match" information theory measures (e.g., not use AIC to compare Model A to Model B and then use BIC to compare Model A to Model C. BIC imposes a greater penalty

than does AIC. There are several "flavors" of AIC such as consistent AIC (CAIC). BIC may be recommended when sample size is large or the number of parameters in the model is small.

- *ECVI and MECVI.* The expected cross-validation index (ECVI) is used for comparing non-nested models in multi-sample analysis of a development sample model with the same model using a validation dataset, in cross-validation. ECVI may be thought of as the AIC value per unit of analysis. Lower ECVI is a better fit. Modified expected cross-validation index (MECVI) is a variant which imposes a greater penalty for model complexity. Like ECVI, lower is better between models.
- *Relative chi square* is easily calculated as model chi-square divided by degrees of freedom. This norming is an attempt to make model chi-square less dependent on sample size. Scholars vary on cutoff criteria but lower relative chi-square is still better. While some authorities consider 3 or less to be acceptable, some set a cutoff as high as 5 or less. High relative chi-square means there is higher error in the model that expected based on the number of parameters. Very low relative chi square may signal lower than expected error, perhaps due to the researcher's model being too saturated.
- *Standardized root mean square residual* (SRMR). Root mean square error is normally reported in its standardized form. Since error (residual) is the difference between observed and model-implied standardized covariance matrices, SRMR is the average difference. Note that SRMR tends to be lower when sample size is larger or there are more parameters in the model. The smaller the SRMR, the better the model fit.
- *Root mean square error of approximation (RMSEA).* RMSEA is a popular measure of fit, partly because it penalizes for model complexity. This penalty is less for large sample size. Nonetheless, RMSEA is one of the fit criteria which is less sensitive to sample size than some others. A value less than 0.05 is widely considered a good fit (or for confirmatory purposes) and below 0.08 adequate fit (or for exploratory purposes).
- *Comparative fit index (CFI).* CFI, compares the covariance matrix predicted by the model to the observed covariance matrix, and compares the null model (covariance matrix of 0's) with the observed covariance matrix, to gauge the percent of lack of fit which is accounted for by going from the null model to the researcher's default model. CFI = 0.90 means that 90% of the covariation in the data can be reproduced by the given model. As such CFI has a "proportionate improvement of fit" interpretation. CFI is among the most widely used GOF measures. CFI is independent of model complexity and is among the measures least affected by sample size. By convention, CFI of 0.95 or higher is a good fit for confirmatory purposes and 0.90 or higher is an adequate fit for exploratory purposes. However, there is no universal agreement on these cut-offs, nor on cut-offs for other CFA measures.
- *Tucker–Lewis index (TLI),* also called the non-normed fit index (NNFI) or rho2, is a criterion less affected by sample size. Though CFI is more widely reported, TLI is preferred when the researcher wishes to penalize more strictly for lack of parsimony. The same cutoffs as for CFI are suggested. Note that if a path is dropped from the model, there is a possibility that TLI will increase due to diminished complexity even when the model chi-square increase was small and the path was trivial or non-significant. Also, TLI is not guaranteed to vary from 0 to 1.
- *Incremental fit index (IFI).* IFI, also called the Bollen non-normed index Delta 2, is an adjustment to NFI to reduce the influence of sample size. Overall, however, IFI

is among the less-reported GOF measures. For small samples, NFI is biased downward. IFI attempts to compensate for this by subtracting model degrees of freedom in the denominator of its equation. However, it has been found that for small sample sizes, IFI is biased upward, especially when the model is misspecified. Cutoffs are the same as for CFI.

- *Parsimony comparative fit index (PCFI), parsimony normed fit index (PNFI), and other parsimony measures.* These measures penalize their "regular" counterparts for model complexity/lack of parsimony. As a result, parsimony measures tend to run lower. As a rule of thumb, values above 0.6 are considered good fit for confirmatory purposes and above 0.50 for exploratory purposes. There is controversy surrounding use of parsimony measures, with some researchers opposed to penalizing models with more parameters for purposes of model acceptance. This school suggests using other GOF measures to assess acceptable models, then using parsimony measures to select among the set of acceptable models.
- *Other GOF measures.* There are a number of other GOF measures for CFA and SEM. Some are older metrics which are no longer recommended. These include the goodness-of-fit index (GFI), the adjusted goodness-of-fit index (AGFI), the normed fit index (NFI – TLI is recommended instead), and the relative fit index (RFI – TLI is recommended instead).

The researcher should be aware that the correspondence of GOF measures to true causality is a subject of great controversy, with positions ranging from rejection of any use of fit measures to embrace of their use. As a practical matter, however, the great majority of CFA studies finding their way to publication report GOF measures. Meeting GOF criteria, however, does not assure that the causal model is true or even plausible, nor that the model comports with theory and the literature, nor that the true model was the one tested, nor that better-fitting models might not exist, nor that researcher bias and subjectivity might not have infiltrated the reported results.[1] However, failing to meet fit criteria is reason to strongly suspect the model is problematic.

Modification indices and parameter change coefficients

Modification indexes (MI) are used in conjunction with parameter change coefficients to judge whether the researcher's model would be improved significantly by adding causal connections (arrows) to the model. There are various types of MIs:

- MI's for regression weights (paths): In a CFA model, there are regression paths from each latent variable to its indicators. In the case of MI for estimated regression weights, the MI may suggest that a path should be added between a latent variable and an indicator variable for a different construct. That is, an indicator for latent variable B may be suggested as an indicator for latent variable A, something not contemplated by the researcher. To add such an arrow raises questions about construct specification as generally the researcher wishes each construct to be correspond to a unique set of indicators.
- MI's for covariances: In CFA, latent variables are allowed to covary. In graphical SEM, a curved two-headed arrow signifying covariance connects each latent variable to each other in the measurement model. In equation-based SEM, covariance equations connect each latent variable. In the case of modification indexes for

covariances, the MI has to do with the decrease in deviance (model chi-square) if two error term variables are allowed to correlate (recall lower is less error).

- MI's for error terms: MI's also may suggest adding covariance arrows connecting error terms associated with indicators. Generally, the researcher does not want to add arrows between the error term for an indicator variable for one latent variable and the error term for an indicator variable for a different latent variable. Rather, the researcher generally wishes the indicator variables for different latent variables to be uncorrelated, and for their error terms to be uncorrelated as well, so as to preserve the conceptual distinction between latent variables. If the suggested covariance is between error terms of two indicator variables for the same latent variable, this means that once the latent cause of correlation of the two indicator variables is accounted for (by the latent variable), there is still an unmeasured influence connecting the two indicator variables. This influence might be that one indicator causes the other, or they may share an unmeasured common outside influence connecting them. Further investigation is warranted.

MI coefficients may be used to alter models to achieve a better fit but this must be done only with theoretical justification. Some oppose use of MI's altogether on the ground that all model specifications should be theory-driven, not data driven (e.g., Gorsuch, 1988: 235). Others recommend thorough model diagnosis with EFA and only then running a CFA model, not using MIs for re-specification of the model (Reise et al., 2018: 699). Blind use of MI runs the risk of capitalization on data noise and model adjustments which make no substantive sense. MI-driven modifications may not generalize to other data (Kline, 2012, 2013; Bandalos & Finney, 2019) and therefore use of cross-validation may be an appropriate check. The researcher should look at MI's as just flags to consider adding a direct path or covariance. These flags may represent good or may represent mistaken suggestions.

If MI's are used, the researcher should also take into account the effect size of the arrow to be added as indicated by the parameter change coefficient. "Par change" is the estimated path coefficient change when adding arrows. Since absence of an arrow corresponds to a 0 path coefficient weight, "Par change" is the regression coefficient for the added arrow. Most SEM software will list the expected parameter change and its significance level along with the MI for that path. If a given arrow is added, the actual new parameter value may differ somewhat from the "Par Change" estimate. The researcher may decide not to add an arrow flagged by MI if the parameter change is trivial. Likewise, the researcher may wish to add an arrow where the parameter change is large in absolute size even if the corresponding MI is not the largest one.

How large is a "large" MI? The minimum MI value would be 3.84, since chi-square must drop that amount simply by virtue of having one less parameter (path) in the model. This is why the default threshold is set to 4 in most SEM software. The researcher can set a higher value if desired, causing MI's below the specified level not to be output.

To summarize, the researcher may consider adding an arrow to the model if the MI is high and the parameter change coefficient is high and if it makes theoretical sense. A variety of criteria have been used. Although a number of criteria have also been suggested as rules-of-thumb for a "large" MI (>4, >10, >30, and >100 all have been mentioned), such criteria are beside the point, which is to consider adding arrows to the model if justified by theory and if adding the arrow will cause model fit to reach acceptable levels. The less certain the theoretical justification for adding an MI-flagged path or covariance, the greater the need for cross-validation of the model on other data than the model development data.

While the discussion above deals with modification indexes, MIs are univariate. The multivariate version is called the "Lagrange multiplier coefficient" (LM). Some statistical packages, such as SAS and Stata, output LM rather than MI coefficients, though they may still be called "modification indices". In general, LM is preferred to MI since LM is based on fuller model information. However, the difference may be moot since MI and LM coefficients often carry the same modeling implications and in any event are only advisory with no clear cut-offs. For both MI and LM coefficients, implied recommendations for adding arrows may not be justified by theory.

Path significance and critical ratios

If modification indexes have to do with "growing the model" by adding arrows, then "trimming the model" by deleting arrows is done in the traditional manner: non-significant regression paths and/or covariances are considered for dropping from the model. While not required, the strategy is often to grow the model to its maximum, then trim the model.

Model trimming usually centers on deleting arrows in the model which are not significant, provided it makes theoretical sense to do so. For regression paths, most SEM software will print out the p significance level. In social science, p values above 0.05 flag non-significant paths. If the p value is not available, paths with a critical ratio less than 1.96 indicate that the given path is non-significant at the 0.05 level or better. Significance of covariances is interpreted in the same manner. The significance of the standardized and unstandardized path estimates is identical.

R packages for CFA and SEM

In this chapter we use the "sem" package, authored by John Fox et al., to implement a CFA model. The "sem" package is actually a wrapper for and relies on the "lavaan" package. Among the functions in the sem package are `sem()` and `cfa()` for CFA, `multigroupModel()` for multi-group modeling, `bootSem()` for bootstrap estimates, `miSem()` for multiple imputation of missing data, and the `Diagram()` for visualizing SEM models. Fox is also author of the "polycor" package and the `polychor()` command for polychoric correlations for input using mixed data levels. See Fox (2006).

The "lavaan" package is authored by Yves Rosseel with Terrence D. Jorgensen, Nicholas Rockwood, and others (Rosseel, 2012). For information, see Broc and Gana (2019). See also the tutorial at https://lavaan.ugent.be/tutorial/index.html. Key functions in this package in addition to the `lavaan()` function itself include `cfa()` for CFA, `sem()` for structural equation models, `fitMeasures()` for GOF measures, `modificationIndices()` for modification indices, `bootstrapLavaan()` for bootstrapped estimates, `growth()` for growth curve models. The "semTools" package provides a variety of supplementary commands for lavaan objects, such as the `nullRMSEA()` function, which calculates that GOF measures for the null model. The "semPlot" package and its `semPaths()` program supports creation of path diagrams for sem, lavaan, and other models.

Though not discussed in this chapter, another popular SEM package in R is "OpenMx". This package supports various advanced SEM models and works with lavaan. See Boker et al. (2011). Others include:

- The widely-used "psych" package has the `lavaan.diagram()`, `sem.diagram()`, and `sem.graph()` functions for SEM visualization, as well as `esem()` functions for exploratory SEM.
- The "qgraph" package, among other functions, imports and visualizes lavaan SEM models.
- The "SEMModComp" package supports comparing mean and covariance structure SEM models.
- The "simsem" package provides a framework for Monte Carlo simulation of SEM models.
- Note also that the "MplusAutomation" package in R is an interface to MPlus, a leading commercial SEM package. The same is done for the commercial SEM package EQS by the "REQS" package in R.

Example data

The Wheaton dataset is a classic dataset used in this chapter to illustrate CFA. The data take the form of a covariance matrix of variables based on a sample of 932 individuals who were measured on psychological scales of anomia and powerlessness as well as on their education and a socioeconomic status index. The research purpose was to confirm that the measurement model for the alienation latent variable (factor) was the same in 1971 as in 1967, controlling for changes in socioeconomic status. See Wheaton, Muthén, Alwin and Summers (1977).

The Wheaton data has these latent and indicator variables:

- The latent variable "alien67" measured by the indicators "anomia67" and "powerless67"
- The latent variable "alien71" measured by the indicators "anomia71" and "powerless71"
- The latent variable "ses" (socio-economic status) measured by the indicators "education" and "sei" (socio-economic index)

Creating the CFA model with lavaan

The CFA analysis below is based on the classic Wheaton et al. (1977) model panel data on the stability of the latent variable "alienation" discussed above in the "Example Data" section. Data are input as a covariance matrix, not observation-level data. The example is similar to that on the help page on the `getCov()` function of the "lavaan" package.

The purpose of the model is to test the stability of the latent variable "Alienation" between two points in time (1967 and 1971). The following variables were in the model:

- Anomia67 and Powerless67: observed indicators of the latent variable "Alienation67". The "Powerless" indicators measured perceived powerlessness.
- Anomia71 and Powerless71: observed indicators of the latent variable "Alienation71"
- Education and SEI: observed indicators of "SES", which is socioeconomic status.

The model posits that Alienation71 is caused by Alienation67, and both are caused by SES, which is modeled as stable over time. There is an initial "Model A" and a better-fitting revised "Model B," which adds covariances between Anomia67 and Anomia71. These are presented for instructional purposes only. Actual research on the Wheaton data involved a number of other models.

Setup

```
base::setwd("C:/Data")        # Declare the working directory
# The "lavaan" and "semPlot" packages must be installed first.
base::library(lavaan)         # The "lavaan" package has cfa() and
                                related functions
base::library(semPlot)        # For visualization of CFA
```

Data for the classic Wheaton et al. (1977) model, which was based on panel data for 1967 and 1971, is input as a covariance matrix. The Wheaton model assesses the stability over time of the latent variable "alienation". Missing values, if any, were handled when the covariance matrix was created and are not an issue here. If we had individual-level input (e.g., in a comma-separated values file), we would want to deal with missing values and also to check on the normal distribution of the variables.

Step 1. Read the Wheaton covariance matrix into the object "lowertriangle", then convert this into the covariance matrix "Wheaton's".

```
# The upper triangle in the covariance matrix would be redundant.
lowertriangle <- '
 11.834,
  6.947,    9.364,
  6.819,    5.091,   12.532,
  4.783,    5.028,    7.495,    9.986,
 -3.839,   -3.889,   -3.841,   -3.625,    9.610,
-21.899,  -18.831,  -21.748,  -18.775,   35.522,   450.288 '

# Convert "lowertriangle" to a full symmetric covariance matrix with
names.
wheaton.cov <- lavaan::getCov(lowertriangle, names =
base::c("anomia67", "powerless67", "anomia71",
"powerless71","education","sei"))
```

Step 2. In step 2, we create but do not yet run the CFA model. The CFA model specifies the number and labels of the latent variables and specifies the indicator measures for each. Unlike the example below, typically in SEM, there are three or more indicators per latent variable. As latent variables have no inherent scale of measurement, it is necessary to assign a metric to each one. This is done by constraining the regression path of a latent variable's first indicator variable to have a value of 1.0. While a different indicator than the first might be used, this would make no difference in the results. As assigning the metric is done automatically "behind the scenes", this assignment does not appear explicitly in the model below but its presence will be seen in the later visualization of the worked model. In addition, it is usual to freely calculate rather than constrain to 0 the covariances of each latent variable with each other.

```
# Create the CFA measurement model using formulas.
# Start and end with a single quotation mark.
wheaton.meas.model <- '
   # Specification of latent variables.
   # Note the use of the equal sign followed by a tilde operator
   for latent variables.
   # Line 1 defines ses as a latent variable associated with the
   education and sei indicators.
   ses      =~ education + sei
   alien67 =~ anomia67 + powerless67
   alien71 =~ anomia71 + powerless71

   # Covariances connecting each latent variable to each other.
   # Note the use of the double tilde operator for covariances and
   variances
   # Line 1 says ses and alien67 are allowed to covary
   ses ~~ alien67
   ses ~~ alien71
   alien67~~ alien71
   '
   # End of model specification.
```

Step 3. Next, we run the measurement model using the `cfa()` function in the "lavaan" package. Results (the fit) are put in the object "mm_fit", which is of data class "lavaan".

```
# Run the model, making sure the "lavaan" package is still loaded.
base::library(lavaan)
set.seed(123)
mm_fit <- lavaan::cfa(wheaton.meas.model, sample.cov = wheaton.cov,
sample.nobs = 932)
```

Step 4. Summarize the results, including goodness-of-fit measures. In the "Parameter Estimates" section, we see that all paths and covariances specified in the model are significant. Had any been non-significant, we would consider dropping them from the model

```
# Show the results and note non-significant paths and covariances.
# Simply typing "mm_fit" would give the first seven lines.
lavaan::summary(mm_fit, standardize=TRUE, fit.measures=TRUE)
[Output:]
lavaan 0.6-11 ended normally after 82 iterations
  Estimator                                              ML
```

```
  Optimization method                              NLMINB
  Number of model parameters                           15
  Number of observations                              932

Model Test User Model:
  Test statistic                                   71.546
  Degrees of freedom                                    6
  P-value (Chi-square)                              0.000

Model Test Baseline Model:
  Test statistic                                 2133.722
  Degrees of freedom                                   15
  P-value                                           0.000

User Model versus Baseline Model:
  Comparative Fit Index (CFI)                       0.969
  Tucker-Lewis Index (TLI)                          0.923

Loglikelihood and Information Criteria:
  Loglikelihood user model (H0)                -15246.680
  Loglikelihood unrestricted model (H1)        -15210.906
  Akaike (AIC)                                  30523.359
  Bayesian (BIC)                                30595.919
  Sample-size adjusted Bayesian (BIC)           30548.281

Root Mean Square Error of Approximation:
  RMSEA                                             0.108
  90 Percent confidence interval - lower            0.087
  90 Percent confidence interval - upper            0.131
  P-value RMSEA <= 0.05                             0.000

Standardized Root Mean Square Residual:
  SRMR                                              0.021

Parameter Estimates:
  Standard errors                                Standard
  Information                                    Expected
  Information saturated (h1) model             Structured

Latent Variables:
                   Estimate  Std.Err
  ses =~
    education         1.000
    sei               5.329    0.430
  alien67 =~
    anomia67          1.000
    powerless67       0.889    0.041
  alien71 =~
```

```
    anomia71                1.000
    powerless71             0.849    0.040
  z-value  P(>|z|)        Std.lv  Std.all
                            2.580    0.833
   12.406    0.000         13.751    0.648
                            2.795    0.813
   21.422    0.000          2.483    0.812
                            2.970    0.839
   21.252    0.000          2.521    0.798

Covariances:
                         Estimate  Std.Err
  ses ~~
    alien67                -4.087    0.356
    alien71                -4.040    0.367
  alien67 ~~
    alien71                 6.216    0.439
  z-value  P(>|z|)        Std.lv  Std.all
  -11.469    0.000        -0.567   -0.567
  -11.010    0.000        -0.527   -0.527
   14.143    0.000         0.749    0.749

Variances:
                         Estimate  Std.Err
   .education               2.941    0.499
   .sei                   260.713   18.212
   .anomia67                4.011    0.343
   .powerless67             3.188    0.271
   .anomia71                3.697    0.373
   .powerless71             3.621    0.292
    ses                     6.659    0.640
    alien67                 7.810    0.590
    alien71                 8.821    0.645
  z-value  P(>|z|)        Std.lv  Std.all
    5.894    0.000          2.941    0.306
   14.315    0.000        260.713    0.580
   11.708    0.000          4.011    0.339
   11.763    0.000          3.188    0.341
    9.917    0.000          3.697    0.295
   12.417    0.000          3.621    0.363
   10.404    0.000          1.000    1.000
   13.236    0.000          1.000    1.000
   13.667    0.000          1.000    1.000
```

Discussion of selected parts of the summary output above:

- Estimation is by ML, which is standard for SEM models, though other estimation methods exist and are listed at https://lavaan.ugent.be/tutorial/est.html. Options include estimation methods with robust standard errors. Some methods require individual-level data rather than covariance or correlation matrix input. Different estimation and optimization methods will lead to different results.

- The optimizer is NLMINB, which is used to find the best fitting parameter values for the estimator. Other optimizers are available and are listed at https://rdrr.io/cran/lavaan/man/lavOptions.html.
- The "Test statistic" for the "Model Test User Model" (71.546) is the model chi-square (deviance, −2 log likelihood). It conforms to a chi-square distribution and so may be used to calculate the significance of the model. It tests if the model-implied covariance matrix is the same as the observed covariance matrix (see discussion below in the "Residual analysis" subsection. A finding of non-significance means the researcher's model reproduces the observed covariance matrix. However, this statistic is often discounted because it is highly sensitive to sample size. Even tiny differences from observed may be "significant".
- Using the baseline GOF measures of the comparative fit index (CFI = 0.969) and the Tucker-Lewis index (TLI = 0.923), fit is judged good for confirmatory research if above 0.95. This is a more rigorous rule-of-thumb cutoff than the 0.90 level formerly used, but which is now considered adequate only for exploratory research. On the other hand, some researchers even today consider 0.90 to be "okay". That the two measures here differ illustrates that in CFA and SEM modeling, the researcher should look at multiple fit criteria. See the discussion of TLI, CFI, and other fit measures in the earlier section on "Goodness-of-fit measures" in this chapter.
- Another fit measure, the root mean square error of approximation (RMSEA) is considered "good" if 0.05 or less and "adequate" if 0.08 or less. Here it is estimated to be 0.108, which is considered less than adequate fit. The "P-value RMSEA <= 0.05" is the PCLOSE coefficient, which is the probability that the RMSEA value reflects "close fit", which here is 0.000.
- Yet another fit measure, standardized root mean square residual (RMSR) here is estimated to be 0.021. By rule of thumb, a value less than 0.08 is considered good fit (Hu & Bentler, 1999).
- Akaike' information criterion (AIC), the Bayesian information criterion (BIC), and sample-size-adjusted BIC are reported. These are not directly useful here, but when comparing models, lower is better. BIC is a little more conservative than AIC.
- The latent variables, being standardized, by definition all have variances of 1 and means of 0.0.
- In the estimates section, the first two rows deal with the "ses" latent variable, which is a function of the measured variables "education" and "sei". The first indicator (education) is the reference indicator which assigns the metric to ses. It is constrained to have a factor loading (interpreted like regression estimates) of 1.0. In standardized terms, this is 0.833. The unstandardized estimate for the "sei" variable is 5.39, which is 0.648 in standardized terms. The regression connecting ses to sei is significant at the 0.000 level. Significance for the education indicator is not shown because it is the reference variable. All the regression paths and covariances in the model are significant.

```
                Estimate  Std.Err  z-value  P(>|z|)   Std.lv  Std.all
ses =~
  education        1.000                                2.580    0.833
  sei              5.329    0.430   12.406    0.000   13.751    0.648
```

A convenient table of factor loadings may be obtained using the R code below, which takes values from mm_fit, filters to select the latent variables (the ones for which the operator is "=~"), selects the columns wanted, and displays a simple table with the `kable()` function from the "knitr" package. In the table, "b" represents the unstandardized weights, "Beta" is the standardized weights, and "p-value" is the significance of either one. All loadings are significant.

```
# The "dplyr" and "knitr" packages need to have been installed first.
base::library(dplyr)         # has filter(),  and select()
base::library(knitr)         # has  kable() to generate simple tables
lavaan::parameterEstimates(mm_fit, standardized=TRUE) |>
  dplyr::filter(op == "=~") |>
dplyr::select('Latent Variable'=lhs, Indicator=rhs, "b"= est, SE=se,
Z=z, 'p-value'=pvalue, Beta=std.all) |>
  knitr::kable(digits=2,format="simple",caption="Factor Loadings")
[Output:]

Table: Factor Loadings

Latent Variable   Indicator        b      SE        Z   p-value
Beta
---------------- ------------  -----  -----  ------  --------  -----
ses               education     1.00   0.00      NA           0.83
ses               sei           5.33   0.43   12.41         0   0.65
alien67           anomia67      1.00   0.00      NA            0.81
alien67           powerless67   0.89   0.04   21.42         0    .81
alien71           anomia71      1.00   0.00      NA            0.84
alien71           powerless71   0.85   0.04   21.25         0   0.80
```

Also available to extract the unstandardized and standardized loadings are the `lavaan::parameterEstimates()` and `lavaan::standardizedSolution()` functions respectively. These functions provide some additional information, such as the upper and lower confidence limits on the estimates of the loadings.

Step 5. Visualize the CFA model.

```
# The simple command below, not run, displays a default figure for
the model.
semPlot::semPaths(mm_fit, whatLabels="std")

# However, one may set many visualization parameters (even more than
the ones listed here).
# The result is Figure 5.1.
semPlot::semPaths(mm_fit,
what="stand",                # Calls for standardized estimates
                               ("std" is a synonym);
                             # Use "est" for unstandardized (but set
                               fade = FALSE)
```

```
    fade = TRUE,               # If TRUE, weaker paths are faded
    rotation = 2,              # Place latent variables on left (1=top,
                               3=bottom, 4=right)
    style="lisrel",            # Default is "ram"; "mx" and "OpenMx" are
                               also available.
    layout="tree2",            # tree is the default; others include
                               circle, spring
    curvature = 3,             # Curvature of covariance arrows; default
                               is 1.
    sizeLat = 14,              # Size of latent variables
    sizeMan = 16,              # Size of manifest (indicator) variables
    sizeMan2 = 5,              # Height of manifest variables
    esize=3,                   # Size of largest edge
    fixedStyle = 5,            # Line type (lty) for fixed parameters.
                               Default is 2. Solid
                               # line is 1. The regression arrow from
                               the indicator that
                               # sets the latent variable metric is
                               fixed at 1.0, for ex.
    nCharNodes = 0,            # Do not abbreviate
    edge.color="dodgerblue",          # Color of lines
    edge.label.cex = 1.5,             # Size of values
    edge.label.color = "gray20"       # Color of values
)
```

Figure 5.1 below shows standardized factor loadings, covariances, and error terms. The latent variables are portrayed as circles and the manifest indicators as rectangles.

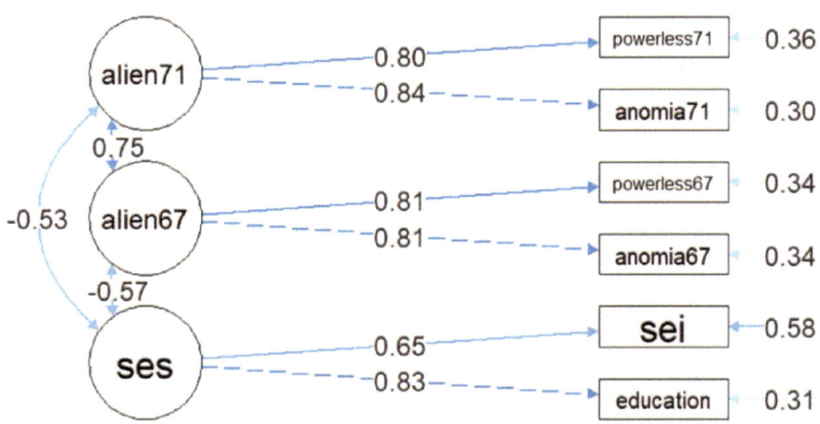

Figure 5.1 The Wheaton measurement model in lavaan.

Residual analysis

A residual is the difference between observed minus expected (O − E). In CFA and SEM generally, this is the difference between the observed covariance matrix and the model-implied covariance matrix. GOF is usually defined in terms of low residuals, which means small differences between these two matrices. Put another way, good fit assesses how well the model reproduces the observed covariance matrix. Observed and implied will differ wherever there is no arrow (causal path) or double-headed curved arrow (covariance) specified by the researcher. Absence of arrows implies covariances of zero. Observed and implied may also differ if the researcher has constrained some paths or covariances, as when they are constrained to be a specific value or when constrained to be equal between groups.

Note that what we here call the "covariance matrix" is short for "variance-covariance matrix", as it is customary for the matrix diagonal to contain variances and the off-diagonal cells to contain covariances. The upper off-diagonal triangle is redundant with the lower off-diagonal triangle.

It follows that it is helpful to compare the observed and implied covariance matrices, and the difference between them (the residual matrix). For the current example, the R commands below summarize these three matrices. For this, we use the `lavInspect()` function from the "lavaan" package. This function is used below for the observed, implied, and residual matrices, but it supports inspection of dozens of other elements in lavaan models. It works with the results object from a `lavaan()`, `cfa()` or `sem()` command, such as the object "mm_fit" for the Wheaton measurement model.

```
# The observed covariance matrix.[2]
observed <- lavaan::lavInspect(mm_fit, what="observed")
observed <- observed$cov
base::round(base::summary(observed[base::lower.tri(observed)]),3)
[Output:]
      Min.  1st Qu.   Median     Mean  3rd Qu.     Max.
   -21.876  -11.320   -3.621   -1.649    5.949   35.484

# The implied covariance matrix.
implied <- lavaan::lavInspect(mm_fit, what="implied")
implied <- implied$cov
base::round(base::summary(implied[base::lower.tri(implied)]),3)
[Output:]
      Min.  1st Qu.   Median     Mean  3rd Qu.     Max.
   -21.781  -11.180   -3.429   -1.634    5.869   35.484

# The residual covariance matrix.
resids <- lavaan::lavInspect(mm_fit, what="resid")
resids <- resids$cov
base::round(base::summary(resids[base::lower.tri(resids)]),3)
[Output:]
      Min.  1st Qu.   Median     Mean  3rd Qu.     Max.
    -0.498   -0.224    0.000   -0.015    0.228    0.596
```

Above we see that model-implied covariances are close to observed ones and that the differences (residuals) are small in relation to the scale of the observed values. The commands below print out the residual matrix so that we may see just where error (high residuals) is concentrated. When reading these values, recall that residuals are observed minus expected (here, implied), not vice versa. Thus, for instance, the highest residual is 0.596, for the covariance of anom67 with anom71. This covariance was underestimated by that amount (observed was higher than implied, resulting in a positive value; negative values represent model underestimates).

The actual residual matrix looks like that below.

```
residsmatrix <- lavaan::lavInspect(mm_fit, what="resid")
base::round(residsmatrix$cov,3)
[Output:]
             eductn sei    anom67 pwrl67 anom71 pwrl71
education     0.000
sei           0.000  0.000
anomia67      0.252 -0.095  0.000
powerless67 -0.253  0.542  0.000  0.000
anomia71      0.203 -0.195  0.596 -0.438  0.000
powerless71 -0.192 -0.482 -0.498  0.335  0.000  0.000
```

Many GOF coefficients are based on residuals between the observed and model-implied matrices. They provide useful cutoffs for assessing when residuals are too high.

Modification indices

The researcher should inspect the modification indices and consider acting on large ones, if any. Below we set the cutoff at 3 for instructional reasons, but this is extremely low. SPSS, for instance, does not list modification indices less than 4. We are particularly looking out for strong suggestions (high MIs) for adding a covariance arrow modeling correlated error connecting indicators. To recall, the indicators are education, sei, anomia67, powerless67, anoma71, and powerless 1. There are four such strong suggestions. If we add these error correlations to the model, fit will improve, but we should only do so if theoretically warranted.

- The "mi" column is the MI coefficient.
- The "lhs" and "rhs" columns are the left- and right-and side of the equation
- The "op" column tells which type of MI pertains to the row. The double-tilde suggests adding a covariance arrow; the equal sign followed by a tilde suggests adding a regression arrow.
- The "epc" and "sepc.lv" columns are the unstandardized and standardize parameter change coefficients. Other formulas for calculating these parameters are in the last two columns.

```
lavaan::modificationIndices(mm_fit, minimum.value = 3, sort = TRUE,
standardized=TRUE, alpha=0.05)
[Output:]
```

```
          lhs op         rhs     mi    epc sepc.lv sepc.all sepc.nox
41   anomia67 ~~    anomia71 63.775  1.952   1.952    0.507    0.507
42   anomia67 ~~ powerless71 49.883 -1.507  -1.507   -0.395   -0.395
43 powerless67 ~~    anomia71 49.805 -1.533  -1.533   -0.447   -0.447
44 powerless67 ~~ powerless71 37.297  1.158   1.158    0.341    0.341
33   education ~~ powerless67  7.715 -0.555  -0.555   -0.181   -0.181
32   education ~~    anomia67  6.077  0.554   0.554    0.161    0.161
34   education ~~    anomia71  3.576  0.425   0.425    0.129    0.129
30     alien71 =~ powerless67  3.483 -0.252  -0.749   -0.245   -0.245
19         ses =~    anomia67  3.483  0.125   0.322    0.094    0.094
20         ses =~ powerless67  3.483 -0.111  -0.286   -0.094   -0.094
29     alien71 =~    anomia67  3.483  0.284   0.843    0.245    0.245
```

In output above, the first four rows show the largest MI values. The double-tildes in the "op" column tell us that these are all covariance arrows suggested to be added. Brief comments on each are given below:

- anomia67 ~~ anomia71: This suggests that anomia in 1967 will covary with anomia in 1971. As this is very plausible, in the revised CFA model further below, we add this covariance arrow.
- powerless67 ~~ powerless71: This suggests that powerlessness in 1967 will covary with powerlessness in 1971. As this is also very plausible, in the revised CFA model further below, we add this covariance arrow.
- anomia67 ~~ powerless71: This suggests anomia in an earlier time period covaries with powerlessness in a later time period. While this is not implausible, in the interest of model parsimony we do not add this covariance arrow in the revised model but rather look to see if we do not get a good model fit without adding this complexity.
- powerless67 ~~ anomia71: For similar reasons, this covariance arrow is also not added to the revised model.

Goodness-of-fit measures

In the next phase, we consider whether the initial model has a good fit to the data. Below, we look at selected GOF measures for the measurement model. These are discussed in Garson (2018) and other SEM texts.

```
# The fitmeasures() function is the same as fitMeasures().
lavaan::fitMeasures(mm_fit, fit.measures="all", output="text")
 [Output:]
  Model Test User Model:

    Test statistic                                   71.546
    Degrees of freedom                                    6
    P-value                                           0.000

  Model Test Baseline Model:

    Test statistic                                 2133.722
    Degrees of freedom                                   15
```

```
  P-value                                        0.000

User Model versus Baseline Model:

  Comparative Fit Index (CFI)                    0.969
  Tucker-Lewis Index (TLI)                       0.923
  Bentler-Bonett Non-normed Fit Index (NNFI)     0.923
  Bentler-Bonett Normed Fit Index (NFI)          0.966
  Parsimony Normed Fit Index (PNFI)              0.387
  Bollen's Relative Fit Index (RFI)              0.916
  Bollen's Incremental Fit Index (IFI)           0.969
  Relative Noncentrality Index (RNI)             0.969

Loglikelihood and Information Criteria:

  Loglikelihood user model (H0)             -15246.680
  Loglikelihood unrestricted model (H1)     -15210.906

  Akaike (AIC)                               30523.359
  Bayesian (BIC)                             30595.919
  Sample-size adjusted Bayesian (BIC)        30548.281

Root Mean Square Error of Approximation:

  RMSEA                                          0.108
  90 Percent confidence interval - lower         0.087
  90 Percent confidence interval - upper         0.131
  P-value RMSEA <= 0.05                          0.000

Standardized Root Mean Square Residual:

  RMR                                            0.283
  RMR (No Mean)                                  0.283
  SRMR                                           0.021

Other Fit Indices:

  Hoelter Critical N (CN) alpha = 0.05         165.024
  Hoelter Critical N (CN) alpha = 0.01         220.000

  Goodness of Fit Index (GFI)                    0.975
  Adjusted Goodness of Fit Index (AGFI)          0.913
  Parsimony Goodness of Fit Index (PGFI)         0.279

  McDonald Fit Index (MFI)                       0.965

Expected Cross-Validation Index (ECVI)           0.109
```

The researcher should select at least three GOF measures to examine, including the p value of model chi-square (deviance, −2LL, labeled 'Test Statistic' in output below). We also choose to look at the comparative fit index (CFI) and root mean square error of approximation (RMSEA).

1 Model chi-square: The p value is 0, which means the researcher's model-implied covariance matrix is significantly different than the observed covariance matrix. Put another way, the researcher's model does not reproduce observed covariation. However, sample size is 932. This compares to Hoelter's critical N at the 0.05 level being 165. If sample size is much larger than 165, and it is, then it is too large to use model chi-square as a reason to reject the researcher's model. This is because, with a large sample, even small and trivial differences may be significant. Instead, the researcher should look at other GOF measures.
2 CFI: CFI is 0.969. Values above 0.95 are considered sufficient to judge that the researcher has a good model.
3 RMSEA: For confirmatory research, RMSEA should be 0.05 or lower. For exploratory research, it should be 0.08 or lower. RMSEA here is 0.108, which is considered a reason to judge the researcher's model as inadequate.

For the example data, the GOF measures present a mixed picture. We therefore seek improvements through revision of the model, in part based on MI coefficients.

Revised model

Two of the highest modification indices for the example data suggest that the measures of anomia and powerless as measured in 1967 and 1971 have correlated errors. Correlated error may be due to unmeasured mutual causes or to reciprocal causation. Here, it seems theoretically justified to believe for anomia and powerless that whatever unmeasured causes accounted for error in 1967 may still to be accounting for error in 1971. On this basis, we add these two error covariances to the model.

Step 1: Specify the revised model with added covariance arrows.

```
wheaton.meas.model2 <- '
   # Specification of latent variables, same as earlier.
   ses      =~ education + sei
   alien67 =~ anomia67 + powerless67
   alien71 =~ anomia71 + powerless71

   # Specify the covariances. The last two lines add the error
   covariances for anomia and powerless
   ses ~~ alien67
   ses ~~ alien71
   alien67~~ alien71
   anomia67 ~~ anomia71
   powerless67 ~~ powerless71
   '
   # End of model specification.
```

Step 2. Next, we run the measurement model as before, except that results (fit) are put in the object "mm_fit2".

```
# Make sure lavaan is still in the R environment.
base::library(lavaan)
# Run the revised CFA model.
mm_fit2 <- lavaan::cfa(wheaton.meas.model2, sample.cov = wheaton.
cov, sample.nobs = 932)
```

```
Step 3. We look to see if goodness of fit has reached satisfactory
levels in the revised model.
lavaan::summary(mm_fit2, standardize=TRUE, fit.measures=TRUE)
[Partial output:]

  Estimator                                         ML
  Optimization method                           NLMINB
  Number of model parameters                        17

  Number of observations                           932

Model Test User Model:

  Test statistic                                 4.735
  Degrees of freedom                                 4
  P-value (Chi-square)                           0.316

Model Test Baseline Model:

  Test statistic                              2133.722
  Degrees of freedom                                15
  P-value                                        0.000

User Model versus Baseline Model:

  Comparative Fit Index (CFI)                    1.000
  Tucker-Lewis Index (TLI)                       0.999

Loglikelihood and Information Criteria:

  Loglikelihood user model (H0)             -15213.274
  Loglikelihood unrestricted model (H1)     -15210.906

  Akaike (AIC)                               30460.548
  Bayesian (BIC)                             30542.783
  Sample-size adjusted Bayesian (BIC)        30488.792

Root Mean Square Error of Approximation:
```

```
  RMSEA                                             0.014
  90 Percent confidence interval - lower            0.000
  90 Percent confidence interval - upper            0.053
  P-value RMSEA <= 0.05                             0.930

Standardized Root Mean Square Residual:

SRMR                                              0.007
...
```

Summarizing the output above for the revised model, we see a finding of non-significance by the model chi-square (likelihood ratio) test, specifically p = 0.316. The model-implied covariance matrix is not significantly different from that for the observed covariance matrix. That is, the model reproduces the observed covariance matrix well. Baseline fit measures are above the 0.95 good fit level. CFI is 1.00. RMSEA at 0.014 is well below the 0.05 cutoff for good fit, as is SRMR. The previous model without correlated error connecting the 1967 and 1971 measurements of anomia and powerlessness, fit was marginal. Adding these correlated errors to the measurement model results in good fit.

In this section, we have treated the CFA model for the classic Wheaton dataset. This tests the measurement model, which is that portion of a full SEM model which deals with the latent variables and their indicators but not with hypothesized causal connections among latent variables. Having defined a well-fitting measurement model by CFA, we may then proceed to the second step of SEM, which is creating a structural model. Structural modeling goes beyond CFA and the scope of this book. Some structural models for the Wheaton data are presented in Garson (2018) and for R, in Chapter 3 of Garson (2022).

Visualization

Figure 5.1 presented the visualization for the original Wheaton measurement model without correlated error. Figure 5.2 presents the visualization for the revised model with the correlated error arrows added. It is created in the same manner, using `semPaths()`, with minor adjustments for better appearance. We observe that once correlated errors are added to the model, all indicators load heavily on their respective factors (latent variables).

```
semPlot::semPaths(mm_fit2,
   what="stand",          # Calls for standardized estimates ("std"
                            is a synonym);
                          # Use "est" for unstandardized (but set
                            fade = FALSE)
   fade = FALSE,          # If TRUE, weaker paths are faded
   rotation = 2,          # Place latent variables on left (1=top,
                            3=bottom, 4=right)
   style="lisrel",        # Default is "ram"; "mx" and "OpenMx"
                            are also available.
   layout="tree2",        # tree is the default; others include
                            circle, spring
   curvature = 1.1,       # Curvature of covariance arrows; default
                            is 1.
```

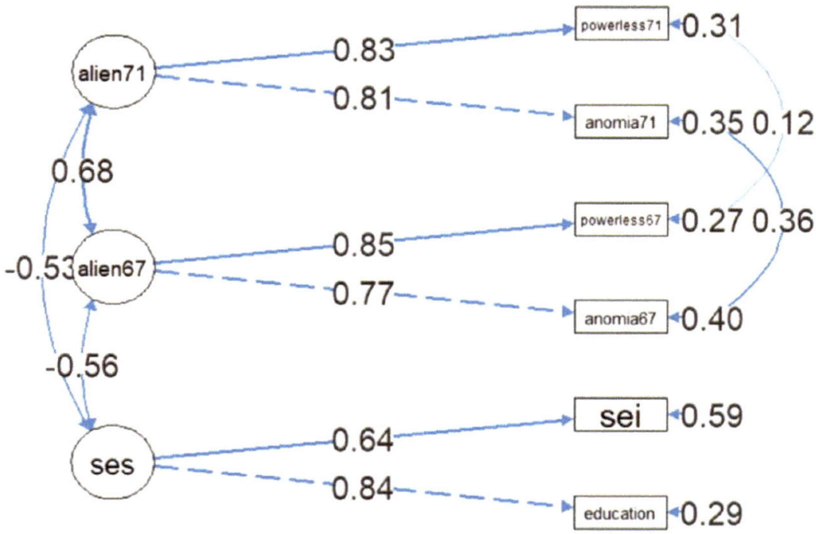

Figure 5.2 The revised Wheaton measurement model in lavaan.

```
    sizeLat = 14,           # Size of latent variables
    sizeMan = 16,           # Size of manifest (indicator) variables
    sizeMan2 = 5,           # Height of manifest variables
    esize=3,                # Size of largest edge
    fixedStyle = 5,         # Line type (lty) for fixed parameters.
                              Default is 2. Solid
                            # line is 1. The regression arrow from
                              the indicator that
                            # sets the latent variable metric is
                              fixed at 1.0, for ex.
    nCharNodes = 0,         # Do not abbreviate
    edge.color="dodgerblue",      # Color of lines
    edge.label.cex = 1.5,         # Size of values
    edge.label.color = "gray20"   # Color of values
)
```

Notes

1 CFA lends itself to "confirmatory bias", which is when the researcher keeps revising a model and uses as a stopping criterion the revision when a theoretically predicted result is finally achieved (Greenwald et al., 1986).

2 The observed covariance matrix is adjusted slightly by lavaan as compared to the input covariance matrix. See https://lavaan.ugent.be/tutorial/cov.html for discussion of handling covariance matrix input. See also Cheung (2015) on the difference between raw data input and meta-analysis of covariance or correlation matrices.

6 Quick start

Principal component analysis (PCA) in R

Introduction

Principal component analysis (PCA) is one of the most widely used procedures for exploring multivariate data. One of the most powerful R programs for PCA is the `PCA()` function from the "FactoMineR" package, treated in this chapter. The companion "factoextra" package provides additional visualization tools. As an example, we use a data frame consisting of measures of preference for various genres of music, based on the U. S. General Social Survey (GSS). The data are contained in the file "musicvars.csv", which contains 18 variables on 1,500 observations, taken from the 1993 version of the GSS. For purposes of demonstrating PCA, we further reduce this dataset to ten variables which we place in a data frame called "dat".

Setup and data for principal components analysis with PCA()

We start by loading the "FactoMineR", "factoextra", and "FactoInvestigate" packages for PCA and its visualization.[1] In addition, we load the "psych" package from R's System library because it supports the `KMO()` command to assess sampling adequacy for factor analysis and more. A variety of other packages are also loaded, as indicated below and later explained in the appropriate subsection.

```
base::library(EFAtools)           # Has the BARTLETT() command
base::library(EGAnet)             # Creates network graphs with the
                                    EGA() function
base::library(FactoMineR)         # Has the PCA() command
base::library(factoextra)         # Has visualization tools and more
base::library(FactoInvestigate)   # Automates reporting of PCA
                                    results
base::library(nFactors)           # Has the eigenComputes() command
base::library(psych)              # Supports the KMO() command
base::library(sna)                # Needed by EGA()
```

We also need to install the archived package "REdaS", which supports the `bart_spher()` command. This requires a set of more elaborate installation steps:

1 Go to https://cran.r-project.org/src/contrib/Archive/REdaS/
2 Click on REdaS_0.9.3.tar.gz to download it.

DOI: 10.4324/9781003279693-7

3 In RStudio, click on the "Packages" tab, then "Install". This brings up the "Install Packages" window.
4 In the "Install from" drop-down menu, select "Package Archive File" rather than the usual "Repository (CRAN)".
5 Click the "Browse" button and go to your local Downloads directory, then select REdaS_0.9.3.tar.gz and click "Open".
6 In the "Install Packages" window, click "Install".
7 Type `base::library(REdaS)`.

The commands explained below set up the data and put it in a data frame labeled "dat". Ten music preference variables are included in the data frame: bigband, blues, blugrass, classicl, country, hvymetal, jazz, musicals, opera, and rap.

```
# Set the working directory (yours may differ)
base::setwd("C:/Data")

# Read in the data
music <- utils::read.csv("musicvars.csv", header=TRUE, sep = ",",
stringsAsFactors=TRUE)

# Create music2 as a complete cases version of music, dropping cases
with missing values.
# Imputing missing values would be better, but we skip that step
here for instructional reasons.
music2 <- music[stats::complete.cases(music),]

# Create a subset of music2 to analyze, then copy it into the object
"dat".
# Column index numbers below refer to bigband, blues, blugrass,
classicl, country, hvymetal,
# jazz, musicals, opera, and rap. Use names(music2) to see these
column index numbers.
musicColumns <- base::c(2,3,5,6,8,10,11,13,14,15)
musicsubset <- music2[,musicColumns]
dat <- musicsubset

# musicsubset has 10 variables, so this is the maximum number of
factors
maxfactors <- base::length(musicsubset)

# View the variables (columns) in the dat data frame.
base::names(dat)
[Output:]
   [1] "bigband"  "blues"    "blugrass" "classicl" "country"
   [6] "hvymetal" "jazz"     "musicals" "opera"    "rap"
```

Testing factor adequacy with `KMO()`

The Kaiser-Meyers-Olkin (KMO) output provides a measure of sampling adequacy (MSA) for any data frame. Sampling adequacy may be thought of as a measure of factorability, helping assess if the factor analysis is likely to have an interpretable factor structure. MSA measures for the example data are shown below. The `KMO()` command is from the "psych" package in R's system library. The data object "dat" was created in the previous section.

```
psych::KMO(dat)
[Output:]
   Kaiser-Meyer-Olkin factor adequacy
   Call: psych::KMO(r = dat)
   Overall MSA =  0.72
   MSA for each item =
    bigband    blues blugrass classicl  country hvymetal
         0.80     0.68     0.65     0.75     0.52     0.52
         jazz musicals    opera      rap
         0.69     0.79     0.79     0.53
```

The usual cutoff for judging data to be suitable for factor/component analysis is that the overall MSA should be 0.60 or higher. For the "dat" data used here it is 0.72, which meets the criterion. We infer that the factor structure of "dat" should be strong enough to lend itself to interpretation. If MSA were lower, one strategy would have been to drop the variable with the lowest MSA, then compute KMO again, and continue one variable at a time until overall MSA rises to 0.60 or higher. Another rule-of-thumb is to drop variables with MSA < 0.60. Yet another strategy used by some researchers is to ignore KMO and simply compute the PCA solution and see how interpretable it is.

Bartlett's test for sphericity

For Bartlett's test of sphericity (Bartlett, 1950), a finding of significance means that there is at least minimal adequate correlation among variables to justify proceeding with factor analysis. Specifically, if Bartlett's test returns a finding of significance, the researcher may conclude that correlations in the data are greater than would be expected by chance. If the data pass the KMO test in the previous section and also pass Bartlett's test, the researcher concludes that the data and correlations based upon it are suitable to proceed with factor analysis.

As noted in Chapter 2, Bartlett's test is extremely sensitive to sample size. It is recommended only for small samples, such as when the sample size is five times the number of variables or less (Tabachnick & Fidell, 2013: 619). The example data file here, "dat", contains ten variables. By the foregoing rule of thumb, applying Bartlett's test would only be appropriate if the sample were 5 * 10 = 50 or less. The actual sample size is 773.

In spite of Bartlett's test being inappropriate for the example data, we proceed simply to illustrate the mechanics of the test. Bartlett's test below returns a finding of strong significance, implying that by this criterion the researcher may proceed with factor analysis.

```
base::library(EFAtools)
EFAtools::BARTLETT(dat, N = NA, use = "pairwise.complete.obs", cor_
method = "pearson")
[Output:]
   i 'x' was not a correlation matrix. Correlations are found from
   entered raw data.
   √ The Bartlett's test of sphericity was significant at an alpha
   level of .05.

      These data are probably suitable for factor analysis.
      x²(45) = 1698, p < .001
```

A second method of calculating Bartlett's test of sphericity is to use the `bart_spher()` command from the "REdaS" package. The findings is one of significance, the same as for the `BARTLETT()` command above.

```
base::library(REdaS)
REdaS::bart_spher(dat, use = "pairwise.complete.obs")
[Output:]
      Bartlett's Test of Sphericity
  Call: REdaS::bart_spher(x = dat, use = pairwise.complete.obs")
        X2 = 1698.456
        df = 45
  p-value < 2.22e-16
```

Determining the number of factors to request

The number of factors to extract issue was discussed in Chapter 2. Here we use the most popular method, which is applying the Kaiser rule: select all factors with an eigenvalue of 1.0 or greater. The `eigen()` command from R's "base" package will return the eigenvalues from a correlation matrix. The command below applies `eigen()` to the correlation matrix of the example data, "dat". The package prefixes "base::" and "stats::" are not really needed but are included to specify what packages are being used.

```
base::eigen(stats::cor(dat))
[Partial output:]
   eigen() decomposition
   $values
     [1] 2.9590359 1.5174835 1.3416894 1.0199456 0.7703126
     [6] 0.6109687 0.5346705 0.4621152 0.4108412 0.3729376
```

Using the Kaiser criterion, we conclude that we should ask for four components since the first four have eigenvalues of 1.0 or higher.

Creating the model with `PCA()`

In this example, we perform PCA on the "dat" data frame, which is based on "musicvars.csv" but stripped of cases with missing values and filtered to have just 10 music preference items. The PCA model is created with the PCA() command. The PCA result from this command is put in the object "PCAresult".

Parameters for the `PCA()` command below include these:

- ncp: number of components extracted and retained in the results (default is 5).
- Scale = TRUE. Variables are placed on the same scale.
- graph: If TRUE a plot is displayed (the default).
- axes: a vector of length 2 specifying the components to plot. By default this is the first two.

We now run the `PCA()` command to create the PCA factor model. Results are placed in the object "PCAresult", which is a list of class PCA.

```
PCAresult <- FactoMineR::PCA(dat, ncp=4, scale=TRUE, graph=TRUE,
axes=base::c(1,2))
```

```
base::class(PCAresult)
[Output:]
   [1] "PCA"   "list"
```

The PCAresult object contains 15 elements, listed below.

	Element	*Description*
1	*"$eig"*	*"eigenvalues"*
2	"$var"	"results for the variables"
3	"varcoord"	"coord. for the variables"
4	"varcor"	"correlations variables - dimensions"
5	"varcos2"	"cos2 for the variables"
6	"varcontrib"	"contributions of the variables"
7	"$ind"	"results for the individuals"
8	"indcoord"	"coord. for the individuals"
9	"indcos2"	"cos2 for the individuals"
10	"indcontrib"	"contributions of the individuals"
11	"$call"	"summary statistics"
12	"$call$centre"	"mean of the variables"
13	"$call$ecart.type"	"standard error of the variables"
14	"$call$row.w"	"weights for the individuals"
15	"$call$col.w"	"weights for the variables"

For instance, the call PCAresult$eig returns the eigenvalues and percent of variance explained for all components (all 10, not just the extracted 4). This is illustrated in the next section. Other result elements are returned in the same way: name of results object ("PCAresult"), dollar sign ("$"), name of element (e.g., "eig"). Alternative code for the same results is: `factoextra::get_eigenvalue(PCAresult).`

Eigenvalues and the scree plot

Below is the output for the "$eig" element. It shows not only the eigenvalues but the percent of variance explained by each component and the cumulative percentage, for all ten components, not just the four components which are extracted. Some texts refer to this as the "inertia" of the model. We see that the first two components explain about 45% of the variance. Often the first two components explain the majority of the variance but not for this example data. Here, three components are needed to explain over half the variance in the 10 input variables; four components explain about two-thirds; and eight components would be needed to explain over 90%.[2]

```
PCAresult$eig
[Output:]
        eigenvalue percentage of variance cumulative percentage of
variance
comp 1   2.9590359              29.590359                 29.59036
comp 2   1.5174835              15.174835                 44.76519
comp 3   1.3416894              13.416894                 58.18209
comp 4   1.0199456              10.199456                 68.38154
comp 5   0.7703126               7.703126                 76.08467
comp 6   0.6109687               6.109687                 82.19436
comp 7   0.5346705               5.346705                 87.54106
comp 8   0.4621152               4.621152                 92.16221
comp 9   0.4108412               4.108412                 96.27062
comp 10  0.3729376               3.729376                100.00000
```

The next set of commands creates a "scree plot" of eigenvalues, shown in Figure 6.1. Eigenvalues are always in a descending order as the first component is always the largest and the last component will have the smallest eigenvalue. The scree plot is an alternative to the Kaiser rule for selecting the number of components to retain. In principle, one selects the number at the "elbow" of the scree plot, where it begins to level off. Finding the "elbow" is subjective in many instances (Mulaik, 2018).

The scree plot may be created with the `fviz_screeplot()` command from the "factoextra" package. A synonym is `fviz_eig()`

```
factoextra::fviz_screeplot(PCAresult,
    addlabels=TRUE,
    barfill="dodgerblue",
    barcolor="black",
    linecolor="red",
    main="Scree Plot for PCAresult",
    xlab="Components",
    ylab="Eigenvalues with Percent Explained Variance",
    ggtheme = theme_classic())
```

Empirical scree tests

To reduce the subjectivity of the scree test, Gilles Raiche and David Magis authored the "nFactors" package, which reports the recommended number of components or factors to retain based on any of three criteria: standard error, acceleration factor, or optimal coordinates. These are implemented below for the example data. Results reveal three

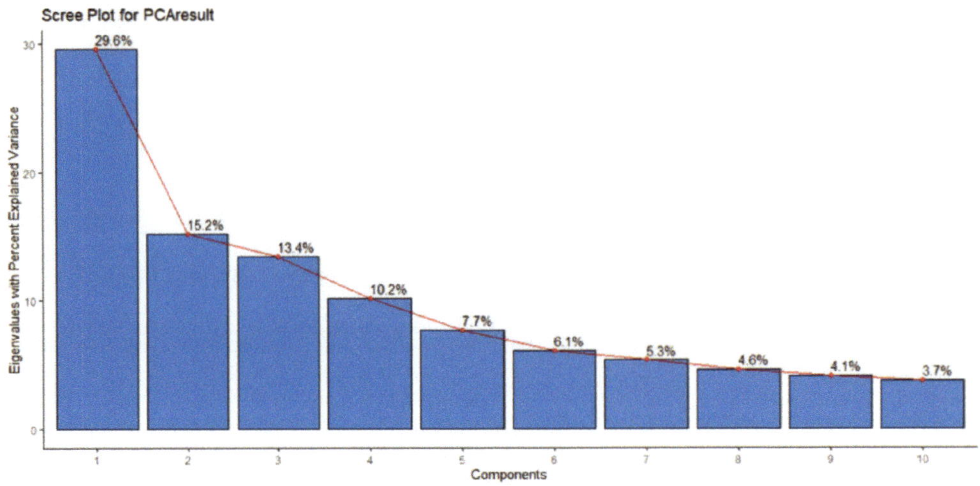

Figure 6.1 Scree plot for PCA result.

different recommendations depending on the criterion selected. For these data, the optimal coordinates method corresponded to the Kaiser criterion.

```
# Setup
base::library(nFactors)
```

Step 1. We calculate the eigenvalues for the "dat" example data, which contains ten music genre preference items. The "dat" data object was created in the setup section. We see that by the Kaiser criterion, extracting the first four components is recommended since these have eigenvalues of 1.0 or greater.

```
evalues <- nFactors::eigenComputes(dat)
evalues
[Output:]
   [1] 2.9590359 1.5174835 1.3416894 1.0199456 0.7703126
   [6] 0.6109687 0.5346705 0.4621152 0.4108412 0.3729376
```

Step 2. We next ask for recommendations using an empirical scree test based on standard error (Zoski & Jurs, 1996). The "model" option is here "components" but could be "factors". By this test, extracting three factors are recommended.

```
se_scree <- nFactors::nSeScree(evalues, model="components")
se_scree$nFactors[1]
[Output:]
   se
    3
```

Step 3. In this step, we ask for recommendations using an empirical scree test based on the acceleration factor. The acceleration factor indicates where the elbow of the scree plot curve appears. By this test, 1 component is recommended.

```
scree2 <- nFactors::nScree(evalues, model="components")
scree2$Components$naf
[Output:]
    [1] 1
```

Step 4. In this step, we ask for recommendations using an empirical scree test based on optimal coordinates. The optimal coordinates (OC) criterion involves extrapolation of the preceding eigenvalue by a regression line between the eigenvalue coordinates and the last eigenvalue coordinates. By this test, four factors are recommended.

```
scree2$Components$noc
[Output:]
    [1] 4
```

Step 5. Somewhat redundantly we may also ask for the number of factors recommended by the Kaiser rule. This is also 4.

```
scree2$Components$nkaiser
[Output:]
    [1] 4
```

Exploratory graph analysis of factor memberships

An additional method of determining the desired number of factors is to examine the network graph and network membership table from the `EGA()` command from the "EGAnet" package. See Golino and Epskamp (2017). Note that this method, like scree plots and empirical scree tests, is not unique or limited to the `PCA()` command.

Step 1. Load needed libraries.

```
base::library(EGAnet)       # Has the EGA() function to create
                              network graphs
base::library(sna)          # Needed by EGA()
```

Step 2. Create the "dat" data object, same as before.

```
# Read in the data
music <- utils::read.csv("musicvars.csv", header=TRUE, sep = ",",
stringsAsFactors=TRUE)
# Create music2 as a complete cases version of music, dropping cases
with missing values.
music2 <- music[stats::complete.cases(music),]
# Create a subset of music2 to analyze and copy it into the object
"dat".
# Column index numbers below refer to bigband, blues, blugrass,
classicl, country, hvymetal,
# jazz, musicals, opera, and rap.
musicColumns <- base::c(2,3,5,6,8,10,11,13,14,15)
musicsubset <- music2[,musicColumns]
dat <- musicsubset
```

Step 3. Create the network graph of component memberships. This is Figure 6.2. The cor=cor_auto option automatically selects Pearson or Spearman correlations. The algorithm may be set to "walktrap" (the default) or "Louvain".

```
g <- EGAnet::EGA(dat, n=base::nrow(dat), cor="cor_auto",
plot.EGA=TRUE, algorithm="walktrap")
```

Step 4. Finally, list the group memberships of each variable in dat.

```
g$wc
[Output:]
   bigband    blues blugrass classicl  country
         1        2        3        1        3
  hvymetal     jazz musicals    opera      rap
         4        2        1        1        4
```

Based on the graph and group membership table above, we see that for these data, exploratory graph analysis suggests that there should be four factors. Factor 1 would be bigband, classicl, musicals, and opera. Factor 2 would be blues and jazz. Factor 3 would be blugrass and country. Factor 4 would be hvymetal and rap.

PCA variable plot

By default, the PCA() command displays the solution graphically in a correlation circle plot, as shown in Figure 6.3. To recall, this command was:

```
PCAresult <- FactoMineR::PCA(dat, ncp=4, scale=TRUE, graph=TRUE,
axes=base::c(1,2))
```

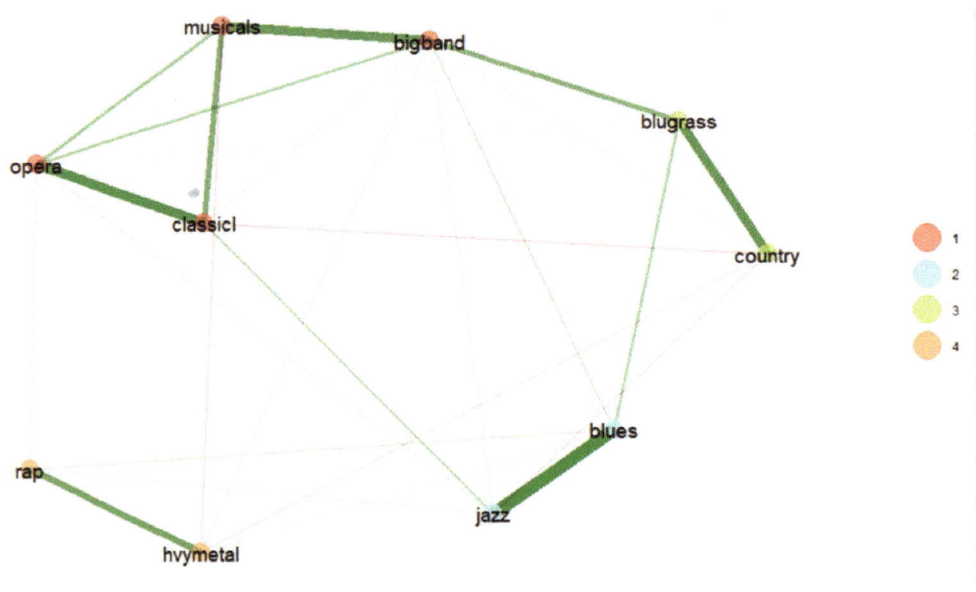

Figure 6.2 Network graph for "dat" (musicsubset).

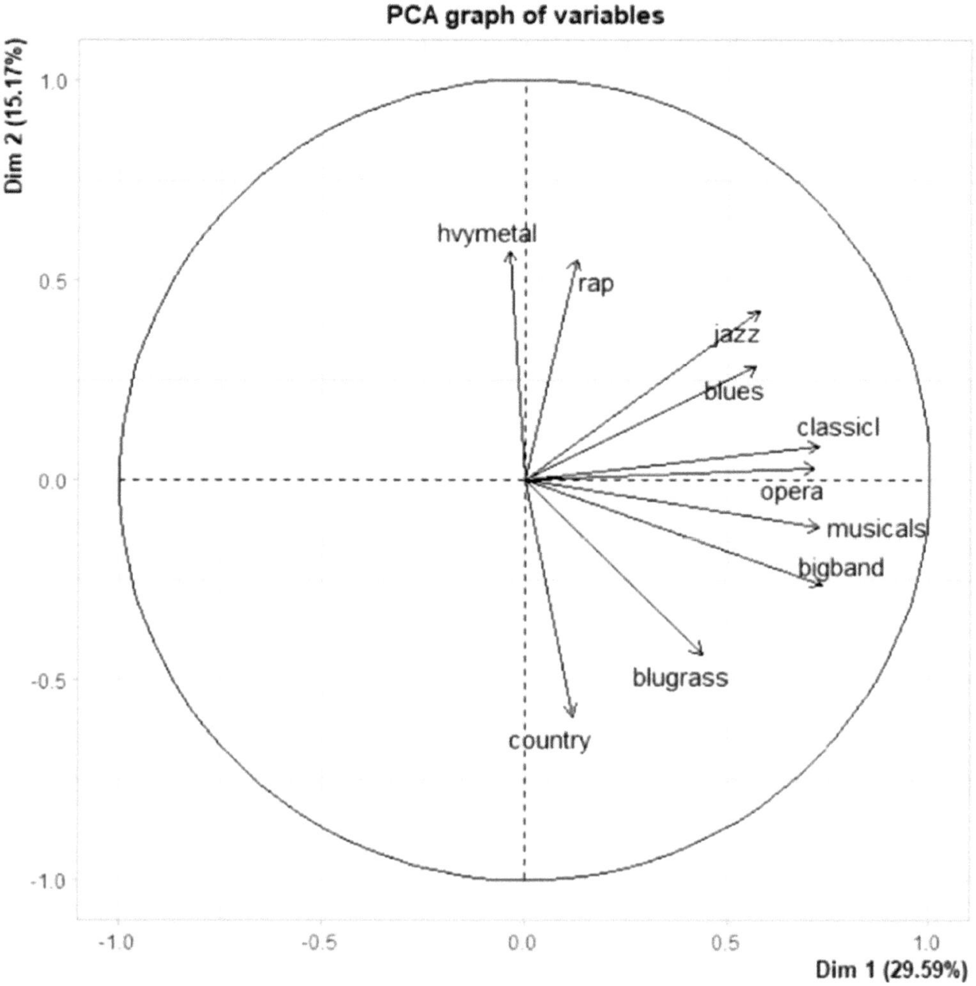

Figure 6.3 Component solution from PCA().

In Figure 6.3, for instance, we can observe that jazz and blues are located close to each other. The same is true of hvymetal and rap.

In Figure 6.3, a variable is placed close to the circumference of the variable correlation plot if it is well represented by the principal components. If representation is perfectly based on components 1 and 2, the variable is located on the circumference of the circle. Placement on or near the circumference reflects a high cos2 value, discussed below. Variables which are not as well represented have lower cos2 coefficients and are located closer to the center of the circle and may be considered less important for the components. Variables close to each other have a similar profile (same direction and length) on the components. Variables on opposite sides of the plot, in opposite quadrants, are negatively correlated in projected component space, though they may not be negatively correlated in raw terms.[3] The angle of the variable arrows shows the relation of the variables to the components. The more the arrow is horizontal and further from the origin, the more that variable is associated with PCA 1. The more it is vertical and further from the origin, the more that variable is associated with PCA 2.

The "cos2" coefficient, which is the squared cosine of a PCA component with an observation, is a measure of the importance of a given component to a given observation. Abdi and Williams (2010: 438) thus write, "Components with a large value of cos2 contribute a relatively large portion to the total distance and therefore these components are important for that observation". The cos2 measure is considered a measure of the observation's "quality of representation" in the PCA analysis. The cos2 coefficients are located in the varcos2 element of the result object (PCAresult, created in the earlier section on "Creating the model with PCA()".)

```
PCAresult$var$cos2
[Output:]
                Dim.1        Dim.2        Dim.3        Dim.4
bigband  0.534231019 0.0685303263 0.0001481734 0.0003307955
blues    0.323156867 0.0793931070 0.1519972694 0.2378254323
blugrass 0.187993403 0.1881123457 0.2878682518 0.0065466525
classicl 0.522129577 0.0073796332 0.1342744818 0.0333059662
country  0.013645863 0.3484784805 0.3350819592 0.0261430199
hvymetal 0.001648183 0.3257640982 0.1243807252 0.1976693130
jazz     0.331321862 0.1801836287 0.0206091659 0.2411139114
musicals 0.522125924 0.0139323241 0.0839178907 0.0235186046
opera    0.507240707 0.0009029411 0.0586146639 0.0938242814
rap      0.015542449 0.3048065967 0.1447968268 0.1596675761
```

It is also possible to obtain a bar chart showing the importance of component 1 (or components 1 and 2 considered jointly) to each of the music variables in the PCA analysis using the `fviz_cos2()` command from the "factoextra" package. Though Figure 6.4 is for dimension 1, note that the `axes=` option may be set to a range of dimensions such as 1:2, or a collection of dimensions (using the `c()` combinations function). In Figure 6.4, the height of the bar corresponds to the importance of that variable to component 1.

```
factoextra::fviz_cos2(PCAresult,
    choice = "var",
    axes = 1,
    fill = "dodgerblue",
    title="Variable importance (contribution)  to dimension 1",
    subtitle="Height of bars corresponds to magnitude of the cos2
    coefficient")
```

Eigenvectors

The `PCA()` function calls eigenvectors "coordinates" and places them in the varcoord element of the results object (here, PCAresult). Eigenvectors are used to plot variables in component space, as in Figure 6.5. They are also the unrotated component loadings.[4]

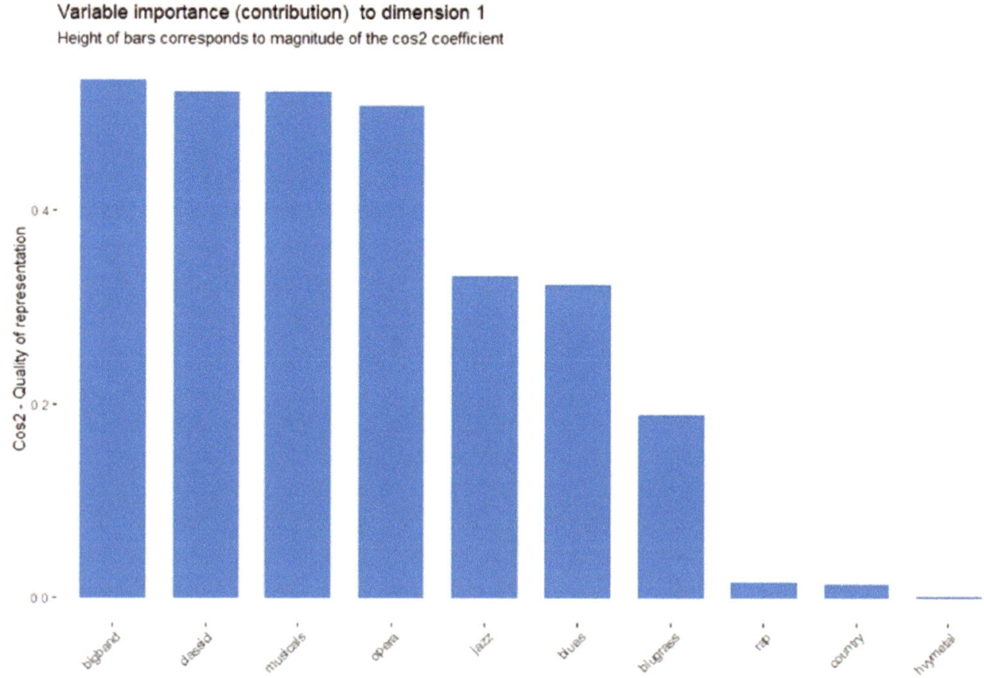

Figure 6.4 Plot of the importance of PCA1 to variables as measured by cos2.

For convenience, the largest standardized eigenvector for each variable is shown in red (red highlights were added manually for instructional reasons). Determining the meanings of the components is subjective, but dimension 2, for example, may be considered rap and country (in opposite directions). Most other music variables fall primarily in component/dimension 1.

```
PCAresult$var$coord
[Partial output:]
                Dim.1       Dim.2       Dim.3       Dim.4
bigband    0.73091109 -0.26178298  0.01217265  0.01818779
blues      0.56846888  0.28176782  0.38986827 -0.48767349
blugrass   0.43358206 -0.43371920  0.53653355  0.08091139
classicl   0.72258534  0.08590479 -0.36643483  0.18249922
country    0.11681551 -0.59032066  0.57886264  0.16168803
hvymetal  -0.04059782  0.57075748  0.35267652  0.44460017
jazz       0.57560565  0.42448042  0.14355893 -0.49103351
musicals   0.72258281 -0.11803527 -0.28968585  0.15335777
opera      0.71220833  0.03004898 -0.24210465  0.30630749
rap        0.12466936  0.55209292  0.38052178  0.39958425
```

Similar results with added information on variables may be had using the `facto_summarize()` command. Coefficients in the Dim.1 and Dim.2 columns are the eigenvectors, which are the coordinates used for plotting variables in component space. The "coord" column contains coefficients which are the same as the "cos2" column and are the sum of cos2 coefficients for Dim.1 and Dim.2 for the given variable. These cos2 coefficients are listed by the command `PCAresult$var$cos2`. Recall that cos2 measures the quality of representation of variables.

```
summary <- factoextra::facto_summarize(PCAresult, "var", axes=1:2)
summary
[Output:]
              name       Dim.1       Dim.2     coord      cos2   contrib
bigband    bigband  0.73091109 -0.26178298 0.6027613 0.6027613 13.464956
blues        blues  0.56846888  0.28176782 0.4025500 0.4025500  8.992477
blugrass  blugrass  0.43358206 -0.43371920 0.3761057 0.3761057  8.401745
classicl  classicl  0.72258534  0.08590479 0.5295092 0.5295092 11.828592
country    country  0.11681551 -0.59032066 0.3621243 0.3621243  8.089418
hvymetal  hvymetal -0.04059782  0.57075748 0.3274123 0.3274123  7.313992
jazz          jazz  0.57560565  0.42448042 0.5115055 0.5115055 11.426411
musicals  musicals  0.72258281 -0.11803527 0.5360582 0.5360582 11.974890
opera        opera  0.71220833  0.03004898 0.5081436 0.5081436 11.351311
rap            rap  0.12466936  0.55209292 0.3203490 0.3203490  7.156208
```

We now turn to visualization of these data.

Step 1. Put the variable coordinates (eigenvectors) into the object "coords":

```
coords <-  PCAresult$var$coord
```

Step 2. Create "musicLabels" as a vector of variable names:

```
musicLabels <- base::rownames(coords)
```

Step 3. Save the x coordinates to x and the y coordinates to y.

```
x <- coords[,1]
y <- coords[,2]
```

Step 4A. Plot the variables in coordinate space. The xlim= and ylim= options expand the plot for better labeling. The pch= option sets the symbol to use and cex= sets its size.

```
base::plot(PCAresult$var$coord, pch=16, cex = 1.4, col="blue",
xlim = base::c(-0.1, 0.8), ylim = base::c(-0.2, 0.8), main = "No
Rotation")
```

Step 4B. Add the variable labels and reference lines to create Figure 6.5:

```
graphics::text(x = x, y = y, labels = musicLabels, pos = 1)
graphics::abline(v=0,col="red")
graphics::abline(h=0,col="red")
```

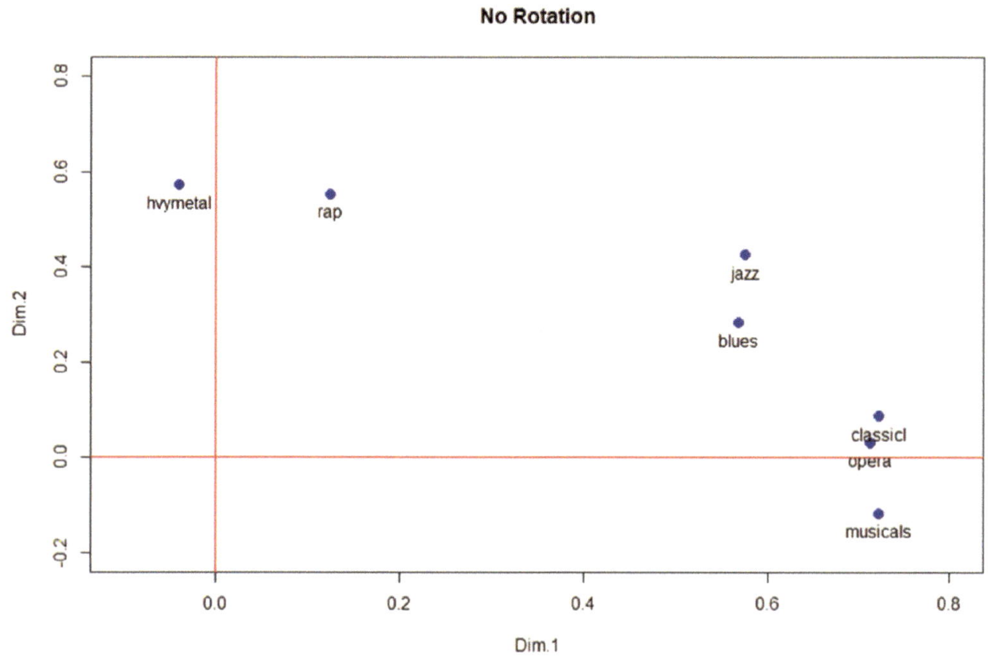

Figure 6.5 Variables plotted in coordinate space for the first two dimensions.

Component loadings

The correlation coefficient between a component and a variable is the component loading. These are found in the varcor element of the result object, as shown below. These coefficients also go under the labels "standardized regression coefficients" or "regression weights". For the unrotated components, the correlation table below is the same as the eigenvector table in the previous section (that is, the varcor element is identical to the varcoord element).

The varcor element contains the correlations of the variables with the components (a.k.a., the "component matrix"). Note this is for the unrotated solution. The rotated solution will differ. The rotated loadings are illustrated in the next section on "Component rotation".

```
PCAresult$var$cor
[Output:]
                Dim.1       Dim.2       Dim.3       Dim.4
bigband    0.73091109 -0.26178298  0.01217265  0.01818779
blues      0.56846888  0.28176782  0.38986827 -0.48767349
blugrass   0.43358206 -0.43371920  0.53653355  0.08091139
classicl   0.72258534  0.08590479 -0.36643483  0.18249922
country    0.11681551 -0.59032066  0.57886264  0.16168803
hvymetal  -0.04059782  0.57075748  0.35267652  0.44460017
jazz       0.57560565  0.42448042  0.14355893 -0.49103351
musicals   0.72258281 -0.11803527 -0.28968585  0.15335777
opera      0.71220833  0.03004898 -0.24210465  0.30630749
rap        0.12466936  0.55209292  0.38052178  0.39958425
```

Component rotation

The `PCA()` function, unlike `principal()` and most other R programs for PCA, does not have an option for rotation. Other PCA programs which do have a rotation option are discussed in Chapter 10. However, we can rotate the loadings from the `PCA()` command by using the `varimax()` command from the "stats" package.[5]

Step 1: Rotate the loadings.

```
unrotatedLoadings <- PCAresult$var$cor
varimaxLoadings <- stats::varimax(unrotatedLoadings)

varimaxLoadings$loadings
[Output: Loadings less that 0.1 appear as blank cells.]
   Loadings:
            Dim.1  Dim.2  Dim.3  Dim.4
   bigband   0.623 -0.137  0.367 -0.248
   blues     0.118         0.168 -0.862
   blugrass  0.181         0.782 -0.162
   classicl  0.817               -0.141
   country                 0.842
   hvymetal         0.802
   jazz      0.221               -0.843
   musicals  0.785               -0.106
   opera     0.801  0.116
   rap              0.780        -0.121

                  Dim.1 Dim.2 Dim.3 Dim.4
   SS loadings    2.418 1.306 1.513 1.601
   Proportion Var 0.242 0.131 0.151 0.160
   Cumulative Var 0.242 0.372 0.524 0.684
```

Step 2A. Plot the rotated loadings, but without labels yet. The command below plots the example variables in rotated coordinate space.

```
base::plot(varimaxLoadings$loadings, pch=16, cex = 1.4, col="blue",
xlim = base::c(-0.1, 0.9), ylim = base::c(-0.2, 0.8), main= "Varimax
Rotation")
```

Step 2B. Save the x coordinates to object x and the y coordinates to object y:

```
x <- varimaxLoadings$loadings[,1]
y <- varimaxLoadings$loadings[,2]
```

Step 2C. Add the variable labels and reference lines to create Figure 6.6.

```
graphics::text(x = x, y = y, labels = musicLabels, pos = 1)
graphics::abline(v=0,col="red")
graphics::abline(h=0,col="red")
```

Compare Figure 6.6 with the unrotated version in Figure 6.5. It will be seen that after varimax rotation, variables are drawn into closer association with one axis or the other, making the dimensions more interpretable.

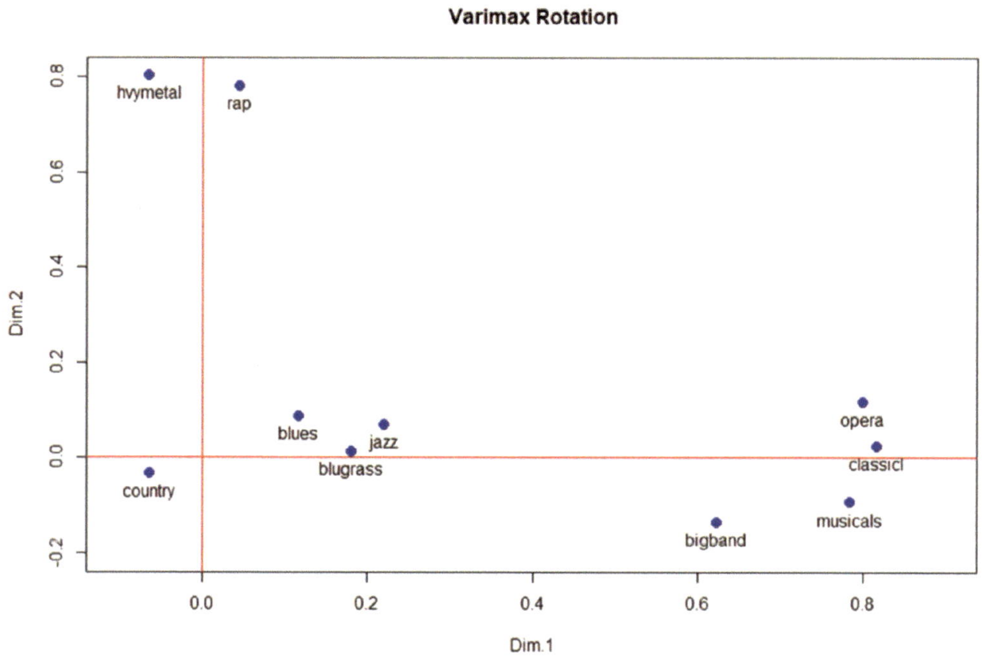

Figure 6.6 Music variables in varimax-rotated component space.

If instead an oblique rotation is wanted, R's built-in "stats" package has the `promax()` command. The "m" option is the power used by the target for promax. Values of 2 to 4 are recommended but 4 is the default.

```
promaxLoadings <- stats::promax(unrotatedLoadings, m=4)
promaxLoadings$loadings
[Output: Loadings less that 0.1 appear as blank cells.]
   Loadings:
            Dim.1  Dim.2  Dim.3  Dim.4
   bigband   0.558 -0.126  0.301 -0.140
   blues    -0.107         0.102 -0.904
   blugrass                0.778 -0.101
   classicl  0.858        -0.174
   country  -0.142         0.876  0.135
   hvymetal         0.807
   jazz                   -0.178 -0.878
   musicals  0.805
   opera     0.849  0.133
   rap              0.781

                  Dim.1 Dim.2 Dim.3 Dim.4
   SS loadings    2.456 1.306 1.540 1.652
   Proportion Var 0.246 0.131 0.154 0.165
   Cumulative Var 0.246 0.376 0.530 0.695
```

Biplots and outliers

Biplots are plots of both variables and observations simultaneously, as shown in Figure 6.7. Biplots may be created with the command `fviz_pca_biplot()` from the "factoextra" package, To identify outlying observations, an ellipse may be drawn at some level, such as 0.95. Observations outside the ellipse may be considered outliers. The smaller the ellipse level (e.g., 0.80), the smaller the ellipse and the greater the number of outliers.

A simple biplot may be created with this simple command:
`factoextra::fviz_pca_biplot(PCAresult)`
However, to get the outlier ellipse and to control other parameters, we use the full syntax below. The resulting plot is shown in Figure 6.7.

```
factoextra::fviz_pca_biplot(
     PCAresult,
     axes = base::c(1, 2),
     geom = base::c("text"),
     geom.ind = base::c("text"),
     geom.var = base::c("arrow", "text"),
     col.ind = "dodgerblue",
     fill.ind = "white",
     col.var = "darkred",
     fill.var = "white",
     label = "all",
     addEllipses = TRUE,
     ellipse.level = 0.95,
     title ="PCAresult biplot with outlier ellipse level = 0.95",
      # The remaining options are not used in this example
     # gradient.cols = NULL,
     # invisible = "none",
     # repel = FALSE,
     # habillage = "none",
     # palette = NULL
)
```

Variable contributions

The contribution coefficient for a variable is a measure of its importance in contributing to the definition of the factors in the solution. The command below plots variables color coded by the magnitude of their unrotated contributions to the dimensions. The resulting plot is Figure 6.7.

```
fig <- factoextra::fviz_pca_var(PCAresult,
    col.var="contrib",
    gradient.cols = base::c("blue", "yellow", "red"),
    repel = TRUE,
    title = "Variables color-coded by contribution")
fig
```

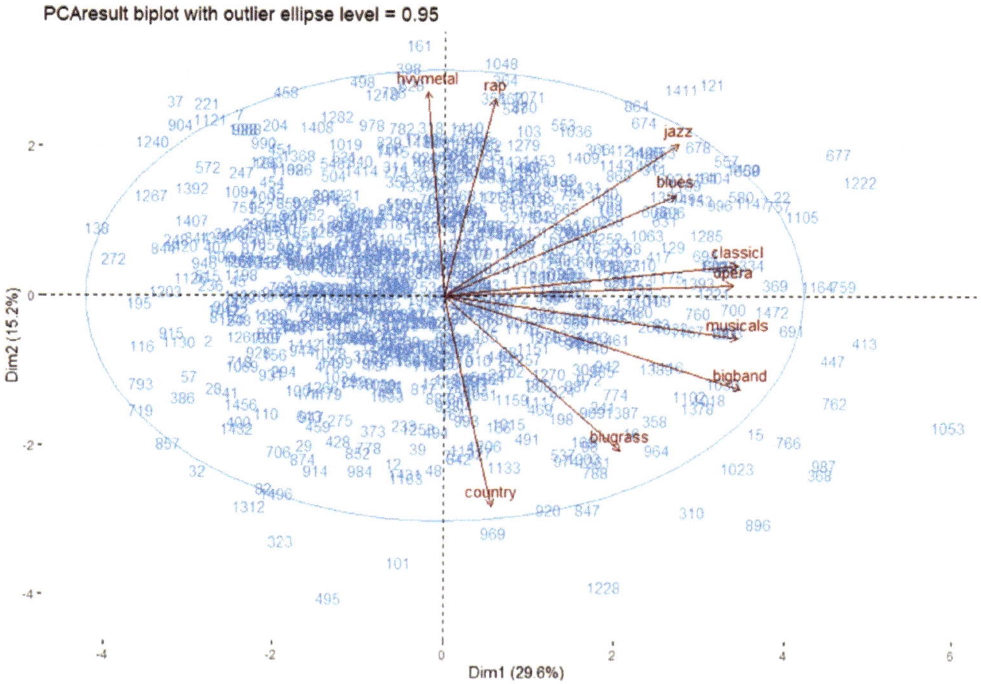

Figure 6.7 Biplot with outlier ellipse.

To see the data behind the plot in Figure 6.8, we print out the "data" element. The contributions are the same as the unrotated component loadings printed earlier.

```
fig$data
[Output:]
              name           x           y     coord      cos2   contrib
bigband    bigband  0.73091109 -0.26178298 0.6027613 0.6027613 13.464956
blues        blues  0.56846888  0.28176782 0.4025500 0.4025500  8.992477
blugrass  blugrass  0.43358206 -0.43371920 0.3761057 0.3761057  8.401745
classicl  classicl  0.72258534  0.08590479 0.5295092 0.5295092 11.828592
country    country  0.11681551 -0.59032066 0.3621243 0.3621243  8.089418
hvymetal  hvymetal -0.04059782  0.57075748 0.3274123 0.3274123  7.313992
jazz          jazz  0.57560565  0.42448042 0.5115055 0.5115055 11.426411
musicals  musicals  0.72258281 -0.11803527 0.5360582 0.5360582 11.974890
opera        opera  0.71220833  0.03004898 0.5081436 0.5081436 11.351311
rap            rap  0.12466936  0.55209292 0.3203490 0.3203490  7.156208
```

Taking bigband as an example, we see it is located at 0.731 on the x axis (the PCA1 axis) and at −0.262 on the y axis (the PCA2 axis) in Figure 6.8. Both the coord and the cos2 values for bigband are the sum of x-squared plus y-squared. The contrib contribution value for bigband is 13.46, which is the highest of any variable, causing it to be shown in red in Figure 6.8. The contribution for rap and hvymetal is the lowest, causing them to be shown in blue.

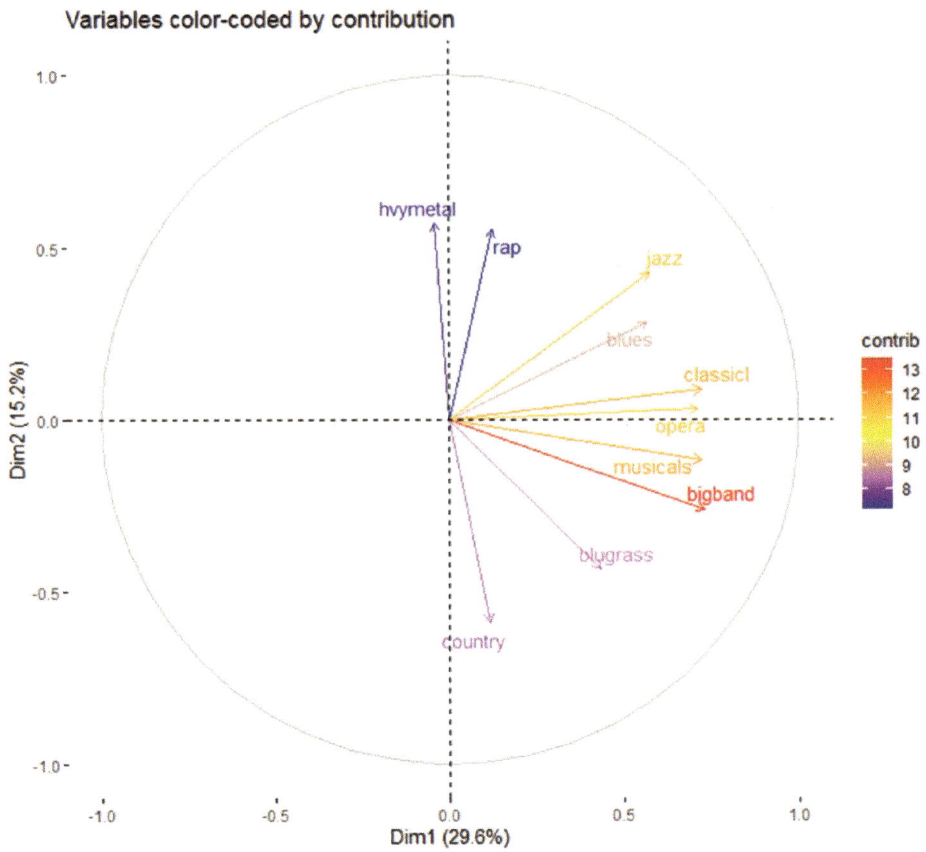

Figure 6.8 Unrotated variable contributions color-coded by contribution.

It is also possible to create a bar chart of variables' contributions to a particular component. The code below creates such a bar chart, shown in Figure 6.9 , for the first dimension. The horizontal dotted red reference line shows the average contribution. The variables bigband, classicl, musicals, and opera make above-average contributions to the definition of the first dimension. Of course, similar bar charts could be created for any dimension by changing the `axes =` option ".

```
factoextra::fviz_contrib(
    PCAresult,
    choice =  "var",
    axes = 1,
    fill = "steelblue",
    color = "steelblue",
    sort.val = "desc",
    top = Inf,
    xtickslab.rt = 45,
    ggtheme = theme_minimal() +
    ggplot2::theme(plot.title= ggplot2::element_text(size=22)) +
```

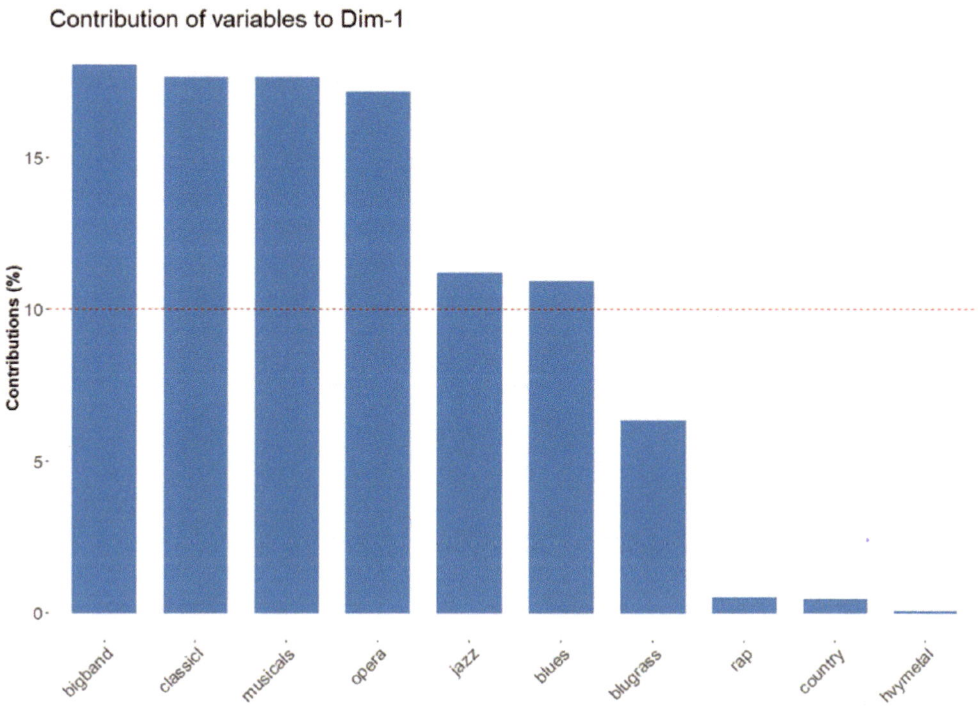

Figure 6.9 Bar chart of unrotated variable contributions to dimension 1.

```
    ggplot2::theme(axis.text=ggplot2::element_text(size=16),
        axis.title=ggplot2::element_text(size=16,face="bold"))
)
```

To see rotated contributions for the first two dimensions, recall that we had created the object "varimaxLoadings" earlier in the subsection titled "Component rotation". From this object, we may retrieve the rotated loadings for components 1 and 2, which we label "contrib1" and "contrib2".

```
rotLoadings <- varimaxLoadings$loadings

contrib1 <- rotLoadings[1:10,1]
contrib1
[Output:]
       bigband        blues     blugrass     classicl      country
    0.62295340   0.11793895   0.18094048   0.81676235  -0.06331763
      hvymetal         jazz     musicals        opera          rap
   -0.06496988   0.22102042   0.78520880   0.80052255   0.04565800

contrib2 <- rotLoadings[1:10,2]
contrib2
[Output:]
       bigband        blues     blugrass     classicl      country
   -0.13659604   0.08628587   0.01205433   0.02360968  -0.03406720
```

```
  hvymetal       jazz    musicals       opera         rap
0.80170668 0.06942545 -0.09517599 0.11570741 0.77959645
```

Below we create barplots for the rotated contribution to the first two components. This is Figure 6.10.

```
# This sets up display for two rows, one column, of barplots, then
resets the display.
graphics::par(mfrow=base::c(2,1))
graphics::barplot(contrib1,
   col="red",
   main = "Rotated contributions to first dimension")
graphics::barplot(contrib2,
   col="blue",
   main = "Rotated contributions to second dimension")
graphics::par(mfrow=base::c(1,1))
```

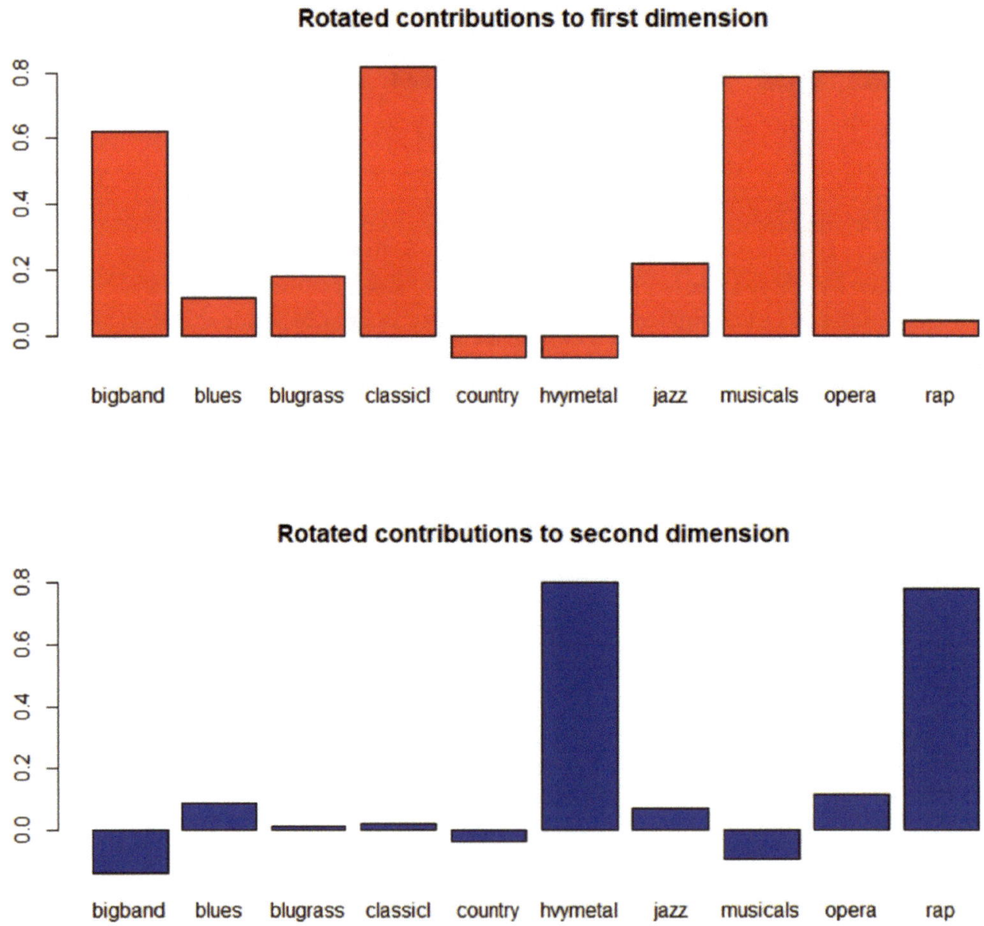

Figure 6.10 Contributions to first and second dimensions.

Residual analysis

Residual analysis is useful to assess the goodness of fit of the PCAresult model to the data. Residuals could be defined in different ways but here we define residuals as the observed correlations minus the correlations reproduced from factor loadings.

The `factor.model()` command from the "psych" package serves to reproduce the correlation matrix of the original data from the factor loading matrix. Recall that factor loadings are the correlations of the variables with the factors. Specifically, the reproduced correlation matrix is the product of the factor loading matrix times its transpose. Residuals are the difference between the observed correlation matrix and the reproduced one. The lower the residuals, the better the fit of the factor model. The initial command is that below, giving us the reproduced correlation matrix. We assume that the PCAresult object and the "dat" data frame created earlier are still in the R environment.

```
reproduced <- psych::factor.model(PCAresult$var$cor)
reproduced
[Output:]
            bigband      blues    blugrass    classicl     country
bigband   0.6032403 0.33761422  0.43845287  0.50451600  0.24990469
blues     0.3376142 0.79237268  0.29398886  0.20311114  0.04690182
blugrass  0.4384529 0.29398886  0.67052065  0.09420317  0.63034415
classicl  0.5045160 0.20311114  0.09420317  0.69708966 -0.14890969
country   0.2499047 0.04690182  0.63034415 -0.14890969  0.72334932
hvymetal -0.1667087 0.05842017 -0.03995496 -0.02839836 -0.06563460
jazz      0.3024115 0.74225192  0.10276095  0.31017088 -0.17963324
musicals  0.5583064 0.18977946  0.22147532  0.64612661  0.01119533
opera     0.5153187 0.16956814  0.19065443  0.66182911 -0.02516064
rap      -0.0415068 0.17991939  0.05109471  0.07099906 -0.02647071

            hvymetal        jazz    musicals       opera         rap
bigband  -0.16670868  0.30241148  0.55830641  0.51531865 -0.04150680
blues     0.05842017  0.74225192  0.18977946  0.16956814  0.17991939
blugrass -0.03995496  0.10276095  0.22147532  0.19065443  0.05109471
classicl -0.02839836  0.31017088  0.64612661  0.66182911  0.07099906
country  -0.06563460 -0.17963324  0.01119533 -0.02516064 -0.02647071
hvymetal  0.64946232  0.05122332 -0.13068730  0.03903632  0.62190619
jazz      0.05122332  0.77322857  0.24892830  0.23754281  0.16453106
musicals -0.13068730  0.24892830  0.64349474  0.62819159 -0.02403493
opera     0.03903632  0.23754281  0.62819159  0.66058259  0.13564994
rap       0.62190619  0.16453106 -0.02403493  0.13564994  0.62481345
```

The output above shows the reproduced correlation matrix. The diagonal contains the communalities. To make the diagonal be 1's instead, add the option `U2=FALSE`. To take one coefficient as an example, above the reproduced correlation of bigband with blues is 0.33761422.

For the bigband-blues correlation, the observed correlation is `cor(dat$bigband,-dat$blues)` = 0.2919103. The residual is observed – expected = 0.33761422 – 0.2919103

= 0.04570392. A rule of thumb is that residuals less than or equal to absolute 0.05 are not large. By this rule, this correlation would not be considered a large residual. In a well-fitting factor model, the researcher wants most absolute residuals to be 0.05 or less.

To calculate residuals, first convert the "dat" data frame to a correlation matrix. This gives the observed correlations, from which the reproduced correlations above may be subtracted to get the residuals. Then run the `factor.residuals()` command from the "psych" package.

```
observed <- stats::cor(dat)

# The two commands below return the same residuals matrix.
residuals <- observed - reproduced

residuals2 <-
psych::factor.residuals(observed,PCAresult$var$cor)
```

Make residuals just the upper triangle of the matrix. This avoids double-counting.

```
# as.matrix() and upper.tri() are in R's "base" package.
residuals <- base::as.matrix(residuals[base::upper.tri(residuals)])

# Get vector for large residuals. This returns TRUE or FALSE for
each correlation.
largeResiduals <- base::abs(residuals) > 0.05

# Count how many large residuals there are.
# Note that sum() counts, not adds the residual values.
residcount <-base::sum(largeResiduals)
residcount
[Output:]
   [1] 19

# Count how many residuals there are in all.
residtotal <- base::nrow(residuals)
residtotal
[Output:]
   [1] 45

# Calculate the percentage of residuals which are large.
pctlarge <- round(residcount/residtotal,2)
pctlarge
[Output:]
   [1] 0.42
```

Discussion: While having 42% of all correlation residuals be absolute 0.05 or greater does show imperfect fit, this is not unusual. A common rule of thumb is to accept PCA results as long as the percent of large residuals does not exceed 50%.

Compute mean square residual as another fit measure. Note that the residuals object is the upper triangle.

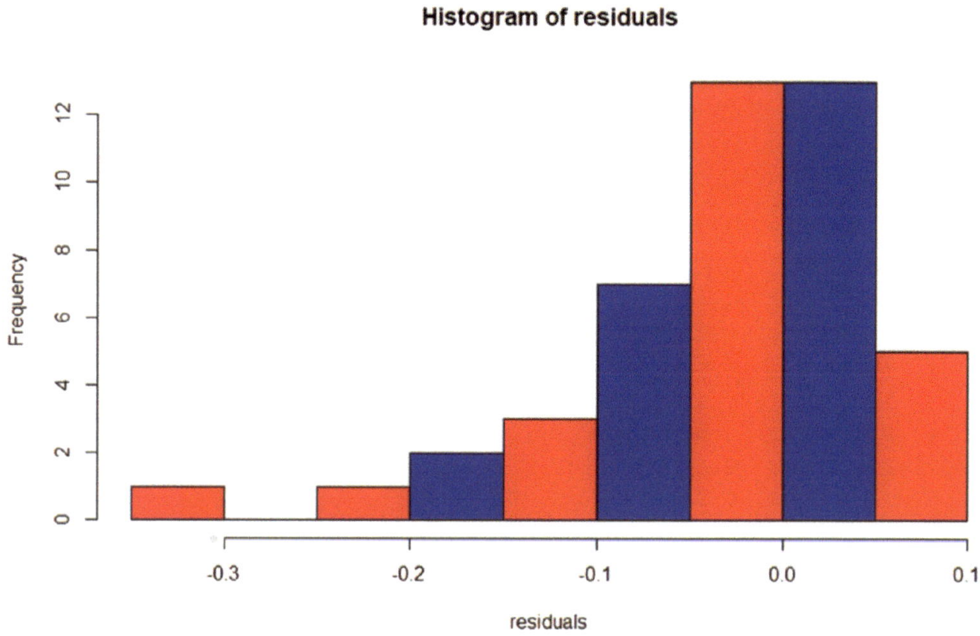

Figure 6.11 Histogram of residuals.

```
MSR <- base::sqrt(base::mean(residuals^2))
MSR
[Output:]
[1] 0.09062859
```

Discussion: By rule of thumb, a well-fitting model would have an MSR or 0.05 or less. An MSR of 0.08 or higher suggests the need to reconsider the PCA model, possibly requesting more components. This is the case for the present model.

In any well-fitting model, one expects most estimates to be close to the mark, with residuals approaching 0. Then the number of low and high estimates will trail off to either side. The result, displayed in the histogram in Figure 6.11, shows residuals are skewed. It is more likely that expected (reproduced) correlations will be higher than observed than lower than observed. Nonetheless, some researchers would consider this histogram not to represent an extreme departure from a normal distribution of residuals.

```
graphics::hist(residuals,col=c("red","blue"))
```

Saving component scores

One possible use of factor analysis is to derive observation-level scores on each component or factor. These scores may be saved and used as predictor variables in other forms of analysis such as regression. By default, these variables will be orthogonal (uncorrelated), with no mathematical possibility of multicollinearity, assuming varimax or some other orthogonal rotation has been used.

The `PCAresult$ind$coord` element contains the individual level scores as defined by FactoMineR. Specifically, these are the unrotated and unstandardized component scores for each observation on each component. The ind.coord element contains the "coord" values for individuals, just as the var.coord element contained "coord" values for variables. Since there were 14 variables, there are 14 components. The scores for the first two components are in `PCAresult$ind$coord[,1:2]`. The command below lists these for the first five observations. Due to differences in algorithms, component scores may differ for the `PCA()` method compared to other methods.

```
utils::head(PCAresult$ind$coord[,1:2],5)
[Output:]
            Dim.1       Dim.2
   1 -0.5912952 -0.57902508
   2 -2.7645081 -0.81060698
   3 -1.8243841 -0.00988843
   7 -2.3989843  2.20787971
   9 -0.2161190 -0.94908495
```

To visualize component scores, use the `plot()` command with the option choice = "ind" (for individuals, as opposed to "var" for variables). This is Figure 6.12, which shows individuals plotted in component space for the first two components. The plot is divided into quadrants (e.g., the upper left quadrant is high on component 2 and low on component 1). While having 773 individuals necessarily creates a very cluttered plot, the more extreme values in each of the quadrants are readily identifiable. Note that while there were only 773 individuals, id numbering was not sequential but rather had values up to 1499. Type `View(dat)` to see this and to identify individuals shown in Figure 6.12.

```
base::plot(PCAresult,choice="ind", col.ind="dodgerblue")
```

Unrotated but standardized (so variance approaches 1.0) component scores are found in the element svdU. These coefficients approach the values of the component scores as defined and computed in SPSS. The svdU element should be selected if the researcher prefers standardized scores, but be aware that when `PCA()` documentation discusses "scores" it is referring to the unstandardized ones. Also be aware that if later using scores as variables in regression, most regression models use unstandardized variables as predictors.

```
utils::head(PCAresult$svd$U[,1:2],5)
[Output:]
                [,1]         [,2]
   [1,] -0.3437393 -0.470040621
   [2,] -1.6070994 -0.658034026
   [3,] -1.0605745 -0.008027224
   [4,] -1.3946084  1.792311213
   [5,] -0.1256371 -0.770447588
```

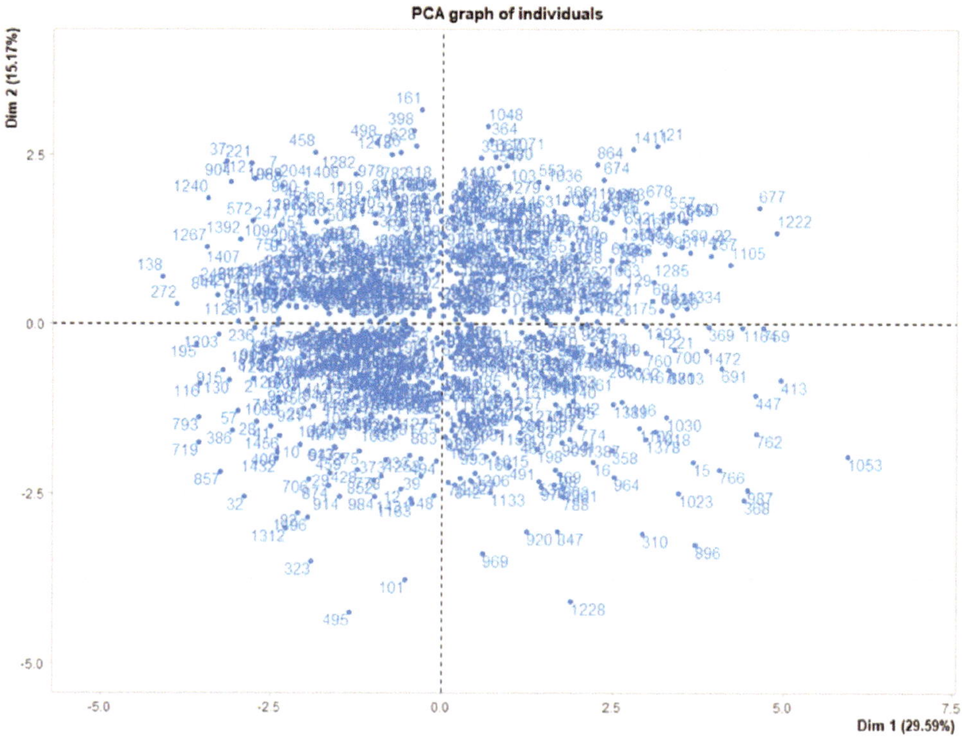

Figure 6.12 Component score plot for the first two components.

We add the unstandardized, unrotated component scores for all observations to the music3 data frame created below by copying the "dat" data object into music3. If we preferred to add PCA1 and PCA2 scores to the "music2" data frame, it would work in the same way except we would start with `music3 <- music2`.

```
music3 <- dat
music3$PCA1 <- PCAresult$ind$coord[,1]
music3$PCA2 <- PCAresult$ind$coord[,2]
```

To save music3, with component scores, as a csv file in the default directory, use the command below. Note column names are added to the csv file by default.

```
utils::write.csv(music3, file = "music3.csv", row.names = FALSE)
```

Unfortunately, the `PCA()` program does not directly support factor rotation and therefore also does not support computation of rotated component scores. If rotated component score output is important to the researcher's project, the researcher may select one of the other R programs which does directly support rotation, as discussed and illustrated in Chapter 10.

Automatic PCA reporting with "FactoInvestigate"

Using its `Investigate()` command, the supplementary package "FactoInvestigate" will automatically create a report on a `PCA()` result object such as PCAresult. The report covers detection of outliers, evaluation of inertia (percent and cumulative percent of variance explained) distribution, dimensions description, classification, and visualization. There are a large number of options (type `help(Investigate)`) including subsetting variables and/or observations.

All the results may be written as Word, html, or PDF documents. When a report is generated, it automatically includes the warning, "This interpretation of the results was carried out automatically, it cannot match the quality of a personal interpretation". If Word output is chosen, as below, by default the file "Investigate.docx" will be written to the default working directory, but there is an option to specify a different filename and path.

```
FactoInvestigate::Investigate(PCAresult, document = "word_document")
```

The report is shown below. It is presented "as is", without editing changes.

Principal Component Analysis

DATASET DAT

This dataset contains 773 individuals and 10 variables.

1. Study of the outliers

The analysis of the graphs does not detect any outlier.

2. Inertia distribution

The inertia of the first dimensions shows if there are strong relationships between variables and suggests the number of dimensions that should be studied.

The first two dimensions of analyse express **44.77%** of the total dataset inertia ; that means that 44.77% of the individuals (or variables) cloud total variability is explained by the plane. This is an intermediate percentage and the first plane represents a part of the data variability. This value is strongly greater than the reference value that equals **23.74%**, the variability explained by this plane is thus highly significant (the reference value is the 0.95-quantile of the inertia percentages distribution obtained by simulating 2949 data tables of equivalent size on the basis of a normal distribution).

From these observations, it may be interesting to consider the next dimensions which also express a high percentage of the total inertia.

An estimation of the right number of axis to interpret suggests to restrict the analysis to the description of the first 3 axis. These axis present an amount of inertia greater than those obtained by the 0.95-quantile of random distributions (58.18% against 34.7%). This observation suggests that only these axis are carrying a real information. As a consequence, the description will stand to these axis.

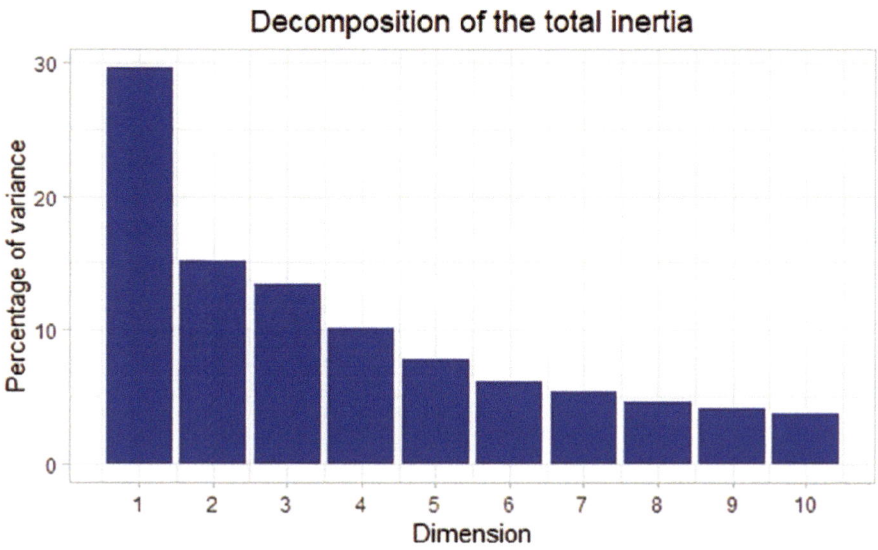

Figure 6.13 Decomposition of the total inertia

3. Description of the plane 1:2

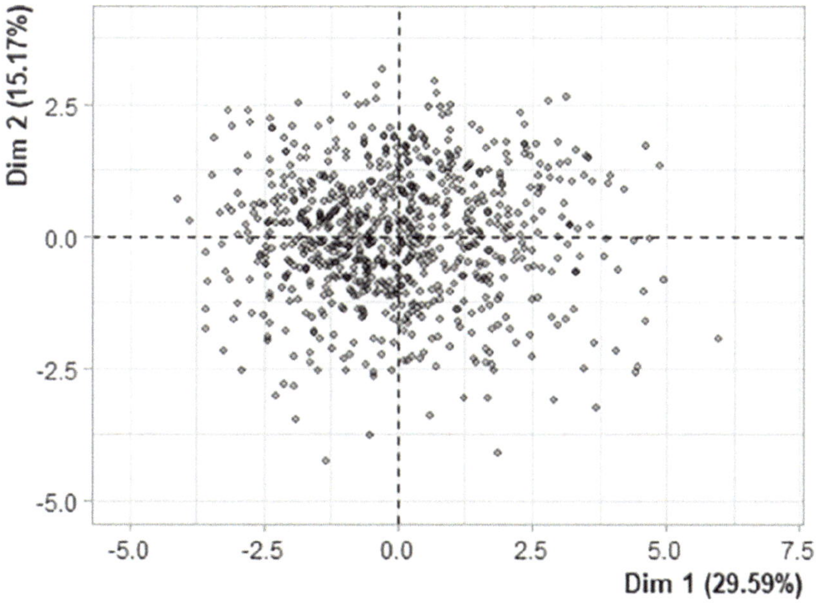

Figure 6.14 **Individuals factor map (PCA)** The labeled individuals are those with the higher contribution to the plane construction.

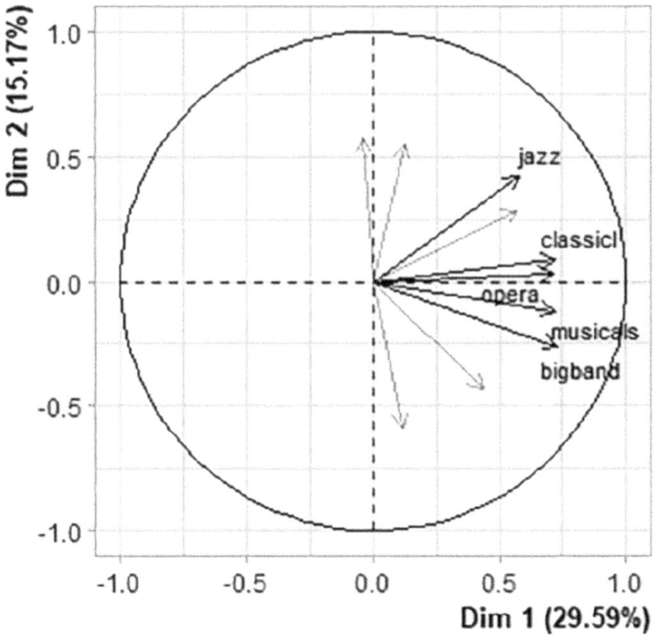

Figure 6.15 **Variables factor map (PCA)** The labeled variables are those the best shown on the plane.

The **dimension 1** opposes individuals characterized by a strongly positive coordinate on the axis (to the right of the graph) to individuals characterized by a strongly negative coordinate on the axis (to the left of the graph).

The group 1 (characterized by a positive coordinate on the axis) is sharing :

- high values for the variables *bigband, musicals, classicl, opera, blugrass, blues, jazz* and *country* (variables are sorted from the strongest).
- low values for the variable *hvymetal.*

The group 2 (characterized by a negative coordinate on the axis) is sharing :

- high values for the variable *country.*
- low values for the variables *jazz, rap, blues, opera, classicl, hvymetal, musicals* and *bigband* (variables are sorted from the weakest).

The group 3 (characterized by a negative coordinate on the axis) is sharing :

- high values for the variables *hvymetal* and *rap* (variables are sorted from the strongest).
- low values for the variables *bigband, blugrass, musicals, country, classicl, opera* and *blues* (variables are sorted from the weakest).

The **dimension 2** opposes individuals characterized by a strongly positive coordinate on the axis (to the top of the graph) to individuals characterized by a strongly negative coordinate on the axis (to the bottom of the graph).

The group 1 (characterized by a positive coordinate on the axis) is sharing :

- high values for the variables *hvymetal* and *rap* (variables are sorted from the strongest).
- low values for the variables *bigband*, *blugrass*, *musicals*, *country*, *classicl*, *opera* and *blues* (variables are sorted from the weakest).

The group 2 (characterized by a negative coordinate on the axis) is sharing :

- high values for the variable *country*.
- low values for the variables *jazz*, *rap*, *blues*, *opera*, *classicl*, *hvymetal*, *musicals* and *bigband* (variables are sorted from the weakest).

The group 3 (characterized by a negative coordinate on the axis) is sharing :

- high values for the variables *bigband*, *musicals*, *classicl*, *opera*, *blugrass*, *blues*, *jazz* and *country* (variables are sorted from the strongest).
- low values for the variable *hvymetal*.

4. Description of the dimension 3

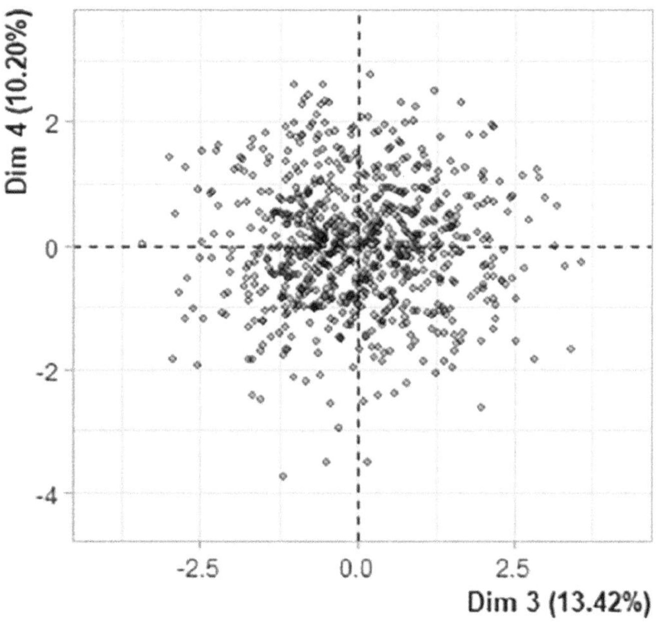

Figure 6.16 **Individuals factor map (PCA)** The labeled individuals are those with the higher contribution to the plane construction.

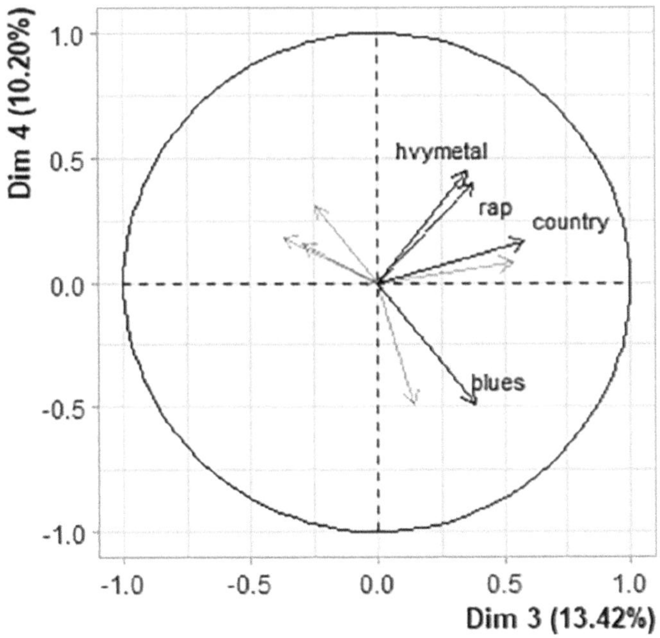

Figure 6.17 **Variables factor map (PCA)** The labeled variables are those the best shown on the plane.

The **dimension 3** opposes individuals characterized by a strongly positive coordinate on the axis (to the right of the graph) to individuals characterized by a strongly negative coordinate on the axis (to the left of the graph).

The group 1 (characterized by a positive coordinate on the axis) is sharing :

- high values for the variables *blues, blugrass, country, jazz, hvymetal* and *rap* (variables are sorted from the strongest).
- low values for the variables *classicl, opera* and *musicals* (variables are sorted from the weakest).

The group 2 (characterized by a negative coordinate on the axis) is sharing :

- high values for the variables *musicals, classicl* and *jazz* (variables are sorted from the strongest).
- low values for the variables *hvymetal, rap, blugrass* and *country* (variables are sorted from the weakest).

The group 3 (characterized by a negative coordinate on the axis) is sharing :

- high values for the variables *opera, hvymetal, classicl, rap* and *musicals* (variables are sorted from the strongest).
- low values for the variables *jazz* and *blues* (variables are sorted from the weakest).

5. Classification

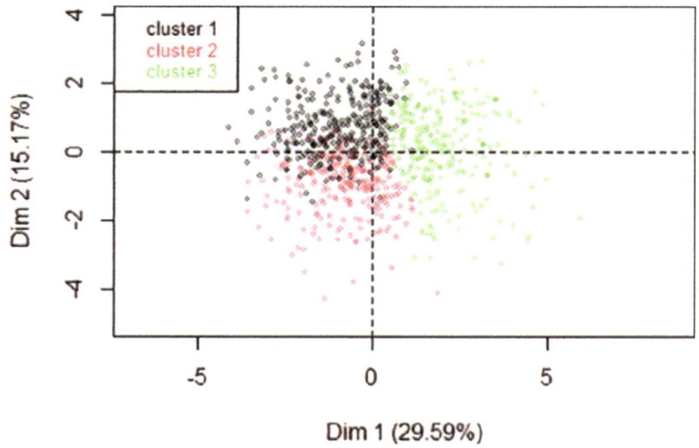

Figure 6.18 **Ascending Hierarchical Classification of the individuals.** The classification made on individuals reveals 3 clusters.

The **cluster 1** is made of individuals sharing :

- high values for the variables *hvymetal* and *rap* (variables are sorted from the strongest).
- low values for the variables *musicals*, *bigband*, *classicl*, *opera*, *blugrass* and *country* (variables are sorted from the weakest).

The **cluster 2** is made of individuals sharing :

- low values for the variables *rap*, *hvymetal*, *blues*, *jazz*, *classicl*, *opera*, *blugrass* and *bigband* (variables are sorted from the weakest).

The **cluster 3** is made of individuals sharing :

- high values for the variables *classicl*, *bigband*, *musicals*, *opera*, *jazz*, *blues*, *blugrass* and *rap* (variables are sorted from the strongest).

Relating factor analysis to theory

Factor analysis may be used to test theory. Consider the following output for a different example in which code letters reflect the four grid-group Cultural Theory categories of Egalitarian, Fatalist, Hierarch, and Individualist respectively. The "E" items were intended to load on Dimension 1, the "F" items on Dimension 2, the "H" items on Dimension 3, and the "I" items on Dimension 4. The PCA output below is for a four-factor solution in which E1–E3 were expected to load on Dimension 1, which was labeled "Egalitarian". Other dimensions were other cultural categories. There were twelve items in all, three for each of the four cultural groups.[6]

Component loadings

	Dimension			
	1	*2*	*3*	*4*
E1: What society needs is a fairness revolution to make the distribution of goods more equal.	0.842	−0.148	0.244	0.293
E2: Society works best if power is shared equally.	0.911	−0.140	0.206	0.277
E3: It is our responsibility to reduce differences in income between the rich and the poor.	0.654	0.300	−0.444	−0.110
I1: Even if some people are at a disadvantage, it is best for society to let people succeed or fail on their own.	−0.609	0.330	−0.224	0.572
I2: Even the disadvantaged should have to make their own way in the world.	−0.283	−0.761	−0.184	0.400
I3: We are all better off when we compete as individuals.	−0.306	−0.781	−0.156	0.417
H1: The best way to get ahead in life is to work hard and do what you are told to do.	−0.511	0.088	0.389	0.525
H2: Society is in trouble because people do not obey those in authority.	−0.579	0.602	0.319	0.112
H3: Society would be much better off if we imposed strict and swift punishment on those who break the rules.	−0.098	−0.110	0.907	−0.147
F1: The most important things that take place in life happen by chance.	0.226	0.733	−0.286	0.411
F2: No matter how hard we try, the course of our lives is largely determined by forces beyond our control.	0.868	−0.091	0.165	0.415
F3: For the most part, succeeding in life is a matter of chance.	0.146	0.771	0.145	0.209

The ideal outcome is shown below. Egalitarians were expected to load strongly on three items (E1 through E3) and to load not strongly on the other nine items, with no cross-loadings. Ideal results for the three other cultural categories were similar.

	Loadings expected > 0.6	*Loadings expected < 0.4*	*Expected > 0.6 but not (cross-loadings)*	*Expected < 0.4 but not (cross-loadings)*	*Percent as expected*
Egalitarians	3	9	0	0	100
Individualists	3	9	0	0	100
Hierarchs	3	9	0	0	100
Fatalists	3	9	0	0	100

The actual outcome differed from the ideal:

	Expected > 0.6	*Expected < 0.4*	*Expected > 0.6 but not*	*Expected < 0.4 but not (cross-loadings)*	*Percent as expected*
Egalitarians	3	5	0	4	67
Individualists	0	6	3	3	50
Hierarchs	1	8	2	1	75
Fatalists	2	6	1	3	67

Discussion

1 Only for the Egalitarian dimension were all its three indicators most highly loaded on the intended dimension. The Fatalist dimension had 2, the Hierarch dimension had 1, and the Individualist dimension had 0.
2 Even the Egalitarian dimension had four unexpected high cross-loadings with items intended for other dimensions.
3 The percent of loadings which were as expected varied from 50% for the Individualist dimension to 75% for the Hierarch dimension.
4 However, the Hierarch dimension had only 1 of its 3 intended indicators load >0.6.
5 Overall, only four of the 12 items (33%) conformed to the ideal pattern of having the item load >0.6 on its intended dimension and having no high cross-loadings. These were Items E1, E2, H3, and F3. There were no ideal items for the Individualist dimension. By common rule of thumb, these four items would be retained.
6 Overall, all dimensions had at least 1 high cross-loading with an item for a different dimension. Some 11/48 = 22.9% of coefficients were high cross-loadings. By one rule of thumb, we would like no more than 5% of coefficients to be high cross-loadings.
7 Overall, only 50% (6/12) of the indicators loaded >0.6 on their intended dimensions, though two of these had high cross-loadings. By common rule of thumb, the other 50% of indicators might be dropped from the model.
8 Most researchers would fail to agree that the factor structure expected by Cultural Theory was upheld by these results. Put another way, Cultural Theory dimensions fail to differentiate responses to the 12 survey items in the manner it would predict.

Notes

1 The reader may find the FactoMineR user group helpful. It is located at https://groups.google.com/g/factominer-users.
2 The eigenvalues and variance explained are the same as for unrotated or varimax rotated output in SPSS.
3 The true variables are in p-dimensional space, but the biplot shows them projected in two-dimensional space.
4 The eigenvectors are the same as in SPSS in the unrotated "Component Matrix" table.
5 The rotated loadings are the same as in SPSS for varimax rotation, but SPSS switches the labels, labeling component 2 in R as component 4 in SPSS, and vice versa. Also, SPSS reverses the sign of all component 4 loadings, but not the others. As the importance of variables is assessed in terms of the magnitude of their absolute loadings, this difference may not matter to the researcher.
6 A PCA analysis of these variables with different data is presented in Ripberger, Gupta, Silva, and Jenkins-Smith (2014).

7 Oblique and higher-order factor models

Oblique PFA with fa()

Introduction

Oblique factor analysis, discussed in Chapters 2 and 3, is used when the theory guiding the researcher suggests that the dimensions (factors) are correlated. This means, of course, that factors are not independent (not orthogonal). The meanings of orthogonal factors are imputed from unique and separate sets of measured indicator variables, making the resulting factors unique and separate from each other. The meanings of oblique factors are not unique and separate but rather overlap in imputed meaning. By removing independence (orthogonality) constraints in an oblique factor solution it is possible to rotate the axes so they align even better with underlying dimensions of the indicator variables, increasing their loadings compared to what they would be if constrained to be orthogonal. Not only are results stronger in this sense but many researchers believe that the correlation of factors is truer to real life, particularly in social science settings.

The cost of oblique rotation is that the resulting factors are no longer independent. If the researcher had wanted to establish divergent (discriminant) validity for the factor model, this is done in a well-fitting orthogonal model because in such a model, by definition, each factor is independent of each other factor. Oblique models do not establish divergent validity of factors. Indeed, lack of divergent validity of the factors can be assessed by how much better the fit of the oblique solution is than the fit for the orthogonal solution. Some researchers use this principle to run an oblique solution first, then proceed to an orthogonal solution only if the improvement in strength in the oblique model is weak and does not change the substantive interpretation and labeling of the factors.

In orthogonal solutions, the strength of the model is reflected in the size of indicator loading in the factor loadings table. In oblique solutions, two tables are generated. The one called the "pattern matrix" is the one primarily used to assess model strength. As discussed further below, however, the researcher must also examine the second table, called the "structure" matrix.

There are a variety of oblique approaches to rotation, all of which allow factors to be correlated. The most widely used package for PFA is the `fa()` program from the "psych" package. This program supports a variety of oblique as well as orthogonal rotations. Below we use the "oblimin" method, which is perhaps the most widely-used oblique rotation approach and also is the `fa()` default. We could have used "promax" to get a slightly different oblique rotation. Supported oblique rotations are "Promax",

DOI: 10.4324/9781003279693-8

"promax", "oblimin", "simplimax", "bentlerQ", "geominQ", "biquartimin", and "cluster".

Factor, component, pattern, and structure matrices: Oblique rotation returns both a "structure matrix" as in ordinary (orthogonal) factor or component analysis, and it returns a "pattern matrix". These represent two sets of factor or component loadings. They are only the same if the factors are perfectly uncorrelated.

- *The pattern matrix* contains coefficients which represent just unique contributions of an item, without common contributions (unique contributions are those after the effects of other variables in the model are controlled by partialling them out). The more factors/components, the lower the pattern coefficients as a rule since there will be more common contributions to variance explained. Pattern coefficients sometimes exceed plus or minus 1.0. Some researchers prefer to use pattern loadings since, because they represent unique contributions, they may be more helpful for evaluation of causal models. In this view, the oblique model should be developed using pattern coefficients, then structure coefficients may be used to help confirm the model further.
- *The structure matrix*, in contrast, is simply the factor or component loading matrix as in orthogonal rotation, representing the variance in a measured variable explained by a factor or component on both a unique and common contributions basis. That is, structure coefficients reflect the simple or whole correlation of a variable with its factor. Structure coefficients always range from −1.0 to +1.0. Some researchers prefer to use structure loadings since, when squared, they are the percent of variance in a variable explained by a factor (some call this the reliability of the item), and vice versa. That is, structure coefficients, when squared are also the percent of variation in the factor explained by the variable (some call this the contribution of the item).

For oblique rotation, the researcher looks at both the structure and pattern coefficients when attributing a label to a factor. In oblique rotation, the sum of squared loadings for a factor reflects its importance even though assignment of proportions of variance to particular factors is ambiguous and cannot be calculated.

In the example below we use the "ability.cov" data on six psychological tests, the correlations for which are placed in the correlation matrix labeled "corrmatrix". That is, we use covariance matrix input, not raw data input, though either is acceptable. If we wanted individual-level residuals, we would have had to have used raw data input.

Setup and data

```
# The ability.cov object is from R's built-in "datasets" package.
utils::data(ability.cov)
base::library(psych)        # Has the fa() command
base::library(GPArotation) # Needed by fa() for oblique rotation

# Create the corrmatrix object.
covmatrix <- ability.cov$cov
corrmatrix <- stats::cov2cor(covmatrix)
```

```
# For comparison purposes, we create the orthogonal solution,
faTestsVarimax.
# faTestsVarimax is of data class "psych" and  "'fa ".
faTestsVarimax <- psych::fa(r = corrmatrix, nfactors = 2, rotate
="varimax", fm="pa")

faTestsVarimax
[Output not shown]
```

The oblimin model

The command below runs an oblique factor model with maximum likelihood estimation using the `fa()` program with oblique rotation (oblimin). Bootstrapped confidence intervals are requested by virtue of adding the options `n.obs=112` and `n.iter=1200`. (The number of observations was given in Appendix 1.) Bootstrapping is an iterative process and we ask for 1,200 iterations. The resulting b coefficients for variables are not significant at the 0.05 level if they are outside the lower and upper confidence limits (shown near the bottom of the output below), which none do for our example. Warning: Bootstrapping involves random processes. Set the random seed to assure replicability of coefficients.

```
# Oblimin oblique rotation with a maximum likelihood factoring
method.
# The n.obs= and n.iter- options are needed for bootstrapped confi-
dence estimate,
# however these must be removed to run the residuals() command, as
discussed below.
base::set.seed(123)

faTestsOblimin <- psych::fa(corrmatrix,
    nfactors = 2,
    rotate ="oblimin",
    fm="ml",
    residuals = TRUE,
    SMC = TRUE,
    missing=FALSE,
    n.obs=112,
    n.iter=1200)
```

Typing the name of the `fa()` results object prints out the factor pattern matrix and associated statistics, as shown below. The pattern matrix, along with the structure matrix, is used to impute labels for the factors. Recommended practice is to examine both matrices. The results object, faTestsOblimin, is of data class "psych" and "fa.ci".

In the pattern matrix below, the squared loadings represent the unique contribution of each factor to each measured variable. For instance, the contribution of factor ML1 to the "general" indicator is $0.39^2 = 0.1521$. The sum of all squared loadings for ML1 is 1.82. The proportion of variance explained by the factor ML1 is 51% because that is the percentage that 1.82 is of the sum of sums of squared loadings for all factors (for ML1 and ML2 = 1.82 + 1.76). Also, in the output below, we see that the correlation of the two factors is 0.46, a substantial level. The correlation represents the correlation of the factor scores for ML1 with the factor scores for ML2.[1]

```
faTestsOblimin
[Output:]
   Factor Analysis with confidence intervals using method =
   psych::fa(r = corrmatrix, nfactors = 2, n.obs = 112,
   n.iter = 1200,
       rotate = "oblimin", residuals = TRUE, missing = FALSE,
       fm = "ml")
   Factor Analysis using method =  ml
   Call: psych::fa(r = corrmatrix, nfactors = 2, n.obs = 112,
   n.iter = 1200, rotate = "oblimin", residuals = TRUE, missing =
   FALSE, fm = "ml")
   Standardized loadings (pattern matrix) based upon correlation
   matrix
             ML1   ML2   h2    u2 com
   general  0.39  0.47 0.54 0.455 1.9
   picture -0.01  0.65 0.41 0.589 1.0
   blocks  -0.03  0.90 0.78 0.218 1.0
   maze    -0.02  0.49 0.23 0.769 1.0
   reading  0.99 -0.04 0.95 0.052 1.0
   vocab    0.79  0.05 0.67 0.334 1.0

                         ML1  ML2
   SS loadings           1.82 1.76
   Proportion Var        0.30 0.29
   Cumulative Var        0.30 0.60
   Proportion Explained  0.51 0.49
   Cumulative Proportion 0.51 1.00

    With factor correlations of
       ML1  ML2
   ML1 1.00 0.46
   ML2 0.46 1.00

   Mean item complexity =  1.2
   Test of the hypothesis that 2 factors are sufficient.

   The degrees of freedom for the null model are  15  and the
   objective function was  2.48 with Chi Square of  268
   The degrees of freedom for the model are 4  and the objective
   function was  0.06

   The root mean square of the residuals (RMSR) is  0.04
   The df corrected root mean square of the residuals is  0.07

   The harmonic number of observations is  112 with the empirical
   chi square  4.23  with prob <  0.38
```

```
The total number of observations was  112  with Likelihood Chi
Square =  6.11  with prob <  0.19

Tucker Lewis Index of factoring reliability =  0.968
RMSEA index=0.068 and the 90 % confidence intervals are  0 0.171
BIC =  -12.8
Fit based upon off diagonal values = 0.99
Measures of factor score adequacy
                                                  ML1  ML2
Correlation of (regression) scores with factors  0.98 0.92
Multiple R square of scores with factors         0.95 0.85
Minimum correlation of possible factor scores    0.91 0.69

 Coefficients and bootstrapped confidence intervals
          low   ML1 upper   low   ML2 upper
general  0.24  0.39  0.56  0.28  0.47  0.67
picture -0.17 -0.01  0.17  0.44  0.65  0.84
blocks  -0.16 -0.03  0.12  0.72  0.90  1.06
maze    -0.20 -0.02  0.20  0.25  0.49  0.69
reading  0.83  0.99  1.09 -0.12 -0.04  0.11
vocab    0.67  0.79  0.97 -0.06  0.05  0.16

 Interfactor correlations and bootstrapped confidence intervals
        lower estimate upper
ML1-ML2  0.28     0.46  0.61
```

To compare the varimax and oblimin solutions, we can run a congruence check using the `congruence()` command from the "psych" package. This check reveals that the two solutions are highly correlated.

```
psych::factor.congruence(faTestsVarimax$loadings,faTestsOblimin$-
loadings)
[Output:]
       ML1  ML2
  PA1 0.97 0.32
  PA2 0.32 0.97
```

Also, for comparison, we print out the "fit.off" element of the results object. This is a measure of how well the off-diagonal elements in the correlation matrix are reproduced by the factor model. Again, the fit is high and very similar for both the oblimin and varimax rotations.

```
# Set subsequent output to three digits.
base::options(digits=3)

# Off-diagonal fit measures
faTestsVarimax$fit.off
[Output:]
    [1] 0.995
```

```
faTestsOblimin$fit.off
[Output:]
    [1] 0.993
```

We may also print out the "fit" measure for the varimax and oblimin solutions. This is a similar fit measure, but one which describes how well the factor model reproduces the correlation matrix. Both models are very similar by this measure also.

```
faTestsVarimax$fit
[Output:]
   [1] 0.883

faTestsOblimin$fit
[Output:]
    [1] 0.886
```

Discussion of oblique output above:

1 By allowing factors to correlate rather than be constrained to be orthogonal as in the varimax solution, the solution is just a bit closer to simple factor structure as shown by the fact that mean item complexity has drops from 1.3 to 1.2. This comes at the price of allowing factors to correlate, which means they overlap in meaning.

```
# numItems is the number of items in corrmatrix, which was based
on ability.cov = 6
numItems <- base::dim(corrmatrix)[1]

# Mean item complexity is the mean of the complexity values for
the six items.
# For varimax:
base::sum(faTestsVarimax$complexity)/numItems
[Output:]
   [1] 1.270

# For oblimin:
base::sum(faTestsOblimin$complexity)/numItems
[Output:]
   [1] 1.16
```

2 However, the goodness of fit measures fit based upon off-diagonal values (0.99, shown above) and RMSR (shown below) are similar for both oblimin (0.035) and varimax (0.032) solutions. As the fit is similar, researchers might revert to a varimax solution so that factors would be independent (orthogonal) and thus more easily interpretable.

```
faTestsOblimin$rms
[Output:]
   [1] 0.0355

faTestsVarimax$rms
[Output:]
   [1] 0.0322
```

There are a large number of output elements in the results object created by the `fa()` command. Below we show the 45 elements for faTestsOblimin. This is done using the `attributes()` command from R's "base" package, which is loaded by default. Each element may be accessed by typing faTestsOblimin, then a dollar sign, then the name of the element (without double quote marks). Examples follow.

```
base::attributes(faTestsOblimin)
[Output:]
$names
 [1] "residual"                 "dof"
 [3] "ENull"                    "chi"
 [5] "rms"                      "nh"
 [7] "EPVAL"                    "crms"
 [9] "EBIC"                     "ESABIC"
[11] "fit"                      "fit.off"
[13] "sd"                       "factors"
[15] "complexity"               "n.obs"
[17] "PVAL"                     "objective"
[19] "criteria"                 "Call"
[21] "null.model"               "null.dof"
[23] "r.scores"                 "R2"
[25] "valid"                    "score.cor"
[27] "weights"                  "rotation"
[29] "communality"              "communalities"
[31] "uniquenesses"             "values"
[33] "e.values"                 "loadings"
[35] "model"                    "fm"
[37] "rot.mat"                  "Phi"
[39] "Structure"                "communality.iterations"
[41] "method"                   "R2.scores"
[43] "r"                        "fn"
[45] "Vaccounted"

$class
[1] "psych" "fa"
```

The "loadings" element of the `fa()` result object gives the oblique standardized factor loadings. These are the rounded pattern coefficients, not displaying low loadings. These are the same as those above produced by simply typing the name of the results object, here "faTestsOblimin". We see that the verbal tests (reading, vocab) have high loadings on factor ML1, meaning that ML1 strongly explains the unique variance in these tests. Likewise, ML2 explains the unique variance in the spatial test called blocks, but is less explanatory of the picture and especially the maze tests. The general test cross loads on both ML1 and ML2.

```
faTestsOblimin$loadings
[Output:]
   Loadings:
```

```
         ML1    ML2
general  0.386  0.475
picture         0.646
blocks          0.896
maze            0.488
reading  0.990
vocab    0.791

                 ML1   ML2
SS loadings     1.756 1.688
Proportion Var  0.293 0.281
Cumulative Var  0.293 0.574
```

The structure matrix is a second type of loadings matrix which is produced for oblique solutions. Correlations in the structure matrix represent the unique association of an indicator variable with its factor *plus* the relationship of the variable to the overlapping variance of the correlated factors in an oblique solution. The pattern loadings, in contrast, reflected just the unique association of a variable with its factor. Factor structure coefficients are the product of the pattern loadings and Phi matrix, which is the factor correlation matrix. To verify this, run this command: `faTestsOblimin$loadings %*% faTestsOblimin$Phi`. The "%*%" is the matrix multiplication operator in R. However, it is simpler just to call the "Structure" element of the result:

```
faTestsOblimin$Structure
[Output:]
   Loadings:
            ML1   ML2
   general 0.607 0.654
   picture 0.289 0.641
   blocks  0.390 0.884
   maze    0.209 0.480
   reading 0.973 0.423
   vocab   0.815 0.420

                    ML1   ML2
   SS loadings     2.258 2.205
   Proportion Var  0.376 0.368
```

Patten coefficients, which represent unique contributions of a factor to an indicator, are usually used to interpret factors. However, as Tabachnick and Fidell (2013: 632) note, "If factors are very highly correlated, it may appear that no variables are related to them because there is almost no unique variance once overlap is omitted". If a variable has a low coefficient in the pattern matrix but a high coefficient in the structure matrix, this means that the variable's relation to the factor is mostly due to shared variance reflected in the high correlation of factors in an oblique solution.

Structure and pattern coefficients may be used to check for suppression effects (Horst, 1941). Suppression effects cause the correlation of two variables to be lower than expected because a third variable, which is a prior cause of both, acts positively on the first variable and negatively or has zero correlation with the second. In partial correlation terms, the partial correlation controlling for the third variable goes up rather than down as is normally the case. Not the case here, but if a structure coefficient approaches zero but the pattern coefficient is high, or vice versa, or if a structure and pattern coefficient are of different signs, this can be evidence of a suppression effect (Graham, Guthrie and Thompson, 2003). For more on suppression effects, see Lancaster (1999) and Thompson (2004).

The "r.scores" element of the `fa()` result object contains the factor correlation matrix. Here there is a moderate (r = 0.498) correlation between the two factors. Some authors will revert to an orthogonal method such as PCA if the correlation is less than 0.32 (one factor then explains 10% or less of the variance of the other factor).

```
faTestsOblimin$r.scores
[Output:]
          [,1] [,2]
     [1,] 1.00 0.51
     [2,] 0.51 1.00
```

The input correlation or covariance matrix may be reproduced by the product of a factor loading matrix times its transpose. Residuals reflect the original matrix minus the reproduced matrix. Thus, output is the matrix of residual correlations after the factor model is applied. Residuals may be retrieved from the "residual" element of the faTestsOblimin results object. These are not the usual observation-level residuals because for this example no individual-level data were entered, only a correlation matrix. The residual matrix below reflects error in reproducing the observed correlation matrix.

```
resids <- faTestsOblimin$residual
resids
[Output:]
          general   picture    blocks    maze   reading     vocab
general  4.55e-01  0.050543 -0.018440  0.0319 -6.34e-05  2.81e-04
picture  5.05e-02  0.589332  0.005843 -0.1146  5.02e-04 -2.29e-02
blocks  -1.84e-02  0.005843  0.218182  0.0205  6.54e-04  1.65e-03
maze     3.19e-02 -0.114570  0.020518  0.7694 -5.00e-03  2.85e-02
reading -6.34e-05  0.000502  0.000654 -0.0050  5.25e-02  1.05e-05
vocab    2.81e-04 -0.022937  0.001647  0.0285  1.05e-05  3.34e-01
```

The "resids" object is a residuals object of class "matrix" and "array". It takes the form of a matrix with variables (here, tests) being rows and columns. The diagonal is not suppressed and both upper and lower triangles are included.

The residual matrix is seen better formatted and rounded by using the `residuals()` command from R's built-in "stats" package. First, however, we create the results object faTestsOblimin2, which is the same as faTestsOblimin but without the bootstrapped confidence limits.

```
base::set.seed(123)
faTestsOblimin2 <- psych::fa(corrmatrix, nfactors=2,ro-
tate="oblimin", fm="ml", residuals=TRUE, SMC=TRUE, missing=FALSE)

resids2 <- stats::residuals(faTestsOblimin2, diag=FALSE)
```

```
resids2
[Output:]
          genrl pictr blcks maze  redng vocab
   general   NA
   picture  0.05    NA
   blocks  -0.02  0.01    NA
   maze     0.03 -0.11  0.02    NA
   reading  0.00  0.00  0.00  0.00    NA
   vocab    0.00 -0.02  0.00  0.03  0.00    NA
```

We now compute the mean absolute residual. Both `mean()` and `abs()` are functions in R's "base" package. If the mean residual is high, the model may need respecification, such as adding additional factors. Note that this must be computed on the resids2 object, which does not include the diagonal as did the resids object.

```
base::mean(base::abs(resids2), na.rm=TRUE)
[Output:]
   [1] 0.0201
```

Visualization

As faTestsOblimin is of classes "psych" and "fa", visualization like those illustrated previously for other factor analysis packages via commands from the "factoextra" package will result in the error message, "An object of class : psychfa can't be handled by factoextra."

One way to visualize the results of faTestsOblimin, which is an object of class "'psych' 'fa'", is to use the `factor.plot()` command from the "psych" package. This command plots variables by their factor loadings. The result is Figure 7.1.

```
psych::factor.plot(faTestsOblimin, pos=3,
cex=1.5,
   xlim= base::c(-0.1, 1.1),
   ylim = base::c(-0.1,1.1),
   labels = base::rownames(faTestsOblimin$loadings),
   title="Ability variables plotted by factor loadings")
```

The `fa.diagram()` command, also from the "psych" package, produces a factor diagram (left side, Figure 7.2). We also display the factor loadings plot (right side, Figure 7.2). The diagram shows standardized loadings above the 0.30 cutoff level. The curved arrow connecting PA1 and PA2 is the inter-factor correlation calculated by the oblimin method, which allows an oblique (correlated) solution. We see that vocab and readings load on factor ML1. In descending order of strength, blocks, picture, and maze load on ML2. The variable general loads moderately on both dimensions (cross-loads).

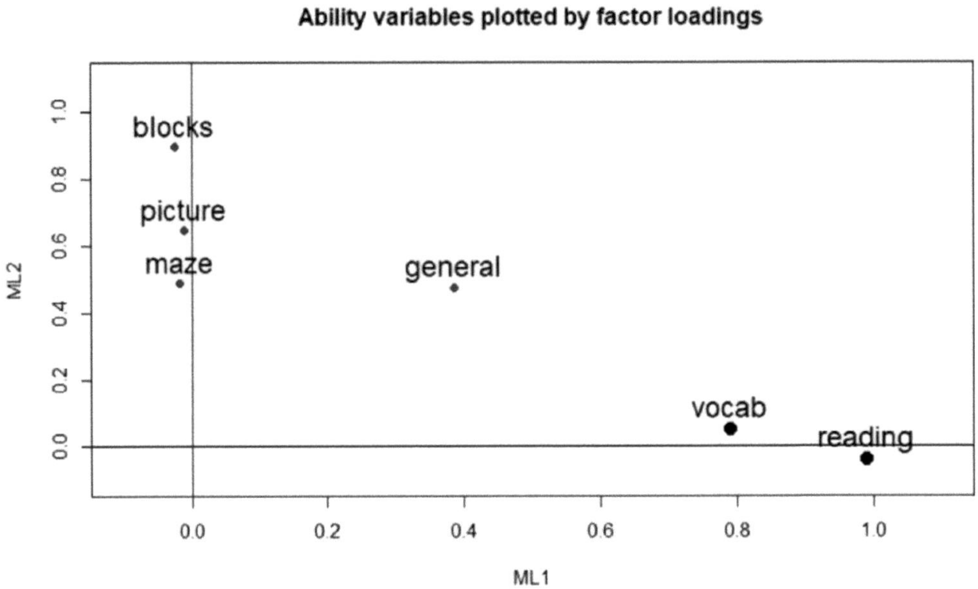

Figure 7.1 Plot of ability tests variables using principal axis factoring with oblimin rotation.

```
# Create vector of labels as ordered in RStudio's "Environment" tab
for faTestsOblimin.
labelvector <- base::c("general", "picture", "blocks", "maze",
"reading", "vocab")
# Set for side-by-side display.
graphics::par(mfrow=base::c(1,2))

# Create the factor diagram.
psych::fa.diagram(faTestsOblimin, simple= FALSE, cut=.30, digits=2,
rsize=.5, marg=base::c(4,4,4,1), main="Diagram for faTestsOblimin")

# Plot the factor loadings.
base::plot(faTestsOblimin$loadings, ylim= base::c(-.2,1),
xlim=base::c(-0.1, 1.4), main="Loadings for faTestsOblimin")
graphics::abline(h=0.0, col="black")

graphics::abline(v=0.0, col="black")

# Add labels to the plot.
graphics::text(faTestsOblimin$loadings[,1:2], labels=labelvector,
cex=0.9, font=2, pos=4)

# Reset the display for single images.
graphics::par(mfrow=base::c(1,1))
```

Next, we plot the correlation of residuals, shown in Figure 7.3. Since the input was a correlation matrix, this is simply the contents of the "resids2" object in graphic format.

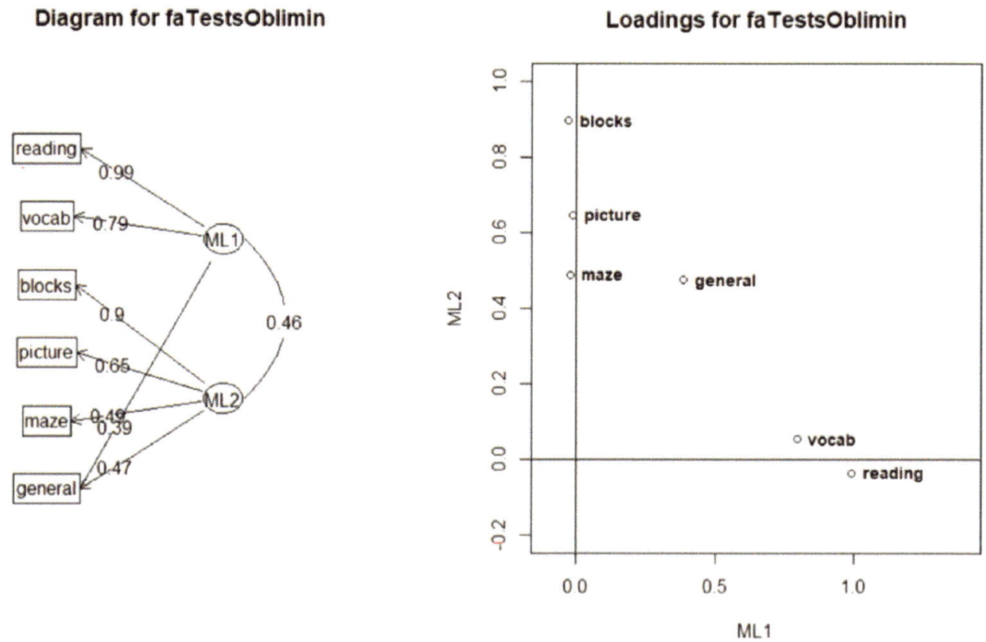

Figure 7.2 Diagram of ability tests (left) and factor loadings plot (right).

Note that we must use resids2, from the `residuals()` command, not the earlier "resids" object. Type `help(plot.psych)` for more options and information.

```
base::plot(resids2, col="red", type="cor")v
```

We may calculate the percent of entries in the resids table greater than 0.05 (similar to Figure 7.3), but resids has a redundant upper triangle as well. The method below works

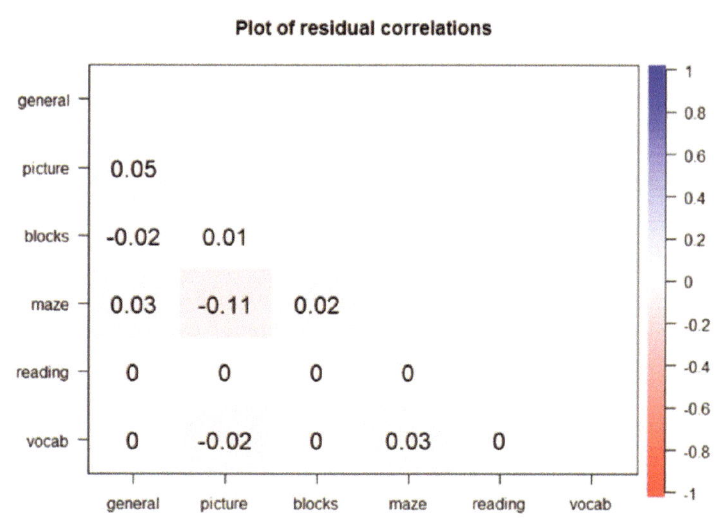

Figure 7.3 Plot of correlations of faTestsOblimin residuals.

for a residual matrix based on correlation input and with the only NAs being those on the main diagonal. Such calculation is useful when the residuals table is much larger than the instructional example here.

```
# Count total off diagonal entries (15 for these data)
# We divide by 2 due to the redundant upper triangle.
offDiagCount <- (base::length(resids2)-base::sqrt(base::length(res
ids2)))/2

offDiagCount
[Output:]
   [1] 15

# Count total number of cells greater than .05  in resids_gt05.
# We use resids_gt05 as a copy of resids2, so as not to alter it.
# Note 0.05 is a measure of size of the residual, not its
significance.
# For these data, the total is 2 since the 0.05 for picture/general
is fractionally above .05.
# Note that resids2 must be used, not resids.
resids_gt05 <- resids2
resids_gt05[base::is.na(resids_gt05)] = 0
total_gt05 <- base::sum(base::abs(resids_gt05)>.05)/2

# Percent greater than .05. For these data, it is 13% (2/15).
base::round(total_gt05/offDiagCount,2)
[Output:]
   [1] 0.13
```

Finally, we may create a histogram of residuals to see if they are roughly normally distributed, which they are, as shown in Figure 7.4.

```
graphics::hist(resids2, xlab="Residuals", col="dodgerblue")
```

Oblique PCA with `principal()`

Introduction

In this section, we use the `principal()` program from the "psych" package to implement another type of oblique rotation (promax rather than oblimin) on the same data as in the previous section. The `principal()` program is only one of several capable of implementing oblique rotation in R. Oblique rotation is requested simply by setting the rotate= parameter to "promax", "oblimin", or another oblique rotation method. Results are similar to those in the previous section (which used the `fa()` command with oblimin rotation) with regard to which variables load on which factors. However, promax rotation is able to create a simpler, more interpretable factor structure for these data.

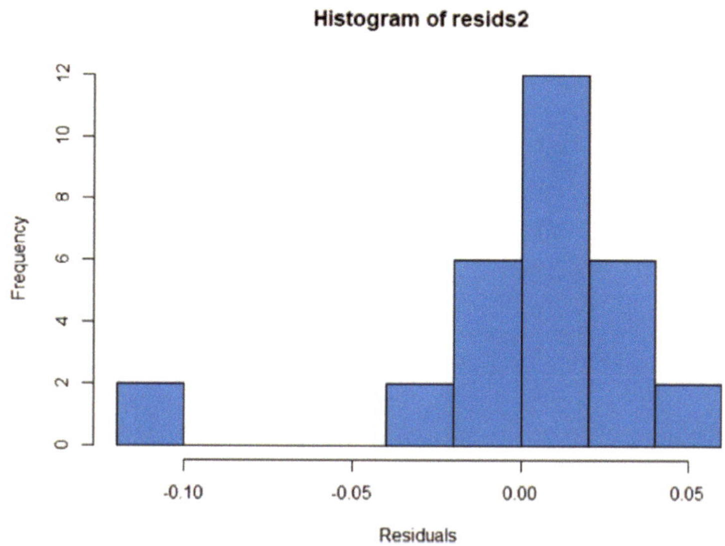

Figure 7.4 Histogram of standardized residuals.

In comparing output, note that the `fa()` command implements principal factor analysis (PFA) whereas `principal()` implements principal components analysis (PCA). The factors created by the former are not the same in meaning as the components created by the latter, as was discussed in Chapter 3. PCA components are independent or correlated (oblique), with each representing the total variance of the indicator variables. PFA factors also are independent or oblique, but each represents the shared (common) variance of the indicator variables.

Setup and data

We use the same ability.cov data object as in the previous oblimin section. This object is from R's built-in "datasets" package.

```
utils::data(ability.cov)
base::library(psych)                    # Has the principal() command
```

We again create the corrmatrix object for data input in the form of a correlation matrix.

```
covmatrix <- ability.cov$cov
corrmatrix <- stats::cov2cor(covmatrix)
```

The promax model

As with the oblimin model in the previous section, we ask for the two-factor solution, but with a promax rotation.

```
principalTestsPromax  <- psych::principal(corrmatrix,
nfactors = 2,
```

```
    n.obs = 112 ,
    rotate = "promax",
    scores=TRUE,
    oblique.score=FALSE)
```

Discussion of option in principal():

- *corrmatrix*: The name of the data object goes first.
- *nfactors = 2*: This specifies that a two-factor solution is wanted.
- *n.obs = 112*: This specifies the number of observations, used to compute goodness of fit statistics.
- *rotate = "promax"*: This specifies that a promax rotation is wanted. Other choices are "none", "varimax", "quartimax", "oblimin", "simplimax" or "cluster".
- *scores = TRUE*: Normally this would compute factor scores. However, since input was a correlation matrix, not observation-level data, no "scores" element is available for the result object.
- *oblique.score = FALSE*: This specifies that factor scores should be based on the pattern matrix. The default is TRUE, which computes scores based on the structure matrix.
- The `principal()` command has other options. Type `help(principal)`.

Results from `principal()`:

```
principalTestsPromax
[Output:]
    Principal Components Analysis
    Call: psych::principal(r = corrmatrix, nfactors = 2, rotate =
    "promax",
        n.obs = 112, scores = TRUE, oblique.scores = FALSE)
    Standardized loadings (pattern matrix) based upon correlation
    matrix
              RC2   RC1   h2   u2 com
    general  0.51  0.46 0.68 0.32 2.0
    picture  0.79 -0.07 0.58 0.42 1.0
    blocks   0.87  0.01 0.77 0.23 1.0
    maze     0.71 -0.14 0.44 0.56 1.1
    reading -0.03  0.96 0.89 0.11 1.0
    vocab   -0.04  0.94 0.86 0.14 1.0

                           RC2  RC1
    SS loadings           2.17 2.05
    Proportion Var        0.36 0.34
    Cumulative Var        0.36 0.70
    Proportion Explained  0.51 0.49
    Cumulative Proportion 0.51 1.00
```

```
 With component correlations of
     RC2  RC1
RC2 1.00 0.45
RC1 0.45 1.00

Mean item complexity =  1.2
Test of the hypothesis that 2 components are sufficient.

The root mean square of the residuals (RMSR) is  0.1
 with the empirical chi square  36.72  with prob <  2.1e-07

Fit based upon off diagonal values = 0.94
```

Discussion of results

The principalTestsPromax results object produces many output elements, listed below. These are invoked by typing the results object name (principalTestsPromax), a dollar sign, and the element name (e.g., principalTestsPromax$loadings).

```
base::attributes(principalTestsPromax)
[Output:]
   $names
    [1] "values"         "rotation"    "n.obs"      "communality" "loadings"
    [6] "fit"            "fit.off"     "fn"         "Phi"          "Call"
   [11] "uniquenesses"  "complexity"  "chi"        "EPVAL"        "R2"
   [16] "objective"      "residual"    "rms"        "factors"      "dof"
   [21] "null.dof"       "null.model"  "criteria"   "STATISTIC"    "PVAL"
   [26] "weights"        "r.scores"    "rot.mat"    "Vaccounted"   "Structure"
```

- *Loadings (principalTestsPromax$loadings[,]).* This gives the pattern matrix shown at the top of the output. The "[,]" part at the end asks for all rows, all columns, in the pattern matrix, to eight decimal places. Leaving this out will give a "prettified" listing in which smaller loadings are omitted and coefficients are listed to three decimal places. Except for the "general" (general intelligence) test, which cross-loads on rotated components RC1 and RC2, we see an easily-interpretable simple factor structure in which all the verbal tests (reading, vocab) load above 0.90 on RC1 while all the spatial tests (picture, blocks, maze) load above 0.70 on RC2.
- *Variance accounted for (principalTestsPromax$Vaccounted).* This element contains the next block in the output above, except that eight rather than two decimal places are displayed. We see that variance explained by components RC1 and RC2 are relatively balanced, at 34% and 36% respectively for a cumulative variance explained by both factors of 70%.
- *Component correlations (principalTestsPromax$r.scores).* This element gives the third block in the output above, except to eight decimal places. Since and oblique correlation (promax) was requested, the components are correlated. The correlation of the two components rounds to 0.45. Since this is higher than 0.32 (which would explain about 10%) reversion to an orthogonal solution such as varimax rotation is not indicated.
- *Mean item complexity (principalTestsPromax$complexity).* This gives Hoffman's index of complexity for each item. Since this example was a two-component solution, items

with a greater split of loadings in the pattern matrix have higher complexity, approaching 2.0 for these data, whereas those loading mainly on just one component are closer to 1.0. Thus the "general" test, which is cross-loaded, has higher complexity than the other tests. Taking the average of the complexities of all variables gives the rounded mean of 1.2 which appears in the output above.

```
principalTestsPromax$complexity
[Output:]
     general  picture   blocks     maze  reading    vocab
    1.979076 1.016459 1.000156 1.072308 1.002460 1.003237
```

```
base::mean(principalTestsPromax$complexity)
[Output:]
    [1] 1.178949
```

Since mean item complexity is low, we may consider this a test of the hypothesis that two components are sufficient, as stated in the output above. See Hofmann (1978) and Pettersson and Turkheimer (2010).

- *Root mean square of residuals (principalTestsPromax$rms).* RMSR and its chi-square probability is returned in three elements: rms, chi, and EPVAL. These provide the coefficients in the output statement above regarding root mean square of the residuals. For these data, RMSR is highly significant, meaning only that average residuals for the correlation matrix differ significantly from 0.

```
principalTestsPromax$rms
[Output:]
   [1] 0.1045395
```

```
principalTestsPromax$chi
[Output:]
   [1] 36.71982
```

```
principalTestsPromax$EPVAL
[Output:]
   [1] 2.057294e-07
```

RMSR is a goodness of fit measure reflecting average residuals in the correlation matrix, which should be low. Researchers have suggested such cutoffs as 0.05 or 0.10 as criteria for RMSE being low enough to use the retained factors to describe the correlation structure of the data. For these data, RMSR = 0.10, meeting the higher cutoff. However, some researchers disparage comparison of RMSR to an absolute cut-off value and instead recommend RMSR only be used for comparisons, with lower being better fit.

If we would like to investigate just which correlations are contributing most to error, we can print out the residual correlation matrix. The highest residual correlation is the maze test with the picture test, at −0.311.

```
# Make rcor as copy of the residual correlation matrix
rcor <- principalTestsPromax$residual
# Put rounded version of rcor into "upper".
upper <- base::round(rcor,3)
# Hide the upper triangle and diagonal.
upper[base::upper.tri(rcor,diag=TRUE)] <- ""
# Make the result into a data frame called "lower".
lower <- base::as.data.frame(upper)
# Print the lower triangle of the residual correlation matrix.
base::print(lower)
[Output:]
        general picture blocks  maze reading vocab
general
picture  -0.053
blocks   -0.079  -0.093
maze     -0.078  -0.311 -0.125
reading   -0.06   0.015      0 0.028
vocab    -0.108  -0.001  0.012 0.068   -0.08
```

- *Fit based on off-diagonal values (principalTestsPromax$fit.off).* This is another goodness-of-fit measure. It assesses how well the off-diagonal coefficients in the estimated correlation matrix reproduce the off-diagonal observed coefficients. That is, the ratio of the estimated (1 − resid2) to the observed (cor2) is 0.94. For these data, coefficients may be said to be reproduced well by the promax-rotated PCA model.

Visualization

Of the many visualizations possible for oblique (promax) PFA with the principal() program, we provide three figures, along with the R code which created them: a scree plot, a plot of variable loadings on components, and a correlation table in graphic format.

The scree plot shows descending eigenvalues for the solution. Though we chose to extract only two components, the x axis of the scree plot shows all six. Only the first two are above the Kaiser criterion, having eigenvalues of 1.0 or greater, justifying the choice to extract only the first two components (Figure 7.5).

```
x <- principalTestsPromax$values
base::plot(x,
   pch=16,
   col="red",
   type="b",
   main="Scree plot for principalTestsPromax",
   xlab="Component",
   ylab="Eigenvalue")
graphics::abline(h=1.0)
graphics::text(x=4, y=1.1, "Kaiser criterion line", cex=1.2)
```

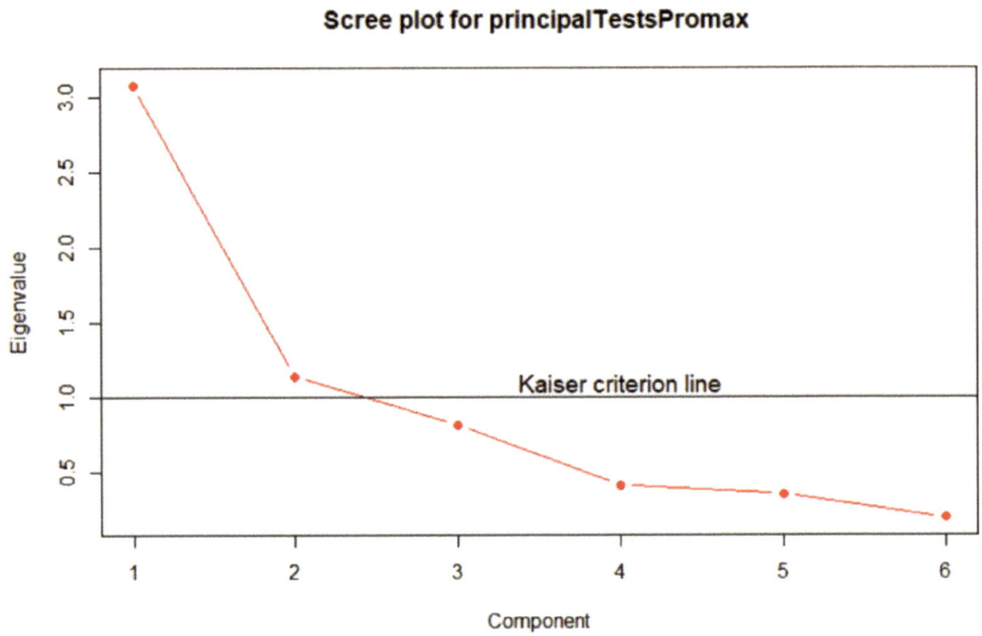

Figure 7.5 Scree plot for PCA on ability tests variables using promax rotation.

Figure 7.6 shows the variables in the ability.cov example data (data of six ability tests) plotted in component space. We see, for example, that the picture, blocks, and maze tests are close together and are heavily loaded on component RC2.

```
base::set.seed(123)
psych::factor.plot(principalTestsPromax,
    main="PCA, promax rotation, ability.cov test variables",
    xlim=base::c(-0.2,1),
    ylim=base::c(-.3,1.1),
    jiggle=TRUE,
    amount=0.15,
    labels=base::rownames(principalTestsPromax$loadings))
```

The third visualization is a graphic version of the table of residual correlations as produced for the ability.cov data by PCA with promax rotation. We see, for instance, that the largest error is reflected in the residual correlation of the maze and picture tests (Figure 7.7).

```
psych::corPlot(r=principalTestsPromax$residual,
    numbers=TRUE,
    diag=FALSE,
    main="Residual correlations for principalTestsPromax")
```

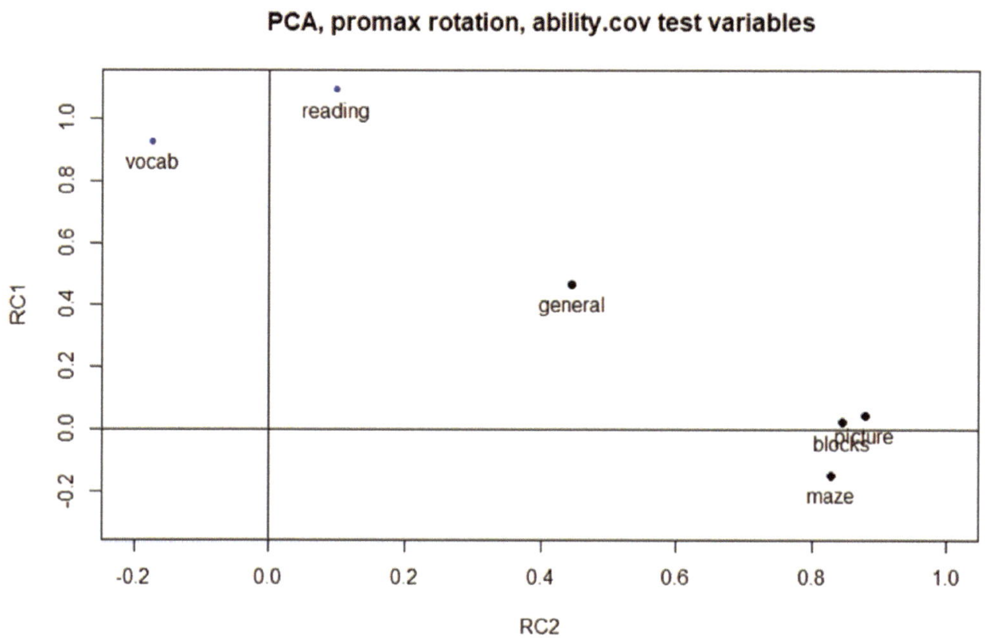

Figure 7.6 Plot of ability tests variables using principal components analysis with promax.

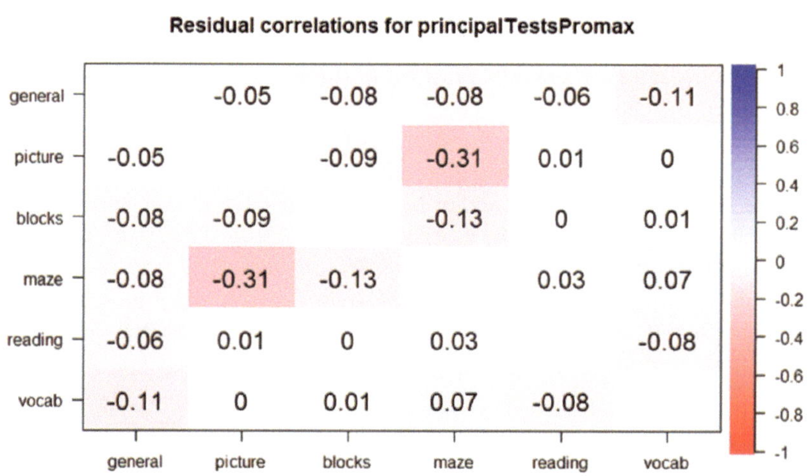

Figure 7.7 Residual correlation chart of ability tests variables using PCA with promax rotation.

Second-order oblique factor analysis

Introduction

It is possible to have factors (or components) which are based on lower factors. In a second-order factor analysis, there are two levels of factors. This is also labeled higher-order or hierarchical factor analysis (HFA). The lowest level is ordinary oblique factor analysis. It must be oblique because the second-level factor or factors are based on the intercorrelations of a minimum of three level 1 factors. Second-order factor models may be based on PCA as well as PFA. While conceptually one might have third- or fourth-level higher factor models, most examples in the literature are limited to two levels.

A second-order factor model of general intelligence was famously set forth by Carroll (1993) and discussed by Jensen (1998). Carroll called his theory a "three stratum theory", with the indicator level being level 1 (narrow abilities), the factor level being level 2 (nine broad abilities, such as quantitative knowledge, reading and writing ability, and visual processing ability), then the third and highest level being a second-order factor called general intelligence (g). This model was subsequently extended and revised, but also encountered objections including opposition to the idea of a "g" general intelligence factor at all. Nonetheless, Cattell's model remains a classic example of second-order factor analysis.

When factors are correlated, as they are in oblique models, this implies there are one or more exogenous factors causing the intercorrelations. Those exogenous factors are the second-order factors. Thus Gorsuch (1988: 250) wrote, "correlated factors imply the existence of higher-order factors." Factor scores or a correlation matrix of factors from the first stage can be input to a second-stage orthogonal factor analysis (e.g., varimax) to generate the hierarchical factor solution.

HFA thus seeks to base higher-order factors on a set of correlated lower-order factors. For instance, HFA has been used prominently in trait theory, in the debate over whether personality is best characterized in terms of three, five, or some other number of higher-order traits (see Matthews, Deary, & Whiteman, 2003).

HFA is a two-stage process. First, an oblique (e.g., oblimin) factor analysis is conducted on the raw dataset. Rotation must be oblique since this generates a correlation matrix for the factors, needed as input for the higher stage. As it is critical in HFA to obtain the simplest factor structure possible, it is recommended to explore different values of tuning parameters. For instance, for oblimin one may run the model for several different values of delta, not just the default delta = 0. A delta of 0 gives the most oblique solutions but the more the researcher specifies an increasingly negative delta, the factors become less and less oblique. To override the default delta of 0, the researcher enters a value less than or equal to 0.8. When the researcher feels the simplest factor structure has been obtained, the result is a correlated set of lower-order factors supporting one or more general factors at level 2.

The second-order model

To illustrate second-order factor models we use the classic Holzinger.9 dataset, which comes with the "psych" package.[2] The data measure nine cognitive abilities scored for middle school students for the tests "vis_perc". "cubes", "lozenges", "par_comp", "sen_comp", "wordmean". "addition", "count_dot", and "s_c_caps". For convenience, these data are copied into "dat", which is an object of class "matrix" "array". Holzinger.9 is a nine-by-nine correlation matrix of these ability tests based on a sample size of 145. Type `help(Holzinger.9)` for more information on this dataset or see Holzinger and Swineford (1937, 1939).

```
base::library(psych)
utils::data("Holzinger.9")
dat <- Holzinger.9
```

We put variable names in the "varLabels" vector and list it out.

```
varLabels <- base::colnames(dat)
varLabels
[Output:]
   [1] "vis_perc"  "cubes"     "lozenges"  "par_comp"
   [5] "sen_comp"  "wordmean"  "addition"  "count_dot"
   [9] "s_c_caps"
```

We then create the model result object, labeled "Level1Result". This result is for an oblique three-factor model using `fm="pa"` (principal axis factoring) and oblimin rotation. The result is printed out below simply by listing the name of the object.

```
level1Result <- psych::fa(dat,
   nfactors=3,
   fm="pa",
   rotate="oblimin",
   missing=FALSE,
   residuals=TRUE)

level1Result
[Output:]
   Factor Analysis using method =  pa
   Call: fa(r = dat, nfactors = 3, rotate = "oblimin", residuals =
   TRUE,
       missing = FALSE, fm = "pa")
   Standardized loadings (pattern matrix) based upon correlation
   matrix
               PA1   PA2   PA3   h2   u2 com
   vis_perc   0.10  0.09  0.60 0.47 0.53 1.1
   cubes      0.01 -0.02  0.52 0.27 0.73 1.0
   lozenges   0.09 -0.03  0.67 0.50 0.50 1.0
   par_comp   0.86 -0.04  0.04 0.76 0.24 1.0
   sen_comp   0.83  0.09 -0.04 0.70 0.30 1.0
   wordmean   0.81 -0.05  0.05 0.68 0.32 1.0
   addition   0.14  0.77 -0.19 0.59 0.41 1.2
   count_dot -0.11  0.80  0.16 0.69 0.31 1.1
   s_c_caps   0.09  0.49  0.37 0.54 0.46 2.0

                          PA1  PA2  PA3
   SS loadings           2.23 1.55 1.42
   Proportion Var        0.25 0.17 0.16
   Cumulative Var        0.25 0.42 0.58
   Proportion Explained  0.43 0.30 0.27
   Cumulative Proportion 0.43 0.73 1.00
```

```
 With factor correlations of
      PA1  PA2  PA3
PA1 1.00 0.26 0.41
PA2 0.26 1.00 0.30
PA3 0.41 0.30 1.00

Mean item complexity =  1.2
Test of the hypothesis that 3 factors are sufficient.

The degrees of freedom for the null model are  36  and the ob-
jective function was  3.49
The degrees of freedom for the model are 12  and the objective
function was  0.07

The root mean square of the residuals (RMSR) is  0.02
The df corrected root mean square of the residuals is  0.04

Fit based upon off diagonal values = 1
Measures of factor score adequacy
                                                  PA1  PA2  PA3
Correlation of (regression) scores with factors  0.94 0.90 0.86
Multiple R square of scores with factors         0.89 0.80 0.73
Minimum correlation of possible factor scores    0.77 0.61 0.46
```

Next, we display the level 1 factor loadings. These show that the verbal items (par_comp, sen_comp, wordmean) tend to load on factor PA1, the numeric items (addition, count_dot, s_c_caps) tend to load on factor PA2, and visual items (vis_perc, cubes, lozenges) tend to load on factor PA3.

```
level1Result$loadings
[Output:]
  Loadings:
            PA1    PA2    PA3
  vis_perc                 0.601
  cubes                    0.524
  lozenges                 0.675
  par_comp   0.860
  sen_comp   0.826
  wordmean   0.813
  addition   0.136  0.767 -0.195
  count_dot -0.109  0.796  0.159
  s_c_caps          0.486  0.375

                   PA1   PA2   PA3
  SS loadings    2.140 1.479 1.301
  Proportion Var 0.238 0.164 0.145
  Cumulative Var 0.238 0.402 0.547
```

Next, we create and display the level 1 factor correlation matrix. The interfactor correlations are in the "Phi" results element. We see that the correlation of factors PA1/

PA2 and PA2/PA3 are weak (the latter is below 3.0 at 0.2997409) and the correlation of factors PA1.PA1 is moderate in strength.

```
level1Corrmatrix <- level1Result$Phi
level1Corrmatrix
[Output:]
         PA1   PA2   PA3
   PA1 1.000 0.256 0.413
   PA2 0.256 1.000 0.300
   PA3 0.413 0.300 1.000
```

Having created the level1 model, we may use its factor correlation matrix to create a single-factor level 2 model. We switch to ML estimation but the results are almost the same as for the "pa" method. Follow creation of the "level2Result" model, we list out its results in the usual manner, simply by typing the object name.

```
level2Result <- psych::fa(level1Corrmatrix,
   nfactors=1,
   fm="ml",
   rotate="oblimin",
   missing=FALSE,
   residuals=TRUE)

level2Result
[Output:]
   Factor Analysis using method =  ml
   Call: fa(r = level1Corrmatrix, nfactors = 1, rotate = "oblimin",
       residuals = TRUE, missing = FALSE, fm = "ml")
   Standardized loadings (pattern matrix) based upon correlation
   matrix
        ML1   h2   u2 com
   PA1 0.59 0.35 0.65   1
   PA2 0.43 0.19 0.81   1
   PA3 0.70 0.48 0.52   1

                    ML1
   SS loadings     1.02
   Proportion Var 0.34

   Mean item complexity =  1
   Test of the hypothesis that 1 factor is sufficient.

   The degrees of freedom for the null model are  3  and the
   objective function was  0.3
   The degrees of freedom for the model are 0  and the objective
   function was  0
```

```
The root mean square of the residuals (RMSR) is  0
The df corrected root mean square of the residuals is  NA

Fit based upon off diagonal values = 1
Measures of factor score adequacy
                                                   ML1
Correlation of (regression) scores with factors   0.79
Multiple R square of scores with factors          0.63
Minimum correlation of possible factor scores     0.26
```

Visualization

The second-order factor model may be diagrammed using the steps below.

```
# Step 1. Set for side-by-side display.
graphics::par(mfrow=base::c(1,2))


# Step 2. Diagram the level 1 model.
psych::fa.diagram(level1Result, cut=0.0001, digits=2, main="Level 1
Model")


# Step 3. Diagram the level 2 model.
psych::fa.diagram(level2Result, cut=0.0001, digits=2, main="Level 2
Model")


# Step 4. Reset the display for single images.
graphics::par(mfrow=base::c(1,1))
```

In Figure 7.8, note that the second-order general factor (ML1) affects the measured indicator variables on the left only as mediated by the level 1 factors (PA1, PA2, and PA3). By the same token, each indicator reflects not only its level 1 factor but also the general ability factor. A challenge for the researcher is to provide a theoretical rational for the mediation of the general ability factor in relation to the indicator variables. Also, in reporting results, the researcher must be careful to attribute loadings (e.g., for the three verbal ability items associated with factor PA1) not just to the level 1 factor (PA1), but also to the general factor.

Schmid and Leiman (1957) set forth a method to partition the variance of measured indicators and to estimate the percent of variance associated with the general factor in a second-order factor model. That is, their orthogonal partition method separated indicator variance into a component associated with the level 1 factors and a component associated with the level 2 general factor.

The "psych" package has operationalized the Schmid-Leiman method in the `schmid()` command. For more information, type `help(schmid)`. In the output below for the same Holzinger.9 model, the "p2" column on the left gives the percent of variance in any indicator attributable to the general factor. The "mean percent general =

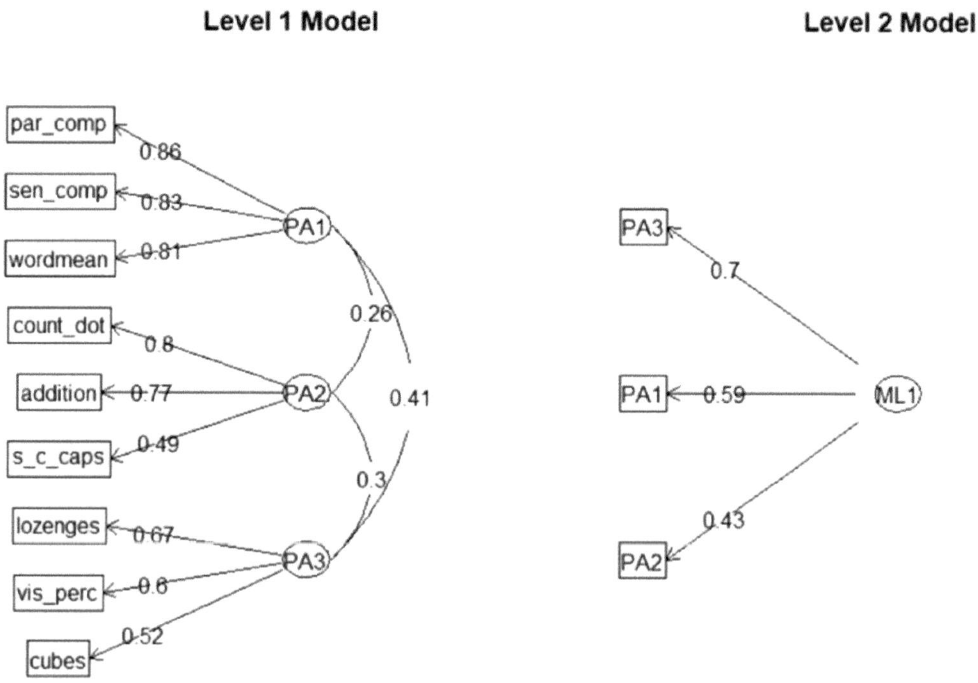

Figure 7.8 Second-order factor analysis of Holzinger ability data.

0.39" line in output means that 39% of the common variance in measured indicators is accounted for by the general factor. However, this varies by the method used. With promax rotation, the corresponding percentage was 45%. Gorsuch (1983: 253) suggested that when the general factor accounted for 40% or more, a second-order factor analysis was of interest and was justified.

```
directeff <- psych::schmid(Holzinger.9,nfactors=3,fm="pa", digits=3,
rotate="oblimin")
directeff
[Partial output:]
   …
   Schmid Leiman Factor loadings greater than  0.2
                 g    F1*   F2*   F3*   h2   u2   p2
   vis_perc  0.51               0.44 0.47 0.53 0.56
   cubes     0.36               0.38 0.27 0.73 0.47
   lozenges  0.51               0.49 0.50 0.50 0.51
   par_comp  0.53  0.69              0.76 0.24 0.37
   sen_comp  0.51  0.66              0.70 0.30 0.37
   wordmean  0.50  0.65              0.68 0.32 0.37
   addition  0.28        0.69        0.59 0.41 0.13
   count_dot 0.39        0.72        0.69 0.31 0.22
   s_c_caps  0.52        0.44  0.27 0.54 0.46 0.50
```

```
With eigenvalues of:
   g  F1*  F2*  F3*
1.94 1.37 1.20 0.69

general/max  1.41   max/min =    1.99
mean percent general =  0.39    with sd =  0.14 and cv of  0.36
...
The root mean square of the residuals is  0.02
The df corrected root mean square of the residuals is  0.04
```

Bifactor models

Introduction

A bifactor model is one in which indicator variables reflect (1) the factor to which they belong, as in ordinary factor analysis; and (2) a general factor which directly affects all indicator variables but does not directly affect the other domain-specific factors. This is similar to the second-order factor model described in the previous section but the general factor acts directly on the measured indicators and is not mediated through the factors which directly affect the indicators. The bifactor model was first set forth by Holzinger and Swineford (1937).

The theoretical assumption of a bifactor model is that there is a general factor underlying all measured indicators, but that sets of indicators are also associated with unique domain factors which are not directly influenced by the common general factor. In the "psych" package, bifactor models utilize the `rotate="bifactor"` option of the `fa()` command.

Online documentation for the `omega()` function used below explains,

> Many scales are assumed by their developers and users to be primarily a measure of one latent variable. When it is also assumed that the scale conforms to the effect indicator model of measurement (as is almost always the case in psychological assessment), it is important to support such an interpretation with evidence regarding the internal structure of that scale. In particular, it is important to examine two related properties pertaining to the internal structure of such a scale. The first property relates to whether all the indicators forming the scale measure a latent variable in common. The second internal structural property pertains to the proportion of variance in the scale scores (derived from summing or averaging the indicators) accounted for by this latent variable that is common to all the indicators.

The bifactor model

We ask for the maximum likelihood method, but the method could be set to "pa" or others.

```
base::library(psych)

bifactorResult <-  psych::fa(Holzinger.9, nfactors = 4, fm="ml",
rotate="bifactor")
```

The loadings element of bifactorResult has 9 rows (the tests) and 4 factors, with ML1 being the general factor and the others being domain-specific factors. Below, fload contains the untransformed standardized loadings for these other factors.

```
fload <- bifactorResult$loadings[1:9,2:4]
base::round(fload,3)
[Output:]
                ML2    ML3    ML4
    vis_perc  -0.007 -0.141  0.192
    cubes      0.022 -0.060  0.461
    lozenges   0.070 -0.124  0.461
    par_comp   0.705  0.007  0.046
    sen_comp   0.637  0.038 -0.107
    wordmean   0.675 -0.012  0.063
    addition   0.025  0.927 -0.021
    count_dot -0.230  0.413 -0.015
    s_c_caps  -0.123  0.144 -0.039
```

Below, gload contains the untransformed standardized loadings for the general factor, which is represented by ML1. As this is a single column, gload would be a numeric vector but `as.matrix()` converts it to a matrix.

```
gload <- bifactorResult$loadings[1:9,1]
base::round(gload,3)
[Output:]
               [,1]
    vis_perc  0.650
    cubes     0.357
    lozenges  0.537
    par_comp  0.504
    sen_comp  0.546
    wordmean  0.474
    addition  0.368
    count_dot 0.569
    s_c_caps  0.781
```

Visualization

We now proceed to visualization of the bifactor model. Coefficient omega estimates the general factor saturation of a test, as proposed by McDonald (1999). The Schmid-Leiman transformation of the Holzinger higher-order model is applied (this is the default). This transformation was discussed in the previous section on second-order oblique models. Therefore, coefficients in Figure 7.9 are not the loadings above, but are transformed. We ask for oblimin rotation but could have asked for promax or another oblique rotation. We ask for the maximum likelihood method, but the method could be set to "pa" or others.

```
holzBifact <- psych::omega(Holzinger.9,
    nfactors=3,
    digits=3,
    fm="ml",
    rotate="oblimin",
    sl=TRUE,
    title = "Nine ability tests,  Holzinger-Swineford data")
```

The complete listing of bifactor results is below.

```
holzBifact
    Nine ability tests,  Holzinger-Swineford data
    Call: omegah(m = m, nfactors = nfactors, fm = fm, key = key,
    flip = flip, digits = digits, title = title, sl = sl, labels =
    labels, plot = plot, n.obs = n.obs, rotate = rotate, Phi = Phi,
    option = option, covar = covar)
```

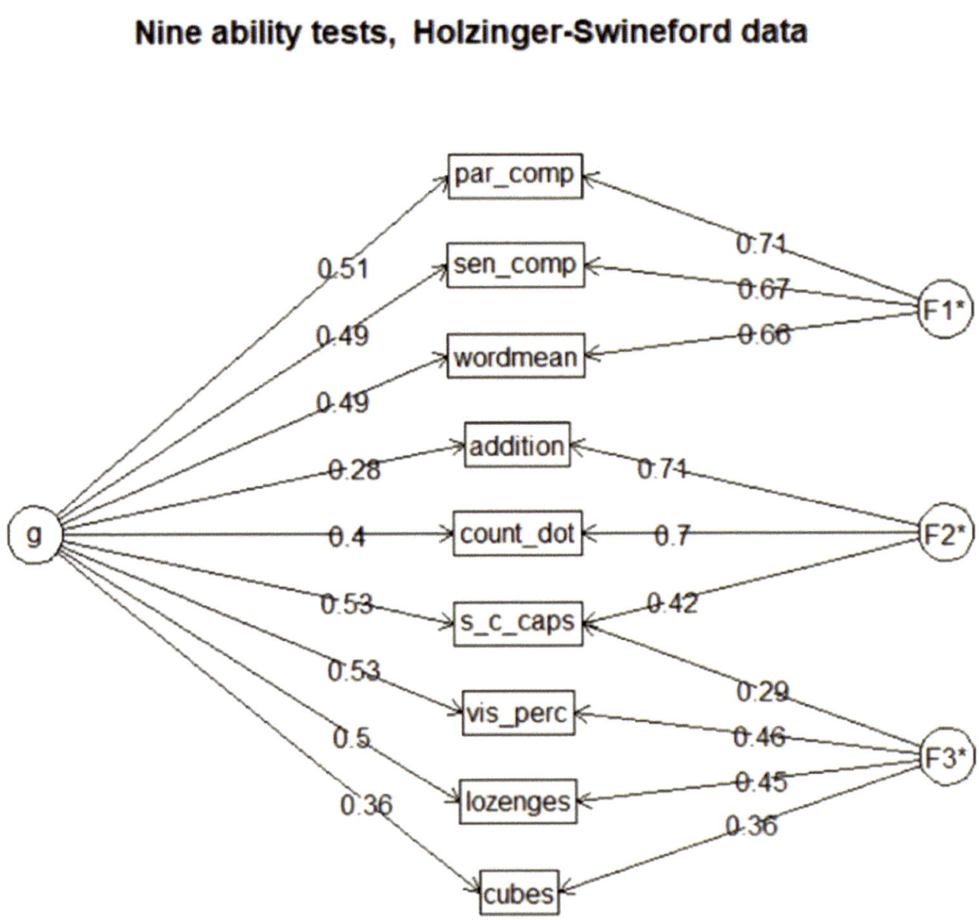

Figure 7.9 Biplot of Holzinger.9 dataset.

```
Alpha:                   0.81
G.6:                     0.85
Omega Hierarchical:      0.53
Omega H asymptotic:      0.6
Omega Total              0.88

Schmid Leiman Factor loadings greater than  0.2
             g   F1*  F2*   F3*   h2   u2   p2
vis_perc  0.53              0.46 0.50 0.50 0.57
cubes     0.36              0.36 0.26 0.74 0.50
lozenges  0.50              0.45 0.46 0.54 0.54
par_comp  0.51  0.71             0.76 0.24 0.34
sen_comp  0.49  0.67             0.70 0.30 0.35
wordmean  0.49  0.66             0.68 0.32 0.35
addition  0.28       0.71        0.61 0.39 0.13
count_dot 0.40       0.70        0.68 0.32 0.24
s_c_caps  0.53       0.42   0.29 0.54 0.46 0.52

With eigenvalues of:
   g  F1*  F2*  F3*
1.93 1.43 1.18 0.66

general/max  1.36   max/min =    2.17
mean percent general =  0.39 with sd =  0.15 and cv of  0.38
Explained Common Variance of the general factor =  0.37

The degrees of freedom are 12  and the fit is  0.07

The root mean square of the residuals is  0.02
The df corrected root mean square of the residuals is  0.04

Compare this with the adequacy of just a general factor and no
group factors
The degrees of freedom for just the general factor are 27  and
the fit is  1.7

The root mean square of the residuals is  0.19
The df corrected root mean square of the residuals is  0.22

Measures of factor score adequacy
                                                 g  F1*  F2*   F3*
Correlation of scores with factors            0.74 0.82 0.83  0.63
Multiple R square of scores with factors      0.55 0.67 0.69  0.40
Minimum correlation of factor score estimates 0.09 0.33 0.39
-0.21
```

```
  Total, General and Subset omega for each subset
                                                    g  F1*  F2*  F3*
Omega total for total scores and subscales     0.88 0.88 0.79 0.66
Omega general for total scores and subscales 0.53 0.31 0.24 0.36
Omega group for total scores and subscales     0.29 0.57 0.55 0.30
```

Notice that above the "Explained Common Variance of the general factor" is 0.37. This ECV coefficient is the ratio of the general factor eigenvalue to the sum of all eigenvalues. Online documentation states of the ECV that "it is a better indicator of unidimensionality than of the amount of test variance accounted for by a general factor". The bottom three lines of output above give the total, general, and group (subset) omega coefficients.

Omega total (the omega_t coefficient) is the amount of variance in the group (column) accounted for by the general (common) factor plus group (unique) factors. This coefficient may be regarded as a model-based measure of reliability alternative to Cronbach's alpha.

Omega general is the amount of variance in the group that is accounted for just by the general factor. The larger the general factor variance, the more erroneous it is to interpret F1, F2, or F3 as being solely and purely measures of verbal, spatial, or mental speed dimensions.

Omega group (aka., omega hierarchical or omegaS or subset omega) is the amount of variance in the group that is accounted for just by the group factor. Put another way, omega group is the amount of group variance after controlling for the general factor. For the F1, F2, and F3 groups, omega group plus omega general equals omega total. This equality is not true for the general (g) factor.[3] Reise (2012) has recommended that omega group should be at least 0.50 and preferably 0.75. By this criterion, factor F3 would be considered inadequate by the bifactor model. For these data, omega general is greater than omega group only for the F3 factor.

The ratio of omega general to omega total reflects the proportion of reliable variance associated with the general factor. For instance, for the F1 group, this is 0.31/0.88 = 35.2%. Similarly, the ratio of omega group to omega total reflects the proportion of the reliable variance which is within group but not general. For instance, for the F1 group, this is 0.57/0.88 = 64.8%.

Be aware that there is some evidence that bifactor models may be biased if factor loadings are small or cross-loadings are high (Dombrowski et al., 2019). Giordano and Waller (2020) found that bifactor models were less accurate than Schmid-Leiman transformations of higher-order models in terms of correctly reproducing factor structure. Such criticisms suggest using bifactor modeling as an exploratory technique rather than confirmatory. For further discussion, see Reise et al. (2013) and Mansolf and Reise (2016), among other sources.

Notes

1 For computation of factor scores, see Tabachnick and Fidell (2013: 630–631).
2 These data appear in the LISREL manual as example NPV.KM.
3 The equality does not hold because coefficients are based on a cluster solution in which every item is assigned to one group) and some of the variance is found in the cross loadings between groups.

8 Factor analysis for binary, ordinal, and mixed data

Polychoric PCA and PFA

Perhaps the most common way of dealing with mixed data in factor analysis is to convert the data to a polychoric or heterogenous correlation matrix, then proceed as usual. The central idea of polychoric PCA is to use a type of correlation (polychoric, polyserial, tetrachoric, and/or others) which is designed for the data level the researcher has. After constructing a polychoric or heterogeneous correlation matrix, this rather than raw data or a Pearsonian correlation matrix is used as input to the PCA or other dimension reduction procedure. In R, there are diverse ways to implement polychoric PCA and polychoric PFA. In this section, we document a few of them and also show how John Fox's heterogeneous PCA works for mixed-level data using a combination of Pearson, polychoric, and polyserial correlation coefficients.

Polychoric correlation is relatively robust against non-normality. Polychoric correlation is appropriate if (1) the researcher can assume that the ordinal variable (or binary for tetrachoric correlation, which is an extension of polychoric correlation for binary variables) has an underlying continuous normal distribution, and (2) skew is not extreme (skew >= 5.0); and kurtosis is not extreme (kurtosis >= 50.0) (Flora & Curran, 2004).

As background, two decades ago it was common practice to run PCA on discrete and categorical data by converting such variables into sets of dummy variables. Kolenikov and Angeles (2004) demonstrated through simulation studies that the dummy variable method was inferior to using polychoric correlation. Factor analysis with dummy variables is now disparaged.

Example data and setup

To illustrate the polychoric approach, we use the "music" and "music2" data frames employed with the `principal()` command from the "psych" base::library. This data frame has mostly ordinal items on preferences for various categories of music but also has one dichotomous variable (sex) and three continuous reference variables (age, educ, rincom91). Reference variables are useful in exploratory research but usually would be removed in the final run of factor analysis. For purposes here, however, these reference variables are retained because they are useful in the discussion of PCA or PFA with mixed data levels.

DOI: 10.4324/9781003279693-9

```
base::setwd("C:/Data")        # Set the working directory
base::library(psych)          # Has the principal() function for  PCA
and others

# Read data from a comma-separated values format file into the data
object music.
music <- utils::read.csv("musicvars.csv", header=TRUE, sep = ",",
stringsAsFactors=TRUE)

# Create music2 as a complete cases version of music
# As there were many missing values, count is reduced from 1500 to 773
music2 <- music[stats::complete.cases(music),]

# View the data.
View(music2)
```

Note that further refinements when reading in the music data are discussed in the later subsection on "Heterogeneous PCA with `hetcor()`".

Polychoric PCA with `principal()`

As seen below, the `principal()` command accepts as input a polychoric correlation matrix based on output from the `polychoric()` command, also from the "psych" package. However, use of continuous variables throws this error message: "You have more than 8 categories for your items, polychoric is probably not needed". This is because when the number of levels is greater than 8, the bias of a Pearsonian correlation in comparison with a polychoric correlation is minimal. The bottom line is that to get polychoric correlations we must bin our continuous variables into 8 levels. We do this below with the `cut()` command from R's "base" package. There are other more complex binning methods, including using researcher-set custom breakpoints, but for simplicity here, `cut()` was used.

The new binned variables will become columns 19–21 in music2

```
music2$agecat <- base::cut(music2[,"age"],8,labels=FALSE)
music2$educat <- base::cut(music2[,"educ"],8,labels=FALSE)
music2$incomecat <- base::cut(music2[,"rincom91"],8,labels=FALSE)
```

We then create a polychoric matrix using the ordinal, dichotomous, and binned variables but leaving out the continuous variables age, educ, and rincom91.

```
# Define the columns (variables) wanted. Use names(music2) if a
listing is wanted.
musicColumns <- base::c(2,3,5,6,8,10,11,13:15,18:21)
```

We now use the `corCi()` command, with `poly=TRUE`, to generate polychoric correlations. Results are sent to the object "polyoutput". Choices other than "pearson" for `method=` are "kendall" and "spearman". Note that execution of this command may take two or three minutes before output appears.

```
polyoutput <- psych::corCi(music2[,musicColumns],
   keys = NULL,
   n.iter = 100,
   p = 0.05,
   overlap = FALSE,
   poly = TRUE,
   method = "pearson", plot=TRUE)

#  Polychoric correlations are in the rho element of polyoutput,
which we convert to a matrix.
polymatrix <- base::as.matrix(polyoutput$rho)

# Optionally, print out one of the polychoric correlations
polymatrix["bigband","blues"]
# Output:
   [1] 0.3063139

# Run a varimax principal components analysis on polymatrix for a
5-component solution.
polyout <- psych::principal(polymatrix, nfactors=5,
rotate="varimax")
polyout
   Principal Components Analysis
   Call: psych::principal(r = polymatrix, nfactors = 5, rotate =
   "varimax")
   Standardized loadings (pattern matrix) based upon correlation
   matrix
                RC1   RC2   RC3   RC5   RC4   h2   u2 com
   bigband     0.63 -0.21  0.36  0.27  0.02 0.64 0.36 2.3
   blues       0.14  0.11  0.15  0.87 -0.04 0.81 0.19 1.2
   blugrass    0.25  0.00  0.76  0.16 -0.19 0.70 0.30 1.4
   classicl    0.82  0.04 -0.13  0.13 -0.06 0.71 0.29 1.1
   country    -0.04 -0.08  0.78 -0.03  0.11 0.62 0.38 1.1
   hvymetal   -0.01  0.83  0.02 -0.03 -0.14 0.71 0.29 1.1
   jazz        0.22  0.09 -0.11  0.84 -0.05 0.78 0.22 1.2
   musicals    0.77 -0.18  0.02  0.17  0.27 0.74 0.26 1.5
   opera       0.82  0.11  0.08  0.05 -0.03 0.69 0.31 1.1
   rap         0.11  0.69  0.02  0.13  0.11 0.52 0.48 1.2
   sex        -0.21  0.14  0.13 -0.01 -0.84 0.78 0.22 1.2
   agecat     -0.29  0.64 -0.27  0.14  0.12 0.60 0.40 2.0
   educat     -0.40  0.00  0.50 -0.12  0.32 0.53 0.47 2.8
   incomecat -0.14  0.30  0.23 -0.12  0.69 0.66 0.34 1.8

                          RC1  RC2  RC3  RC5  RC4
   SS loadings           2.80 1.81 1.76 1.68 1.46
   Proportion Var        0.20 0.13 0.13 0.12 0.10
   Cumulative Var        0.20 0.33 0.45 0.57 0.68
   Proportion Explained  0.29 0.19 0.18 0.18 0.15
   Cumulative Proportion 0.29 0.48 0.67 0.85 1.00
```

```
Mean item complexity =  1.5
Test of the hypothesis that 5 components are sufficient.
The root mean square of the residuals (RMSR) is  0.07
Fit based upon off diagonal values = 0.9>
```

Standard PCA with `principal()`

We also present a standard PCA analysis of the music2 data as a basis for convenient comparison with the polychoric PCA solution. We again ask for a five-factor solution with varimax rotation.

```
# In music2, columns 1:18 include all variables in music2.
principalResult <- psych::principal(music2[,musicColumns], nfac-
tors=5, rotate="varimax")

# List the contents of principalResult:
principalResult
Output:
   Principal Components Analysis
   Call: psych::principal(r = music2[, musicColumns], nfactors = 5,
   rotate = "varimax")
   Standardized loadings (pattern matrix) based upon correlation
   matrix
                RC1   RC2   RC3   RC5   RC4   h2   u2 com
   bigband     0.62 -0.19  0.34  0.26 -0.01 0.61 0.39 2.2
   blues       0.14  0.09  0.15  0.85 -0.03 0.78 0.22 1.1
   blugrass    0.26 -0.01  0.71  0.16 -0.21 0.65 0.35 1.6
   classicl    0.81  0.05 -0.14  0.12 -0.06 0.70 0.30 1.1
   country    -0.03 -0.08  0.75  0.00  0.07 0.57 0.43 1.0
   hvymetal    0.02  0.80  0.03 -0.08 -0.16 0.68 0.32 1.1
   jazz        0.21  0.11 -0.09  0.82 -0.04 0.73 0.27 1.2
   musicals    0.75 -0.16  0.01  0.17  0.26 0.69 0.31 1.5
   opera       0.80  0.07  0.05  0.05 -0.03 0.64 0.36 1.0
   rap         0.07  0.62  0.01  0.18  0.13 0.44 0.56 1.3
   sex        -0.15  0.13  0.12 -0.04 -0.84 0.75 0.25 1.2
   agecat     -0.28  0.65 -0.28  0.14  0.07 0.61 0.39 1.9
   educat     -0.37  0.01  0.53 -0.16  0.30 0.54 0.46 2.7
   incomecat -0.10  0.36  0.27 -0.18  0.62 0.64 0.36 2.3

                          RC1  RC2  RC3  RC5  RC4
   SS loadings           2.64 1.70 1.68 1.65 1.35
   Proportion Var        0.19 0.12 0.12 0.12 0.10
   Cumulative Var        0.19 0.31 0.43 0.55 0.65
   Proportion Explained  0.29 0.19 0.19 0.18 0.15
   Cumulative Proportion 0.29 0.48 0.67 0.85 1.00
   Mean item complexity =  1.5
   Test of the hypothesis that 5 components are sufficient.
```

```
The root mean square of the residuals (RMSR) is  0.08
with the empirical chi square  877.79  with prob <  1.6e-164

Fit based upon off diagonal values = 0.86
```

Summary

When we compare polychoric PCA with standard PCA using `principal()` we see that fit has improved. Fit based upon off-diagonal values has increased from 0.86 to 0.90. Root mean square residuals (error) are a bit less at 0.07 rather than 0.08. The cumulative proportion explained by the first two components has increased from 0.31 to 0.33. While this is not a strong model, we have achieved somewhat better fit with polychoric input. Moreover, in getting the polychoric results we have not violated the data level assumptions of standard PCA.

Polychoric PFA with `fa()`

Polychoric correlation input may be used with principal factor analysis (PFA) as well as principal component analysis (PCA). Polychoric PFA, a.k.a. "factor analysis" or "common factor analysis", is supported by the `fa()` command in the "psych" package, which implements the polychoric version. Below we demonstrate with the same data and variables as for polychoric PCA in the preceding section. These data, from the previous section, are "music2[,musicColumns]" for raw data input or are in the "polymatrix" object for polychoric correlation input. Both have the same 14 variables, which include 10 music preference items and 4 demographic reference variables included for instructional reasons.

PFA with polychoric input in `fa()`

We then run the polychoric PFA model, using the polymatrix object as input.

```
faResultPoly <- psych::fa(r=polymatrix, nfactors = 5, fm="pa", cor =
"poly", rotate ="varimax")

faResultPoly
[Output:]
  Factor Analysis using method =  pa
  Call: psych::fa(r = polymatrix, nfactors = 5, rotate = "varimax",
      fm = "pa", cor = "poly")
  Standardized loadings (pattern matrix) based upon correlation
  matrix
              PA1   PA5   PA2   PA3   PA4   h2   u2 com
  bigband    0.58  0.26 -0.21  0.32 -0.03 0.54 0.46 2.3
  blues      0.16  0.81  0.13  0.14  0.03 0.73 0.27 1.2
  blugrass   0.24  0.16 -0.04  0.65  0.15 0.53 0.47 1.5
  classicl   0.78  0.14  0.02 -0.13  0.04 0.65 0.35 1.1
  country   -0.03 -0.02 -0.09  0.62 -0.09 0.41 0.59 1.1
  hvymetal  -0.01  0.00  0.71  0.00  0.13 0.52 0.48 1.1
  jazz       0.26  0.67  0.12 -0.10  0.06 0.54 0.46 1.4
  musicals   0.72  0.18 -0.18  0.04 -0.29 0.66 0.34 1.6
```

```
opera       0.73  0.10  0.06  0.06  0.02 0.55 0.45 1.1
rap         0.06  0.14  0.49  0.00 -0.08 0.27 0.73 1.3
sex        -0.16  0.01  0.08  0.06  0.78 0.64 0.36 1.1
agecat     -0.24  0.08  0.57 -0.21 -0.07 0.44 0.56 1.7
educat     -0.32 -0.14  0.03  0.38 -0.18 0.30 0.70 2.7
incomecat -0.13 -0.13  0.29  0.23 -0.46 0.39 0.61 2.7

                       PA1  PA5  PA2  PA3  PA4
SS loadings           2.36 1.32 1.28 1.22 1.00
Proportion Var        0.17 0.09 0.09 0.09 0.07
Cumulative Var        0.17 0.26 0.35 0.44 0.51
Proportion Explained  0.33 0.18 0.18 0.17 0.14
Cumulative Proportion 0.33 0.51 0.69 0.86 1.00

Mean item complexity =  1.6
Test of the hypothesis that 5 factors are sufficient.

The degrees of freedom for the null model are  91  and the
objective function was  4
The degrees of freedom for the model are 31  and the objective
function was  0.34

The root mean square of the residuals (RMSR) is  0.03
The df corrected root mean square of the residuals is  0.06

Fit based upon off diagonal values = 0.98
Measures of factor score adequacy
                                                  PA1  PA5
Correlation of (regression) scores with factors  0.91 0.86
Multiple R square of scores with factors         0.82 0.75
Minimum correlation of possible factor scores    0.64 0.49
                                                  PA2  PA3
Correlation of (regression) scores with factors  0.83 0.83
Multiple R square of scores with factors         0.69 0.69
Minimum correlation of possible factor scores    0.38 0.38
                                                  PA4
Correlation of (regression) scores with factors  0.84
Multiple R square of scores with factors         0.70
Minimum correlation of possible factor scores    0.41
```

PFA with raw data input with `fa()`

For comparison, we run the PFA model using the raw data in the music2 data frame, using just the variables in the "musicColumns" vector defined in the previous PCA section.

```
faResultVarimax <- psych::fa(r=music2[,musicColumns], nfactors = 5,
fm="pa", rotate ="varimax")
```

```
faResultVarimax
[Output:]
   Factor Analysis using method =  pa
   Call: psych::fa(r = music2[, musicColumns], nfactors = 5, rotate
   = "varimax",
       fm = "pa")
   Standardized loadings (pattern matrix) based upon correlation
   matrix
               PA1   PA5   PA2   PA3   PA4   h2   u2 com
   bigband    0.55  0.25 -0.20  0.31  0.02 0.51 0.49 2.4
   blues      0.16  0.76  0.11  0.14 -0.04 0.63 0.37 1.2
   blugrass   0.23  0.15 -0.03  0.61 -0.15 0.47 0.53 1.6
   classicl   0.77  0.13  0.04 -0.11 -0.04 0.63 0.37 1.1
   country   -0.03 -0.01 -0.08  0.56  0.07 0.32 0.68 1.1
   hvymetal  -0.02 -0.01  0.61  0.01 -0.10 0.39 0.61 1.1
   jazz       0.25  0.65  0.13 -0.08 -0.05 0.51 0.49 1.4
   musicals   0.69  0.18 -0.18  0.03  0.27 0.61 0.39 1.6
   opera      0.68  0.11  0.02  0.06  0.00 0.48 0.52 1.1
   rap        0.05  0.15  0.41 -0.01  0.08 0.20 0.80 1.4
   sex       -0.13  0.00  0.07  0.06 -0.61 0.39 0.61 1.1
   agecat    -0.22  0.08  0.60 -0.24  0.04 0.47 0.53 1.7
   educat    -0.32 -0.14  0.01  0.39  0.21 0.32 0.68 2.8
   incomecat -0.12 -0.14  0.31  0.23  0.47 0.40 0.60 2.7

                          PA1  PA5  PA2  PA3  PA4
   SS loadings           2.16 1.21 1.11 1.09 0.76
   Proportion Var        0.15 0.09 0.08 0.08 0.05
   Cumulative Var        0.15 0.24 0.32 0.40 0.45
   Proportion Explained  0.34 0.19 0.18 0.17 0.12
   Cumulative Proportion 0.34 0.53 0.71 0.88 1.00

   Mean item complexity =  1.6
   Test of the hypothesis that 5 factors are sufficient.

   The degrees of freedom for the null model are  91  and the ob-
   jective function was  3.17 with Chi Square of  2427.94
   The degrees of freedom for the model are 31  and the objective
   function was  0.23

   The root mean square of the residuals (RMSR) is  0.03
   The df corrected root mean square of the residuals is  0.05

   The harmonic number of observations is  773 with the empirical
   chi square  119.69  with prob <  2.3e-12
   The total number of observations was  773  with Likelihood Chi
   Square =  177.71  with prob <  1.6e-22
   Tucker Lewis Index of factoring reliability =  0.815
   RMSEA index =  0.078  and the 90 % confidence intervals are
   0.067 0.09
```

```
BIC =   -28.44
Fit based upon off diagonal values = 0.98
Measures of factor score adequacy
                                                  PA1  PA5
Correlation of (regression) scores with factors   0.89 0.83
Multiple R square of scores with factors          0.79 0.68
Minimum correlation of possible factor scores     0.58 0.36
                                                  PA2  PA3
Correlation of (regression) scores with factors   0.80 0.80
Multiple R square of scores with factors          0.63 0.64
Minimum correlation of possible factor scores     0.26 0.29
                                                  PA4
Correlation of (regression) scores with factors   0.75
Multiple R square of scores with factors          0.56
Minimum correlation of possible factor scores     0.12
```

Summary

Comparing the PFA model with polychoric input with the same model with raw data input we see that model fit metrics are very similar.

Unlike the PCA models, for these data, the polychoric PFA solution does not improve on model fit. Both models are a good fit by root mean square error or by fit-based off-diagonal values. For corrected RMSR, the raw data model is marginally better (lower is a better fit for this metric). Both perform equally in terms of attaining simple factor structure as measured by mean item complexity, which is higher than it would otherwise be due to the presence of demographic reference variables. Later, in Chapter 9, we will see how these reference variables can be included in the data but not in the calculation of the factors, by treating them as "extension variables".

Metric	*Raw data input*	*Polychoric correlation input*
RMSR	0.03	0.03
Corrected RMSR	0.05	0.06
Fit-based off-diagonal values	0.98	0.98
R-square of scores with first factor	0.89	0.91
R-square of scores with second factor	0.80	0.83
Mean item complexity	1.6	1.6

As discussed in Chapter 3, PFA seeks to reflect common or shared variance in the variables. PCA, in contrast, seeks to reflect the total (common plus unique plus error) variance of the indicator variables. Because of this difference, PFA may not reproduce the observed correlation matrix as well as PCA. For these data, using polychoric correlation to avoid violating the data level assumption of PFA seems to make less difference than it did for use in seeking to avoid violating the data level assumption of PCA.

Heterogeneous PCA with `hetcor()`

Introduction

The `hetcor()` function from the package "polycor" computes a heterogeneous correlation matrix for data of mixed levels. This matrix may be used as input for factor analysis. Between continuous-numeric variables, it computes a Pearson correlation coefficient. Between ordinal variables, it computes a polychoric correlation coefficient. Between a continuous variable and an ordinal variable, it computes a polyserial correlation coefficient. If declared as a factor, a dichotomous variable is given a polyserial correlation with numerics and a polychoric correlation with ordinals or other dichotomies. The "polycor" package was authored by John Fox (McMaster University). In addition to `hetcor()`, this package also contains the related functions `polychor()` and `polyserial()`. For reference, see Olsson (1979) and Drasgow (1986).

The general format of the `hetcor()` command is:

```
polycor::hetcor(data,
    ML = FALSE,
    std.err = TRUE,
    use=c("complete.obs", "pairwise.complete.obs"),
    bins=4,
    digits = 3,
    pd=TRUE)
```

In this format, "data" is a data frame, whose columns (variables) may be numeric-continuous, unordered (nominal) or ordered (ordinal) factors, logical, or character. If character, the order of levels is alphabetical (consider converting to factors with `as.factor()`).

Other settings:

- *ML*: If TRUE, compute maximum-likelihood estimates; otherwise the default is FALSE, corresponding to computing quick two-step estimates.
- *std.err*: The default (TRUE) is to compute standard errors.
- *use*: If "complete.obs", remove observations with any missing data; if "pairwise.complete.obs", compute each correlation using all observations with valid data for that pair of variables.
- *bins*: Number of bins to use for continuous variables when testing for bivariate normality. The default is 4.
- *digits*: Number of significant digits.
- *pd*: If TRUE and the correlation matrix is not positive-definite, an attempt is made to adjust it to a positive-definite matrix, using the `nearPD()` function in the "Matrix" package.

The `hetcor()` function returns an object of data class "hector" with these output elements:

$correlations	The correlation matrix.
$type	The type of each correlation: "Pearson", "Polychoric", or "Polyserial".
$std.errors	The standard errors of the correlations, if requested.
$n	The number of observations on which the correlations are based.
$tests	The p-values for tests of bivariate normality for each pair of variables.
$NA.method	Method for handling NAs: "complete.obs" or "pairwise.complete.obs".
$ML	TRUE for ML estimates, FALSE for two-step estimates.

Data preparation for `hetcor()`

The `hetcor()` function must be told which numerically-coded variables are ordinal factors and which are numeric. This is most easily done by assigning colClasses when reading the data, assigning columns their appropriate data classes. In the case of the "music" data example below, the first column (age) is numeric, then seven columns are ordinal music preferences, then one more is numeric (educ), then seven are more ordinal music preferences, then 1 is numeric (rincom91), then sex is entered as a factor, for a total of 18 columns. The data level types are specified using the `rep()` replicate command nested within the `c()` combine command as shown below.

```
base::setwd("C:/Data")

music <- utils::read.csv("musicvars.csv", header=TRUE, sep = ",",
stringsAsFactors = TRUE, colClasses = base::c("numeric",base::rep("-
factor",7), "numeric", base::rep("factor",7), "numeric","factor"))
```

Then we reduce the data frame to complete cases:

```
music2 <- music[stats::complete.cases(music),]
```

Then we eliminate redundant columns which otherwise would cause `hetcor()` to throw an error message. For instance, we retain "blues" as a five-level ordinal variable but delete its three-level cousin, "blues3":

```
music2[, base::c("blues3", "classic3", "jazz3", "rap3")]
<- base::list(NULL)
```

Optionally, we may view the data classes of columns of music2 with the command below. Note that we have retained the reference demographic variables (age,educ, rincome91, sex) but unlike the polychoric method in the previous section, it has not been necessary to bin age, education, and income into an eight-level categorical format. Again, we remind the reader that later, in Chapter 9, we will see how these reference variables can be included in the data but not in the calculation of the factors, by treating them as "extension variables".

```
base::sapply(music2, class)
[Output:]
        age    bigband      blues   blugrass   classicl
  "numeric"   "factor"   "factor"   "factor"   "factor"
    country       educ   hvymetal       jazz   musicals
```

```
"factor" "numeric"  "factor"  "factor"   "factor"
   opera        rap  rincom91       sex
"factor"  "factor" "numeric"  "factor"
```

It is often recommended to standardize variables prior to PCA, so that they are all on the same scale. This can be controversial for variables of mixed data levels and we omit such a step here. Instead we use the `hetcor()` command to create the object "musicHet", which is of class hetcor. This produces ignorable warnings but the correlation coefficients are correct.

```
library(polycor)
music2Het <- polycor::hetcor(music2)
```

The "type" element shows us what types of correlation have been used for each combination of variables. Below we show only the correlation types of the first six variables with the 14 variables in the data frame:

```
utils::head(music2Het$type)
[Output:]
(music2Het$type)
     [,1]         [,2]         [,3]         [,4]         [,5]
[1,] ""           "Polyserial" "Polyserial" "Polyserial" "Polyserial"
[2,] "Polyserial" ""           "Polychoric" "Polychoric" "Polychoric"
[3,] "Polyserial" "Polychoric" ""           "Polychoric" "Polychoric"
[4,] "Polyserial" "Polychoric" "Polychoric" ""           "Polychoric"
[5,] "Polyserial" "Polychoric" "Polychoric" "Polychoric" ""
[6,] "Polyserial" "Polychoric" "Polychoric" "Polychoric" "Polychoric"
     [,6]         [,7]         [,8]         [,9]         [,10]
[1,] "Polyserial" "Pearson"    "Polyserial" "Polyserial" "Polyserial"
[2,] "Polychoric" "Polyserial" "Polychoric" "Polychoric" "Polychoric"
[3,] "Polychoric" "Polyserial" "Polychoric" "Polychoric" "Polychoric"
[4,] "Polychoric" "Polyserial" "Polychoric" "Polychoric" "Polychoric"
[5,] "Polychoric" "Polyserial" "Polychoric" "Polychoric" "Polychoric"
[6,] ""           "Polyserial" "Polychoric" "Polychoric" "Polychoric"
     [,11]        [,12]        [,13]        [,14]
[1,] "Polyserial" "Polyserial" "Pearson"    "Polyserial"
[2,] "Polychoric" "Polychoric" "Polyserial" "Polychoric"
[3,] "Polychoric" "Polychoric" "Polyserial" "Polychoric"
[4,] "Polychoric" "Polychoric" "Polyserial" "Polychoric"
[5,] "Polychoric" "Polychoric" "Polyserial" "Polychoric"
[6,] "Polychoric" "Polychoric" "Polyserial" "Polychoric"
```

Finally, we create the correlation matrix for mixed data. This matrix is used for input for principal components analysis (PCA) in the next section.

```
music2HetcorMatrix <- music2Het$correlations
```

Using the heterogeneous correlation matrix for PCA

We first "manually" compute the principal components, eigenvalues, and eigenvectors. This step does not involve selecting how many factors to extract, rotating the axes, and computing factor loadings for the rotated axes. For the initial principal components solution, the number of components equals the number of input variables (or sample size minus 1 in the unusual case that there are fewer variables than observations). In this solution, we look to see how many components should be extracted, often using the Kaiser rule of extracting those with eigenvalues greater or equal to 1.0.

Calculations below rely on the `eigen()` function in R's "base" package, using the correlation matrix obtained above from `hetcor()` for mixed-level data. Type `help(eigen)` for more information.

```
# Create a correlation matrix object
music2Cormat <- music2Het$correlations


# Get eigenvalues
# By the Kaiser rule, the first 5 components would be interpreted
since they're greater than 1.0.
music2Eigen <- base::eigen(music2Cormat)
music2Eigen
(Partial output:)
   eigen() decomposition
   $values
    [1] 3.3193596 2.0678461 1.7078766 1.4571441 1.0071116 0.8342436
   0.7624442
    [8] 0.5875917 0.4819254 0.4326306 0.4037563 0.3511209 0.2975513
   0.2893981

   $vectors
                [,1]        [,2]        [,3]        [,4]        [,5]
    [1,]  0.17515198  0.47952823  0.13749075  0.10652568  0.04637138
    [2,] -0.41665984 -0.16507668  0.10986003 -0.04668497  0.02595105
   . . .

# Calculate proportions of variance explained by each component:
proportions <- music2Eigen$values/base::sum(music2Eigen$values)
proportions
(Output:)
    [1] 0.23709711 0.14770329 0.12199119 0.10408172 0.07193655
    [6] 0.05958883 0.05446030 0.04197083 0.03442324 0.03090219
   [11] 0.02883973 0.02508006 0.02125367 0.02067129

# Calculate cumulative proportions of variance explained.
# The first 5 components are associated with about 2/3 of variance
accounted for.
cumulativeProportions <- base::cumsum(proportions)
cumulativeProportions
    [1] 0.2370971 0.3848004 0.5067916 0.6108733
    [5] 0.6828099 0.7423987 0.7968590 0.8388298
```

```
 [9] 0.8732531 0.9041552 0.9329950 0.9580750
[13] 0.9793287 1.0000000
```

The `principal()` function from the "psych" package is among the PCA programs which support correlation matrix input. Below we use the mixed-level correlation matrix (music2Cormat) created above using `hetcor()`.

```
music2Principal5Result <- psych::principal(music2Cormat, nfactors =
5, residuals = TRUE, rotate="varimax", n.obs=base::nrow(music2), co-
var=FALSE, scores=TRUE, missing=FALSE, impute="median", method="re-
gression", use ="pairwise")

music2Principal5Result
[Partial output:]
   Principal component analysis
   . . .
   Standardized loadings (pattern matrix) based upon correlation
   matrix
              RC1   RC2   RC3   RC5   RC4   h2   u2 com
   age      -0.23  0.70 -0.27  0.07  0.06 0.63 0.37 1.6
   bigband   0.62 -0.23  0.37  0.28  0.02 0.66 0.34 2.4
   blues     0.14  0.11  0.15  0.87 -0.05 0.81 0.19 1.1
   blugrass  0.25 -0.01  0.77  0.14 -0.19 0.71 0.29 1.4
   classicl  0.83  0.04 -0.11  0.12 -0.06 0.73 0.27 1.1
   country  -0.04 -0.07  0.78 -0.05  0.11 0.62 0.38 1.1
   educ     -0.44 -0.03  0.49 -0.06  0.34 0.55 0.45 2.8
   hvymetal  0.00  0.83  0.03 -0.02 -0.13 0.70 0.30 1.1
   jazz      0.22  0.09 -0.11  0.85 -0.05 0.79 0.21 1.2
   musicals  0.77 -0.19  0.03  0.17  0.28 0.74 0.26 1.5
   opera     0.82  0.08  0.08  0.08 -0.02 0.69 0.31 1.1
   rap       0.09  0.68  0.03  0.17  0.14 0.51 0.49 1.3
   rincom91 -0.13  0.30  0.22 -0.12  0.68 0.64 0.36 1.8
   sex      -0.21  0.13  0.14  0.00 -0.83 0.77 0.23 1.2

                          RC1  RC2  RC3  RC5  RC4
   SS loadings           2.79 1.87 1.77 1.68 1.45
   Proportion Var        0.20 0.13 0.13 0.12 0.10
   Cumulative Var        0.20 0.33 0.46 0.58 0.68
   Proportion Explained  0.29 0.20 0.19 0.18 0.15
   Cumulative Proportion 0.29 0.49 0.67 0.85 1.00

   Mean item complexity =  1.5
   Test of the hypothesis that 5 components are sufficient.

   The root mean square of the residuals (RMSR) is  0.07
    with the empirical chi square  769.76  with prob <  6.7e-142

   Fit based upon off diagonal values = 0.9
```

Comparison with standard PCA

We now run a standard PCA model for comparison purposes.

```
music3 <- music2
i <- base::c(1:14)
music3[,i] <- base::apply(music3[,i], 2,
# Specify function within apply
function(x) as.numeric(as.character(x)))
base::sapply(music3, class)
[Output shows all variables are now numeric, as needed by the prin-
cipal() command:]
        age    bigband      blues   blugrass   classicl    country       educ
  "numeric"  "numeric"  "numeric"  "numeric"  "numeric"  "numeric"  "numeric"

   hvymetal       jazz   musicals      opera        rap   rincom91        sex
  "numeric"  "numeric"  "numeric"  "numeric"  "numeric"  "numeric"  "numeric"
```

We now run the conventional PCA model on the numeric data in music3.

```
music3Principal5Result <- psych::principal(music3, nfactors = 5,
residuals = TRUE, rotate="varimax", n.obs=base::nrow(music2), co-
var=FALSE, scores=TRUE, missing=FALSE, impute="median", method="re-
gression", use ="pairwise")

music3Principal5Result
[Partial output:]
  Principal component analysis
  . . .
  Standardized loadings (pattern matrix) based upon correlation
  matrix
             RC1   RC2   RC3   RC5   RC4   h2   u2 com
  age      -0.24  0.69 -0.27  0.09  0.05 0.62 0.38 1.6
  bigband   0.62 -0.21  0.34  0.26  0.00 0.61 0.39 2.3
  blues     0.14  0.08  0.14  0.86 -0.03 0.78 0.22 1.1
  blugrass  0.27 -0.01  0.71  0.15 -0.21 0.65 0.35 1.6
  classicl  0.82  0.05 -0.13  0.11 -0.06 0.70 0.30 1.1
  country  -0.02 -0.08  0.75 -0.02  0.07 0.57 0.43 1.0
  educ     -0.40  0.00  0.54 -0.11  0.30 0.55 0.45 2.6
  hvymetal  0.02  0.80  0.04 -0.07 -0.15 0.67 0.33 1.1
  jazz      0.21  0.10 -0.09  0.82 -0.04 0.74 0.26 1.2
  musicals  0.75 -0.17  0.01  0.16  0.27 0.69 0.31 1.5
  opera     0.79  0.06  0.06  0.06 -0.02 0.64 0.36 1.0
  rap       0.07  0.62  0.01  0.20  0.14 0.44 0.56 1.3
  rincom91 -0.11  0.35  0.28 -0.17  0.63 0.63 0.37 2.3
  sex      -0.16  0.12  0.12 -0.03 -0.83 0.75 0.25 1.2
```

```
                           RC1  RC2  RC3  RC5  RC4
SS loadings               2.63 1.74 1.70 1.64 1.35
Proportion Var            0.19 0.12 0.12 0.12 0.10
Cumulative Var            0.19 0.31 0.43 0.55 0.65
Proportion Explained      0.29 0.19 0.19 0.18 0.15
Cumulative Proportion     0.29 0.48 0.67 0.85 1.00

Mean item complexity =  1.5
Test of the hypothesis that 5 components are sufficient.

The root mean square of the residuals (RMSR) is  0.08
 with the empirical chi square  869.71  with prob <  7.8e-163

Fit based upon off diagonal values = 0.86
```

We now turn to visually comparing results from standard PCA with `principal()` with results from `principal()` using `hetcor()` correlation matrix input. The results appear in Figure 8.1.

Step 1: Standard Results: Plot the variables in rotated coordinate space for the standard `principal()` estimates. List out the variable coordinates, which are eigenvectors. Documentation calls these loadings, but they are not in the usual social science definition, which is loadings = eigenvectors * sqrt(eigenvalues). See the section of output below on "Standardized loadings (pattern matrix) based upon correlation matrix".

```
music3Principal5Result
[Output:]
   music3Principal5Result
   Principal Components Analysis
   Call: psych::principal(r = music3, nfactors = 5, residuals =
   TRUE,
       rotate = "varimax", n.obs = base::nrow(music2), covar =
   FALSE,
       scores = TRUE, missing = FALSE, impute = "median",
       method = "regression", use = "pairwise")
   Standardized loadings (pattern matrix) based upon correlation
   matrix
              RC1   RC2   RC3   RC5   RC4   h2   u2 com
   age      -0.24  0.69 -0.27  0.09  0.05 0.62 0.38 1.6
   bigband   0.62 -0.21  0.34  0.26  0.00 0.61 0.39 2.3
   blues     0.14  0.08  0.14  0.86 -0.03 0.78 0.22 1.1
   blugrass  0.27 -0.01  0.71  0.15 -0.21 0.65 0.35 1.6
   classicl  0.82  0.05 -0.13  0.11 -0.06 0.70 0.30 1.1
   country  -0.02 -0.08  0.75 -0.02  0.07 0.57 0.43 1.0
   educ     -0.40  0.00  0.54 -0.11  0.30 0.55 0.45 2.6
   hvymetal  0.02  0.80  0.04 -0.07 -0.15 0.67 0.33 1.1
   jazz      0.21  0.10 -0.09  0.82 -0.04 0.74 0.26 1.2
   musicals  0.75 -0.17  0.01  0.16  0.27 0.69 0.31 1.5
   opera     0.79  0.06  0.06  0.06 -0.02 0.64 0.36 1.0
   rap       0.07  0.62  0.01  0.20  0.14 0.44 0.56 1.3
```

```
rincom91 -0.11  0.35  0.28 -0.17  0.63 0.63 0.37 2.3
sex       -0.16  0.12  0.12 -0.03 -0.83 0.75 0.25 1.2

                        RC1  RC2  RC3  RC5  RC4
SS loadings             2.63 1.74 1.70 1.64 1.35
Proportion Var          0.19 0.12 0.12 0.12 0.10
Cumulative Var          0.19 0.31 0.43 0.55 0.65
Proportion Explained    0.29 0.19 0.19 0.18 0.15
Cumulative Proportion 0.29 0.48 0.67 0.85 1.00

Mean item complexity =  1.5
Test of the hypothesis that 5 components are sufficient.

The root mean square of the residuals (RMSR) is  0.08
 with the empirical chi square  869.71  with prob <  7.8e-163

Fit based upon off diagonal values = 0.86
```

The eigenvectors from the "loadings" component are the coordinates for components 1 and 2 used in mapping the variables.

```
coordinates1 <- music3Principal5Result$loadings[,1:2]
```

Now plot variables in the coordinate space formed by the first two dimensions.

```
base::plot(coordinates1,
pch=16,
   col="red",
   cex=1.5,
   cex.axis=1.4,
   cex.lab = 1.8,
   cex.main = 2,
   xlab = "Dimension 1",
   ylab = "Dimension 2",
   ylim = base::c(-0.5, 0.5),
   main="Music2 PCA with Standard Estimates from principal()")
graphics::abline(h=0, v=0, col="blue")
x <- coordinates1[,1]
y <- coordinates1[,2]
graphics::text(x,y,labels=base::rownames(coordinates1), cex=2,
pos=1, offset=0.5)
```

Step 2: Results with `hetcor()` correlation matrix input to the same model

Recall that conventional PCA results were stored in the object music3Principal5-Result. For the version based on `hetcor()`, results were stored in the object music2-Principal5Result. We list out the eigenvectors, labeled "loadings" by `principal()`. They are similar but not identical to coefficients for the standard results. The "music2Principal5Result" object was created earlier above, using `hetcor()` correlation matrix input.

```
music2Principal5Result
[Output:]
     Principal Components Analysis
   Call: psych::principal(r = music2Cormat, nfactors = 5, residuals
   = TRUE,
       rotate = "varimax", n.obs = base::nrow(music2), covar =
   FALSE,
       scores = TRUE, missing = FALSE, impute = "median",
       method = "regression", use = "pairwise")
   Standardized loadings (pattern matrix) based upon correlation
   matrix
              RC1   RC2   RC3   RC5   RC4   h2   u2 com
   age      -0.23  0.70 -0.27  0.07  0.06 0.63 0.37 1.6
   bigband   0.62 -0.23  0.37  0.28  0.02 0.66 0.34 2.4
   blues     0.14  0.11  0.15  0.87 -0.05 0.81 0.19 1.1
   blugrass  0.25 -0.01  0.77  0.14 -0.19 0.71 0.29 1.4
   classicl  0.83  0.04 -0.11  0.12 -0.06 0.73 0.27 1.1
   country  -0.04 -0.07  0.78 -0.05  0.11 0.62 0.38 1.1
   educ     -0.44 -0.03  0.49 -0.06  0.34 0.55 0.45 2.8
   hvymetal  0.00  0.83  0.03 -0.02 -0.13 0.70 0.30 1.1
   jazz      0.22  0.09 -0.11  0.85 -0.05 0.79 0.21 1.2
   musicals  0.77 -0.19  0.03  0.17  0.28 0.74 0.26 1.5
   opera     0.82  0.08  0.08  0.08 -0.02 0.69 0.31 1.1
   rap       0.09  0.68  0.03  0.17  0.14 0.51 0.49 1.3
   rincom91 -0.13  0.30  0.22 -0.12  0.68 0.64 0.36 1.8
   sex      -0.21  0.13  0.14  0.00 -0.83 0.77 0.23 1.2

                          RC1  RC2  RC3  RC5  RC4
   SS loadings           2.79 1.87 1.77 1.68 1.45
   Proportion Var        0.20 0.13 0.13 0.12 0.10
   Cumulative Var        0.20 0.33 0.46 0.58 0.68
   Proportion Explained  0.29 0.20 0.19 0.18 0.15
   Cumulative Proportion 0.29 0.49 0.67 0.85 1.00

   Mean item complexity =  1.5
   Test of the hypothesis that 5 components are sufficient.

   The root mean square of the residuals (RMSR) is  0.07
    with the empirical chi square  769.76  with prob <  6.7e-142

   Fit based upon off diagonal values = 0.9
```

As the "Standardized loadings (pattern matrix)" are a bit different between the standard and hetcor() versions, the coordinates are somewhat different in Figure 8.1. In this figure, the standard PCA version is in the upper half and the hetcor() version in the lower half.

```
coordinates2 <- music2Principal5Result$loadings[,1:2]

base::plot(coordinates2,
   pch=16,
   col="red",
   cex=1.5,
   cex.axis=1.4,
   cex.lab = 1.8,
   cex.main = 2,
   xlab = "Dimension 1",
   ylab = "Dimension 2",
   ylim = base::c(-0.5, 0.5),
   main="Music2 PCA with hetcor() matrix input to principal()")
graphics::abline(h=0, v=0, col="blue")
x <- coordinates2[,1]
y <- coordinates2[,2]
graphics::text(x,y,labels=base::rownames(coordinates2), cex=2,
pos=1, offset=0.5)
```

The differences between the standard and `hetcor()` versions are small, but the latter has adjusted for the data levels of the variables rather than assuming all to be continuous. Both versions achieve the same level of simple factor structure (mean item complexity = 1.5, which is inflated due to the presence of the demographic reference variables) but the `hetcor()` version has a bit less error (RMSR = 0.07 rather than 0.08 for the standard version) and a bit better fit (fit based on off-diagonal values = 0.9 rather than 0.86).

Mixed data PCA with `PCAmix()`

Introduction

An alternative to using polychoric correlation or `hetcor()` is to use an R package designed for mixed data such as "PCAmixeddata" discussed in this section. The "PCAmixdata" package was authored by Marie Chavent at the University of Bordeaux, France, along with her colleagues. See Chavent et al. (2017 or her tutorial at https://cran.r-project.org/web/packages/PCAmixdata/vignettes/PCAmixdata.html). The package contains three main functions:

1	PCAmix()	Principal component analysis of mixed data
2	PCArot()	Varimax-like rotation in PCAmix
3	MFAmix()	Multiple factor analysis of mixed multi-table data

The PCAmixdata package uses generalized singular value decomposition (GSVD) at the center of its algorithm.

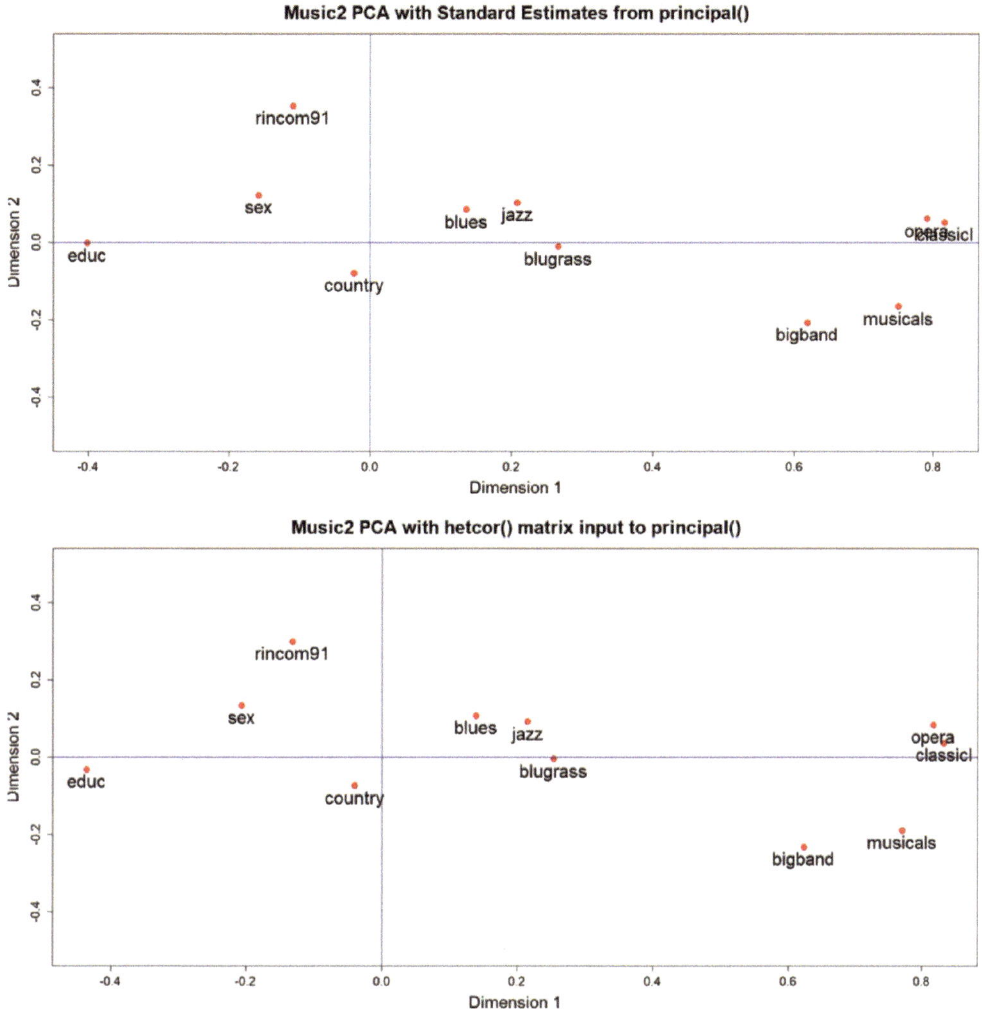

Figure 8.1 Standard (top) and `hetcor()` output (bottom) from `principal()`.

Other R packages for PCA with mixed data include "FactoMineR" and its function `FAMD()`, discussed in the next section. Others include the package "ade4" and its function in `dudi.mix()`. Also, clustering of variables with mixed data is supported by the package "ClustOfVar". Clustering of observations is supported by the package "cluster". And there are numerous other relevant packages.

Example data

The "PCAmixdata" package contains the "gironde" data. The "gironde" object is of data class "list". It contains four datasets: housing, employment, services, and environment. Below, we use the "housing" subset of the data. For simplicity, we put housing data in a data frame labeled "housing". The objects "Gironde$housing" and "housing" are R data frames.

The housing data frame contains data on 542 French cities for 5 variables (2 qualitative and 3 quantitative). The housing variables are population density (density), primary residence (primaryres), percent residential housing (houses, a factor variable coded either "sup 90%" for above 90% or "inf 90%" for below 90%), percent owner-occupied (owners), and the percentage of council housing within the cities (council, a factor variable coded either "sup 5%" for above 5%, or "inf 5%" for below 5%). Type `help(-gironde)` for more information.

```
base::setwd("C:/Data")              # Set the working directory
base::library(PCAmixdata)           # Contains PCA functions for
mixed data
utils::data(gironde)                          # Invoke the Gironde data
housing <- gironde$housing          # Create the housing data frame
```

We then get the names and data types of the five variables in housing. We see that the variables are a mixture of numeric (continuous) and factor (categorical) types. The density, primaryres, and owners variables are numeric and the houses and council variables are factors.

```
base::sapply(housing, class)
[Output:]
      density primaryres      houses      owners      council
    "numeric"  "numeric"    "factor"   "numeric"     "factor"
```

Below, we view the first three observations. Note that the city names are rownames, which are not themselves a variable (column) in the housing data frame.

```
utils::head(housing,3)
[Output:]
                         density primaryres   houses owners council
   ABZAC                  131.70      88.77  inf 90%  64.23   sup 5%
   AILLAS                  21.21      87.52  sup 90%  77.12   inf 5%
   AMBARES-ET-LAGRAVE     531.99      94.90  inf 90%  65.74   sup 5%
```

The `PCAmix()` algorithm handles missing values, if any, by mean replacement for quantitative variables and by zeros in the indicator matrix for qualitative variables. It is preferable to impute missing values by multiple imputations or some other more sophisticated routine prior to running `PCAmix()`. However, there are no missing values for the housing data frame, as shown by the command below.

```
base::sum(base::is.na(housing))
[Output:]
    [1] 0
```

The PCAmix() model

The format for the `PCAmix()` command given below. Note that the researcher must specify which data frame columns are numeric and which are categorical. There is no choice for binary, so binary variables may be listed as either numeric or categorical. True binary variables (e.g., gender) should be entered as qualitative factors. Dichotomizations of continuous variables may be entered as quantitative vectors but by binning into 0, 1, coding has lost information. The `PCAmix()` program makes no provision for ordinal variables. Ordinal variables may be entered as categorical but order information will be ignored, reducing statistical power. If ordinal variables are entered as continuous, this has the same problems as doing so in standard PCA. If all variables are continuous, `PCAmix()` gives the same results as standard PCA. The format of the command is given below.

```
PCAmix(
   # Matrix, vector, or data frame with the numeric variables
   X.quanti = NULL,
   # Matrix, vector, or data frame with the factor variables
   X.quali = NULL,
   # Number of requested dimensions in the solution (default = 5)
   ndim = 5,
   # If rename.level=TRUE, all factor variable levels are renamed
   (FALSE is default)
   # in the format "variable_name=level_name".
   # This is needed when different variables share one or more same
   level names.
   rename.level = FALSE,
   # Optional vectors of possible weights of numeric and factor
   variables (NULL is default)
   weight.col.quanti = NULL,
   weight.col.quali = NULL,
   # Toggles if graphs should be part of output. (TRUE is default)
   graph = TRUE)
```

If graphs are selected, `PCAmix()` produces four types of graphics as appropriate, for the first two dimensions:

- A component map of the individual observations
- A plot of the squared loadings of all the variables (quantitative and qualitative)
- A plot of the correlation circle for quantitative variables
- A component map of the levels for factor variables

We start by using the `splitmix()` function to identify the qualitative and quantitative variables and create separate data frames for the two types of variables. The optional `attributes()` command from R's "base" package reveals the names of these objects.

```
split <- PCAmixdata::splitmix(housing)

base::attributes(split)
[Output:]
   $names
   [1] "X.quanti"  "X.quali"   "col.quant" "col.qual"
```

All of the objects listed above are data frames. The first two contain the quantitative and qualitative data in separate data frames. The last two contain the column index numbers in housing corresponding to quantitative and qualitative variables respectively.

Note that if there are no qualitative variables, results correspond to a standard PCA solution. If there are no quantitative variables, results correspond to a standard multiple correspondence analysis (MCA) solution.

We put the quantitative and qualitative data frames into quantvars and qualvars respectively:

```
quantvars <- split$X.quanti
qualvars <- split$X.quali
```

We then run the `PCAmix()` model with for the default five-dimension solution.

```
PCAmixResult5 <- PCAmixdata::PCAmix(
   X.quanti=quantvars,
   X.quali=qualvars,
   ndim=5,
   rename.level=TRUE,
   graph=FALSE)
```

The object PCAmixResult5 is of data class "PCAmix". It has many numerical results as elements. These are listed below.

```
base::attributes(PCAmixResult5)
[Output:]
   $names
    [1] "call"        "eig"         "ind"         "quanti"      "levels"
    [6] "quali"       "sqload"      "coef"        "Z"           "M"
   [11] "quanti.sup"  "levels.sup"  "sqload.sup"  "rec.sup"     "scores.stand"
   [16] "scores"      "V"           "A"           "categ.coord" "quanti.cor"
   [21] "quali.eta2"  "rec"         "ndim"        "W"           "rename.level"
```

Typing the name of the results object lists definitions for some of these elements.

```
PCAmixResult5
[Partial output. These are described the "Explanation" section
below]
   "name" "description"
   "$eig" "eigenvalues of the principal components (PC) "
   "$ind" "results for the individuals (coord,contrib,cos2)"
   "$quanti" "results for the quantitative variables
   (coord,contrib,cos2)"
   "$levels" "results for the levels of the qualitative variables
   (coord,contrib,cos2)"
   "$quali" "results for the qualitative variables (contrib,rela-
   tive contrib)"
   "$sqload" "squared loadings"
   "$coef" "coef of the linear combinations defining the PC"
```

Explanation of PCAmixResult elements

$eig: A matrix containing the eigenvalues, the % of variance and the cumulative % of variance.

$ ind: A list containing the results for the individuals (observations):

- indcoord: factor coordinates (scores) of the observations
- indcontrib: absolute contributions of the individuals
- indcontrib.pct: relative contributions of the individuals
- indcos2: squared cosines of the individuals

$quanti: A list containing the results for the quantitative variables:

- $quanti$coord: factor coordinates (scores) for the quantitative variables
- $quanti$contrib: absolute contributions for the quantitative variables
- $quanti$contrib.pct: relative contributions for the quantitative variables (in %)
- $quanti$cos2: squared cosines of the quantitative variables.

$levels: A list containing the results for the levels of the qualitative variables:

- $levels$coord: factor coordinates (scores) of the levels
- $levels$contrib: absolute contributions of the levels
- $levels$contrib.pct: relative contributions of the levels (in percentage)
- $levels$cos2: squared cosines of the levels

$quali: A list containing the results for the qualitative variables:

- $quali $contrib: absolute contributions of the qualitative variables (sum of absolute contributions of the levels of the qualitative variable)
- $quali $contrib.pct: relative contributions (in percentage) of the qualitative variables (sum of relative contributions of the levels of the qualitative variable).

$sqload: a matrix of dimension (p, ndim) containing the squared loadings of the quantitative and qualitative variables.

$coef: The coefficients of the linear combinations used to construct the principal components of PCAmix, and to predict coordinates (scores) of new observations in the function predict.PCAmix.

$M: The vector of weights of the columns used in the Generalized Singular Value Decomposition.

Rotation with `PCArot()`

In the foregoing section on `PCAmix()`, the axes (components) being discussed were not rotated. Rotation does not change variance explained as reflected in the sum of eigenvalues, as illustrated further below. Rotation does change the variances associated with each factor and how variables load on the dimensions. Rotation of axes is the norm in social science because the interpretation of the contributions of each variable to the components becomes much clearer. Put another way, the initial axes are arbitrary whereas rotated axes seek to align the variables with the components in a more interpretable way. The most common form of rotation is varimax, which is a type or orthogonal rotation. Orthogonality means that the components are uncorrelated. The `PCArot()` function implements varimax rotation.

In this section, we assume that needed objects created in the `PCAmix()` section are still in the R environment. These include:

- housing, the data frame with the Gironde housing data
- PCAmixResult5 , the result of the PCAmix() command for the unrotated five-component solution
- Quantvars and qualvars, the result of splitting the housing data into quantitative and qualitative variables, using the splitmix() command

The $eig element of the unrotated solution contains the eigenvalues and the proportion of variance explained by each. Eigenvalue proportions from the `PCAmix()` model with all five components, unrotated, sum to 100%. The cumulative proportion for the first two dimensions is 71.92%. For the first three dimensions, it is 84.5%.

```
PCAmixResult5$eig
[Output:]
         Eigenvalue Proportion Cumulative
   dim 1  2.5268771  50.537541   50.53754
   dim 2  1.0692777  21.385553   71.92309
   dim 3  0.6303253  12.606505   84.52960
   dim 4  0.4230216   8.460432   92.99003
   dim 5  0.3504984   7.009968  100.00000
```

The sum of eigenvalues is the same as the number of dimensions:

```
base::sum(PCAmixResult5$eig[,1])
[Output:]
   [1] 5
```

Moreover, had we asked for a two- or three-dimensional solution in `PCmix()` in its `ndim=` option, we would still get the same five eigenvalues and the same sum of eigenvalues as when we requested five dimensions. That is, the `ndim=` option defaults to 5, which is always to solution returned. To get an actual two- or three-dimensional solution, we must so request in the `PCArot()` command. The unrotated solution is used primarily to select how many dimensions to retain and rotate.

There is no statistically "correct" answer to the question, "How many dimensions should be rotated?" For the eigenvalues in the unrotated solution above, the Kaiser criterion would suggest rotating the first two, which are those with eigenvalues greater than 1.0. In the vignette accompanying the package, the first three dimensions are selected on the rationale that these account for the bulk of the variance (85%). Below, for instructional purposes, we create results objects reflecting rotation of the first two, the first three, or all five dimensions

```
PCArotResult2  <-  PCAmixdata::PCArot(PCAmixResult5, dim=2,
graph=FALSE)
```

```
PCArotResult3  <-  PCAmixdata::PCArot(PCAmixResult5, dim=3,
graph=FALSE)
PCArotResult5  <-  PCAmixdata::PCArot(PCAmixResult5, dim=5,
graph=FALSE)
```

Eigenvalues and proportions explained

To list eigenvalues (variances of the principal components), their proportions, and their cumulative variance explained, list the "eig" element. In the previous section, we listed the eigenvalues for PCAmixResult5 (the unrotated five-dimensional solution) and saw the sum of eigenvalues was also 5.

For the rotated five-dimensional solution we see that the variances and proportions explained associated with each factor have changed compared to the unrotated five-dimensional solution, but the eigenvalues still sum to 5 (the same). For both rotated and unrotated solutions, the first dimension always has the highest eigenvalue and largest proportion explained. The last dimension always has the smallest eigenvalue and proportion, and always reflects a cumulative percentage explained of 100% when all dimensions (here, all 5) are represented.

```
PCArotResult5$eig
[Output:]
            Variance Proportion
   dim1.rot 1.0189857   20.37971
   dim2.rot 1.0064724   20.12945
   dim3.rot 1.0286919   20.57384
   dim4.rot 0.9966582   19.93316
   dim5.rot 0.9491918   18.98384

base::sum(PCArotResult5$eig[,1])
   [1] 5
```

As mentioned, in `PCArot()` it is possible to rotate just selected dimensions, not necessarily all of them. Below we look at the eigenvalues and proportions explained when only two or three dimensions are retained and rotated.

For the solution in which the first two dimensions were rotated, the sum of proportions explained in the "housing" data is still 71.92%, the same as it was for the cumulative percent explained for the unrotated results object "PCAmixResult5$eig" earlier.

```
PCArotResult2$eig
[Output:]
            Variance Proportion
   dim1.rot 2.526628   50.53255
   dim2.rot 1.069527   21.39054

base::sum(PCArotResult2$eig[,1])
[Output for sum of variances:]
   [1] 3.596155
```

```
base::sum(PCArotResult2$eig[,2])
[Output for sum of proportions:]
   [1] 71.92309
:
```

For the solution in which the first three dimensions were rotated, the sum of proportions explained in the "housing" data is still 84.53%, the same as it was for the cumulative percent explained for the unrotated results object "PCAmixResult5$eig" earlier.

```
PCArotResult3$eig
[Output:]
            Variance Proportion
   dim1.rot 1.926203   38.52407
   dim2.rot 1.056062   21.12123
   dim3.rot 1.244215   24.88430

base::sum(PCArotResult3$eig[,1])
[Output for sum of variances:]
   [1] [1] 4.22648

base::sum(PCArotResult3$eig[,2])
[Output for sum of proportions:]
   [1] 84.5296
```

Rotated loadings

The main function of rotation is to create a more interpretable component loadings table. Rotation seeks to optimize each variable being heavily loaded on just one component and little loaded on the others. Visually, points wind up closer to one axis or another. The solution comes closer to a simple factor structure. Below, we examine the squared loadings before and after rotation.

Below are the loadings for the unrotated five-dimensional solution.

```
base::round(PCAmixResult5$sqload[,1:5],digit=2)
[Output:]
              dim 1 dim 2 dim 3 dim 4 dim 5
   density     0.50  0.06  0.40  0.02  0.02
   primaryres  0.00  0.95  0.02  0.03  0.00
   owners      0.74  0.02  0.00  0.02  0.23
   houses      0.68  0.03  0.03  0.16  0.09
   council     0.61  0.02  0.18  0.19  0.00
```

Below are the loadings for the rotated five-dimensional solution. Had five-, three-, and two-dimensional solutions been computed, the rotated loadings would differ for each solution for included dimensions (unlike unrotated loadings).

```
base::round(PCArotResult5$sqload[,1:5],digit=2)
[Output:]
             dim1.rot dim2.rot dim3.rot dim4.rot dim5.rot
  density        0.02     0.00     0.90     0.03     0.04
  primaryres     0.00     0.99     0.00     0.00     0.00
  owners         0.07     0.00     0.06     0.09     0.78
  houses         0.07     0.00     0.03     0.82     0.08
  council        0.86     0.00     0.03     0.06     0.05

base::round(PCArotResult3$sqload[,1:3],digit=2)
[Output:]
             dim1.rot dim2.rot dim3.rot
  density        0.04     0.01     0.91
  primaryres     0.00     0.96     0.00
  owners         0.49     0.03     0.24
  houses         0.63     0.03     0.08
  council        0.76     0.03     0.02

base::round(PCArotResult2$sqload[,1:2],digit=2)
[Output:]
             dim1.rot dim2.rot
  density        0.49     0.07
  primaryres     0.00     0.94
  owners         0.74     0.01
  houses         0.69     0.03
  council        0.61     0.02
```

Above, we may observe the following:

1 In the solution rotating 3 axes, dimension 1 is heavily associated with house and council. Dimension 2 is heavily associated with primaryres. Dimension 3 is heavily associated with density. The owners variable loads most heavily on dimension 1, but only at 0.49.
2 For the solution rotating 2 axes, dimension 1 is heavily associated with owners, houses, and council. Dimension 2 is heavily associated with primaryres. The density variable is not heavily associated with either dimension but is most heavily loaded on dimension 1, but only at 0.49.

Rotation does not change the sums of variances proportions explained for the solution with all dimensions. For both unrotated PCAmixResult5 and rotated PCArotResult5, the sum of variances is 5 and the sum of proportions is 100%. However, rotation does affect variable loadings on the components. So which solution should the researcher choose? This is not strictly a statistical question. If the density were a critical construct for the researcher, and it were planned to use component scores in regression or other procedures, the PCArotResult3 solution is preferable as it contains a dimension clearly associated with density. On the other hand, if the researcher felt that owners, houses, and council were closely related on a conceptual basis, then the PCArotResult2 solution might suffice.

Visualization of observations

Below we plot observations, which are French municipalities, on the component space formed by the first two dimensions of the PCArotResult2 solution. For more information on plotting `PCAmix()` models, type `?plot.PCAmix`. The plot is shown in Figure 8.2.

```
# Plotting observations in component space for the 2-dimensional
rotated solution
base::plot(PCArotResult2,
    choice = "ind",
    # choice is the graph type, which may be any of the following.
    # "ind"  = individual observations component map
    # "cor"  = correlation circle map of quantitative variables
    # "levels" = components map of levels of the qualitative
    variables
    # "sqload" =  plot of squared loadings of all variables.
    axes = (base::c(1,2)),
    # Specifies the components (axes) to plot. The first two is the
    default.
    coloring.ind = qualvars$houses,
    # coloring.ind declares the qualitative variable whose levels
    differentiate colors of points.
    # Recall that qualvars was created above as a data frame holding
    the qualitative variables.
    col.ind = base::c("red", "blue"),
    # col.ind sets the colors to be used. If omitted, colors are se-
    lected automatically.
    label=FALSE,
    # If label= is TRUE, then points are labeled.
    posleg="bottomright",
    # posleg = sets the position of the legend.
    main="Observations"
    # main sets the title for the plot
    )
```

Figure 8.2 shows observations, which are French cities, located in `PCAmix()` component space. The color coding of points reflects the qualitative variable "houses". This variable has two levels:

```
base::levels(housing$houses)
[Output:]
    [1] " inf 90%" " sup 90%"
```

In French, "inf" is the abbreviation of inférieur à which means "less than". Likewise, "sup" corresponds to "more than".

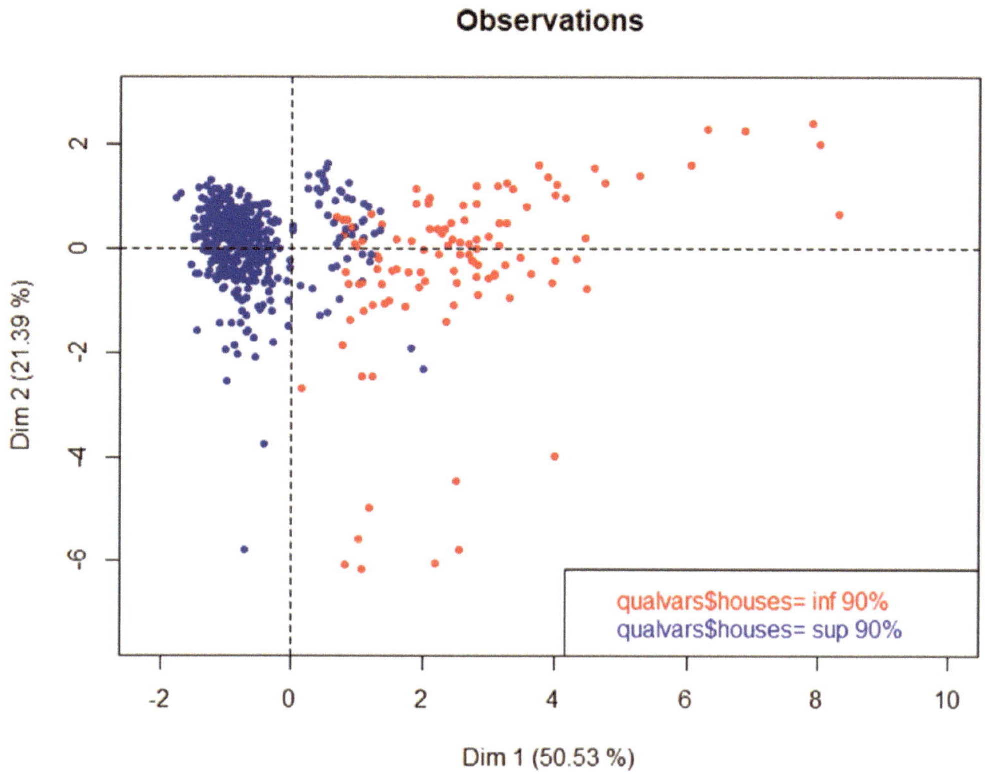

Figure 8.2 Plot of observations in PCAmix component space, Gironde housing data.

The blue points in the upper left quadrant of Figure 8.2 are those for the level "sup 90%". These are cities where houses (privately owned properties) are 90% or more of all units. It may be observed that dimension 1 differentiates the municipalities with a high proportion of privately owned houses. The blue points representing these municipalities tend to be below 2 on dimension 1. Most are below 0 on dimension1 and above −1 on dimension 2. The municipalities with less than 90% private houses are found in the two right-most quadrants of Figure 8.2.

Visualization of variables

For the example data, the most striking change for variables after rotation occurs in the contrast of the correlation of the density variable with component 3. Before rotation, its correlation is 0.39 and after rotation, it is 0.90. Following Chavent et al. (2017), Figure 8.3 shows observations and variables before and after rotation in dimensions 1 and 3. We do this for PCAmixResult3 for prior to rotation and for PCArotResult3, which rotated the first three axes.

First, we create the PCmixResult3 (unrotated0 object, not created above.

```
PCAmixResult3 <- PCAmixdata::PCAmix(
  X.quanti=quantvars,
```

```
  X.quali=qualvars,
  ndim=3,
  rename.level=TRUE,
  graph=FALSE)

# Set graphics for 2x2 panel display of four plots
graphics::par(mfrow=base::c(2,2))

base::plot(PCAmixResult3,
   choice="ind",
   label=FALSE,
   axes=base::c(1,3),
   main="Observations: Before Rotation",
   sub = "PCAmixResult Object")

base::plot(PCArotResult3,
   choice="ind",
   label=FALSE,
   axes=base::c(1,3),
   main="Observations: After Rotation",
   sub = "PCArotResult3 Object")

base::plot(PCAmixResult3,
   choice="sqload",
   coloring.var=TRUE,
   cex = 1.6,
   leg=TRUE,
   axes=base::c(1,3),
   posleg="topright",
   main="Variables: Before Rotation",
   sub = "PCAmixResult Object",
   xlim=base::c(0,1),
   ylim=base::c(0,1))

base::plot(PCArotResult3,
   choice="sqload",
   coloring.var=TRUE,
   cex = 1.6,
   leg=TRUE,axes=base::c(1,3),
   posleg="topright",
   main="Variables: After Rotation",
   sub = "PCArotResult3 Object",
   xlim=base::c(0,1),
   ylim=base::c(0,1))

# Reset graphics to single plots
graphics::par(mfrow=base::c(1,1))
```

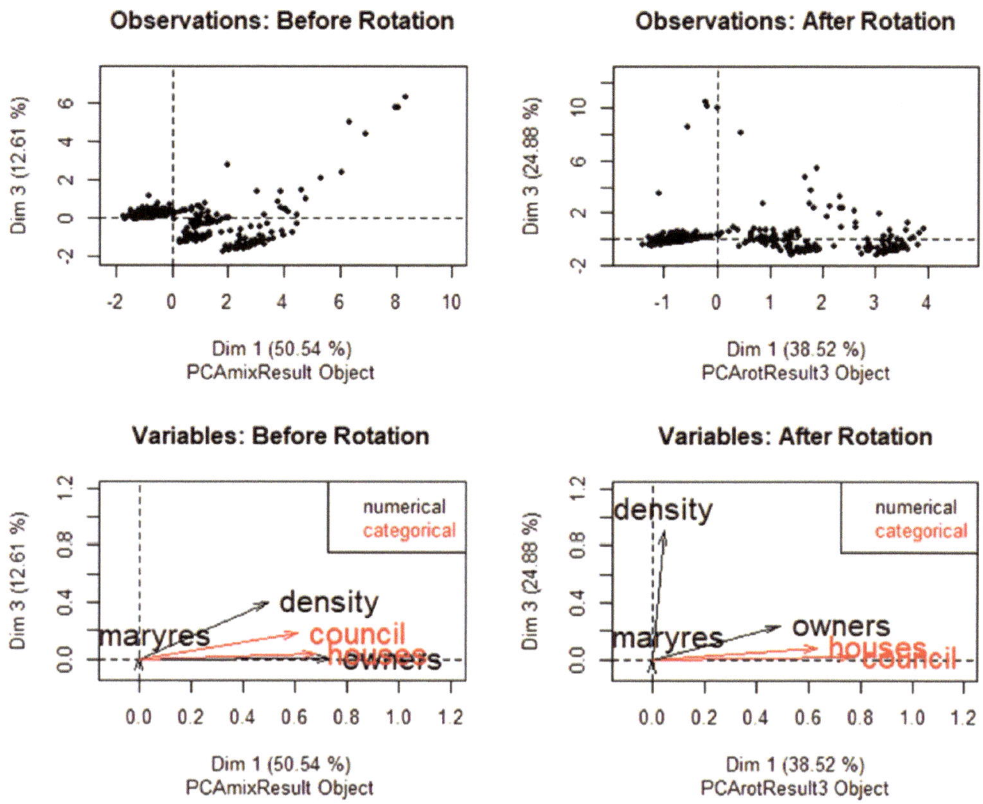

Figure 8.3 Observations and variables before and after rotation, Gironde data.

In Figure 8.3, apart from the striking difference for the density variable, rotation for these data does not have an impact great enough to change most substantive inferences. However, on average, rotation draws both observations and variables closer to the axes representing the components.

A second panel plot in Figure 8.4 again uses components 1 and 3 to plot the numeric variables (top two plots) and the categorical variable levels (bottom two plots) in the component space formed by components 1 and 3.

```
# Set display for 2x2 panel plot
graphics::par(mfrow=base::c(2,2))

base::plot(PCAmixResult3,
   choice="cor",
   coloring.var=TRUE,
   cex = 1.6,
   leg=TRUE,
   axes=base::c(1,3),
   posleg="topright",
   main="Before Rotation",
   sub = "PCAmixResult Object",
   )
```

```
base::plot(PCArotResult3,
    choice="cor",
    coloring.var=TRUE,
    cex = 1.6,
    leg=TRUE,
    axes=base::c(1,3),
    posleg="topright",
    main="After Rotation",
    sub = "PCArotResult3 Object")

base::plot(PCAmixResult3,
    choice="levels",
    cex = 1.6,
    axes=base::c(1,3),
    main="Before Rotation",
    sub = "PCAmixResult Object",
    xlim=base::c(-1.5,2.5))

base::plot(PCArotResult3,
    choice="levels",
    cex = 1.6,
    axes=base::c(1,3),
    main="After Rotation",
    sub = "PCArotResult3 Object",
    xlim=base::c(-1.5,2.5))

# Reset graphics to single plots
graphics::par(mfrow=base::c(1,1))
```

Figure 8.4 in the two upper plots uses a correlation circle to illustrate the effect of rotation on the density variable, for dimensions 1 and 3. The effect on levels of the categorical variables is shown in the two lower plot panels.

Visualization of levels of categorical variables

It is also possible to display values of categorical variables in component space, similar to correspondence analysis. There are two categorical variables, houses, and council, with two levels (values) each. These appear in Figure 8.5.

```
base::plot(PCAmixResult3,
    choice = "levels",
    # A levels plot is selected
    cex = 1.2,
    # cex sets size of point labels
    xlim = base::c(-1.5,2.5),
    # xlim is a numeric vectors of length 2, giving the x axis/coor-
    dinate range.
    # If NULL (the default) the range is defined automatically
    (recommended).
    main = "Levels of Categorical Variables")
```

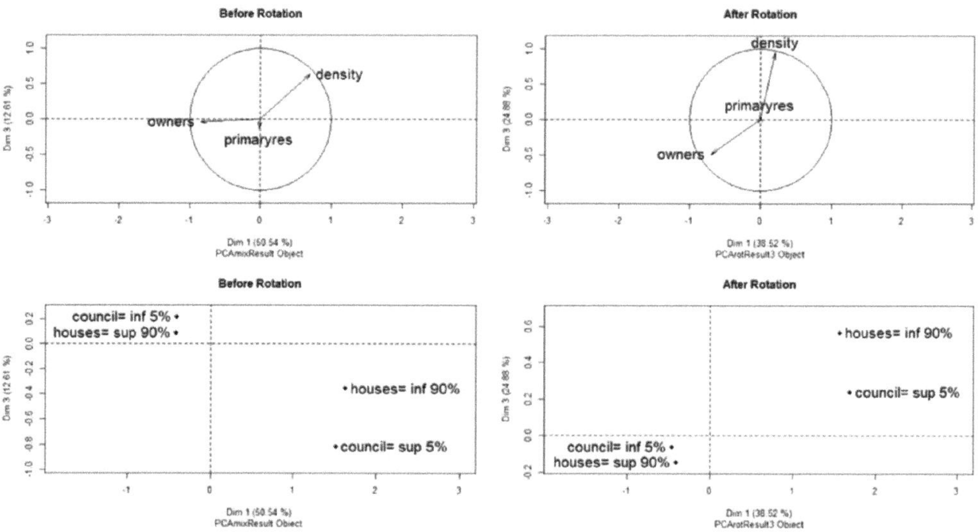

Figure 8.4 Variables and categorical levels before and after rotation, Gironde data.

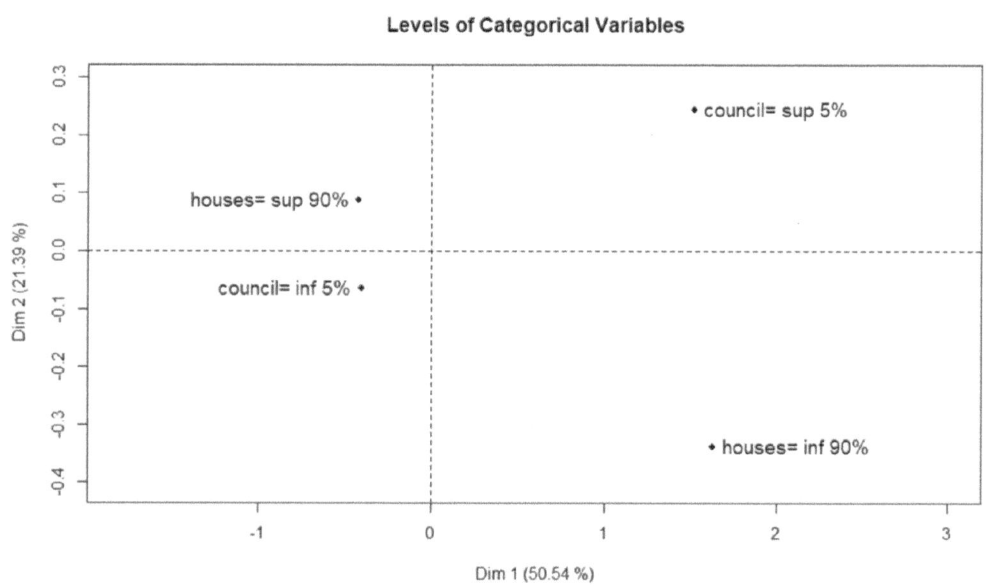

Figure 8.5 Levels of categorical variables in component space, Gironde housing data.

Figure 8.5 is a level map for the qualitative (categorical, factor) variables. It shows that there is a relationship between having a high proportion of private housing and having a low percentage of council (public) housing. That is, "houses=sup 90%" and "council=inf 5%" are both located on the negative side of dimension 1, while "houses=inf 90%" and "council=sup 5%" have positive values on dimension 1.

Visualization of numeric variables

Numeric variables may be visualized in a correlation circle plot, such as shown in Figure 8.6. For this plot, the unrotated three-dimensional solution is used. The rotated result is different.

```
base::plot(PCAmixResult3,
    choice="cor",
    cex = 1.6,
    main="Numeric Variables, PCAmixResult",
    sub = "Gironde Data")
```

Figure 8.6 is a "correlation circle" for the numeric variables, also called simply a "PCA variable plot". Arrows in opposing directions indicate negative correlation. For instance, percentage of home owners = owners and population density = density are negatively correlated along dimension 1. This does not necessarily mean that the variables are negatively correlated in raw terms. Not exemplified here, but arrows close together indicate a positive correlation on the relevant dimension. The more an arrow is horizontal and further from the origin, the more that variable is associated with dimension 1. The more it is vertical and further from the origin, the more that variable is associated with dimension 2.

The closer a variable is to the circumference of the circle, the more it is well represented by the principal components. Here, all three variables are well represented. Placement on or near the circumference reflects a high cos2 value. Variables which are not as well represented have lower cos2 coefficients and are located closer to the center of the circle and may be considered less important for the components. The "cos2"

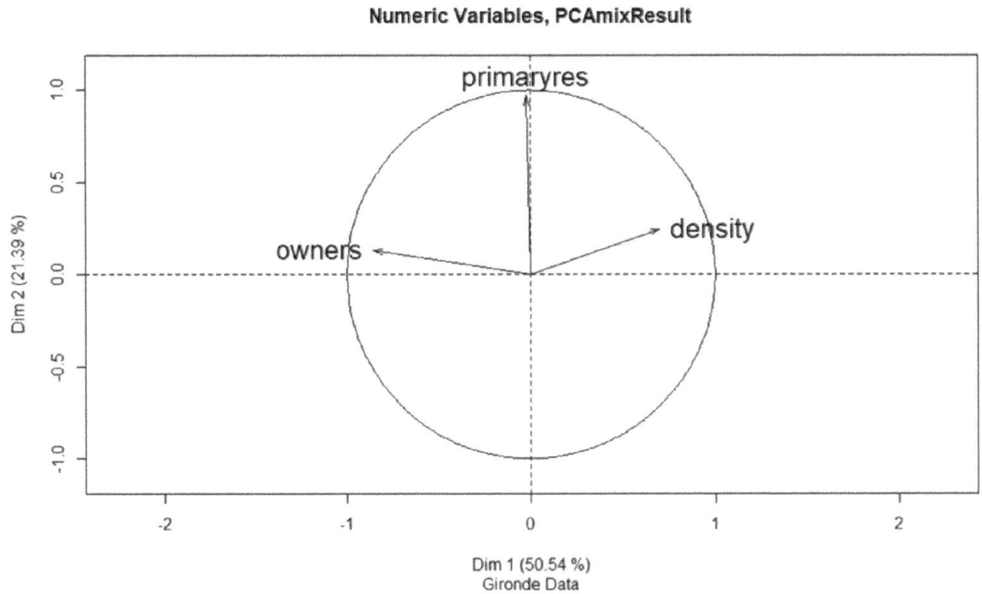

Figure 8.6 Numeric variables in PCAmix component space, Gironde housing data.

coefficient, which is the squared cosine of a PCA component with an observation, is a measure of the importance of a given component to a given observation.

Visualization of component loadings for all variables

It is possible, as in Figure 8.7, to visualize both numerical and categorical variables in component space simultaneously. Again, the three-dimensional unrotated solution is used.

```
base::plot(PCAmixResult3,
    choice="sqload",
    cex = 1.6,
    coloring.var=TRUE,
    leg=TRUE,
    posleg="topright",
    main="All Variables Plotted on Dimensions 1 and 2, PCAmixResult",
    sub = "Gironde Data")
```

Figure 8.7 shows the location of all variables in component space, for the first two dimensions. The coordinates of any variable are its squared loadings. For numeric variables, this is squared correlations. For categorical variables, this is the correlation ratios. Squared loadings reflect the linkage of variables to principal components. In

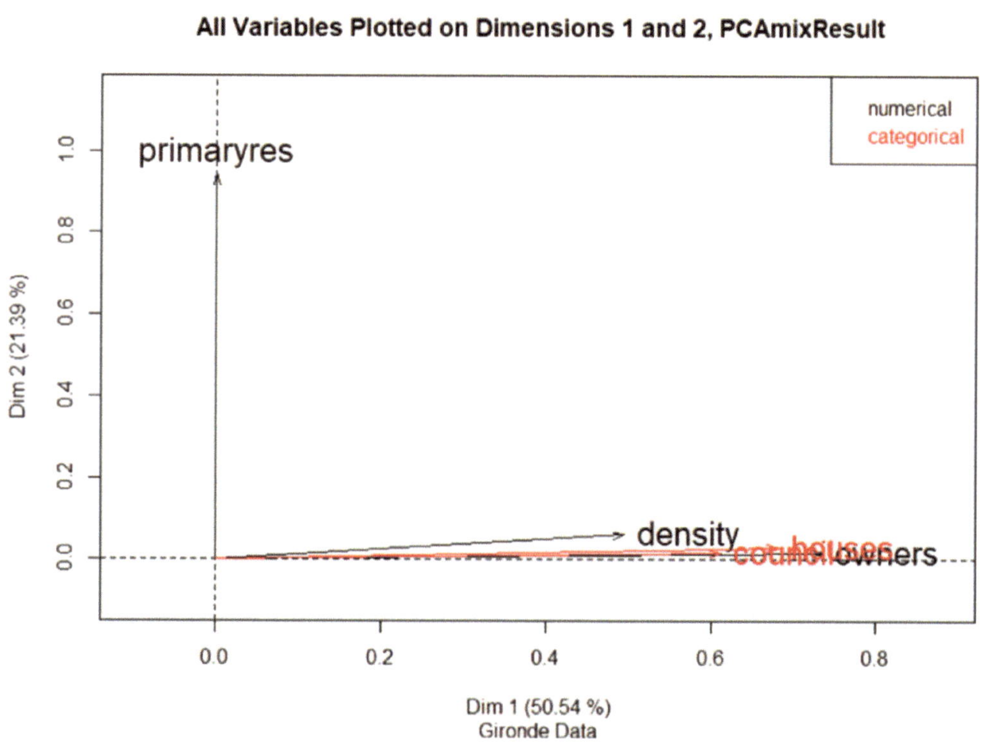

Figure 8.7 Component loadings of all variables, Gironde housing data.

Figure 8.7, primaryres (proportion of houses used as primary residences) is associated with dimension 2 while all other variables are associated with dimension1. Figure 8.7 shows the linkages of variables to dimensions but not whether these linkages show positive or negative association with the dimensions. That is shown in Figures 8.5 and 8.6.

For Figure 8.7, the numerical values of the coordinates are in the "contrib" sub-elements.

```
# For quantitative variables, these values are identical to
PCAmixResult3$quanti$cos2.
PCAmixResult3$quanti$contrib
 [Output:]
                  dim 1       dim 2       dim 3       dim 4       dim 5
density    0.4955009833 0.06107142 0.401118144 0.02102911 0.021280343
primaryres 0.0003542291 0.94553912 0.017772819 0.03327953 0.003054306
owners     0.7365056261 0.01714341 0.001247151 0.01795169 0.227152122

# For categorical variables:
PCAmixResult3$quali$contrib
 [Output:]
             dim 1      dim 2      dim 3     dim 4       dim 5
houses  0.6822569 0.02958542 0.03242219 0.1609776 0.094757858
council 0.6122593 0.01593829 0.17776496 0.1897836 0.004253778
```

Cross-validation and prediction

As with any predictive procedure, it is possible to develop the model on a training subset of the data (here, "train") and then test it on a validation subset (here, "test"). In this section, 100 municipalities are sampled at random to form the test set. The remaining municipalities form the train set. The `PCAmix()` model is developed on the train set (shown as "-test" in R code below). Type `?predict.PCAmix` for more information on using `predict()` with PCAmix objects.

Set a random seed since sampling is being used. Create the test sample. The training sample is simply "-test". Then apply the `PCAmix()` command to the training set to obtain the results object, which we label "trainResult".

```
base::set.seed(123)
test <- base::sample(1:base::nrow(gironde$housing),100)
trainResult <- PCAmixdata::PCAmix(quantvars[-test,],qual-
vars[-test,],ndim=2,graph=FALSE)
```

The "pred" object created below by the `predict()` command contains the predictions for the 100 municipalities in the test set.

```
pred <- stats::predict(trainResult, quantvars[test,],
qualvars[test,])
```

We then plot the train observations as model points and the test observation as supplemental points on the principal component map developed for the training data. This is shown in Figure 8.8.

```
base::plot(trainResult,
   axes=base::c(1,2),
   label=FALSE,
   main="Train and Test Sets Plotted on First Two Dimensions")
   graphics::points(pred, col=2, pch=16)
graphics::legend("bottomright",legend = base::c("Train","Test"),
fill=1:2, col= 1:2)
```

While the train and test points in Figure 8.8 show a general similarity, there are few test points which are very high on Dimension 1 and very few test points which are very low on Dimension 2.

Component scores

Sometimes observation-level component scores are used as variables in regression or other forms of analysis. If the researcher wishes to retain the scores of each observation (here, each municipality), these are returned in the "$scores" element as a matrix. Below we use the rotated results in PCArotResult3 object for scoring, but other PCA result objects could be used. Only the first three cities in the data are shown.

```
scoresMatrix <- PCArotResult3$scores
utils::head(scoresMatrix,3)
```

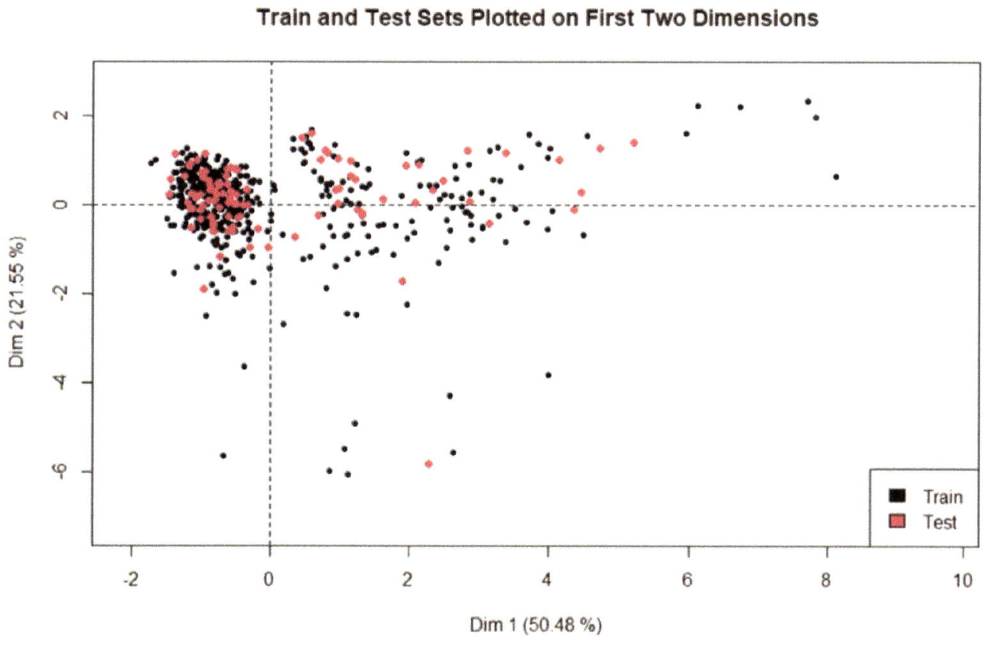

Figure 8.8 Training and test municipalities mapped on the first two PCAmix dimension.

```
                                      dim1.rot       dim2.rot       dim3.
rot
ABZAC                              3.241486176  0.3458916709 -0.8471213893
AILLAS                            -0.731177115  0.1230417153 -0.2246583947
AMBARES-ET-LAGRAVE                 2.855016255  0.9685219705 -0.0140754750
```

If the researcher prefers to retain scores in a data frame, the matrix is easily converted. Rows in the data frame will be the observations (municipalities), the columns will be the components, and cell entries will be the scores. Municipality names are rownames, not a column in the data frame itself. Type `rownames(scoresDF)` to list them.

```
scoresDF <- base::as.data.frame(scoresMatrix)
View(scoresDF)
```

Multiple factor analysis (MFA) with `MFAmix()`

Introduction

The `MFAmix()` module of the "PCAmixdata" package implements multiple factor analysis. In this type of analysis, a set of observations is described by not one but rather by several groups of quantitative and categorical variables. For instance, in the "Gironde" example, there were data not only for housing as utilized in the foregoing sections, but also for employment, services, and environment. MFA allows simultaneous analysis across these sectors. MFA for quantitative variables originated with Escofier and Pages (1994, based on work from 1983) and is extended to categorical variables by the `MFAmix()` function.

Setup

Recall that the "Gironde" data had not just the "housing" data frame used above, but also data frames for "employment", "services", and "environment", each with their own sets of variables. We can use these four data frames to illustrate `MFAmix()`.

```
utils::data(gironde)
base::names(gironde)
   [Output:]
   [1] "employment"  "housing"      "services"     "environment"
```

1 The first step is to concatenate the four data frames into one object, "MFAdat", which is a data frame.

   ```
   MFAdat <- base::cbind(gironde$employment, gironde$housing, gi-
   ronde$services, gironde$environment)
   ```

2 We then create a vector of corresponding names for the groups of variables. We also create a vector of column codes for the concatenated MFAdat data frame. The data frame has 27 variables (columns). The MFAindex object is of length 27. Its numbers

designate from which of the four variable groups each column comes. For instance, the first 9 columns come from set 1, which is the "employment" set.

```
MFAnames <- base::c("employment","housing","servi
ces","environment")

MFAindex <- base::c(base::rep(1,9), base::rep(2,5),
base::rep(3,9),base::rep(4,4))
MFAindex
[Output:]
 [1] 1 1 1 1 1 1 1 1 1 2 2 2 2 2 3 3 3 3 3 3 3 3 3 4 4 4 4
```

Running the MFAmix() program

3 We then run the MFAmix() function on the data, requesting a three-component solution. The results object, which here is MFAmixResults, is of data class "MFAmix".

```
MFAmixResults <- PCAmixdata::MFAmix(
    data    = MFAdat,
    groups = MFAindex,
    name.groups = MFAnames,
    ndim=3,
    rename.level=TRUE,
    graph=FALSE)
```

Elements returned

The `MFAmixResults` object is returned with a $sqload element. This contains the squared loadings of variables on the components. For quantitative variables, this is based on the squared correlations of the variable while for categorical variables, this is based on the correlation ratios. The $sqload element is also returned by `PCAmix()` and `PCArot()`, already discussed.

For the `MFAmix()` results object, the following retrievable elements are also returned.

- $eig: Eigenvalues
- $eig.separate: Eigenvalues of the separate analyses
- $separate.analyses: Separate analyses for each group of variables
- $groups: Results for all the groups
- $partial.axes: Results for the partial axes
- $ind: Results for the individuals
- $ind.partial: Results for the partial individuals
- $quanti: Results for the quantitative variables
- $levels: Results for the levels of the qualitative variables
- $quali: Results for the qualitative variables
- $qqload: Squared loadings
- $listvar.group: List of variables in each group
- $global.pca: Results for the global PCA

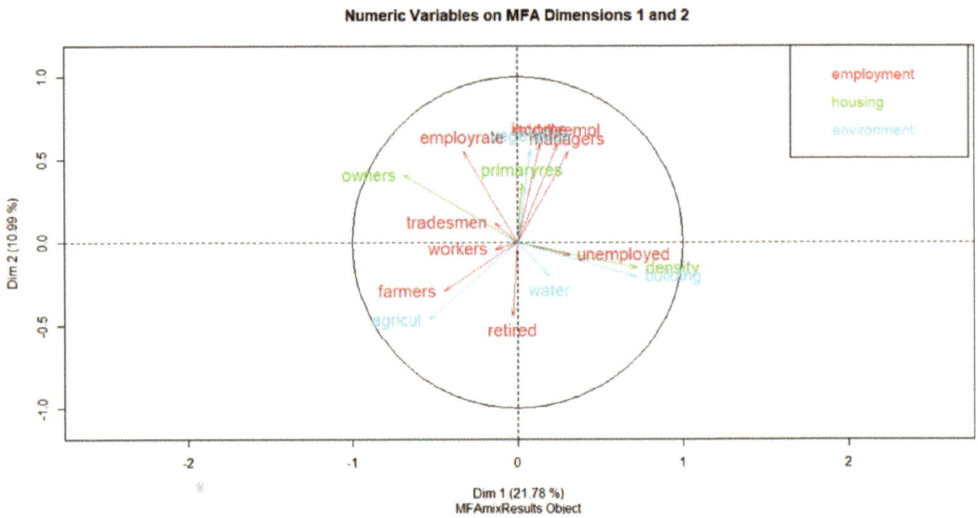

Figure 8.9 `MFAmix()` results for Gironde numeric variables.

MFA visualization

Graphics available for `MFAmix()` results objects largely parallel those for `PCAmix()` and `PCArot()`. Type `?plot.MFAmix` for more information on visualization for `MFAmix()` objects. We start by creating a variable correlation plot as a separate plot for greater readability (Figure 8.9). The first two dimensions explain a lower percent of variance in the numeric variables in the four datasets (housing, employment, services, environment) than was the case earlier when only housing was being analyzed. The services sector does not show up in this figure because all or its variables are qualitative rather than numeric.

```
base::plot(MFAmixResults,
   choice="cor",
   coloring.var="groups",
   leg=TRUE,
   cex = 1.4,
   cex.leg = 1.2,
   posleg="topright",
   main="Numeric Variables on MFA Dimensions 1 and 2",
   sub = "MFAmixResults Object")
```

Below we list numeric variable coordinates (loadings) for the first two dimensions, shown in Figure 8.9.

```
coordinatesNumericvars <- MFAmixResults$quanti$coord[,1:2]
coordinatesNumericvars
[Output:]
   MFAmixResults$quanti$coord[,1:2]
```

```
                 dim 1        dim 2
farmers     -0.44529713 -0.29512901
tradesmen   -0.13670787  0.11802163
managers     0.30664023  0.54634224
workers     -0.13264188 -0.04194179
unemployed   0.31843191 -0.07565052
middleempl   0.24155390  0.59925788
retired     -0.02864241 -0.44465055
employrate  -0.32801079  0.55452095
income       0.13397203  0.60380395
density      0.72379130 -0.15496125
primaryres   0.03135362  0.36008335
owners      -0.68638171  0.41000006
building     0.71618929 -0.20519959
water        0.19150538 -0.20370591
vegetation   0.07645926  0.56424908
agricul     -0.53909672 -0.46935769
```

Next we now list numeric variables with heavy loadings (greater than 0.6) on dimension 1. Three variables are strongly loaded on dimension 1.

```
base::subset(coordinatesNumericvars,base::abs(coordinatesNumeric-
vars[,1]) > 0.6, 1)
[Output:]
              dim 1
   density   0.7237913
   owners   -0.6863817
   building  0.7161893
```

We could do the same for dimension 2. Only income is strongly loaded on dimension 2. Some other variables load in the 0.5–0.6 range but are not listed.

```
base::subset(coordinatesNumericvars,base::abs-
(coordinatesNumericvars[,2]) > 0.6,2)
[Output:]
            dim 2
   income 0.603804
```

We now create a panel plot of three graphs showing MFA results for both numeric and qualitative variables. This is Figure 8.10. If desired, each of the three plots might have been created separately, not as part of a panel plot, for better viewing.

```
# Set display for 3x1 panel of three plots.
graphics::par(mfrow=base::c(3,1))

# The first plot plots observations (municipalities) against the
first two components
# It also highlights the municipality of Bordeaux.
base::plot(MFAmixResults,
```

```
    choice="ind",
    partial=base::c("BORDEAUX"),
    label=TRUE,
    leg=TRUE,
    cex = 1.4,
    cex.leg=1.4,
    posleg="topright",
    main="Municipalities as Observations",
    sub = "MFAmixResults Object")

# The second plot plots squared loadings by group.
base::plot(MFAmixResults,
    choice="sqload",
    coloring.var="groups",
    leg=TRUE,
    cex = 1.4,
    cex.leg=1.4,
    posleg="topright",
    main="Variable Plot (Squared Loadings on the First Two
    Dimensions)",
    sub = "MFAmixResults Object")

# The third plot plots groups on the first two dimensions.
base::plot(MFAmixResults,
    choice="groups",
    coloring.var="groups",
    cex = 1.2,
    main="Groups on the First Two Dimensions",
    sub = "MFAmixResults Object")

# Reset graphics to single plots
graphics::par(mfrow=base::c(1,1))
```

We can also visualize how levels of the categorical variables in "housing" and "services" align themselves on the first two dimensions. By setting `choice=levels`, we filter for just the categorical variables, which are all from either the housing or the services sets, color-coded green or blue respectively. The level "houses = inf 90%" (low proportion of private housing) stands out as an outlier, heavily loaded in a negative direction on dimension 2. This is Figure 8.11.

```
base::plot(MFAmixResults,
    choice="levels",
    coloring.var="groups",
    cex.leg=1.5,
    posleg="bottomleft",
    main="Levels of Categorical Variables on the First Two
    Dimensions",
    sub = "MFAmixResult Object",
    xlim=base::c(-2,4))
```

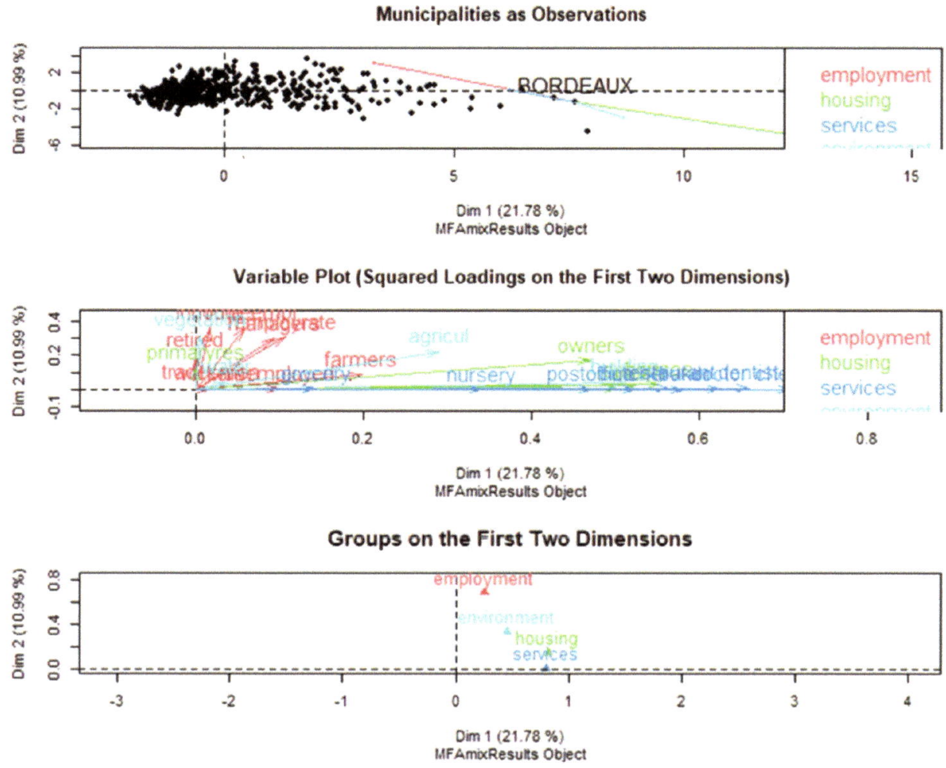

Figure 8.10 Panel Plots of `MFAmix()` Results.

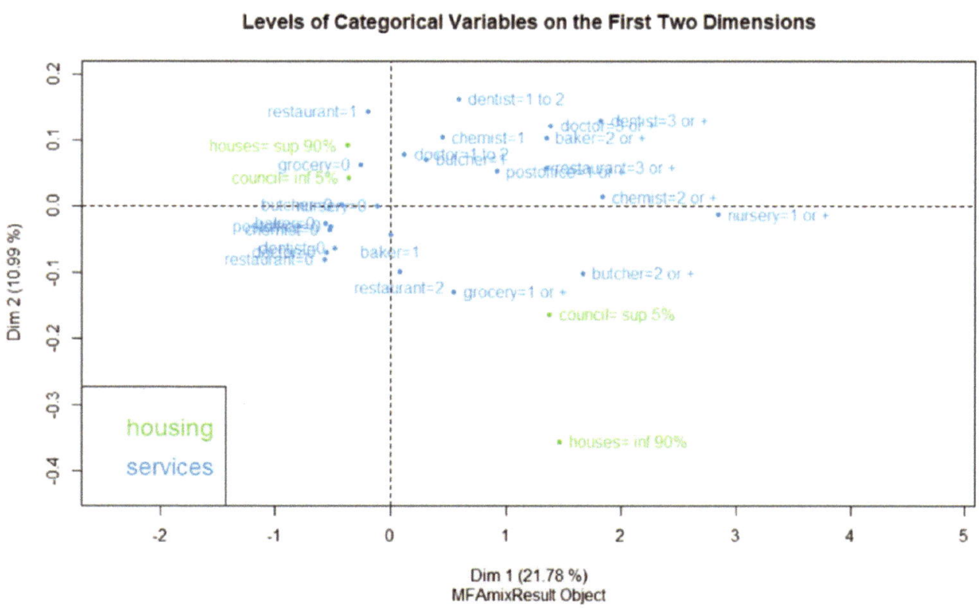

Figure 8.11 Levels of qualitative variables on dimensions 1 and 2.

Mixed data PCA with `FAMD()`

Introduction

The "FactoMineR" package, discussed previously with regard to the `PCA()` command in Chapter 6, also supports a type of principal components analysis for mixed data through its `FAMD()` command (Factor Analysis for Mixed Data). Like `PCAmixdat()`, `FAMD()` bridges aspects of principal components analysis of numeric variables (PCA) with MCA of categorical variables. Numeric variables are scaled to have a variance of 1.0 and categorical variables are scaled using MCA scaling. Scaling serves to treat continuous and categorical variables in a balanced manner so that neither type of variable is advantaged or disadvantaged in the determination of the dimensions of variability. Graphical outputs include plots of observations, correlation circles for continuous variables, visualization of the levels of the categorical variables, and plots show in the associations between continuous and categorical variables.

Setup

```
base::setwd("C:/Data")             # Set the working directory
base::library(FactoMineR)          # Has the FAMD() program
base::library(factoextra)          # Visualization for FactoMineR
base::library(PCAmixdata)          # Contains Gironde housing data
```

Example data

We use the same "housing" data frame as for the PCAmixdata package described in the previous section. This data frame contains three continuous and two categorical variables pertaining to housing in Gironde district, France's largest.

```
utils::data(gironde)
housing <- gironde$housing
```

The `FAMD()` model

The basic syntax of the FAMD() command is shown below.

```
FAMD(df, ncp = 5, sup.var = NULL, ind.sup = NULL, graph = TRUE)
```

Explanation

- df: A data frame (here, housing) in which the rows are observations and the columns are continuous and categorical variables.
- ncp: The number of requested dimensions (default = 5, accepted in this example)
- sup.var: An optional vector (or NULL) indicating the indexes of the supplementary variables not used in training the model. Not used in this example.
- ind.sup: An optional vector (or NULL) indicating the indexes of the supplementary individuals not used in training the model (e.g., the test set). Not used in this example.
- graph: unless FALSE, displays a graph.

Thus, the simplest command for the example data would be that below. It returns five output elements.

```
FAMDresult <- FactoMineR::FAMD(housing)
FAMDresult
[Output:]
  *The results are available in the following objects:
    name             description
  1 "$eig"           "eigenvalues and inertia"
  2 "$var"           "Results for the variables"
  3 "$ind"           "results for the individuals"
  4 "$quali.var"     "Results for the qualitative variables"
  5 "$quanti.var"    "Results for the quantitative variables"
```

Output elements

1 *$eig*: Returns the eigenvalues, which reflect the proportion of variance explained by each component.

```
FAMDresult$eig
[Output:]
          eigenvalue percentage of variance cumulative percent-
   age of variance
   comp 1  2.5268771              50.537541
   50.53754
   comp 2  1.0692777              21.385553
   71.92309
   comp 3  0.6303253              12.606505
   84.52960
   comp 4  0.4230216               8.460432
   92.99003
   comp 5  0.3504984               7.009968
   100.00000
```

The first two components have eigenvalues greater than 1.0, meriting their extraction by the Kaiser rule. Together, they explain 71.9% of the variance in the housing data.

Eigenvalues may be displayed in a scree plot, shown in Figure 8.12. We use the conventional plotting of eigenvalues, but it is possible to set `choice = "variance"` instead, to use percent of variable explained by each component as the basis for the bars.

```
factoextra::fviz_screeplot(FAMDresult, choice="eigenvalue",
addlabels=TRUE)
```

2. $var: Returns the coordinate, contribution, and cos2 values for the variables. Among other purposes, these values are used to plot results. Before viewing the contents of $var, we first list the data class of each variable:

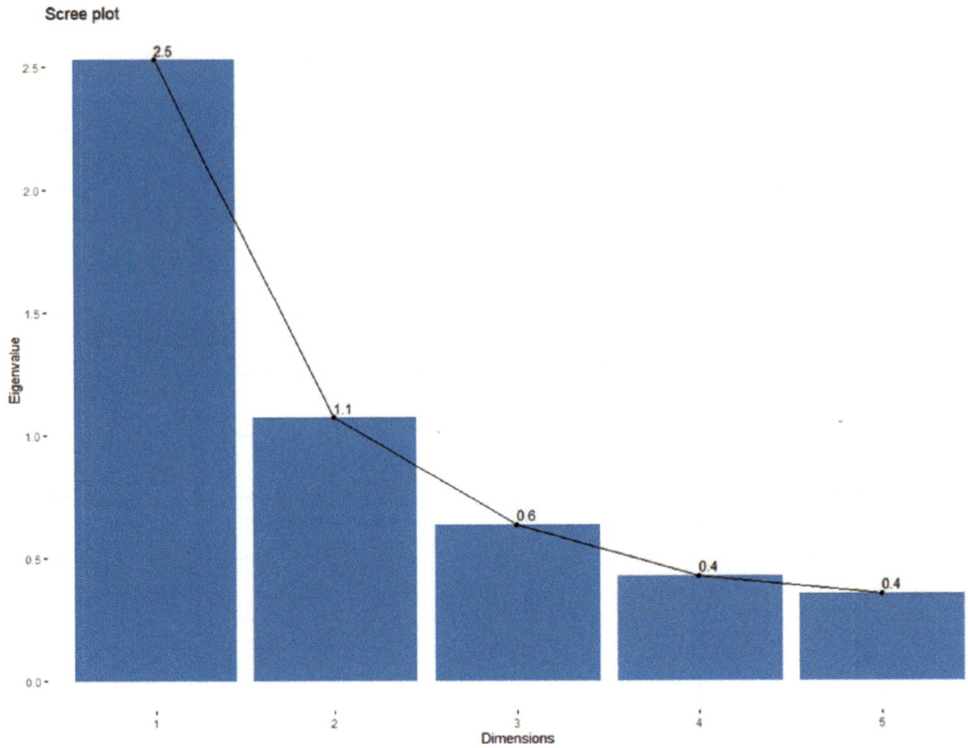

Figure 8.12 Scree plot.

```
base::sapply(housing, class)
[Output:]
      density primaryres      houses      owners     council
    "numeric"  "numeric"    "factor"   "numeric"    "factor"
```

```
FAMDresult$var
[Output:]
   $coord
                     Dim.1      Dim.2       Dim.3      Dim.4       Dim.5
   density    0.4955009833 0.06107142 0.401118144 0.02102911 0.021280343
   primaryres 0.0003542291 0.94553912 0.017772819 0.03327953 0.003054306
   owners     0.7365056261 0.01714341 0.001247151 0.01795169 0.227152122
   houses     0.6822568847 0.02958542 0.032422193 0.16097764 0.094757858
   council    0.6122593496 0.01593829 0.177764958 0.18978363 0.004253778

   $contrib
                    Dim.1     Dim.2      Dim.3     Dim.4      Dim.5
   density    19.60922392  5.711465 63.6366915  4.971168  6.0714522
   primaryres  0.01401845 88.427838  2.8196267  7.867098  0.8714181
   owners     29.14687200  1.603270  0.1978582  4.243681 64.8083179
   houses     27.00000297  2.766861  5.1437241 38.054237 27.0351750
   council    24.22988266  1.490566 28.2020994 44.863815  1.2136369
```

```
$cos2
                  Dim.1        Dim.2        Dim.3        Dim.4
Dim.5
density    2.455212e-01 0.0037297180 1.608958e-01 0.0004422236
4.528530e-04
primaryres 1.254783e-07 0.8940442275 3.158731e-04 0.0011075268
9.328788e-06
owners     5.424405e-01 0.0002938966 1.555384e-06 0.0003222631
5.159809e-02
houses     4.654745e-01 0.0008752972 1.051199e-03 0.0259138015
8.979052e-03
council    3.748615e-01 0.0002540290 3.160038e-02 0.0360178257
1.809463e-05
```

3 $ind: Returns the coord, contrib, and co2 values for the individuals/observations, along with their dist values. As output is too long to show here, instead we show its structure:

```
utils::str(FAMDresult$ind)
[Output:]
  List of 4
   $ coord  : num [1:542, 1:5] 2.364 -0.88 2.623 0.927 1.176 ...
    ..- attr(*, "dimnames")=List of 2
    .. ..$ : chr [1:542] "ABZAC" "AILLAS" "AMBARES-ET-LAGRAVE"
  "AMBES" ...
    .. ..$ : chr [1:5] "Dim.1" "Dim.2" "Dim.3" "Dim.4" ...
   $ cos2   : num [1:542, 1:5] 0.667 0.957 0.741 0.173 0.134 ...
    ..- attr(*, "dimnames")=List of 2
    .. ..$ : chr [1:542] "ABZAC" "AILLAS" "AMBARES-ET-LAGRAVE"
  "AMBES" ...
    .. ..$ : chr [1:5] "Dim.1" "Dim.2" "Dim.3" "Dim.4" ...
   $ contrib: num [1:542, 1:5] 0.4081 0.0565 0.5025 0.0627 0.101
  ...
    ..- attr(*, "dimnames")=List of 2
    .. ..$ : chr [1:542] "ABZAC" "AILLAS" "AMBARES-ET-LAGRAVE"
  "AMBES" ...
    .. ..$ : chr [1:5] "Dim.1" "Dim.2" "Dim.3" "Dim.4" ...
   $ dist   : Named num [1:542] 2.9 0.9 3.05 2.23 3.22 ...
    ..- attr(*, "names")= chr [1:542] "ABZAC" "AILLAS"
  "AMBARES-ET-LAGRAVE" "AMBES"
```

These values may be retrieved by the `get_famd_var()` command and stored in an object (here, "variable_dat") with three result elements: the coordinates needed to plot variables, the cos2 measure of importance of the factors to the variables, and the variables' percent contributions to construction of the dimensions.

```
variable_dat <- factoextra::get_famd_var(FAMDresult)
variable_dat
```

```
[Output:]
   FAMD results for variables
    ===================================================
     Name         Description
   1 "$coord"     "Coordinates"
   2 "$cos2"      "Cos2, quality of representation"
   3 "$contrib"   "Contributions"
```

The coordinates result element may be used to plot variables in component space. These are the same coordinate values as displayed above for FAMDresult$var.

```
variable_dat$coord
[Output:]
                        Dim.1      Dim.2       Dim.3      Dim.4       Dim.5
   density     0.4955009833 0.06107142 0.401118144 0.02102911 0.021280343
   primaryres  0.0003542291 0.94553912 0.017772819 0.03327953 0.003054306
   owners      0.7365056261 0.01714341 0.001247151 0.01795169 0.227152122
   houses      0.6822568847 0.02958542 0.032422193 0.16097764 0.094757858
   council     0.6122593496 0.01593829 0.177764958 0.18978363 0.004253778
```

Likewise, the other result elements contain the cos2 and contrib values. The cosign squared (cos2) is a measure of the quality of representation of the variable. The cos2 coefficient reflects the importance of the column dimension to the row variable. Contributions reflect how much the row variable contributed to the construction of the column dimension. Contribution coefficients are percentages which add to 100% in the columns.

```
variable_dat$cos2
[Output not shown but is the same as for FAMDresult$var:]

variable_dat$contrib
[Output not shown but is the same as for FAMDresult$var:]
```

Visualization of variables

Below we plot both quantitative and qualitative (factor) variables in component space formed by Dim1 and Dim2. This is Figure 8.13. In col.var= color assignment, the order of variables is dictated by `FAMD()`, which puts the three continuous variables first, then the two factor/categorical variables. In Figure 8.13, in addition to points representing variables, the larger unlabeled triangle symbols show the location of the centroids for the numeric and categorical variables as sets.

We use the command `fviz_famd_var()` to visualize variables. The simple command `plot(var$coord)` would generate a bare-bones plot with correct coordinates, but the more complex R code below gives a more communicative figure in which we see "primaryres" is highly associated with Dimension 2 and the other variables with Dimension 1.

```
factoextra::fviz_famd_var(FAMDresult,
    choice="var",
    axes=base::c(1:2),
    labelsize=5,
    pt.cex=24,
    ylim=base::c(-0.2,1),
    repel=TRUE,
    col.var= base::c(base::rep("numeric",3),base::rep("factor",2)))
    +
        ggplot2::theme(text = ggplot2::element_text(size = 14),
              axis.title = ggplot2::element_text(size = 14),
              axis.text = ggplot2::element_text(size = 14))
```

If we let `choice="quali.var"` and adjust the color coding for the two qualitative variables (house, council) we may plot categorical variable levels in component space. This is shown in Figure 8.14.

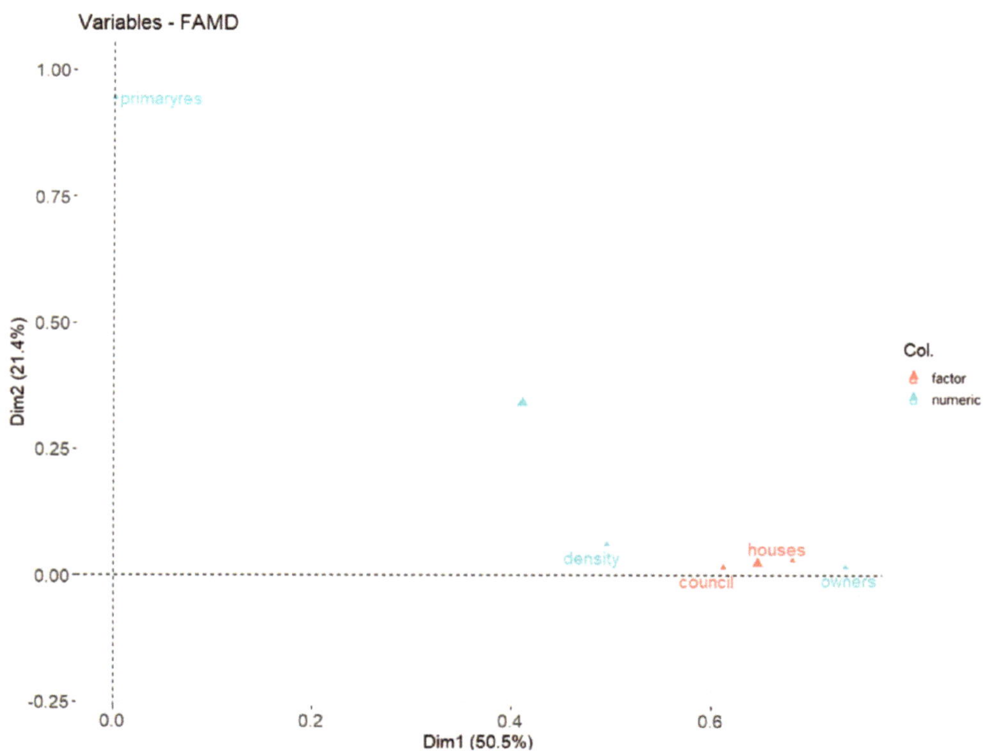

Figure 8.13 Numeric and factor housing variables, first two dimensions.

```
factoextra::fviz_famd_var(FAMDresult,
                choice="quali.var",
                axes=base::c(1:2),
                labelsize=5,
                pt.cex=24,
                ylim=base::c(-0.2,1),
                repel=TRUE,
                col.var=base::c(base::rep("houses",2),
base::rep("council",2))) +
    ggplot2::theme(text = ggplot2::element_text(size = 14),
           axis.title = ggplot2::element_text(size = 14),
           axis.text = ggplot2::element_text(size = 14))
```

One may also print a bar chart of contributions to the first dimension by both quantitative and qualitative variables. This is Figure 8.15.

```
factoextra::fviz_contrib(FAMDresult,
choice =  "var",
axes = 1) +
    ggplot2::theme(text = ggplot2::element_text(size = 14),
           axis.title = ggplot2::element_text(size = 14),
           axis.text = ggplot2::element_text(size = 14))
```

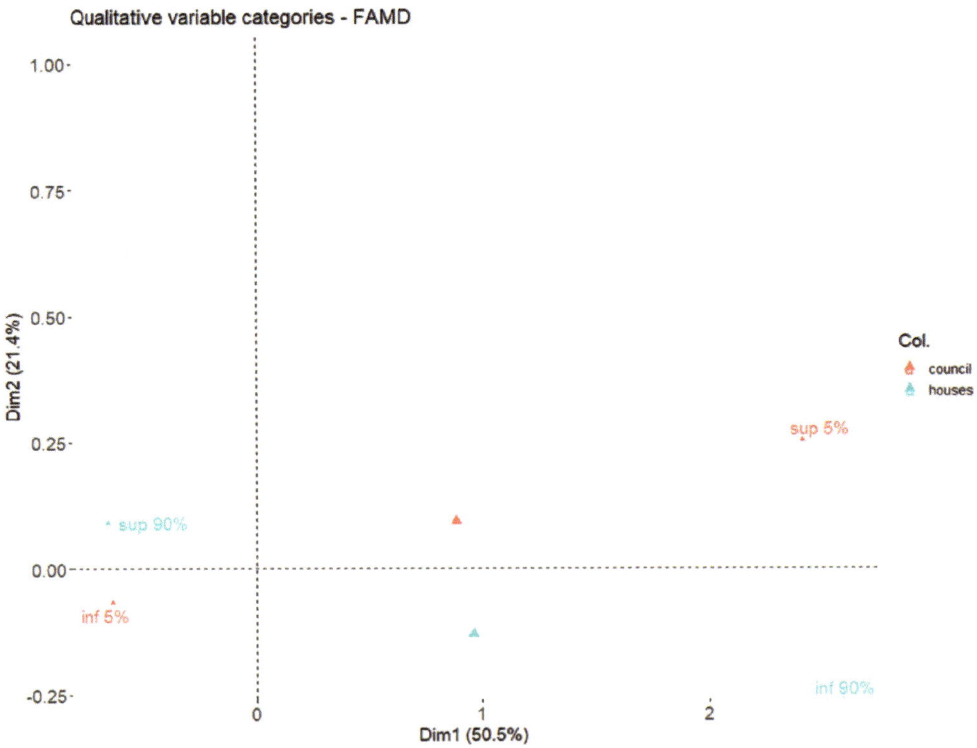

Figure 8.14 Levels of categorical variables plotted on the first two dimensions.

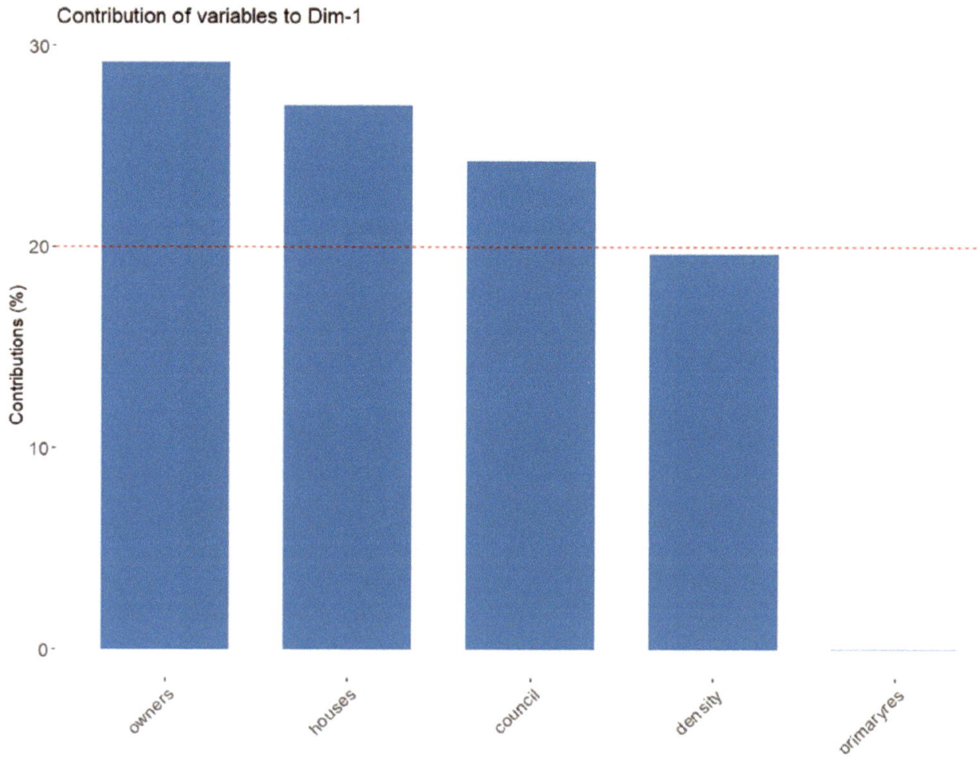

Figure 8.15 Contributions of numeric and categorical variables to the first dimension.

In Figure 8.15, the horizontal red dotted line is the level of contribution if all variables made equal contributions. It is seen that "primaryres" contributes little to dimension 1. The "density" variable is substantial but below average. The left-most three variables make above-average contributions to the construction of the first dimension. Contributions to the second dimension could be visualized simply by using `axes=2`. Contributions to both the first and second dimensions could be made by using `axes = c(1:2)`.

Visualization of observations

It is, of course, possible to visualize observations (here, French cities) as well as variables. This is done with the `fviz_famd_ind()` command and is sown in Figure 8.16. Congestion of points suppresses many labels but those which are high on dimension 1 and/or low on dimension 2 are shown. If labels are not showing, click "Zoom" in the "Plots" tab.

```
p <- factoextra::fviz_famd_ind(FAMDresult,
    col.ind = "cos2",
    axes=base::c(1:2),
    gradient.cols = base::c("blue", "yellow3", "red"),
    repel = TRUE)
```

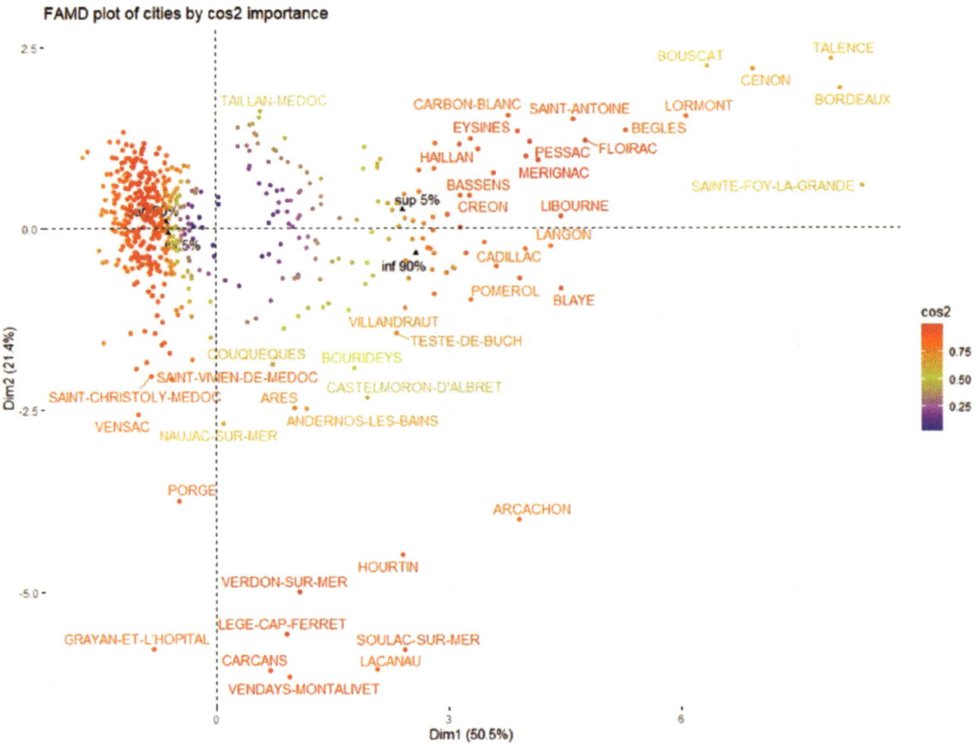

Figure 8.16 Observations plotted on the first two dimensions and color-coded by cos2 importance.

```
# Change plot title
p$labels$title <- "FAMD plot of cities by cos2 importance"
# Display figure
base::print(p)
```

To see actual cos2 importance values of the top and bottom ten cities, ask for the ind$cos2 element as below.

```
# Put individual-level data in "ind".
# The "ind" object is of classes " 'factoextra'  'famd'  famd_ind'  ".
ind <- factoextra::get_famd_ind(FAMDresult)

# List top 10 cities  by cos2 (importance)
utils::head(sort(ind$cos2[,1],decreasing=TRUE), 10)
[Output:]
    PONDAURAT      LIBOURNE SAINT-ANDRONY        AILLAS       LAGORCE
    0.9710840     0.9661642     0.9654504     0.9566490     0.9557928

        BAYAS        ILLATS      MERIGNAC   MONTAGOUDIN         BAGAS
    0.9543805     0.9463752     0.9458114     0.9449562     0.9440060
```

```
# List bottom 10 cities  by cos2 (importance)
utils::tail(sort(ind$cos2[,1],decreasing=TRUE), 10)
[Output:]
                  CARCANS          GRIGNOLS  SAINT-ANTOINE-DU-QUEYRET
             0.0118204143      0.0050680359              0.0043796175

  SAINT-MARTIN-LACAUSSADE SAINT-MICHEL-DE-CASTELNAU           SOUSSAC
             0.0027069362              0.0018629016      0.0008710021

           NAUJAC-SUR-MER           JUILLAC         LESTIAC-SUR-GARONNE
             0.0007948228      0.0005903785              0.0003919649

                    AUROS
             0.0003657487
```

In summary, the `FAMD()` command is a versatile tool for analyzing mixed data. As with other factor analysis and dimension reduction programs, there are many more possibilities, options, and commands associated with the `FAMD()` command than can be illustrated here. Type `help(FAMD)` or view online documentation for the "factoextra" package for more information.

Mixed data with generalized low-rank models (GLRM)

Introduction

A GRLM implements dimension reduction for mixed data, which may include binary, categorical, ordinal, and numeric types. GLRMs are a generalization of PCA and matrix factorization. Using a GRLM one may reduce high-dimensional raw data, even "Big Data", into a much smaller number of low-dimensional constructs called "archetypes". Archetypes are analogous to factors or components in oblique factor analysis. In addition to dimension reduction, GRLMs may be used for cluster analysis and imputation of missing data. For further information, see Udell et al. (2016).

In R, the "h2o" package is used to implement GLRM but also supports other machine learning procedures.[1] The "h2o" package is authored by Erin LeDell and others and according to its online package description, supports

> parallelized implementations of many supervised and unsupervised machine learning algorithms such as Generalized Linear Models (GLM), Gradient Boosting Machines (including XGBoost), Random Forests, Deep Neural Networks (Deep Learning), Stacked Ensembles, Naive Bayes, Generalized Additive Models (GAM), ANOVA GLM, Cox Proportional Hazards, K-Means, PCA, Maximum R GLM (maxrglm), Word2Vec, as well as a fully automatic machine learning algorithm (H2O AutoML).

In this section, however, we discuss only h20's GLRM algorithm.

Let the original dataset "A" have m observations as rows for n variables as columns. GLRM decomposes A into two numeric matrices, X and Y. The X matrix still has the same m rows of observations but has condensed the variables into k columns, loosely analogous to a factor score matrix with observations as rows, factors as columns, and scores as cell entries. Put another way, each row of X is an X archetype which represents an observation (row) of the raw data in A as projected into the k-reduced dimensions in the feature space constituted by the Y matrix, discussed below. The researcher selects the value of k, selecting the number of desired factors to extract in factor analysis. It is typical that k is a number much less than either m (the observations) or n (the original variables). The lower the k, the more the data compression during dimension reduction.

The Y matrix has k rows, each representing a condensed set of variables, where k is set by the user. The Y matrix has n columns, representing the original variables in A. There is one column per numeric variable plus one column for each level of each categorical variable. The Y matrix represents the data after dimension reduction. Each row of the Y matrix may be thought of as representing the most important combination of features from the training data. These condensed features are termed "archetypes". Put another way, each row of Y is a Y archetype based on the original columns of the raw dataset A. Like factors or components, Y archetypes may be manipulated, visualized, and incorporated into various statistical operations.

GLRM uses an iterative estimation process based on "alternating minimization". This means that after the archetype values of X and Y are initialized to random values, a gradient descent method is used to adjust these values, based on minimizing some loss function (quadratic loss is the default, discussed below). Iterations continue until some specified or default stopping condition is reached.

Note that it is also possible to reconstruct the original dataset from the X and Y matrices with minimal loss and doing so will impute missing values automatically. This reconstruction is based on the matrix product XY, giving a matrix of rank k.

Example data and setup

We again use the same "housing" data frame as for the PCAmixdata and FAMD packages described in the previous two sections. This data frame contains three continuous and two categorical variables pertaining to housing in the Gironde district, France's largest, for 542 municipalities.

```
base::setwd("C:/Data")              # Set the working directory
base::library(PCAmixdata)           # Contains the gironde data
utils::data(gironde)                # Invoke the gironde data
housing <- gironde$housing          # Create the housing data frame
```

Optionally, housing may be saved to the working directory as a comma-separated values file.

```
utils::write.csv(housing, file="housing.csv")
# Load the previously-installed package "h2o", which has functions
to implement GLRM.2
install.packages("h2o")
base::library(h2o)
```

Note: One must have 64-bit Java SE SDK installed or the `h2o.init()` command below will generate an error message instructing one to download the latest Java SE JDK from the following URL: https://www.oracle.com/technetwork/java/javase/downloads/index.html. There are numerous online tutorials on the installation of Java. The steps below worked for this author at this writing, when Java 18 was the most recent version:

1 Go to https://www.oracle.com/java/technologies/downloads/
2 Click on the Windows tab
3 Click on "x64 installer" (https://download.oracle.com/java/18/latest/jdk-18_windows-x64_bin.exe). This saves a file to "C:\Users\...\Downloads\jdk-18_windows-x64_bin.exe"
4 Clicking on this .exe file opens the Installation Wizard for Java SE Development Kit 18. Accept the defaults in the wizard and click "Close" at the end.
5 In Java is successfully installed, one should be able to issue the command `h2o::h2o.init()`.

Next we initialize h2o for a single-user local machine. Type `help(h2o.init)` for initialization options for more complex environments. More information on initialization is also at https://docs.h2o.ai/h2o/latest-stable/h2o-docs/starting-h2o.html.

```
h2o::h2o.init()
(Partial output:)
   Connection successful!
   ...

# Convert the housing data frame into an object of class "H2OFrame".
# Note that only the first 6 of the 542 French municipalities are
shown.
housing1 <- h2o::as.h2o(housing)

housing1
[Output:]
      density primaryres   houses owners council
   1  131.70       88.77  inf 90%  64.23  sup 5%
   2   21.21       87.52  sup 90%  77.12  inf 5%
   3  531.99       94.90  inf 90%  65.74  sup 5%
   4  101.21       93.79  sup 90%  66.54  sup 5%
   5  551.87       62.14  inf 90%  71.54  inf 5%
   6   63.82       81.02  sup 90%  80.54  inf 5%

   [542 rows x 5 columns]

base::class(housing1)
[Output:]
   [1] "H2OFrame"
```

Alternatively, import the saved housing.csv file into H2OFrame, as housing2. Again, note that only the first 6 of the 542 French municipalities are shown. Note that this method creates a new column called C1, containing the rownames. Thus, housing1 corresponds to housing2[,2:6].

```
housing2 <- h2o::h2o.importFile("C:/Data/housing.csv")

housing2
[Output:]
                     C1 density primaryres   houses owners council
  1               ABZAC  131.70      88.77  inf 90%  64.23  sup 5%
  2              AILLAS   21.21      87.52  sup 90%  77.12  inf 5%
  3 AMBARES-ET-LAGRAVE  531.99      94.90  inf 90%  65.74  sup 5%
  4               AMBES  101.21      93.79  sup 90%  66.54  sup 5%
  5 ANDERNOS-LES-BAINS  551.87      62.14  inf 90%  71.54  inf 5%
  6             ANGLADE   63.82      81.02  sup 90%  80.54  inf 5%

  [542 rows x 6 columns]

base::class(housing2)
[Output:]
    [1] "H2OFrame"

# Convert categorical variables to factors for housing1, which we
use below.
housing1$houses <- h2o::as.factor(housing1$houses)
housing1$council <- h2o::as.factor(housing1$council)
```

Split the housing1 dataset into train (training) and test (validation) sets. The `ratios=` option sets the approximate split, here 80% for the train object and 20% for the test object. However, the split is probabilistic, not exact, and is affected by the random seed.

```
housing_splits <- h2o::h2o.splitFrame(data = housing1, ratios = 0.8,
seed = 123)
train <- housing_splits[[1]]
test <- housing_splits[[2]]
```

Optionally, view some of train. We observe it has five columns (variables).

```
head(train)
[Output:]
    density primaryres   houses owners council
  1  131.70      88.77  inf 90%  64.23  sup 5%
  2   21.21      87.52  sup 90%  77.12  inf 5%
  3  531.99      94.90  inf 90%  65.74  sup 5%
  4  101.21      93.79  sup 90%  66.54  sup 5%
  5  551.87      62.14  inf 90%  71.54  inf 5%
  6   63.82      81.02  sup 90%  80.54  inf 5%
```

Running the GLRM model

```
# Define and train the model on the "train" subset:
h2o::h2o.init()
```

```
glrm_model1 = h2o::h2o.glrm(training_frame = train,
   seed=123,
   k = 5,
   loss = "Quadratic",
   regularization_x = "None",
   regularization_y = "None",
   gamma_x = 0,
   gamma_y = 0,
   max_iterations = 1000,
   svd_method = "GramSVD",
   recover_svd = FALSE,
   init = "SVD",
   transform = "STANDARDIZE")
```

Optionally, note that the columns of glrm_model1 are the same as for train:

```
glrm_model1@model$names
[Output:]
   [1] "density"   "primaryres" "houses"   "owners"      "council"
```

Explanation of h20.glrm parameters above:

- `training_frame=`: This is a required parameter. However, the entire dataset (e.g., housing for this example) may be the training frame if no validation set is wanted. The training_frame object must be an h2o object such as housing1, not an ordinary data frame such as housing.
- `validation_frame=`: This parameter is optional and may be omitted, as it is here. The named object must have the same number of rows as the training_frame. In this example, test has fewer rows than train and thus may not be entered here. Instead, test is used after the `newdata=` parameter in the `h20_predict()` command.
- `seed=`: Optionally set a random seed.
- `k=`: This sets the number of archetypes to be created in the dimension reduction process. Put another way, this sets the rank of the approximation. The default value is 1.
- `loss=`: This parameter must be one of: "Quadratic", "Absolute", "Huber", "Poisson", "Hinge", "Logistic", "Periodic". Quadratic is the default. There is also an override parameter, `loss_by_col=`, available for specifying the loss function by column (variable). Quadratic loss is the most common criterion and is based on minimizing the squared residuals, which represent error. Quadratic loss is sensitive to outliers. If outliers are thought to be problematic, Huber loss is often recommended.
- `regularization_x` and `regularization_y`: These constrain the X and Y matrices in specified ways, such as disallowing negative values. The default is "None". Other choices are "Quadratic", "L2", "L1", "NonNegative", "OneSparse", "UnitOneSparse", and "Simplex". The h2o package at this writing does not support a built-in tuning command. However, Boehmke and Greenwell (2019: 371–374) give code for doing a grid search for the best regularization values. Regularization may improve predictions, though, for the example data here, near-perfect performance is achieved without regularization.

- `gamma_x=` and `gamma_y=`: These are the regularization weights on the X and Y matrixes respectively. Defaults to 0.
- `max_iterations=`: Maximum number of iterations. The default is 1000.
- `svd_method=`: This specifies the method for singular value decomposition (SVD) of the matrix during initialization. Must be one of: "GramSVD", "Power", or "Randomized". Defaults to Randomized. GramSVD uses the Gram matrix followed by a local SVD. Power computes the SVD using the power iteration method, which in this writing is considered experimental. Randomized uses the randomized subspace iteration method, which in this writing is considered experimental and unstable.
- `recover_svd=`: This is a logical specifying whether to recover singular values and eigenvectors of XY. The default is FALSE.
- `init=`: This specifies the initialization mode. It must be one of "Random", "SVD", "PlusPlus", or "User". The default is "PlusPlus".
- `transform=`: This specifies the transformation of the training data and must be one of: "NONE", "STANDARDIZE", "NORMALIZE", "DEMEAN", "DESCALE". The default is "NONE".

GLRM results

Using the `summary()` command we first check to make sure the model converged on a solution. Recall that the model specification allowed a maximum of 1,000 iterations. In summary output, we see that convergence was reached at 115 iterations. Moreover, both numeric and categorical performance metrics show virtually no error. Not shown here, but convergence may be visualized with the simple command `plot(glrm_model)`, which will show a loss curve with the iteration number on the x axis and the objective function (here, minimizing error, ideally to 0) on the y axis. Note that the exact results depend on the observations selected for the train data subset used for glrm_model1.

```
h2o::summary(glrm_model1)
[Output:]
   Model Details:
   ==============

   H2ODimReductionModel: glrm
   Model Key:  GLRM_model_R_1650815989562_1
   Model Summary:
     number_of_iterations final_step_size final_objective_value
   1                  115         0.00003               0.00000

   H2ODimReductionMetrics: glrm
   ** Reported on training data. **

   Sum of Squared Error (Numeric):  3.830328e-11
   Misclassification Error (Categorical):  0
   Number of Numeric Entries:  1281
   Number of Categorical Entries:  854
```

Most other results are in attributes of glrm_model, listed below.

```
base::attributes(glrm_model1@model)
[Output:]
   $names
    [1] "names"
    [2] "original_names"
    [3] "column_types"
    [4] "domains"
    [5] "cross_validation_models"
    [6] "cross_validation_predictions"
    [7] "cross_validation_holdout_predictions_frame_id"
    [8] "cross_validation_fold_assignment_frame_id"
    [9] "model_summary"
   [10] "scoring_history"
   [11] "cv_scoring_history"
   [12] "reproducibility_information_table"
   [13] "training_metrics"
   [14] "validation_metrics"
   [15] "cross_validation_metrics"
   [16] "cross_validation_metrics_summary"
   [17] "status"
   [18] "start_time"
   [19] "end_time"
   [20] "run_time"
   [21] "default_threshold"
   [22] "help"
   [23] "iterations"
   [24] "updates"
   [25] "objective"
   [26] "avg_change_obj"
   [27] "step_size"
   [28] "archetypes"
   [29] "singular_vals"
   [30] "eigenvectors"
   [31] "representation_name"
   [32] "importance"
```

To retrieve any attribute, enter the model name, "@model", the dollar sign, and the attribute name. For instance, the commands and output below show the relative importance of the five requested dimensions. The first archetype accounts for about two-thirds of variance explained in the example data, the second accounts for about 15%, and so in descending order.

```
glrm_model1@model$importance
[Output:]
   Importance of components:
```

```
                              pc1      pc2      pc3
Standard deviation       1.852563 0.861185 0.636638
Proportion of Variance 0.685754 0.148189 0.080986
Cumulative Proportion  0.685754 0.833943 0.914929
                              pc4      pc5
Standard deviation       0.514162 0.401737
Proportion of Variance 0.052823 0.032248
Cumulative Proportion  0.967752 1.00000
```

The following block of R code uses the "ggplot2" package to visualize the relative importance of the archetypes, shown in Figure 8.17.

```
base::library(ggplot2)      # Leading visualization package in R
base::library(tidyr)            # Has the gather() function and more
base::data.frame(
     Archetype   = glrm_model1@model$importance |> seq_along(),
     Percent_Variance_Explained = glrm_model1@model$importance %>%
.[2,] |> base::unlist(),
     Cumulative_Variance_Explained = glrm_model1@model$importance
%>% .[3,] |> base::unlist()
 ) |>
     tidyr::gather(metric, variance_explained, -Archetype) |>
     ggplot2::ggplot(aes(Archetype, variance_explained)) +
     ggplot2::geom_point(color = "blue",size=3) +
     ggplot2::facet_wrap(~ metric, ncol = 1, scales = "free") +
ggplot2::theme(axis.title=ggplot2::element_text
(size=14,face="bold"))
```

The $archetypes element also may be used to see how the variables load on the five requested archetypes, which are GLRM dimensions. Keep in mind that glrm_model1 is for the training data, not the full data.

```
glrm_model1@model$archetypes
[Output:]
   Archetypes:
         houses_inf_90 houses_sup_90 council_inf_5
   Arch1      3.798553     -3.432970     -2.361855
   Arch2      1.142080     -2.380286     -1.513151
   Arch3      1.119781     -0.498384      1.257642
   Arch4     -0.162993     -0.012722      0.716996
   Arch5     -0.706374      1.114703     -3.516388
         council_sup_5   density primaryres    owners
   Arch1      2.402567  2.014170   0.328172 -1.822090
   Arch2      0.208144 -2.605085  -0.304550  2.678778
   Arch3     -0.664734 -0.598525  -4.257363 -0.996789
   Arch4     -0.884770  2.679414  -0.798622  2.347064
   Arch5      3.754976  0.084164  -0.879249  0.432906
```

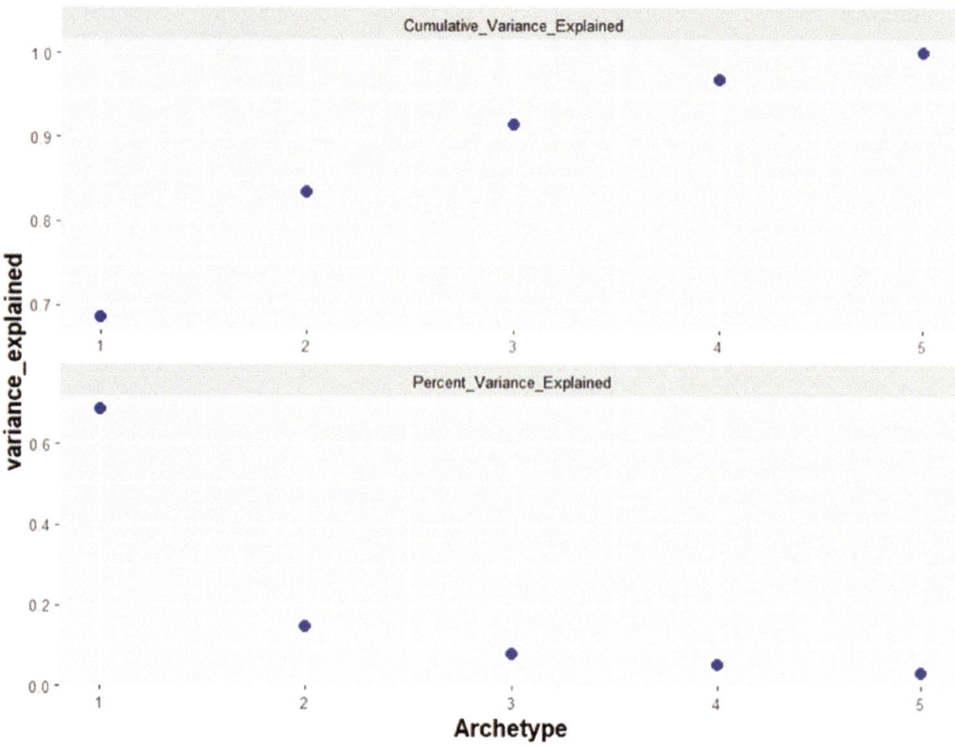

Figure 8.17 Variance explained by model archetype dimensions.

The archetypes are correlated, as shown below. In this sense, GLRM results are more like oblique than orthogonal factor analysis. Results will be orthogonal and approach PCA results only if the data are all numeric, all relationships are linear, and all relationships are orthogonal in reality. By allowing correlated archetypes, the predictions are better but the meaning of each archetype is less clear than compared to orthogonal solutions. By the same token, loadings on the archetypes do not form a simple factor (archetype) structure and therefore the meaning of any given archetype is less clear.

```
stats::cor(glrm_model1@model$archetypes)
[Output:]
                houses_inf_90 houses_sup_90 council_inf_5
   houses_inf_90    1.00000000   -0.91356404   -0.09207861
   houses_sup_90   -0.91356404    1.00000000    0.10126821
   council_inf_5   -0.09207861    0.10126821    1.00000000
   council_sup_5    0.07856617    0.05377224   -0.93722516
   density          0.12731070    0.08260068    0.10270089
   primaryres       0.24294987   -0.41214092   -0.65954630
   owners          -0.60760791    0.24499084    0.12510260
```

```
               council_sup_5     density primaryres     owners
houses_inf_90     0.07856617  0.12731070  0.2429499 -0.6076079
houses_sup_90     0.05377224  0.08260068 -0.4121409  0.2449908
council_inf_5    -0.93722516  0.10270089 -0.6595463  0.1251026
council_sup_5     1.00000000  0.06540957  0.4588262 -0.3810680
density           0.06540957  1.00000000  0.2416753 -0.2711450
primaryres        0.45882622  0.24167532  1.0000000  0.2605053
owners           -0.38106802 -0.27114499  0.2605053  1.0000000
```

Archetype loadings may be used to plot variables in low-dimensional (archetype) space as shown in Figures 8.18 and 8.19. The former figure plots all variables against archetype 1. The latter plots all variables against both archetypes 1 and 2 simultaneously.

```
# Plotting features for a given archetype such as Archetype 1.
# This is R code for creating Figure 8.18:
base::library(dplyr)              # Utilities, including the
mutate() function
base::library(ggplot2)            # Leading R graphics package
varnumber <- base::ncol(glrm_model1@model$archetypes)
p1 <- base::t(glrm_model1@model$archetypes) |>
  base::as.data.frame() %>%
  dplyr::mutate(feature = row.names(.)) |>
  ggplot2::ggplot(aes(Arch1, stats::reorder(feature, Arch1))) +
  ggplot2::geom_point(color = base::c(1:varnumber),size=5) +
  labs(x ="Variable Loadings on Archetype 1", y = "Variables") +

ggplot2::theme(axis.title=ggplot2::element_text(size=14,-
face="bold"))
pic1 <- p1 + ggplot2::theme(axis.text.x = ggplot2::element_text(-
face="bold", color="black", size=12), axis.text.y = ggplot2::ele-
ment_text(face="bold", color="black",size=12, angle=45))
pic1 + ggplot2::labs(title = "Variable Loadings for Archetype 1",
subtitle = "Gironde Housing Data")

# Plotting variables in feature space defined by the first two
archetypes.
# This is R code for creating Figure 8.19:
base::library(dplyr)              # Utilities, including the mu-
tate() function
base::library(ggplot2)            # Leading R graphics package
p2 <- base::t(glrm_model1@model$archetypes) |>
    base::as.data.frame() %>%
    dplyr::mutate(feature = row.names(.)) |>
    ggplot2::ggplot(aes(Arch1, Arch2, label = feature)) +
    ggplot2::geom_text(color="blue", size = 6, vjust=1.0) +
```

```
    ggplot2::geom_point(color = "black",size=3) +
    ggplot2::labs(x =" Archetype 1", y = "Archetype 2") +
    ggplot2::theme(axis.title =
ggplot2::element_text(size=18,face="bold"))
p2 + ggplot2::labs(title = "Gironde Housing Variables in Feature
Space", subtitle = "defined by Archetypes 1 and 2")
```

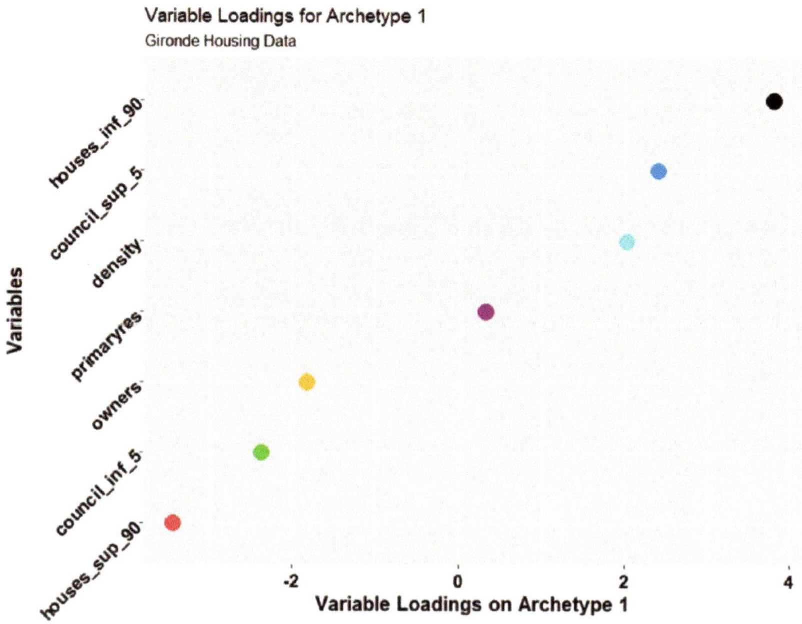

Figure 8.18 Feature loadings on archetype 1.

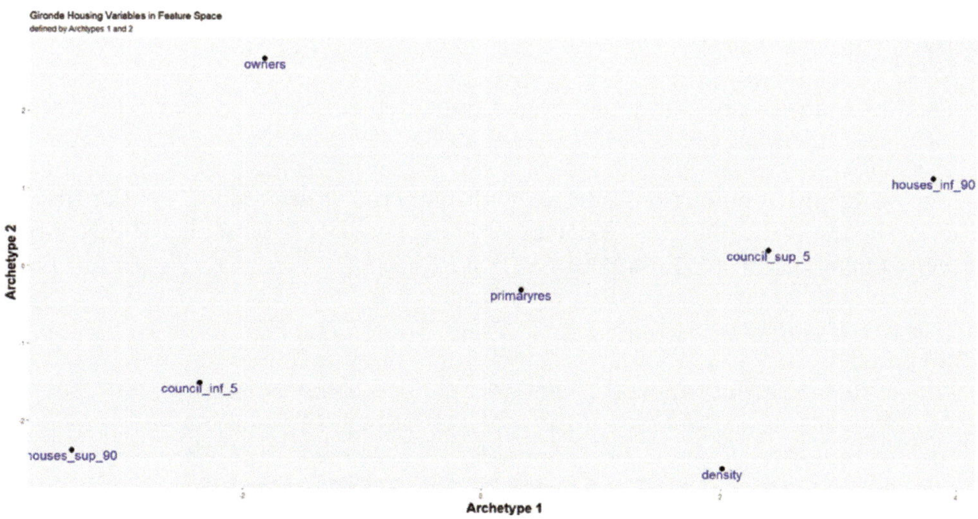

Figure 8.19 Gironde housing variables in Arch1 and Arch2 feature space.

GLRM performance and predictions

In this section, we evaluate performance of the model. For both numeric and categorical variables, error is very low. The metrics are sum of squared error for numeric variables and misclassification error for categorical variables.

```
housing_perf <- h2o::h2o.performance(glrm_model1)

housing_perf
[Output:]
   Sum of Squared Error (Numeric):  3.830328e-11
   Misclassification Error (Categorical):  0
   Number of Numeric Entries:  1281
   Number of Categorical Entries:  854
```

We may then compare predicted values (predicted for the "test" validation set) with observed values.

```
housing_pred <- h2o::h2o.predict(glrm_model1, newdata = test)

housing_pred
[Output for first six cities in the test dataset:]
     reconstr_density reconstr_primaryres reconstr_houses
   1        2.5165147         -4.76628543         inf 90%
   2       -0.2509989          0.24342091         sup 90%
   3       -0.1593622          0.46919330         sup 90%
   4       -0.1774303          0.67971080         sup 90%
   5       -0.2901275         -1.03900696         sup 90%
   6        0.3076841          0.04002236         inf 90%
     reconstr_owners reconstr_council
   1    -1.409069423           sup 5%
   2     0.213904024           inf 5%
   3     0.003090747           inf 5%
   4     0.514240747           inf 5%
   5     0.405464947           inf 5%
   6    -1.293555299           inf 5%
```

Below we compare predictions (above) with observed scaled values, for the first six cities in the test (validation) set. We must scale the raw data because the GLRM model requested `transform = "STANDARDIZE"`.

```
utils::head(scale(test),6)
 [Output:]
       density primaryres    houses      owners council
  1  2.7296239 -4.2003301  inf 90% -1.25195838  sup 5%
  2 -0.2459237  0.2996236  sup 90%  0.21097862  inf 5%
  3 -0.1473986  0.5024229  sup 90%  0.02095299  inf 5%
  4 -0.1668250  0.6915196  sup 90%  0.48170006  inf 5%
  5 -0.2879936 -0.8523134  sup 90%  0.38365031  inf 5%
  6  0.3547555  0.1169214  inf 90% -1.14783475  inf 5%
```

Using GLRM to impute missing values

First, we check to see if our data have missing values. There are none

```
base::sum(base::is.na(housing))
[Output:]
   [1] 0
```

Since the example data have no missing values, for instructional purposes, we add some to a copy of housing1.

```
# List the first three observations.
utils::head(housing,3)
[Output:]
                     density primaryres   houses owners council
   ABZAC              131.70      88.77  inf 90%  64.23  sup 5%
   AILLAS              21.21      87.52  sup 90%  77.12  inf 5%
   AMBARES-ET-LAGRAVE 531.99      94.90  inf 90%  65.74  sup 5%

# Make a copy and assign missing to owners and council for first
three observations
# There are now 6 missing values in housing2.
housing2 < housing[,]
housing2[1:3,4:5]<- NA
utils::head(housing2,3)
[Output:]
                     density primaryres   houses owners council
   ABZAC              131.70      88.77  inf 90%     NA    <NA>
   AILLAS              21.21      87.52  sup 90%     NA    <NA>
   AMBARES-ET-LAGRAVE 531.99      94.90  inf 90%     NA    <NA>

base::sum(base::is.na(housing2))
[Output:]
   [1] 6

# Create an h20 version of housing2, calling it "housing_m" (for
missing values).
housing_m <- h2o::as.h2o(housing2)
```

As the data have now changed, we create a new GLRM model for housing_m. For instructional simplicity, here we train on the entire housing_m dataset in actual research the researcher may prefer to split it into train_m and test_m sets, let train_m be the training frame and let test_m be the validation set for imputation results.

```
glrm_model_m = h2o::h2o.glrm(training_frame = housing_m,
   seed=123,
   k = 5,
   loss = "Quadratic",
   regularization_x = "None",
   regularization_y = "None",
```

```
    gamma_x = 0,
    gamma_y = 0,
    max_iterations = 1000,
    svd_method = "GramSVD",
    recover_svd = FALSE,
    init = "SVD",
    transform = "STANDARDIZE")
```

We are now ready to reconstruct (impute) the data from X and Y archetypes in GLRM.

```
reconstructed_housing_m = h2o::h2o.reconstruct(glrm_model_m, hous-
ing_m, reverse_transform = TRUE)
```

By default, GLRM will have added a "reconstr_" prefix to all names (e.g., reconstr_density). Here we restore the original, shorter names (e.g., density) in reconstructed_housing_m.

```
base::names(reconstructed_housing_m) = base::names(housing1)
```

We also convert from reconstructed_housing_m from an h2o object back into an R data frame.

```
# Before conversion:
base::class(reconstructed_housing_m)
[Output:]
    [1] "H2OFrame"

* Convert to an R data frame:
reconstructed_housing_m = base::as.data.
frame(reconstructed_housing_m)

# After conversion:
base::class(reconstructed_housing_m)
[Output:]
     [1] "data.frame"
```

Finally, we view the results:

```
* Original data, first three observations:
utils::head(housing,3)
[Output:]
                         density primaryres   houses owners council
    ABZAC                 131.70      88.77  inf 90%  64.23  sup 5%
    AILLAS                 21.21      87.52  sup 90%  77.12  inf 5%
    AMBARES-ET-LAGRAVE    531.99      94.90  inf 90%  65.74  sup 5%
```

```
# Data after assigning missing values:
utils::head(housing2,3)
[Output:]
                     density primaryres   houses owners council
   ABZAC              131.70      88.77  inf 90%     NA    <NA>
   AILLAS              21.21      87.52  sup 90%     NA    <NA>
   AMBARES-ET-LAGRAVE 531.99      94.90  inf 90%     NA    <NA>

# Data after GLRM imputation:
utils::head(reconstructed_housing_m,3)
[Output:]
     density primaryres   houses   owners council
   1  131.70      88.77  inf 90% 64.26924  sup 5%
   2   21.21      87.52  sup 90% 72.09100  inf 5%
   3  531.99      94.90  inf 90% 64.40312  sup 5%
```

For these data, GLRM reconstruction has imputed the missing values with a high but not perfect degree of accuracy. Had we split the data into training and test (validation) sets, accuracy of imputation would tend to be lower for the test set than for the training set or for the full dataset used here. As with other forms of imputation, the variance of imputed variables will tend to be lower than for full information data. It may also be necessary to check that imputed values are all within the range of observed data.

Using GLRM for cluster analysis

```
# Load needed libraries (previous installation is assumed).
base::library(h2o)          # Has functions to implement GLRM
h2o::h2o.init()             # Initialize h20 if not already done
base::library(data.table)   # Supports the as.data.table() function.
base::library(factoextra)   # Supports fviz(cluster) for
visualization
base::library(ggplot2)      # Leading R graphics package
base::library(purrr)        # Supports the map_dbl () function

# Get data
base::setwd("C:/Data")      # Set the working directory
base::library(PCAmixdata)   # Contains the gironde data
utils::data(gironde)        # Invoke the gironde data
housing <- gironde$housing # Create the housing data frame

# Add columns needed later
housing$id <- base::seq.int(base::nrow(housing))
housing$town <- base::rownames(housing)
housing$cluster <- 1
```

```
housing1 <- h2o::as.h2o(housing)

# We create a GLRM model for the news housing1,  with all 542 obser-
vations  (not just train).
# We use just the archetype columns, which are the first five.
model_all <- h2o::h2o.glrm(training_frame = housing1,
   seed=123,
   cols=1:5,
   k = 5,
   loss = "Quadratic",
   regularization_x = "None",
   regularization_y = "None",
   gamma_x = 0,
   gamma_y = 0,
   max_iterations = 1000,
   svd_method = "GramSVD",
   recover_svd = FALSE,
   init = "SVD",
   transform = "STANDARDIZE")

# Convert model_all  to data.table format.
glrm_table = data.table::as.data.table(h2o::h2o.getFrame(model_all@
model$representation_name))

# k-mean clustering (k Is the requested number of clusters).
k= 10
tot_withinss = purrr::map_dbl(1:k, function(k)
   {
   model <- stats::kmeans(x = glrm_table, centers = k)
   model$tot.withinss
   })

# Plot error by number of clusters. This is Figure 8.20.
error_df <- base::data.frame(k = 1:k, tot_withinss = tot_withinss)
p1 <- ggplot2::ggplot(error_df,
   ggplot2::aes(x = k, y = tot_withinss)) +
   ggplot2::geom_line() +
   ggplot2::scale_x_continuous(breaks = 1:k) +
   ggplot2::labs(x="Number of Clusters", y="Total Within Sum of
Squares", title="Error by Number of Clusters") +
   ggplot2::theme(axis.title= ggplot2::element_text(size=14,-
face="bold"))

p1
```

We select a two-cluster solution as this gave the best separation of clusters. However, as shown in Figure 8.20, this was not the lowest-error solution. To achieve lower error, as with the 6-cluster solution, one must allow overlapping clusters.

```
# Create the two-cluster kmeans model.
# Only the first five columns (the archetype values) are used.
base::set.seed(123)
model_kmeans = stats::kmeans(glrm_table[,1:5], centers=2, nstart =
25)

# List cluster membership numbers for cities from the glrm_table
object.
clusterMembershipNumber <- model_kmeans$cluster
clusterMembershipNumber
[Partial output:]
     [1] 1 2 1 1 2 2 2 2 1 2 2 2 1 1 2 2 2 2 2 2 2 2 2
    [24] 2 2 2 2 2 2 2 2 1 2 2 2 1 2 2 1 2 1 2 2 2 2 2
  …
   [507] 2 2 1 2 2 2 2 1 2 1 2 2 2 1 2 1 2 2 2 2 2 2 1
   [530] 1 2 2 2 1 1 2 1 2 2 2 1 1
```

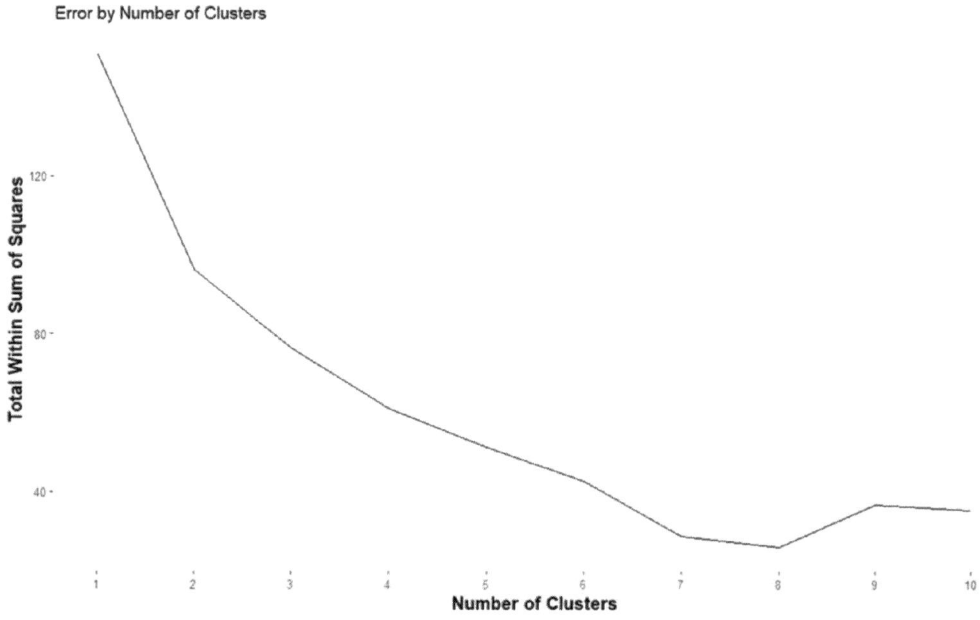

Figure 8.20 Error by a number of clusters.

```
# Setup for listing cases by cluster membership number
# We go back to the original R data frame, housing (not an h2o
object).
# We add variables to glrm_table for observation number ($id), clus-
ter membership number.
# ($cluster), and the name of the municipality ($town).
# Note we recode $cluster with correct cluster membership numbers.
glrm_table$id <- base::seq.int(base::nrow(housing))
glrm_table$cluster <- clusterMembershipNumber
glrm_table$town <- base::rownames(housing)
```

We now visualize the clusters. Points are labeled by observation id numbers from glrm_table2.

Axis values are standardized versions of glrm_table2 archetype values and may be displayed by calling p$data after p0 <- factoextra::fviz_cluster()below is run.[3]

```
# Run fiz_cluster().
# Set point labels to row id numbers.  For no labels, use
geom="point".
p <- factoextra::fviz_cluster(model_kmeans, data = glrm_table[,1:5],
geom = base::c("point","text"),main = "Cluster Plot on First Two
Standardized Archetypes")

# Set custom axis labels.
p$labels$x<-"Archetype 1"
p$labels$y<-"Archetype 2"

# Display Figure 8.21.
p
```

To look up town names corresponding to id numbers in Figure 8.21, we may list observations alphabetically by town grouped by cluster number. The id variable, also shown, was in numerical order in the original data. Output below shows the head (all cluster 1) and tail (all cluster 2) of the townByCluster data frame, which may also be viewed (`View(townByCluster)`).

```
temp1 <- glrm_table
temp2 <-  temp1[, c("id", "town","cluster")]
townByCluster <- temp2[base::order(temp2$cluster, temp2$town)]
townByCluster
[Partial output:]

    id               town cluster
1:   1              ABZAC       1
2:   3 AMBARES-ET-LAGRAVE       1
3:   5 ANDERNOS-LES-BAINS       1
4:   9           ARCACHON       1
```

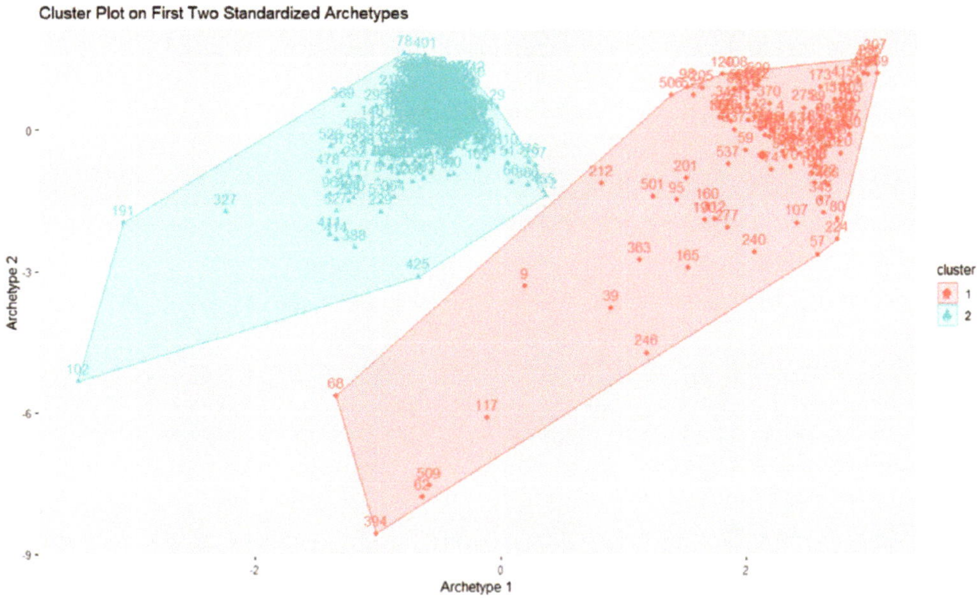

Figure 8.21 Gironde municipalities plotted on first two standardized archetypes.

```
  5:  11                ARES       1
 ---
538: 536 VILLENAVE-DE-RIONS       2
539: 538          VILLENEUVE       2
540: 539            VIRELADE       2
541: 540              VIRSAC       2
542: 542         MARCHEPRIME       2
```

The R code below will list out all 542 rows:

```
base::options(max.print = 10000)
townByCluster |> base::print(n = Inf)
[Output not shown]
```

Above, for instructional simplicity, we displayed the visualization for two clusters. As shown in Figure 8.20, the two-cluster solution has high error even though it has excellent separation. The researcher might prefer the seven-cluster solution, which is low in error but has poor separation for these data. As mentioned earlier, solutions with more clusters than two generate overlapping clusters. Overlapping clusters, in turn, reflect the fact that GLRM returns a solution with correlated archetypes. As with other oblique factor analytic methods, correlated dimensions make improve predictions but make interpretation of dimensions more difficult or even impossible. Figure 8.22 is the seven-cluster visualization for the example data. In this visualization, only cluster 1, with its seven towns, is clearly separated from other clusters when visualized on archetypes (dimensions) 1 and 2. Each other cluster overlaps with at least one additional cluster. Nonetheless, the k-means clustering algorithm assigns each town to one and only one of the seven clusters and these cluster membership numbers may be listed.

```
# R code for visualization in seven clusters.
base::set.seed(123)
model_kmeans7 = stats::kmeans(glrm_table[,1:5], centers=7, nstart =
25)
# Compute  cluster membership numbers for seven cluster output.
clusterMembershipNumber7 <- model_kmeans7$cluster
# Set point labels to row id numbers.  For no labels, use
geom="point".
p7 <- factoextra::fviz_cluster(model_kmeans7, data = glrm_ta-
ble[,1:5], geom = base::c("point","text"),main = "Cluster Plot on
First Two Standardized Archetypes, 7-Cluster Visualization")
# Set custom axis labels.
p7$labels$x<-"Archetype 1"
p7$labels$y<-"Archetype 2"
# Display Figure 8.22.
p7

# List towns by cluster membership number.
glrm_table7 <- glrm_table
glrm_table7$cluster <- clusterMembershipNumber7
townByCluster7 <- glrm_table7[base::order(glrm_table7$cluster,
glrm_table7$town)]
# Just list last three columns (id, cluster, town)
townByCluster7[,6:8]
[Partial output:]
         id cluster          town
     1:  39       1        BEGLES
     2:  62       1      BORDEAUX
     3:  68       1       BOUSCAT
     4: 117       1         CENON
     5: 246       1      LORMONT
    ---
   538: 408       7 SAINTE-HELENE
   539: 506       7 TAILLAN-MEDOC
   540: 522       7       TRESSES
   541: 529       7         VERAC
   542: 535       7    VILLEGOUGE
```

Tailoring loss functions for variable type in GLRM

In the foregoing discussion, the "Quadratic" loss function was applied to all variables. However, it is possible to selectively apply other functions as appropriate to the data type. Binary variables, for instance, may be assigned the "Hinge" loss function.

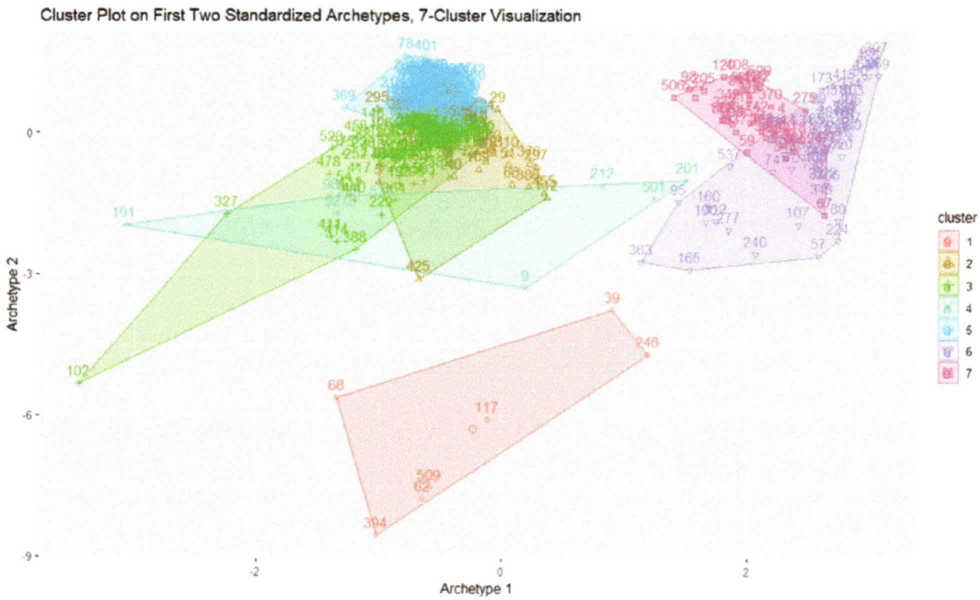

Figure 8.22 The seven-cluster visualization of GLRM results.

Step 1: We list the loss functions associated with each variable.

```
# The index of the first feature is always index = 0.
losses = base::data.frame(
   "index" = base::seq(base::ncol(housing1)) - 1,
   "variable" = base::names(housing1),
   "loss" = "Quadratic",
   stringsAsFactors = FALSE)

losses
[Output:]
      index   variable      loss
   1      0    density Quadratic
   2      1 primaryres Quadratic
   3      2     houses Quadratic
   4      3     owners Quadratic
   5      4    council Quadratic
   6      5         id Quadratic
   7      6       town Quadratic
   8      7    cluster Quadratic
```

Step 2: Assign variables to types. For housing1, there are two binary variables but no ordinal ones.

```
utils::str(housing1)
task = base::list(
  binary = base::c("houses", "council"),
  ordinal = NULL
)
```

```
task
[Output:]
   $binary
   [1] "houses"   "council"
   $ordinal
   NULL
```

Step 3: Customize loss for each variable type.

```
# Customize loss for binary variables.
losses$loss[losses$variable %in% task$binary] = "Hinge"

# Customize loss for ordinal variables. As there are no ordinal
variables, this has no effect.
# Had there been their loss function would be reassigned from "Qua-
dratic" to "Ordinal".
# Likewise, the researcher could make other loss type revisions in
the same manner.
losses$loss[losses$variable %in% task$ordinal] = "Ordinal"

# Review revised losses
losses
[Output:]
     index   variable      loss
   1     0     density Quadratic
   2     1 primaryres Quadratic
   3     2      houses     Hinge
   4     3      owners Quadratic
   5     4     council     Hinge
   6     5          id Quadratic
   7     6        town Quadratic
   8     7     cluster Quadratic
```

Step 4: Run the `h2o.glrm()` model with revised loss types. Using the `loss_by_col` and `loss_by_col_idx` options. This step is discussed in Kurka (2019).

Categorical PCA with `princals()`

Introduction

Categorical principal component analysis (CATPCA) is used when the items used to calculate the components are categorical. CATPCA in R is also called "nonlinear PCA" or "ordinal Princals". Results will differ from ordinary PCA, which assumes items are continuous. Categorical PCA in R is usually implemented using the `princals()` command from the "Gifi" package, named by Dutch researchers after Francis Galton's manservant (see Gifi, 1990).

Before proceeding, it might be asked if there is not also a CATPFA, using PFA rather than PCA, for categorical variables. If the researcher wishes to use PFA rather than

PCA but the variables are categorical, the usual procedure is to compute a polychoric correlation matrix and use this as input to `fa()` or some other function supporting PFA. Polychoric correlation input was discussed earlier in this chapter.

CATPCA implements PCA factor analysis for categorical variables, use of which would otherwise violate the data assumptions of ordinary PCA. When all variables are declared continuous numeric, CATPCA generates the same results as PCA. When all variables are declared multiple nominal, CATPCA yields "homogeneity analysis." When variables are different in measurement level, CATPCA is considered a type of "nonlinear principal component analysis". That is, in this context "nonlinear" is equivalent to "categorical".

The `princals()` program can emulate PCA of continuous variables, implement CATPCA for categorical or ordinal variables, or can handle mixed levels of measurement. Categorical variables may be transformed using various spline regression options in `princals()`. However, in the treatment below, we simply present default `princals()` procedure for ordinal data.

While CATPCA parallels PCA in logic, there are some differences. Perhaps the most important is that in CATPCA the researcher must specify the number of dimensions for the solution in advance, whereas this is not needed by some implementations of PCA. Of course, the researcher may explore alternatives by requesting multiple runs of CATPCA with different numbers of dimensions specified.

Example data

Example data, following Mair (2018), consist of the "Adult Self-Transcendence Inventory" (ASTI) items from Oregon State University psychologist Michael Levenson et al. (2005). These items were administered to 1,215 individuals but there are 1,129 observations in the ASTI data frame used below. There are 25 categorical (ordinal) personality items (listed below) plus two covariates (gender and group, where group is student or non-student). The items are grouped into five dimensions of 3 or more items each: self-knowledge and integration (SI), peace of mind (PM), non-attachment (NA, but labeled "NAT" below because NA is a reserved term in R), self-transcendence (ST), and presence in the here-and-now and growth (PG).

ASTI items are listed below:

ASTI1	I often engage in quiet contemplation. (PM; reversed)
ASTI2	*I feel that my individual life is a part of a greater whole. (ST)*
ASTI3	I don't worry about other people's opinions of me. (NA)
ASTI4	I feel a sense of belonging with both earlier and future generations. (ST)
ASTI5	My peace of mind is not easily upset. (PM)
ASTI6	My sense of well-being does not depend on a busy social life. (NA)
ASTI7	I feel part of something greater than myself. (ST)
ASTI8	My happiness is not dependent on other people and things. (NA; reversed)
ASTI9	I do not become angry easily. (PM)
ASTI10	I have a good sense of humor about myself. (SI; reversed)
ASTI11	I find much joy in life. (PG; reversed)
ASTI12	Material possessions don't mean much to me. (NA)
ASTI13	I feel compassionate even toward people who have been unkind to me. (ST)
ASTI14	I am not often fearful. (PG)

ASTI15	I can learn a lot from others. (PG)
ASTI16	I often have a sense of oneness with nature. (ST)
ASTI17	I am able to accept my mortality. (PG)
ASTI18	I often "lose myself" in what I am doing. (PG)
ASTI19	I feel that I know myself. (SI; reversed)
ASTI20	I am accepting of myself, including my faults. (SI; reversed)
ASTI21	I am able to integrate the different aspects of my life. (SI; reversed)
ASTI22	I can accept the impermanence of things. (PM; reversed)
ASTI23	I have grown as a result of losses I have suffered. (PG; reversed)
ASTI24	Whatever [good] I do for others, I do for myself. (ST; reversed)
ASTI25	Whatever [bad] I do to others, I do to myself. (ST)
gender	gender
group	student vs. non-student

Steps in `princals()` analysis

1 Setup (assumes library packages are already installed)

```
base::setwd("C:/Data")          # Declare the working directory
base::library(Gifi)                   ## Has the princals() function
base::library(MPsychoR)         ## Has the ASTI data

# Invoke the ASTI data frame, supplied as part of the MPsychoR
package.
utils::data(ASTI)
```

2 Check if the data seem factorable. Using Spearman rank correlation for ordinal data, we see the mean intercorrelation of the 25 personality items in the ASTI data frame is low (0.14).

```
base::mean(stats::cor(ASTI[,1:25], method="spearman"))
   [1] 0.1358729
```

Nonetheless, the overall measure of sampling adequacy (MSA) from the Kaiser-Meyers-Olkin (KMO) output is above the 0.60 cutoff discussed earlier for proceeding with factor analysis.

```
ASTIcorrmatrix <- stats::cor(ASTI[,1:25], method="spearman")

psych::KMO(ASTIcorrmatrix)
[Output:]
   Kaiser-Meyer-Olkin factor adequacy
   Call: psych::KMO(r = ASTIcorrmatrix)
   Overall MSA =  0.77
   MSA for each item =
    ASTI1  ASTI2  ASTI3  ASTI4  ASTI5  ASTI6  ASTI7
     0.81   0.65   0.76   0.82   0.82   0.66   0.63
    ASTI8  ASTI9 ASTI10 ASTI11 ASTI12 ASTI13 ASTI14
     0.66   0.74   0.85   0.82   0.70   0.70   0.83
```

```
ASTI15 ASTI16 ASTI17 ASTI18 ASTI19 ASTI20 ASTI21
  0.84   0.83   0.79   0.63   0.82   0.83   0.86
ASTI22 ASTI23 ASTI24 ASTI25
0.82   0.84   0.67   0.70
```

3 Obtain an initial `princals()` CATPCA solution for the entire ASTI data frame and estimate the number of needed factors to extract. We use only first 25 items in ASTI, which are the personality items without the covariates. By asking for ndim = 25, we get all possible dimensions/factors, which is 25 dimensions.

```
princalsResultAll <- Gifi::princals(ASTI[,1:25], ndim = 25,
ordinal=TRUE)
```

Eigenvalues are in the "evals" element of princalsResultAll. The Kaiser criterion (eigenvalues >= 1.0) would suggest an eight-factor solution is needed since the first eight dimensions all have eigenvalues of 1.0 or higher.

```
princalsResultAll$evals
[Output:]
    [1] 3.5160357 2.0920167 1.5893514 1.2975164
    [5] 1.2156333 1.0825345 1.0243047 1.0232990
    [9] 0.9497345 0.9442674 0.8840936 0.8686792
   [13] 0.8506978 0.8009058 0.7854826 0.7449582
   [17] 0.7194315 0.6787812 0.6510093 0.6410886
   [21] 0.5958521 0.5848159 0.5717594 0.5090014
   [25] 0.3787497
```

Alternatively, we can use a scree plot to estimate the number of needed dimensions. This requires a subjective judgment about where the "elbow" is in the scree curve. Based on Figure 8.23, the researcher might choose a seven-factor solution since this is a point where the curve levels.

```
base::plot(princalsResultAll$eval,
    pch=16,
    col="blue",
    cex=1.3,
    ylab = "Eigenvalue",
    xlab = "Number of Factors to Extract",
    main="Scree plot for ASTI variables",
    sub="The red line is the Kaiser criterion line at eig =
    1.0.")

graphics::abline(h=1.0, col="red")
```

4 As a preliminary step, we create separate data frames for each of the five subscales of the ASTI data.

```
SI <- ASTI[ ,base::c(10,19,20,21)]          ## self-knowledge and
integration
```

```
PM <- ASTI[ ,base::c(1,5,9,22)]              ## peace of mind
NAT <- ASTI[ ,base::c(3,6,8,12)]             ## non-attachment
# Note that NA is a reserved term in R.
ST <- ASTI[ ,base::c(2,4,7,13,16,24,25)]## self-transcendence
PG <- ASTI[ ,base::c(11,14,15,17,18,23)]## Here-and-now pres-
ence,  growth
```

5 In spite of the scree plot suggesting seven factors, in order to simplify analysis for instructional purposes, we create a new data frame with just two of the ASTI dimensions and subsequently ask for a two-dimensional CATPCA solution.

To do this, we create the "SIPM" data frame with items from the SI (self-knowledge/integration) and PM (peace of mind) subscales. Column names in SIPM will be labeled as "SI.ASTI10", for instance, for the ASTI10 variable, which is the first SI dimension item. All the SI items will begin with "SI" and all the PM items will begin with "PM."

```
SIPM <- base::data.frame(SI=SI, PM=PM)
# List the first three rows in SIPM:
utils::head(SIPM,3)
[Output:]
     SI.ASTI10 SI.ASTI19 SI.ASTI20 SI.ASTI21
  1          2         1         0         1
  2          2         1         2         1
  3          1         2         1         0
     PM.ASTI1 PM.ASTI5 PM.ASTI9 PM.ASTI22
  1         1        3        3         1
  2         2        3        3         1
  3         1        2        2         2
```

As SI and PM are supposed to be two ASTI subscales, we ask `princals()` for a 2-dimensional CATPCA solution and put the results in the object "princals_SIPM". This object is of data class "'princals' 'gifi'".

```
princals_SIPM <- Gifi::princals(SIPM, ndim=2,ordinal=TRUE)
```

6 Next we display the factor loadings for the princals_SIPM two-dimensional solution. These loadings are interpreted as in PCA.

```
princals_SIPM$loadings
[Output not shown because loadings also appear in summary() out-
put below:]
```

In the summary below, we see that all the SI items load heavily on component 1 except item 10, which loads moderately but without cross-loadings. However, the situation is more complex for the PM items. PM item ASTI9 loads heavily on component 2. PM item ASTI1 loads moderately on component 2. However, PM items ASTI5 and ASTI22 load more on component 1 than on component 2, and do so at a moderate level. Overall, the model has a cumulative variance accounted for (cumulative VAF) of 44.6% of the variation in the eight items, which is less than half the variance. We judge that two dimensions are insufficient to capture the variability of the two subscales, SI and PM.

```
base::summary(princals_SIPM)
[Output:]
   Loadings (cutoff = 0.1):
             Comp1  Comp2
   SI.ASTI10 -0.531  0.119
   SI.ASTI19 -0.658  0.292
   SI.ASTI20 -0.721  0.243
   SI.ASTI21 -0.601  0.275
   PM.ASTI5  -0.590 -0.343
   PM.ASTI22 -0.550
   PM.ASTI1  -0.249 -0.571
   PM.ASTI9  -0.336 -0.684

   Importance (Variance Accounted For):
                     Comp1   Comp2
   Eigenvalues      2.4228  1.1470
   VAF             30.2851 14.3375
   Cumulative VAF 30.2900 44.6200
```

7 Visualize the `princals()`_ SIPM result in Figure 8.23.

In the next step, we create Figure 4.2, which is a factor loadings plot for the SIPM data. While containing the same information as the loadings table above, it graphically shows that all eight items are negatively loaded on Component 1. The

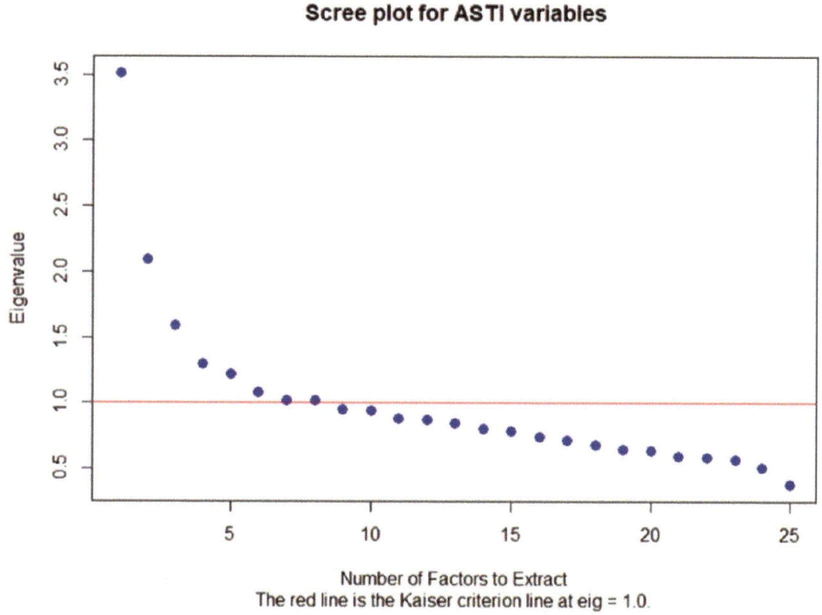

Figure 8.23 Scree plot for ASTI variables.

PM items are differentiated from the SI items because they are negatively loaded on Component 2 while the SI items are positively loaded.

As a preliminary to creating Figure 8.24, we create a color vector for the 8 variables in SIPM. The first four are SI variables and will be colored red. The last four are PM variables and will be colored blue.

```
colvec <- base::c(base::rep("red",4),base::rep("dodgerblue",4))
```

We then call for the loadings plot. Available plot.type settings are "loadplot", "biplot", "screeplot", and "transplot". See `help(princals)` for details. The default is "loadplot".

```
base::plot(princals_SIPM,
    plot.type = "loadplot",
    main="SIPM Loadings Plot",
    col.loadings = colvec,
    xlim=base::c(-1, 1))
```

Princals biplots

It is also possible to create biplots using `princals()`. The R code below will display the variable loadings as red arrows, similar to loading plots. The observations will show in black, with row numbers for each, from 1 to 1,129. With 1,129 observations, display of the points is very cluttered in Figure 8.25. However, it can be used to spot

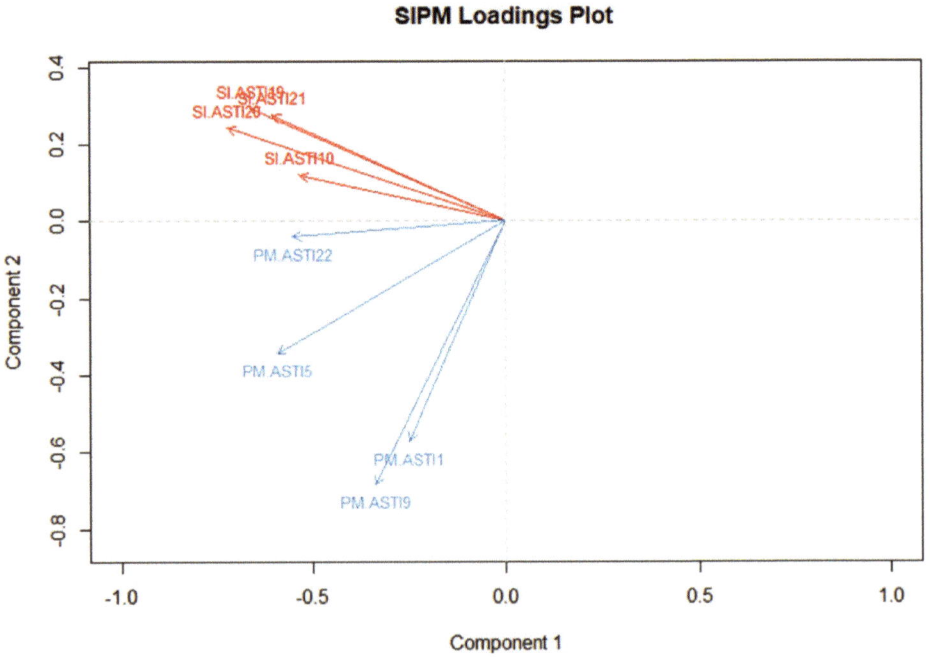

Figure 8.24 Plot of princals_SIPM variables in two CATPCA dimensions.

outlier observations around the periphery. Type `help(plot.princals)` for more information.

```
base::plot(princals_SIPM,
    plot.type = "biplot",
    main="SIPM Biplot",
    col.loadings = "red",
    col.lab="black",
    cex.scores=0.6,
    cex.loadings=0.8,
    expand=.9)
```

Beyond basic `princals()`

As illustrated above, the `princals()` function provides a way of doing factor analysis on ordinal variables. If the researcher has a mixture of ordinal and continuous variables, `princals()` can support that also. However, its default variable transformations

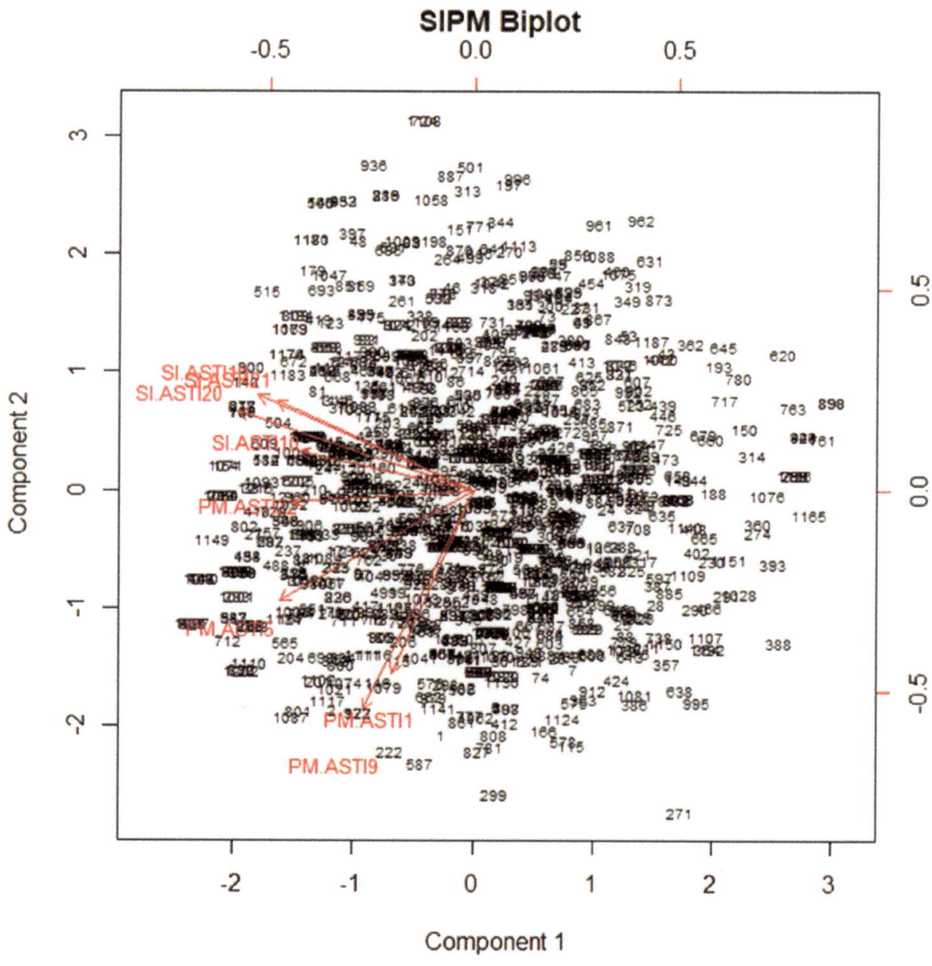

Figure 8.25 Biplot for the princals two-dimension solution, SI and PM ASTI variables.

apply to ordinal variables. For continuous variables, it is necessary to transform using spline regression. This advanced topic is outside the scope here but is treated in the Gifi package documentation at https://cran.r-project.org/web/packages/Gifi/Gifi.pdf and by Mair (2018: Section 8.2.3).

In addition, the `homals()` function from the Gifi package provides a type of factor analysis for nominal variables. This is also treated by Mair (2018: Section 8.3), who also sets forth the strategy for combined homals-princals models. See Mair (2018), discussion of `princals()` on pp. 235–244 and of `homals()` on pp. 244–256.

PCA for binary variables with `logisticPCA()`

Why binary factor analysis is problematic

An internet search for "Can I do PCA with binary variables?" will yield a variety of web pages explaining why it is problematic to do so.[4] Conventional PCA assumes continuous variables and seeks to minimize variance, which is defined in terms of variation around the means of variables. Taking the mean of a binary variable gives a value such as 0.5, which is usually meaningless (think gender equals 0 or 1, so what does gender = 0.5 mean?). Even if one conceptualizes a binary variable (e.g., low and high income, code 0 or 1) as a binned version of an underlying continuous variable, binning in that case is a form of measurement error which, in turn, inflates standard error and attenuates effect size. Attenuated effect size, in turn, inflates the number of components in the solution.

Social scientists often use binary variables in multiple regression and so may be inclined to feel that their use in PCA is acceptable. Indeed, it is true that one may get useful results when binary variables are used as inputs for PCA. However, the operative word is "may". The bottom line is that conventional PCA results will not be the same as those from logistic PCA designed for binary variables and may well differ enough as to affect substantive interpretation of the components.

Dimensionality reduction for binary variables is approached through three methods under the "logisticPCA" package, authored by Andrew J. Landgraf. The mathematical foundation for logisticPCA is treated by Landgraf and Lee (2015). The tutorial in this section is adapted from that by Landgraf (2016) in the "vignettes" section of online package documentation, though it goes beyond that.[5] The three methods Landgraf describes are:

1 Exponential PCA applied to Bernoulli data (Collins et al., 2001, using the algorithm of de Leeuw, 2006). In the "logisticPCA" package, this method corresponds to the command `logisticSVD()`. Results are returned as objects of data class "lsvd".
2 Logistic PCA (Landgraf & Lee, 2015). Where PCA seeks to minimize the Gaussian deviance (squared error), logistic PCA seeks to minimize the Bernoulli deviance. Principal component scores are linear functions of the data as in ordinary PCA but not as in exponential family PCA, which is nonlinear. In the "logisticPCA" package, this method corresponds to the command `logisticPCA()`. Results are returned as objects of data class "lpca".

3 Convex logistic PCA (Landgraf & Lee, 2015). This method is faster for large datasets but is not assured to give a low-rank solution. It seeks to minimize over the convex hull of projection matrices. Cross-validation is part of the algorithm. In the "logisticPCA" package, this method corresponds to the command `convexLogisticPCA()`. Results are returned as objects of data class "cv.lpca".

Setup for logisticPCA

It is assumed that library packages are already installed.

```
base::setwd("C:/Data")           # Declare the working directory
base::library(logisticPCA) # Has the binary logistic PCA functions
base::library(ggplot2)           # Leading visualization package for R
base::library(rARPACK)           # Needed by logisticPCA for partial
                                   decomposition
```

Example data

The example data, which are used in the documentation for the "logisticPCA" package, are 1984 votes by members of the U. S. Congress contained in the object "house_votes84". Type `help(house_votes84)` for further information. As these data are part of the "logisticPCA" package, they may be called with a simple `data()` statement issued following a `base::library(logisticPCA)` statement. The object house_votes84 is of data class matrix, not data.frame. There are 435 observations (one for each U. S. Representative) and 16 columns (the 16 roll-call votes). All data are binary, coded 0 or 1 for voted against or voted for respectively. Rownames indicate the political party, with labels such as "democrat.1" and "republican.1" (party is not a column in the matrix itself).

While the data do contain missing values, the commands in the "logisticPCA" package can accept missing values.

```
utils::data(house_votes84)
View(house_votes84)
```

Parameters and values for logisticPCA functions

`LogisticPCA()` and `convexLogisticPCA()` require input of the "m" parameter. The exponential family's `LogisticPCA()` command, in contrast, does not require or accept the "m" parameter. Larger values of this parameter yield fitted probabilities closer to 0 or 1, while smaller values yield probabilities closer to 0.5. The "logisticPCA" package has the functions `cv.lpca()` and `cv.clpca()` to optimize selection of m.

Below we use the `cv.lpca()` function to help determine the best value for m, given ks, where ks is the number of dimensions to try. The ks parameter can be a range, such as `ks = 2:4`, rather than a single value, in which case the plot will show lines for k = 2, k = 3, and k = 4, where k is the number of dimensions. Because a random element is involved in cross-validation, it is necessary to set a random seed if one wants replicable results.v-

```
set.seed(123)
logpca_cv = logisticPCA::cv.lpca(house_votes84, ks = 2:4, ms = 1:10)
base::plot(logpca_cv)
```

In the resulting Figure 8.26, the x axis is m, representing ascending alternative values of m, the tuning parameter. The y axis is NegLogLikelihood. Negative log likelihood is a measure of model error, with higher values being more error and lower values corresponding to a better model by this criterion. The k value in the legend is the number of requested principal components to try, here 2–4.

Based on Figure 8.26, we pick a value where the negative log likelihood starts to increase (moves toward greater error). For these data, for k = 2 (the two-dimensional solution) this point is m = 5. This may be verified by looking at the logpca_cv table below, where the lowest value for k = 2 is 1964.363 in column m = 5.

```
# List all error values for k rows by m columns
logpca_cv
[Output:]
      m
```

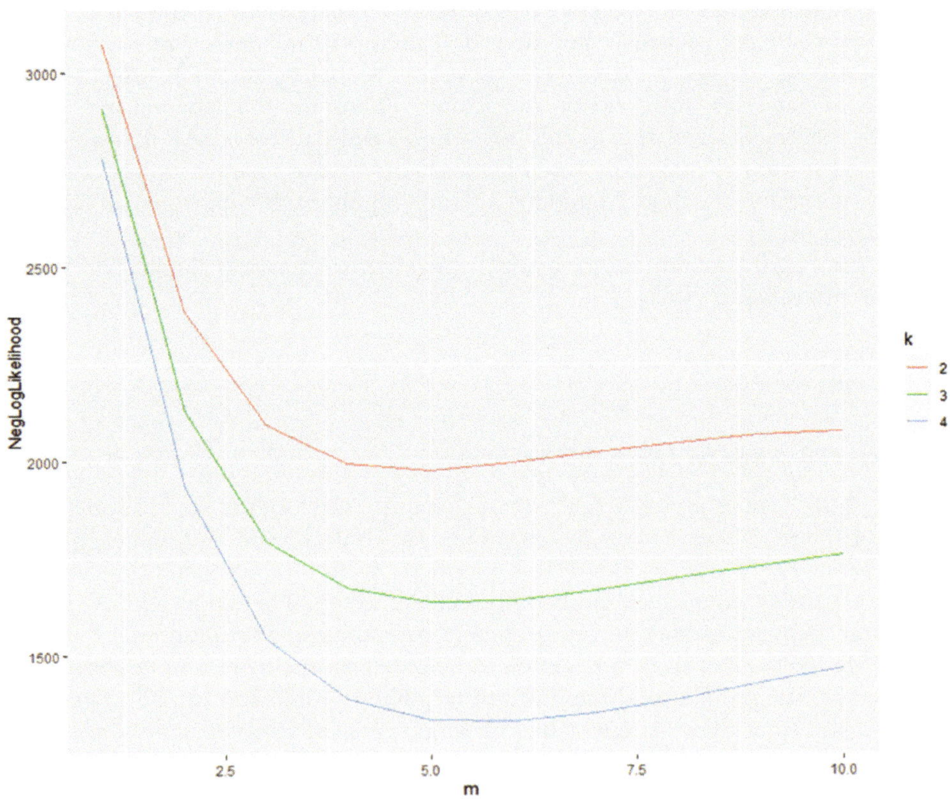

Figure 8.26 Negative log likelihood (error) for m = 1 to 10, for k = 2 to 4.

```
k          1        2        3        4        5
  2 3071.084 2384.407 2094.887 1993.486 1976.777
  3 2905.292 2129.419 1794.883 1674.840 1638.736
  4 2778.072 1931.573 1546.356 1389.265 1336.103
   m
k          6        7        8        9       10
  2 1996.152 2024.657 2048.078 2069.283 2078.073
  3 1641.838 1669.100 1700.820 1731.565 1759.273
  4 1332.556 1354.574 1390.248 1431.609 1467.469
attr(,"class")
[1] "matrix"   "cv.lpca"
```

More simply, we may also list m for the lowest error in row 1, which is for k = 2 dimensions.

```
base::which.min(logpca_cv[1,])
[Output:]
   [1] 5
```

Note that the lowest error value in the entire logpca_cv table is the value of 1299.686, which corresponds to k = 4 and m =6. However, here we choose to use the k=2 row, where the minimum is in row 1, column 5. We choose the "m" for the two-dimensional solution because our objective is to visualize political party, which is binary.

```
m <- logpca_cv[1,5]
m
[Output:]
   [1] 1976.777
```

There is a corresponding function to asses optimal m for convex logistic PCA: `cv.clpca()`. No such function is needed for the exponential family (SVD) model, which does not require the m parameter.

On the output side, the three logistic PCA methods return various values summarized in the table below.

The explanation of returned values in the table above is given below, along with the

Method	*Function*	*Class*	*Returns*
Exponential family PCA	logisticSVD	lsvd	mu, A, B
Logistic PCA	logisticPCA	lpca	mu, U, (PCs)
Convex logistic PCA	convexLogisticPCA	clpca	mu, H, (U, PCs)

meaning of other elements of the result object. A listing of any one of them is obtained by adding "$" and the element name to the model name. For instance, `logsvd_model$prop_deviance_expl` returns the proportion of deviance explained by the logsvd model.

Elements returned by the `logisticSVD()` *command*: mu, A, B, iters, loss_trace, and prop_deviance_expl.

Elements returned by the `logisticPCA()` *command*: mu, U, PCs, m, M, iters, loss_trace, and prop_deviance_expl.

Elements returned by the `convexLogisticPCA()` *command*: mu, H, U, PCs, m, M, iters, loss_trace, proj_loss_trace, and prop_deviance_expl.

Meaning of elements, listed alphabetically:

- A: A k-dimensional orthogonal matrix with the scaled left singular vectors. This applies only to `logisticSVD()` models.
- B: A k-dimensional orthogonal matrix with the scaled right singular vectors. This applies only to `logisticSVD()` models.
- H: This is the rank k Fantope matrix, where k is the number of variables. That is, it is a k-by-k matrix of coefficients. See Landgraf & Lee (2015) for further discussion. The Fantope matrix is used only for convex logistic PCA models.
- iters: The number of iterations required for convergence.
- loss_trace: The trace of the average negative log likelihood of the algorithm, which is a measure of error. Higher loss-trace is higher error. This element should be non-increasing. There is one loss-trace coefficient per iteration, plus one. For `logisticPCA()` models this is based on the projection matrix. For `convexLogisticPCA()` models it is based on the Fantope matrix.
- m: A single-value tuning parameter which is solved for, as per discussion further above.
- M: A deprecated equivalent of m.
- mu: The main effects. There is one main effect for each variable.
- proj_loss_trace: The trace of the average negative log likelihood using the projection matrix rather than the Fantope matrix. This element only pertains to `convexLogisticPCA()` models.
- PCs: The principal component scores for observations on each component. In this matrix element, each observation is a row and there are as many columns of scores as there are components.
- prop_deviance_expl: Proportion of deviance explained by the model.
- U: A k-dimensional orthonormal matrix, where k is the number of variables. The U matrix contains the component coordinates, which logisticPCA documentation calls "loadings". There is one coefficient per variable per component. No U matrix is returned by logistic SVD models, which is why, below, observations are not plotted in component space for SVD models.

The logistic exponential family (lsvd) model

Because the logistic exponential family model uses SVD to handle the binary data matrix, this is also called the "SVD model" in logisticPCA documentation. The percent of variance explained by the model is contained in the element "prop_deviance_expl" (e.g., `logsvd_model$prop_deviance_expl`).

```
logsvd_model <- logisticPCA::logisticSVD(house_votes84, k = 2)
logsvd_model
```

```
[Output:]
   435 rows and 16 columns
   Rank 2 solution
   63.6% of deviance explained
   549 iterations to converge
```

The trace plot in Figure 8.27 shows how quickly a model converged (549 iterations for the logistic SVD model compared to only 12 iterations for the convex logistic model) and at what level of percent of deviance explained (0.636). For exponential (SVD) models, error (deviance) does decline with each iteration. The plot below shows the relation of iterations to NegativeLogLikelihood. Actual values for NegativeLogLikelihood are contained in the loss_trace element of the result object. The next-to-last trace value (0.2440518) is the negative log likelihood at the last (549th) iteration (at convergence).

```
logsvd_model$loss_trace
[Partial output:]
 [1] 0.3085641 0.2946666 0.2895314 0.2865603 0.2843550
. . .
 [546] 0.2440819 0.2440718 0.2440618 0.2440518 0.2440418
```

The decline in error (NegativeLogLikelihood) is visualized in Figure 8.27.

```
base::plot(logsvd_model, type = "trace")
```

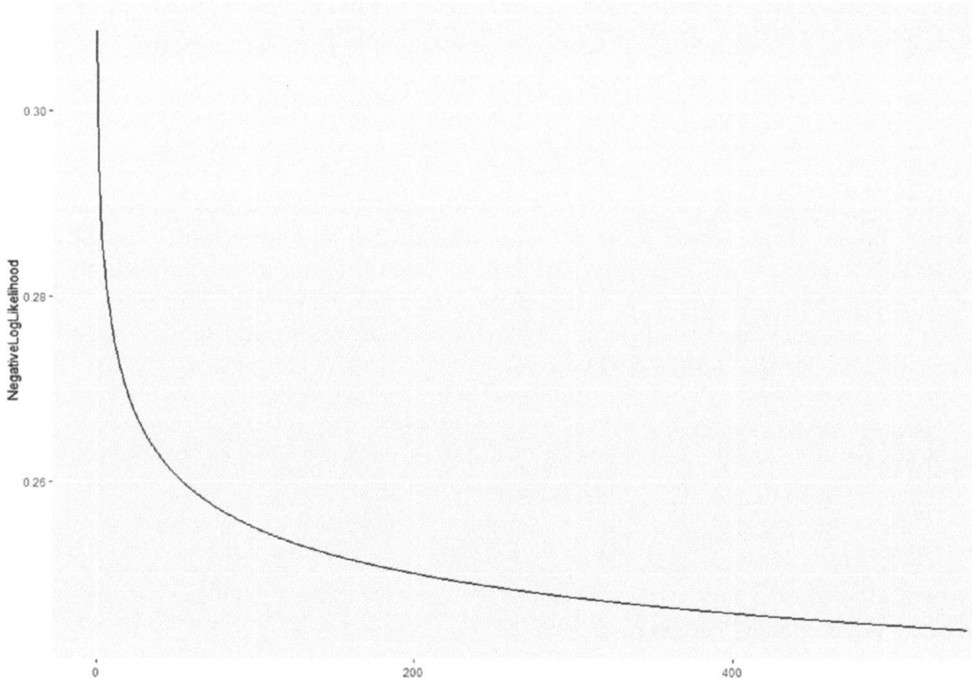

Figure 8.27 Plot of negative log likelihood by iteration for the exponential (SVD) logistic model.

The logistic PCA (lpca) model

```
logpca_model = logisticPCA::logisticPCA(house_votes84, k = 2, m = 5)
logpca_model
[Output:]
    435 rows and 16 columns
    Rank 2 solution with m = 5
    57.1% of deviance explained
    73 iterations to converge
```

For space reasons, we omit the loss_trace listing and plot, but these are created in the same manner as for the logisticSVD() model.

The convex logistic PCA (clpca) model

```
clogpca_model = logisticPCA::convexLogisticPCA(house_votes84, k = 2,
m = 5)
clogpca_model
[Output:]
    435 rows and 16 columns
    Rank 2 Fantope solution with m = 5
    65.7% of deviance explained
    12 iterations to converge
```

The trace plot shows how quickly a model converged (12 iterations for the convex logistic model) and at what level of error as measured by negative log likelihood (0.23). Though it does so here, error (deviance) does not decline with each iteration for convex logistic models, whereas it always does for logistic and exponential (SVD) models.

```
base::plot(clogpca_model, type = "trace")
```

The actual values for the y axis in Figure 8.28 are contained in the loss_trace element of clogpca_model. Trace (error = NegativeLogLikelihood) at convergence is in the next-to-last (12th) position. This value, which is 0.230084, is slightly lower (which is less error) than for the `logisticSVD()` model, where it was 0.2440518. This translates into percent of deviance explained being 65.7% for the convex logistic model, but slightly lower at 63.8% for the logistic SVD model.

```
clogpca_model$loss_trace
[Output:]
     [1] 0.3348437 0.3027998 0.2805234 0.2626397 0.2501345
     [6] 0.2418283 0.2364088 0.2330970 0.2312943 0.2304638
    [11] 0.2301590 0.2300824 0.2300824
```

Selecting the number of components

Although in the remainder of this section we request two components, a scree-type plot of percent of deviance explained by a number of components requested may

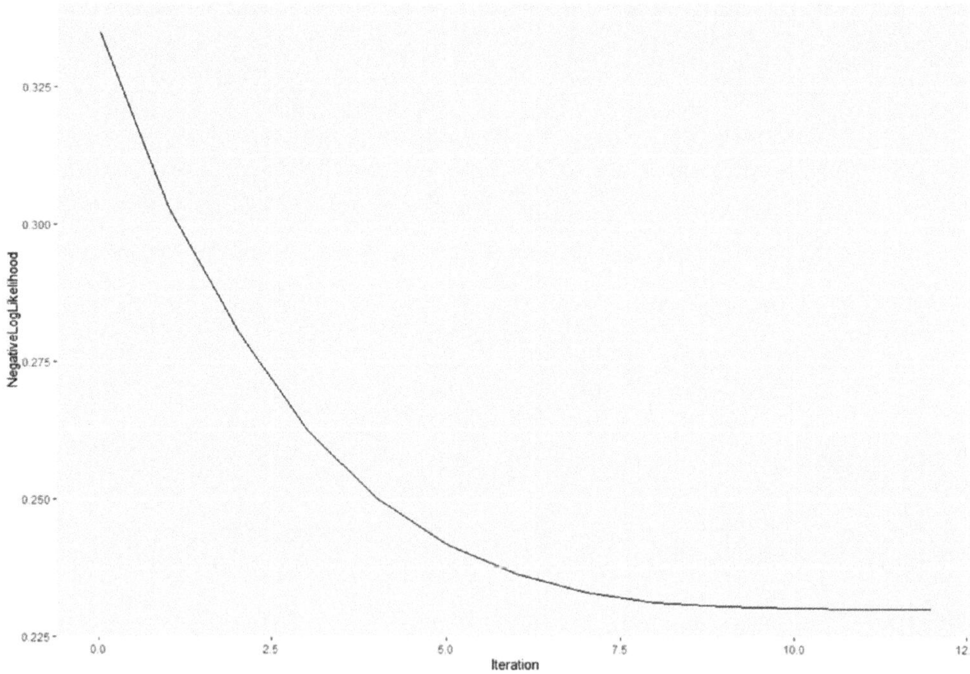

Figure 8.28 Plot of negative log likelihood (error) by iteration for the convex logistic model.

be created based on the $prop_deviance_expl element of the solution for any of the three types of logistic models. This scree plot may aid the researcher in selecting the desired number of components to request. However, this is not a matter of selecting the number of components explaining the highest proportion of deviance explained. Given k variables, that would ways be k components, which defeats the purpose of dimension reduction! Rather, the researcher must balance proportion of deviance explained with such other considerations as model parsimony, theoretical justification for the number of components, and interpretability of the dimensions in the solution.

```
# The number of columns is also the maximum number of components.
varnum <- base::ncol(house_votes84)

# Initialize the pde vector (percent deviance explained).
pde <- 1:varnum

# Loop through 1 to varnum requested components
for(i in 1:varnum){
temp1 = logisticPCA::logisticPCA(house_votes84, k = i, m = 5)
temp2 <-temp1$prop_deviance_expl
pde[i] <- temp2
}
```

```
# Print list of percent deviance explained by number of requested
components
round(pde,2)
[Output:]
    [1] 0.47 0.57 0.65 0.73 0.79 0.83 0.87 0.90
    [9] 0.92 0.94 0.96 0.97 0.98 0.98 0.99 0.99

# Plot the results (Figure 8.29)
base::plot(pde,
   col="red",
   pch=16,
   xlab = "Number of Components",
   ylab= "Percent of deviance explained",
   main ="Percent Deviance Explained by Number of Components
   Requested",
   sub = "Method = logisticPCA()")

graphics::abline(h=.50)

graphics::text(12,.50,"50% explained")
graphics::abline(h=.67)
graphics::text(12,.67,"67% explained")
graphics::abline(h=.80)
graphics::text(12,.80,"80% explained")
graphics::abline(h=.90)
graphics::text(12,.90,"90% explained")
```

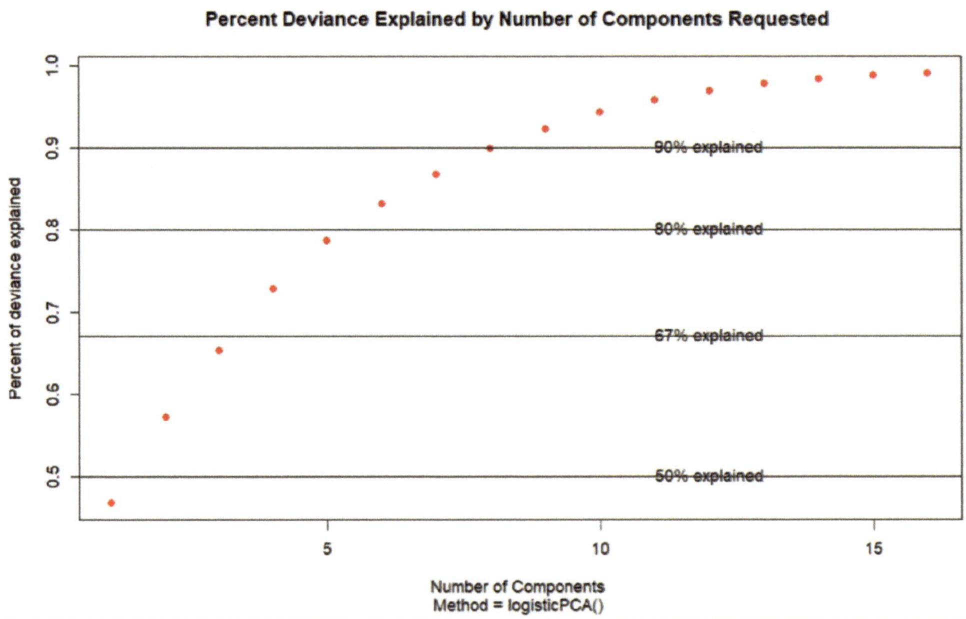

Figure 8.29 Plot of percent of deviance explained by a number of requested components.

Visualization

Regardless of which of the three algorithms is used, the observations (U.S. Representative) are plotable in two-dimensional latent space. We may represent observations color-coded in this case by political party (blue = Democrat, red = Republican). The research question is whether the algorithm separated members by party based on the 1984 roll call votes. As mentioned above, party is stored in the rownames of the house_votes84 data matrix.

```
party = base::rownames(house_votes84)
```

For each model below, we plot the models in descending order of performance based on variance explained.

The convex logistic PCA model

Convex logistic PCA models do not necessarily yield a two-dimensional space based on the Fantope matrix (H) even though k = 2 is specified in the model. However, the algorithm uses the first k eigenvectors of H as an estimate (Figure 8.30).

```
base::plot(clogpca_model, type = "scores") +
   geom_point(aes(colour = party)) +
   ggtitle("Convex Logistic PCA") +
   scale_colour_manual(values = base::c("blue", "red"))
```

The exponential family PCA model

The plot (Figure 8.31) of observations by party is constructed in a similar manner but the exponential SVD model reveals separation by party in a different configuration of points.

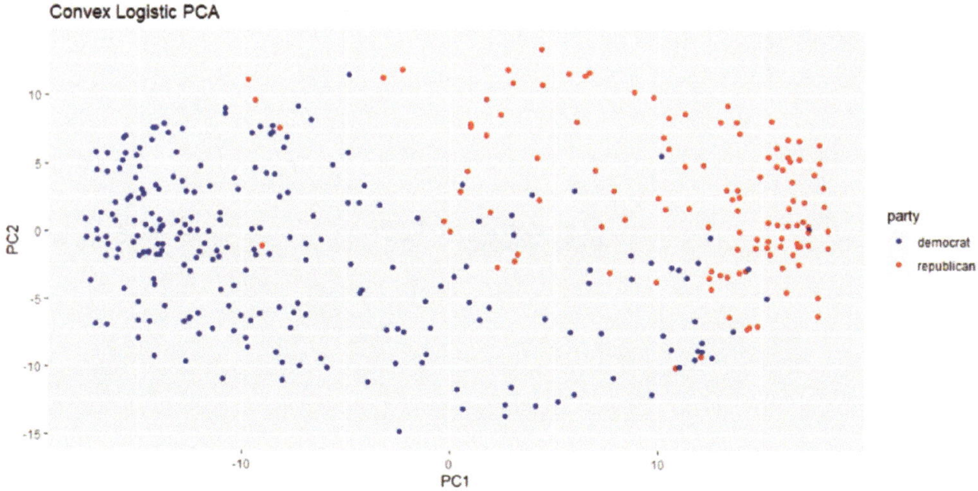

Figure 8.30 U. S. Representative by party, convex logistic PCA solution.

```
base::plot(logsvd_model, type = "scores") +
   ggplot2::geom_point(ggplot2::aes(colour = party)) +
   ggplot2::ggtitle("Exponential Family (SVD) PCA") +
   ggplot2::scale_colour_manual(values = base::c("blue", "red"))
```

The logistic PCA model

The plot of observations by party based on the logistic PCA algorithm is constructed in a similar manner, but is not shown here for space reasons. Like the other two models, it also reveals separation by party.

```
party = base::rownames(house_votes84)
base::plot(logpca_model, type = "scores") +
   ggplot2::geom_point(ggplot2::aes(colour = party)) +
   ggplot2::ggtitle("Logistic PCA") +
   ggplot2::scale_colour_manual(values = base::c("blue", "red"))
```

Component coordinates

Component coordinates for each of the 16 variables (the votes) may be plotted by the bare-bones plot command below. Though the logisticPCA documentation calls these coefficients "loadings", they are not factor loadings in the PCA sense. They are not based on the usual factor loadings metric which multiplies the eigenvectors by the square root of the eigenvalues. Indeed, logisticPCA does not produce eigenvalues or eigenvectors in the PCA sense. These values are better called "coordinates" and are useful for mapping countries in component space.

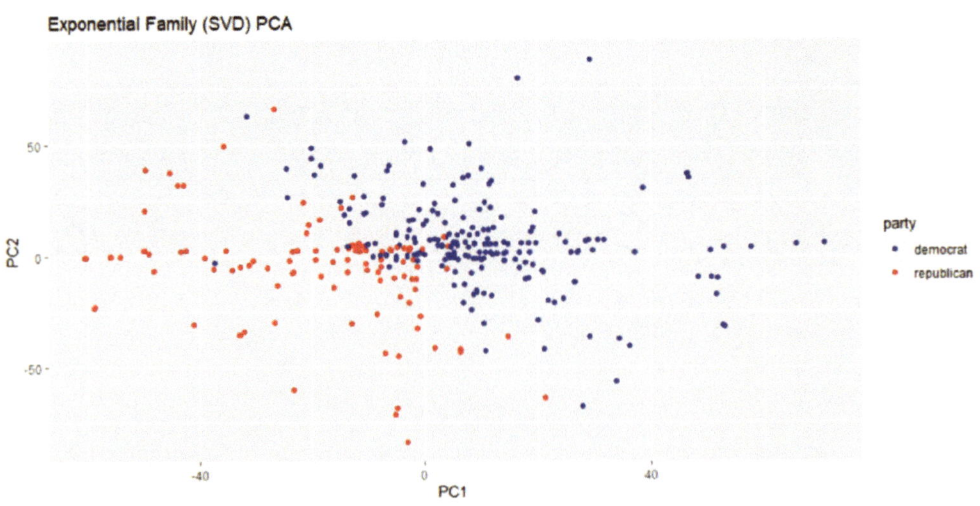

Figure 8.31 U. S. Representatives by party, logistic SVD solution.

Listing the coordinates

The coordinates are in the "U" element of the model. In this section, we use the convex logistic PCA model as an example, but the same listing of coordinates and plotting of variables in component space may be done in an identical manner for the logistic PCA model (logpca_model) and the exponential family logistic model (logsvd_model).

```
# Create a matrix from the clogpca_model result, add column and row
names, then list it.
clogpca_matrix <- clogpca_model$U
base::colnames(clogpca_matrix) <- base::c("PC1", "PC2")
base::rownames(clogpca_matrix) <- base::colnames(house_votes84)

clogpca_matrix
[Output:]
                                                PC1         PC2
handicapped-infants                   -0.1391954014 -0.11948887
water-project-cost-sharing             0.0450812491 -0.63307805
adoption-of-the-budget-resolution     -0.2869655105 -0.17766941
physician-fee-freeze                   0.3216078504  0.23598985
el-salvador-aid                        0.3931800482 -0.07128734
religious-groups-in-schools            0.2300276539 -0.17287340
anti-satellite-test-ban               -0.2875025275  0.24304966
aid-to-nicaraguan-contras             -0.3581028008  0.06058586
mx-missile                            -0.3086494643  0.18761452
immigration                           -0.0008922307  0.13405821
synfuels-corporation-cutback          -0.0484303156 -0.49714303
education-spending                     0.2736391185  0.22100415
superfund-right-to-sue                 0.2456360458 -0.12517516
crime                                  0.2752306565  0.18152765
duty-free-exports                     -0.2068493747 -0.02322416
export-administration-act-south-africa -0.1518160757  0.04050653
```

To plot the coordinate matrix (clogpca_matrix), we first convert it to a data frame, then use the "ggplot2" package to visualize it. A simpler but equivalent figure may be created using R's default `plot()` command, as shown in this endnote.[6] The "|>"operator is the pipe operator, in this case piping the contents of clogplca_df to the `ggplot()` command. The plot is shown in Figure 8.32.

```
clogpca_df <- base::as.data.frame(clogpca_matrix)
clogpca_df |>
  ggplot2::ggplot(ggplot2::aes(PC1,PC2,size=4,label= base::-
colnames(house_votes84))) +
  ggplot2::geom_point(color = "red", alpha = 0.6, size = 4, pch=16) +
  ggplot2::geom_text(check_overlap = TRUE, hjust="inward", size=5,
color="darkblue") +
  ggplot2::theme(axis.text=ggplot2::element_text(size=16, hjust=1)) +
```

```
  ggplot2::theme(axis.title=ggplot2::element_text(size=16,-
face="bold")) +
  ggplot2::theme(plot.title = ggplot2::element_text(color = "dark-
blue", size = 18, face = "bold"), plot.subtitle = ggplot2::element_
text(color = "darkblue", size=16) ) +
  ggplot2::labs(
     title = "Convex logistic PCA model",
     subtitle = "16 House Votes, 1984")
```

Component loadings

This section discusses how to get from component coordinates to component loadings as widely and conventionally used in social science. Where component coordinates show the location of variables in component space, as illustrated in the previous section, component loadings show the relative importance of variables to each component. The conventional definition of a loading is: loadings = eigenvectors * sqrt(eigenvalues). However, the "logisticPCA" package does not output eigenvectors or eigenvalues.

With continuous-normal data, we could compute a Pearsonian correlation matrix and then derive the eigenvector matrix, and then from that matrix, we could compute traditionally-defined component loadings based on the formula, loadings = eigenvectors * sqrt(eigenvalues). However, with binary data, we need an alternative. Below, our alternative is to interpret the absolute coordinate value expressed as a percentage of the column sum of such values as a "loading". These might be called "pseudo-loadings" since they are not based on eigenvectors and eigenvalues. However, these component coordinate-based loadings still reflect variable importance to the dimensions and also they sum to 1.0.

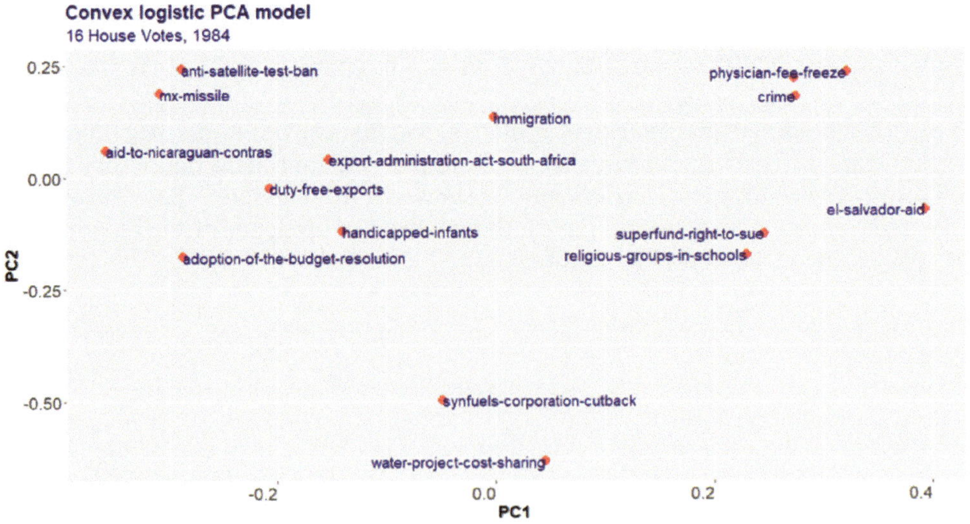

Figure 8.32 Vote variables in convex logistic component space.

Larger coordinates (which the "logisticPCA" package misleadingly calls "loadings") contribute more to the meaning of the component. Large positive and large negative coordinates help define the end-points of the dimension that a given component is supposed to represent. Labels for components may be imputed by the researcher based on which variables load most heavily on them.

Component labels are more easily imputed when there is "simple factor structure". The cutoffs for simple factor structure vary by author, but by common rule of thumb, when there is simple factor structure, any given variable should load strongly (>= 0.6) on a single component and weakly (<0.4) on all others. Cross-loadings in the middle (>=0.4–0.6) should be few in number or absent. The loadings of all variables for a given component sum to 1 (or 100%, depending on the metric used).

We use a convex logistic PCA model as an example. For convenience, we start by copying clogpca_matrix into object x. Note that logsvd_model has no U element containing component coordinates and so this cannot be done for SVD models. However, there is a clogpca_model$U element and so component loadings may be calculated for convex logistic PCA models.

```
clogpca_matrix <- clogpca_model$U

x <- clogpca_matrix

# Convert x to absolute values of component coordinates.
x2 <- base::abs(x)

# Compute sums of absolute component coordinate values for the two
components.
s <- base::colSums(x2)
s1 <- s[1]
s2 <- s[2]

# Convert component coordinate values to percentages of the column
sums.
L1 <- x2[,1]/s1
L2 <- x2[,2]/s2

# Create the array of component loadings.
temp <- x2
temp[,1] <- L1
temp[,2] <- L2
loadings <- temp
loadings

# Add labels to the loadings matrix
base::colnames(loadings) <- base::c("PC1", "PC2")
base::rownames(loadings) <- base::colnames(house_votes84)

# List the component loadings
loadings
```

```
[Output:]
                                                 PC1         PC2
   handicapped-infants                    0.0389596829 0.038245302
   water-project-cost-sharing             0.0126178821 0.202631933
   adoption-of-the-budget-resolution      0.0803193581 0.056867389
   physician-fee-freeze                   0.0900154728 0.075534256
   el-salvador-aid                        0.1100479602 0.022817236
   religious-groups-in-schools            0.0643829061 0.055332311
   anti-satellite-test-ban                0.0804696649 0.077793920
   aid-to-nicaraguan-contras              0.1002301184 0.019391968
   mx-missile                             0.0863885239 0.060050561
   immigration                            0.0002497283 0.042908571
   synfuels-corporation-cutback           0.0135552592 0.159122643
   education-spending                     0.0765894072 0.070737719
   superfund-right-to-sue                 0.0687515705 0.040065334
   crime                                  0.0770348660 0.058102312
   duty-free-exports                      0.0578954905 0.007433452
   export-administration-act-south-africa 0.0424921090 0.012965093

# We verify that component loadings add to 1.0.
base::colSums(loadings)
[Output:]
   PC1 PC2
     1   1
```

We now create a barplot for component 1, shown in Figure 8.33. The barplot for component 2 would be done identically. Bars in the plot are sorted in descending order of loading magnitude. We handle the very long labels in three ways:

1. We take only the first eight letters of each label.
2. The `las=2` parameter makes puts bar labels at 90 degrees.
3. the `cex.names=0.8` prints the labels at 80% of full size.

The `col= parameter` sets the bar colors, here to a rainbow of as many colors as there are rows in loadings.

```
# Sort the loadings for PC1 descending
loadingsSorted1 <- base::sort(loadings[,1], decreasing=TRUE)

graphics::barplot(loadingsSorted1,
   xlab = "Vote",
   ylab = "Loadings",
   main = "Component Loadings for PC1, data = house_votes84",
   sub= "Loadings are based on absolute coordinate values divided by
the sum of absolute values." ,
   names.arg = substr(base::names(loadingsSorted1),1,8),
   las=2,
```

```
   cex.names=0.8,
   col= grDevices::rainbow(base::nrow(loadings)))
```

For PC1, the first six variables (votes, listed below) account for over 50% of the loadings for PC1. It is difficult to impute a label for PC1, but the content leans in the foreign policy and military direction, but physician fee-freeze is also included as is the adoption of the budget resolution, though the military sector is the largest budget component.

- el-salvador-aid
- aid-to-nicaraguan-contras
- physician-fee-freeze
- mx-missile
- anti-satellite-test-ban
- adoption-of-the-budget-resolution

Fitted values and imputing NAs

The "logisticPCA" package contains commands to produce fitted values for logistic PCA models (`fitted.lpca()`) and exponential family PCA models (`fitted.lsvd()`) but not for convex logistic PCA models. For this subsection, we use exponential family logistic (SVD) models.

Fitted values may be either of two types. The `type=` option may be set to "response" to obtain output as probabilities, while "link" gives output on the logit scale. Below we ask for probabilities. The svd_fitted object, which is of data class "matrix", contains 435 rows (one per observation = U. S. Representative) and 16 columns (one per variable = vote).

```
svd_fitted <- stats::fitted(logsvd_model, type = "response")
```

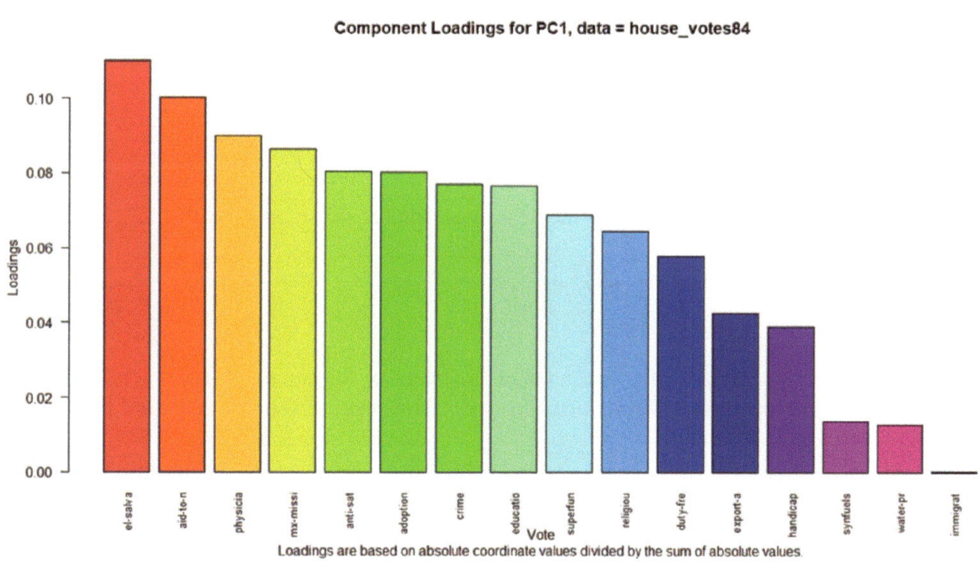

Figure 8.33 Barplot of PC1 component loadings.

```
# List fitted values for the first three observations.
utils::head(svd_fitted, 3)
[Partial output:]
              [,1]        [,2]         [,3]        [,4]
   [1,] 0.17918897 0.9435995 0.04987579 0.9959116
   [2,] 0.06532547 0.9880254 0.00121354 0.9999939
   [3,] 0.44859660 1.0000000 0.50638666 0.3784252
   . . .
             [,13]       [,14]        [,15]        [,16]
   [1,] 0.9495671 0.9999961 0.061040398 0.73621678
   [2,] 0.9957879 1.0000000 0.008984933 0.05403459
   [3,] 0.9323690 0.7497420 0.163600392 0.07389330
```

These fitted probabilities may be used to impute missing values. For instance, the U. S. Representative for observation 3 had missing values on votes 1 and 4 in the original data. As both of the fitted values for these cells in the matrix above were less than 0.50, the probability of a 1 value is less than 50% and we could impute values of 0 to these two cells

Plot supplemental observations

A supplemental observation is one which is added to the plot of observations on the PC1 and PC2 axes without recalculating the solution. For instance, if the solution has been calculated for the 435 U.S. Representatives for the logsvd model and then plotted, suppose a representative dies and is replaced by a new person. Suppose that new person has previously announced how they would have voted (0, 1) on the 16 votes which are variables in the model. How can this new individual be plotted in existing coordinate space, along with the 435 individuals used for the original solution?

The `stats::predict()` command is able to tell if the logisticPCA object passed to it is a `logisticSVD()`, `logisticPCA()`, or `convexLogisticPCA()` model and compute appropriate component coordinates for the new individual.

To demonstrate plotting supplemental points, we follow the steps below, using a `convexLogisticPCA()` model. However, the process is the same for `logisticSVD()` and `logisticPCA()` models.

1 Run the original `convexLogisticPCA()` model as done previously:

```
clogpca_model = logisticPCA::convexLogisticPCA(house_votes84, k
= 2, m = 5)
```

2 Create an array for the party and votes of the new replacement representative. For instructional simplicity, we do not eliminate the representative that died, but could have.

```
newvotes <- base::c(1,0,0,0,1,0,0,1,1,1,0,1,0,1,1,0)
newparty <- "democrat"
datanew <- base::matrix(newvotes,nrow=1,ncol=16,byrow=FALSE,
dimnames = base::list(NULL, base::colnames(house_votes84)))
```

3 Predict the new representatives coordinates on PC1 and PC2 using the original model from step 1.

```
newcoordinates <- stats::predict(clogpca_model, newdata=datanew,
type = "PCs")

newcoordinates
[Output:]
              [,1]      [,2]
   [1,] -0.3552478 8.087313
```

4 Plot the new (supplemental) individual on the plot using newparty from step 2 and newcoordinates from step 3. Normally we would add a point with the `points()` function but this plot uses ggplot2 commands, not just the generic `plot()` command in R's "base" package. The supplemental observation appears as "N" in Figure 8.40.

```
x <- newcoordinates[1,1]
y <- newcoordinates[1,2]
party = base::rownames(house_votes84)
newcolor = base::ifelse(newparty=="democrat","blue","red")
```

The last line adds the new point with the symbol "N" and in blue, denoting a Democrat (Figure 8.34).

```
base::plot(clogpca_model, type = "scores") +
   ggplot2::geom_point(ggplot2::aes(color = party)) +
   ggplot2::ggtitle("Convex Logistic PCA") +
   ggplot2::scale_colour_manual(values= base::c("blue","red")) +
   ggplot2::geom_point(ggplot2::aes(x=x, y=y), pch ="N", cex=5,
color=newcolor)
```

We note that Landgraf (2016) discusses predictions for adding multiple supplemental points rather than just one. Let the object morevotes contain the votes of multiple new observations, where morevotes is a matrix like datanew above, but with multiple rows. The command to obtain principal component coordinates for these new observations is essentially the same as before. If the reader tries this, however, the error message appears, "object 'morevotes' not found" since we have not actually created it here.

```
stats::predict(clogpca_model, newdata = morevotes, type = "PCs")
```

Model fit by classification rate

The fitted values discussed above may be used to compute the classification rate (percent correctly classified) and the misclassification rate for the model. Recall that the model illustrated for fitted values was the logsvd_model (the logistic exponential family model).

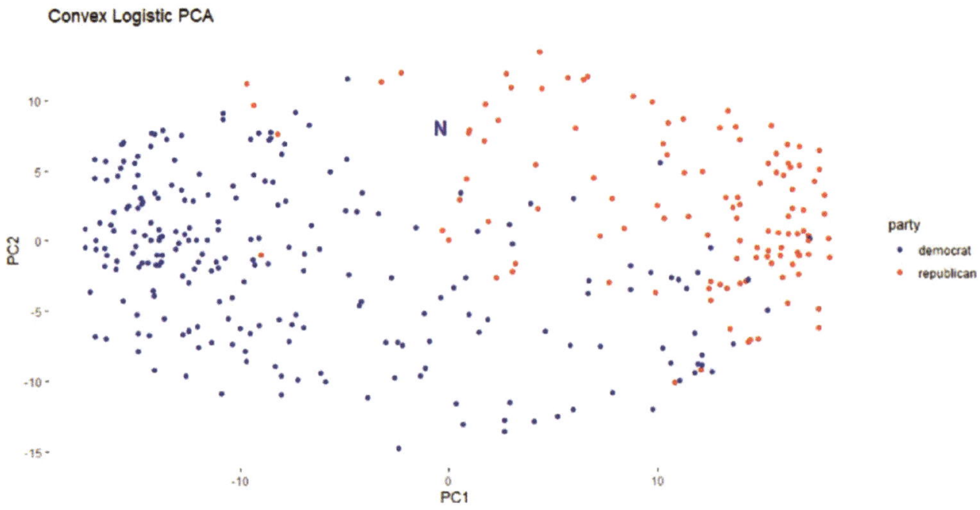

Figure 8.34 Convex logistic model with a supplemental point added.

```
# Copy original observed data into matrix O, for observed values
O <- base::as.matrix(house_votes84)

# Copy fitted values into matrix E, for expected values
E <- svd_fitted

# Re-create E as the binary version of E (all 0 or 1 predicted
values)
E <- base::as.matrix((E >= 0.5) + 0)

# Create the residual matrix, R, as observed minus expected
R <- O - E
```

We now count the matches and misses in the residual matrix. Results are put in the array M. The column value −1 corresponds to O was 0, E was 1; column value 0 corresponds to both O and E matching; and column value 1 corresponds to O being 1 and E being 0.

```
M <- base::tapply(base::unlist(R), base::factor(base::unlist(R)),
length)
M
[Output:]
     -1    0   1
    388 5770  410

# Get the total of all types (-1, 0, 1). The total is 6,568 for
these data.
total <- base::sum(M)

# Get the correct classification rate. The unname() function drops
the "0" column header.
```

```
ClassificationRate  <- M[2]/total
base::unname(ClassificationRate)
[Output:]
   [1] 0.8785018

MisclassificationRate  <- 1 - ClassificationRate
base::unname(MisclassificationRate)
[Output:]
   [1] 0.1214982
```

Earlier, the logistic PCA package computed a model fit measure for the logsvd_model as 63.6% of deviance explained. The classification rate is a supplementary model fit measure, here 87.9%. The former is a percent explained measure while the latter is a correct classification measure. It is not that one is "correct" and the other not. Rather, the two model fit measures provide two perspectives on how well the model (here, the `logisticSVD()` model) functioned.

Notes

1 The h2o manual is at https://docs.h2o.ai/h2o/latest-stable/h2o-r/h2o_package.pdf.

2 Below is the recommended procedure for a more thorough installation of h20, as of this writing:

```
# The following two commands remove any previously installed H2O
packages for R.
if ("package:h2o" %in% search()) { detach("package:h2o",
unload=TRUE) }
if ("h2o" %in% rownames(installed.packages())) { remove.
packages("h2o") }

# Next, download packages that H2O depends on.
pkgs <- c("RCurl","jsonlite")
for (pkg in pkgs) {
  if (! (pkg %in% rownames(installed.packages()))) { install.
packages(pkg) }
}

# Then download and install the H2O package for R.
install.packages("h2o", type="source", repos=(c("http://h2o-
release.s3.amazonaws.com/h2o/latest_stable_R")))
```

3 This endnote illustrates that the coordinates used by `fviz_cluster()`, which are point coordinates in p$data used for the axis tick marks, are standardized versions of the archetype values in glrm_table2. For simplicity, a model based only on the first three rows in housing1 is presented here.

```
p$data
[Output:]
    name          x          y cluster
  1    1  0.4286375 -0.3569149       2
  2    2 -1.1428673 -0.7725730       1
  3    3  0.7142298  1.1294878       2
```

```
# The upper right value in glrm_table2 is 0.9799537.
glrm_table2
[Output:]
            Arch1      Arch2      Arch3        Arch4
   1:  0.9799537 -0.5690748  1.1151615 3.218047e-05
   2: -1.4278762 -0.6782328 -0.4946441 4.754293e-06
   3:  1.4175328 -0.1787233 -0.5632053 1.052853e-05
               Arch5 density primaryres   houses
   1: -4.696333e-05  131.70      88.77  inf 90%
   2: -6.938289e-06   21.21      87.52  sup 90%
   3: -1.536505e-05  531.99      94.90  inf 90%
      owners council id              town cluster
   1:  64.23  sup 5%  1             ABZAC       2
   2:  77.12  inf 5%  2            AILLAS       1
   3:  65.74  sup 5%  3 AMBARES-ET-LAGRAVE      2

# Below we print the upper-right value in standardized form. It
is the same as in p$data.
Arch1_standardized <- scale(glrm_table2[,1:5])
Arch1_standardized[1]
[Output]
   [1] 0.4286375
```

4 E.g., https://stats.stackexchange.com/questions/16331/doing-principal-component-analysis-or-factor-analysis-on-binary-data.

5 This is also found at https://cran.r-project.org/web/packages/logisticPCA/vignettes/logisticPCA.html.

6
```
plot(clogpca _ model, type = "loadings")
Or:
plot(clogpca_matrix, pch=16, cex=2, col="darkblue")
text(clogpca_matrix, y = NULL, labels = colnames(house_votes84),
adj = NULL, pos = 4, offset = 0.3, cex = 1, col = "dodgerblue")
```

9 PFA in greater detail

Introduction

In this chapter, we return to the common factor analysis (PFA) procedure introduced in Chapter 4. Five important topics are covered:

1 *Extension variables with* `fa()`: We show how it is possible to add variables to factor analytic visualizations without their being used to create the factor model. This may be useful, for example, for seeing the placement of socio-demographic variables (e.g., age, family size, income) in the factor space created by, say, test items pertaining to personality constructs.
2 *Orthogonal PFA with* `factanal()`: The `factanal()` program performs maximum likelihood (ML) factor analysis on a data matrix or covariance matrix. For researchers wishing an ML solution, `factanal()` supports more options than does the more widely used `fa()` program.
3 *Oblique PFA with* `fa.promax()`: The `fa.promax()` function provides enhanced output for promax solutions, which are oblique solutions.
4 *Bayesian factor analysis*: The `befa()` command supports a Bayesian approach to factor analysis, incorporating prior information through an MCMC (Markov Chain Monte Carlo) process.
5 *Regularized factor analysis with* `fareg()`: Regularization is an alternative to factor rotation. In some circumstances regularization shows how variables are arrayed on dimensions more clearly than does varimax rotation.

Extension variables with fa()

Dwyer (1937) originated a method to compute factor loadings for variables not present in the original analysis. The `fa.extension()` command from the "psych" package implements this method. The variables not initially present are called "extension variables". The process involves finding the unattenuated correlation of the extension variables with the factor scores from the original analysis. An alternative, which does not correct for factor reliability, was proposed by Gorsuch (1997). Both options implement exploratory factor analysis with extensions to new variables.

A use of the factor extension method would be to visualize the locations in factor space of relevant covariates (e.g., age, income, gender) without having included the covariates in the original computation of factor dimensions. Factor extension is also useful for establishing criterion validity, showing that the location in factor space of an extension variable

DOI: 10.4324/9781003279693-10

conforms to expectations based on the literature or theory. Either the factor dimensions are assumed to be true and validation is for the extension variable, or the extension variable is assumed to be correct and validation is for the factor dimensions.

The format for the `fa.extension()` command in R is:

```
fe <- psych::fa.extension(Roe, fo, correct=TRUE)
```

where `Roe` is the correlation matrix of the original variables with the extension variables;

`fo` is the output results from the `fa()` function as applied to the original variables; and `correct=TRUE` yields Dwyer's solution, whereas `correct=FALSE` produces Gorsuch's solution. Type `help(fa.extension)` for more information.

Example

An example of the extension method is given below. However, since the "ability" data do not contain extension variables, we use simulated data adapted from Revelle (2021). To start, we create 16 simulated variables for 500 cases using the `sim.item()` function from the "psych" package. We label this data object "simdata". Since simulation involves random processes, we must set the random seed in order to assure replicable results. A different seed will lead to slightly different coefficients than those in the output shown below. We also load the needed packages "psych" and "GPArotation", which are assumed to have been installed.

```
base::set.seed(123)
base::library(GPArotation)          # Needed by the fa() command
base::library(psych)                # Supports the fa() command and more

simdata <- psych::sim.item(nvar=16, nsub=500)
```

Second, we create a vector (the "s" vector) containing the column index numbers of the odd-numbered columns. These columns constitute the training set and are used below by the `fa()` command to compute the factors. For example purposes, even-numbered columns will be used as the extension variables.

```
s <- base::c(1,3,5,7,9,11,13,15)
```

Third, we use the `fa()` command from the "psych" package to factor analyze variables (columns) specified in the "s" vector. We request a solution with two dimensions (factors). The "fo" object holds the `fa()` results, which are for the odd-numbered variables (the analysis variables). This is, of course, "fo" in the template above, though it did not have to have this label. The `fa()` command will call on the "GPArotation" package, which must have been installed.

```
fo <- psych::fa(simdata[,s],2)
```

There are, of course, numerous other options for the `fa()` command. By not specifying them, we accept the defaults shown on the `help(fa)` page. One of these is the "`fm =`" option, which specifies the factor method. The default method is "`fm = minres`", which is the minimum residual method. For this reason, the factors generated carry the labels "MR1", "MR2", and so on, just as when performing orthogonal factor analysis using the principal axis method, the factors are labeled "PA1", "PA2", and so on.

Fourth, we create the correlation matrix "Roe" as an object using variables designated as extension variables. The "Roe" object is a list object of class psych-extension. The first term ("`cor(simdata)[s,-s]`") is the correlation matrix of the odd numbered items (s, the analysis variables) as rows with the even numbered items (-s, the extension variables) as columns), based on the fo results object from the `fa()` command. This is "Roe" in the template above, though again it needn't have been labeled "Roe".

```
Roe <- psych::fa.extension(stats::cor(simdata)[s,-s],fo)
```

Fifth, we create the first visualization of the results, shown in Figure 9.1.

```
# 5A. Plot the analysis variables in factor space.
# We use xlim= and ylim= to extend the range of the axes for better
viewing of variables.
psych::factor.plot(fo,
   pos=3,
   cex=1.4,
   xlim=base::c(-.7,.8),
   ylim=base::c(-.8,.8),
   labels=base::rownames(fo$loadings),
   title="Analysis variable  loadings",
   sub="Extension variable points and labels in green",
   pch=16,
   jiggle=TRUE)

# 5B. Add the extension variable points. This assumes we are
plotting factors 1 and 2
# points() is from the "graphics" package in the R System library.
graphics::points(Roe$loadings[,1], Roe$loadings[,2], col = "green",
pch = 16, cex=1.4)

# 5C. Add the text labels for the extension variables.
# text() is from the "graphics" package in the R System library.
graphics::text(Roe$loadings[,1], Roe$loadings[,2], pch=16,cex=1.4,
col="green", pos=1,labels=base::rownames(Roe$loadings))
```

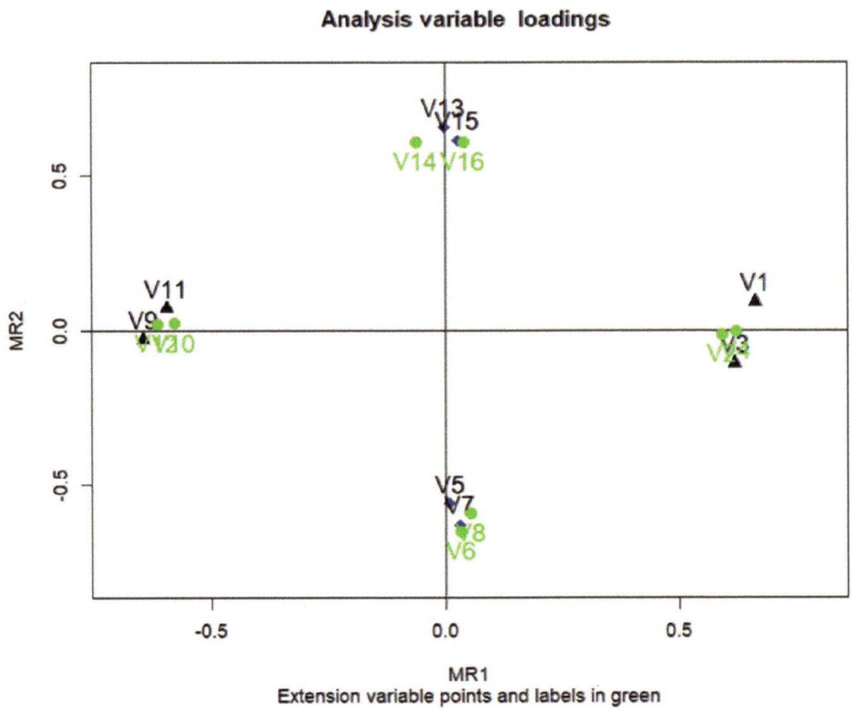

Figure 9.1 Factor analysis diagram with analysis and extension variables.

Alternatively, one could plot the extension variables separately using the code below, but the resulting figure is not shown here for space reasons.

```
psych::factor.plot(Roe, pos=1, cex=1.4,  labels=base::rownames(Roe$-
loadings), title="Extension variable  loadings", pch=16,
xlim=base::c(-.7,.8),
ylim=base::c(-0.8,0.8))
```

Sixth, we create the second visualization of results using the `fa.diagram()` command from the "psych" package. This command provides a different dual view of analysis and extension variables. This view more clearly separates the analysis and extension variables and also shows their loadings on the two factors in this model. To illustrate we use both the "fo" and "Row" objects created above. The resulting visualization is shown in Figure 9.2.

```
psych::fa.diagram(fo,fe=Roe, digits=2, cex=1.4)
```

In Figure 9.2, analysis variables are on the left and extension variables are on the right. The values on the arrows for extension variables are the standardized loadings from the pattern matrix for the largest loading for each variable. The values on the arrows for analysis variables are the standardized loadings from the pattern matrix for the largest loading for each variable. Negative loadings are shown as dashed red arrows, but only for the analysis variables.

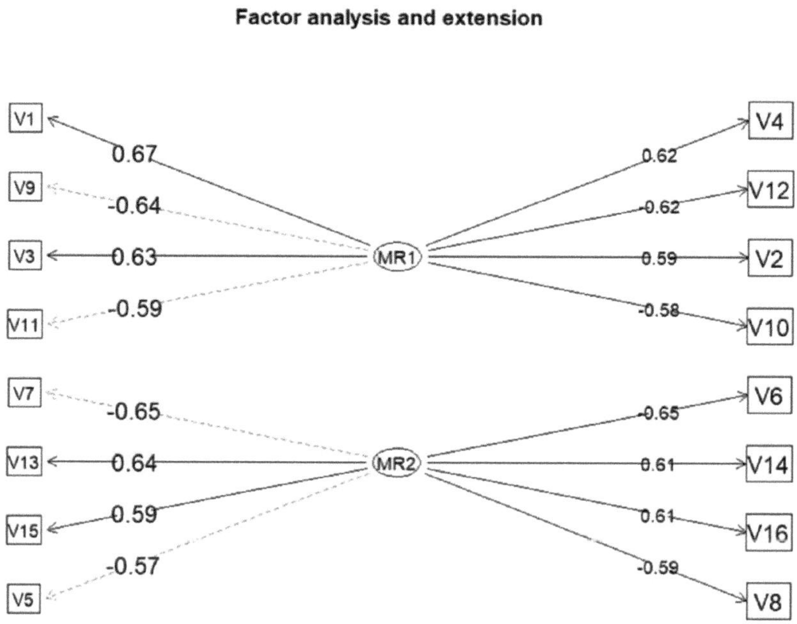

Figure 9.2 Factor analysis plot with analysis and extension variables.

A second extension method example

We now apply the factor extension method to the music2 data, considered earlier in different form (the "dat" data frame) in Chapter 6. Figure 9.3 shows the music variables which were part of the analysis but adds extension variables (the covariates age, educ, rincom91, and sex) on dimensions determined by the analysis (music) variables.

```
# Set the working directory.
base::setwd("C:/Data")

# Load needed libraries
base::library(GPArotation) # Needed by the fa() command
base::library(psych)               # Supports the fa() command and more

# Read in the "music" data.
music <- utils::read.csv("musicvars.csv", header=TRUE, sep = ",",
stringsAsFactors=TRUE)

# Create music2 as a complete cases version of music
music2 <- music[stats::complete.cases(music),]

# View the columns and column numbers in music2
base::names(music2)
[Output:]
    [1] "age"      "bigband"
    [3] "blues"    "blues3"
    [5] "blugrass" "classicl"
```

```
     [7] "classic3" "country"
     [9] "educ"     "hvymetal"
    [11] "jazz"     "jazz3"
    [13] "musicals" "opera"
    [15] "rap"      "rap3"
    [17] "rincom91" "sex"

# Create a vector of index numbers for the analysis variable columns
(the music items).
s2 <- base::c(2,3,5,6, 8,10,11,13:15)
base::names(s2)<- base::c("bigband","blues","blugrass","classicl",
"country", "hvymetal", "jazz", "musicals","opera", "rap")

# Create a vector of index numbers for the extension variable col-
umns (the covariates).
# The covariates are age, educ, rincom91, and sex.
s3 <- base::c(1,9,17,18)
base::names(s3) <- base::c("age", "educ", "rincom91","sex")

# Perform PFA on the analysis variables, putting results in the ob-
ject "fo2".
# We specify principal axis factoring (pa) as the method, with vari-
max rotation.
fo2 <- psych::fa(music2[,s2],nfactors = 2, fm="pa", rotate
="varimax",)

# Process coordinates for the music2 extension variables using fo2
dimensions.
Roe2 <- psych::fa.extension(stats::cor(music2)[s2,s3],fo2)

# 5A. Plot the analysis variables in factor space, for Figure 9.3.
# Note we use xlim= and ylim= to extend the range of the axes for
better viewing of variables.
psych::factor.plot(fo2, pos=3, cex=1.2,  labels=base::rownames(-
fo2$loadings), xlim=base::c(-.6,0.9), ylim=base::c(-.4,0.8),  ti-
tle="Analysis variable and extension covariate loadings, music2
data", sub="Extension covariates in green",pch=16, jiggle=TRUE)

# 5B. Add the extension variable points. This assumes we are plot-
ting factors 1 and 2
graphics::points(Roe2$loadings[,1], Roe2$loadings[,2], col = "char-
treuse", pch = 16)

# 5C. Add the text labels for the extension variables:
graphics::text(Roe2$loadings[,1], Roe2$loadings[,2], cex=1.0,
col="chartreuse4", pos=2,labels=base::rownames(Roe2$loadings))
```

Finally, the command below and the corresponding Figure 9.4 show the factor diagram for the same data. The "cut=0.1" option dictates that only standardized loadings of 0.1

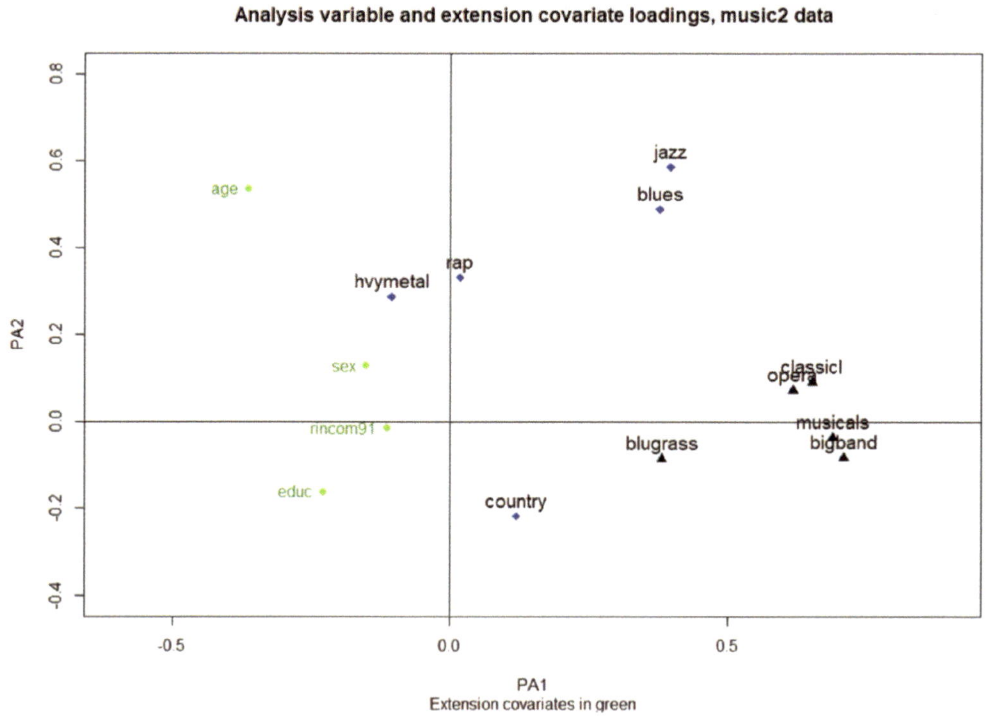

Figure 9.3 Analysis and extension variables plotted by analysis dimensions, music2 data.

or greater are shown in the arrows. The variables or covariates below the cut point do not display an arrow.

```
psych::fa.diagram(fo2,fe=Roe2, digits=2, cex=1.1, cut=0.1)
```

In Figure 9.4, the strongest covariate relationship is 0.54 for age with dimension PA2, which is the dimension most associated with jazz, blues, rap, hvymetal, and country (which is loaded negatively on PA2). As higher values on the music preference items correspond to disliking, those of older age tend to dislike the set of music types represented by dimension PA2. This does not necessarily apply to each music item considered individually but rather to PA2 items considered as a set.

Orthogonal PFA with `factanal()`

Introduction

The `factanal()` program from R's built-in "stats" package is less widely used than `fa()` from the "psych" package but it is an alternative way to create a principal factor analysis (PFA) model. The `factanal()` program performs ML factor analysis on a data matrix or covariance matrix. While `fa()` supports more factoring methods (in its `fm =` option), its ML implementation has fewer options than does the `factanal()` program. If the researcher prefers the ML methods then the `factanal()` program may be more appealing for this reason. However, note that results using ML do not correspond to the default output for `fa()` or for SPSS.

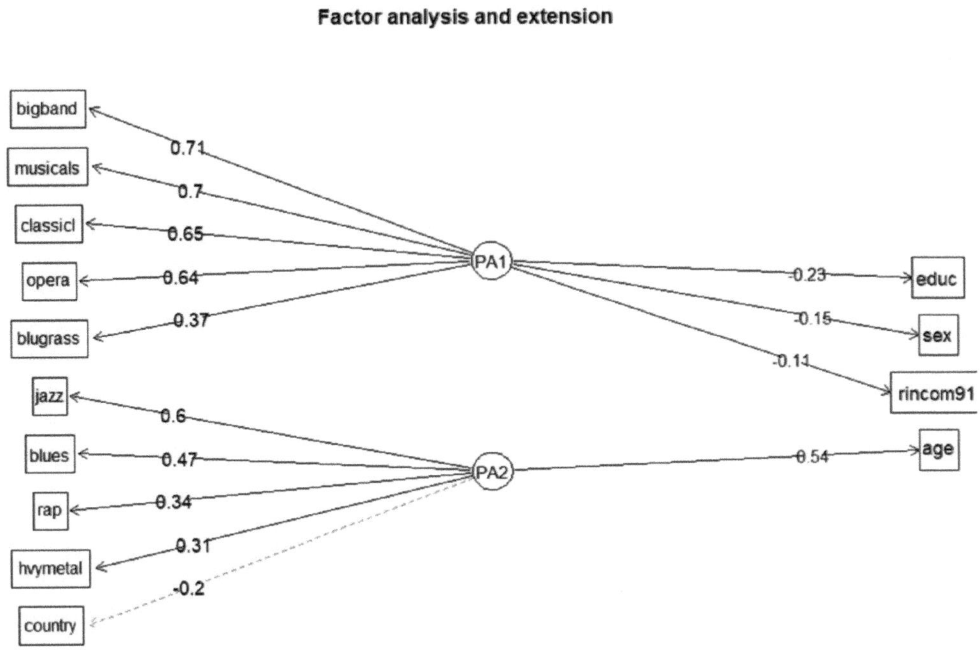

Figure 9.4 Analysis and extension diagram, music2 data.

Setup and data

```
# Set the working directory
base::setwd("C:/Data")

# Read in the "music" data.
base::setwd("C:/Data")
music <- utils::read.csv("musicvars.csv", header=TRUE, sep = ",",
stringsAsFactors=TRUE)

# Delete cases with missing values
# Imputation of missing values would be better (see Garson, 2022).
music2 <- music[stats::complete.cases(music),]

# Limit the dataset fo 10 music preference items. The data are
stored in the data frame "dat".
musicColumns <- base::c(2,3,5,6,8,10,11,13,14,15)
dat <- music2[,musicColumns]
base::names(dat)
[Output:]
    [1] "bigband"   "blues"
    [3] "blugrass"  "classicl"
    [5] "country"   "hvymetal"
    [7] "jazz"      "musicals"
    [9] "opera"     "rap"
```

Determining the number of factors

Prior to running the `factanal()` model, we use the `eigen()` command to display the eigenvalues for the "dat" data. This involves creating a correlation matrix based on dat. We see from the results that by the usual Kaiser criterion, we would request four factors since the first four eigenvalues are 1.0 or higher.

```
corrmatrix <- stats::cor(dat)
base::eigen(corrmatrix)
[Partial output:]
   eigen() decomposition
   $values
      [1] 2.9590359 1.5174835 1.3416894 1.0199456 0.7703126
      [6] 0.6109687 0.5346705 0.4621152 0.4108412 0.3729376
```

The factanal() model

In the `factanal()` model below, we ask for four factors, as discussed in the previous section. We also ask for varimax rotation, which gives an orthogonal solution. We also ask for Bartlett scores. In comparison with regression scores, Bartlett scores, discussed in Chapter 3, are thought to better conform to the original factor structure. Other `scores=` choices are "regression" and "none". Not the case in the example, but if there had been a `covmat=` option, then scores could not be requested. This is because the `covmat=` option bases calculation on either the correlation or the covariance matrix as specified. Scores require input observation-level data, not correlation or covariance matrices.

```
# Run a factanal() model:
factanalResultVarimax <- stats::factanal(dat, factors = 4, rota-
tion="varimax", scores="Bartlett")

factanalResultVarimax
[Output:]
   Call:
   factanal(x = dat, factors = 4, scores = "Bartlett", rotation =
   "varimax")

   Uniquenesses:
    bigband    blues blugrass classicl  country hvymetal
      0.518    0.343    0.511    0.400    0.624    0.585
       jazz musicals    opera      rap
      0.484    0.498    0.489    0.812

   Loadings:
            Factor1 Factor2 Factor3 Factor4
   bigband   0.551   0.232   0.332  -0.121
   blues     0.148   0.769   0.165   0.126
   blugrass  0.189   0.132   0.659
   classicl  0.757   0.141
```

```
country                              0.604
hvymetal                                      0.640
jazz        0.261   0.655                     0.117
musicals    0.676   0.147                    -0.125
opera       0.701   0.102
rap                 0.135                     0.411

                  Factor1 Factor2 Factor3 Factor4
SS loadings         1.956   1.165   0.958   0.657
Proportion Var      0.196   0.116   0.096   0.066
Cumulative Var      0.196   0.312   0.408   0.474

Test of the hypothesis that 4 factors are sufficient.
The chi square statistic is 65.76 on 11 degrees of freedom.
The p-value is 7.75e-10
```

Model appropriateness (p value)

The "p-values" at the bottom of the output above provide a test of the hypothesis that the model has an adequate number of factors. We had specified four factors. If the p-value is <0.05, the model has too few factors to be adequate. That is the case here. We can see in the loadings above, for instance, that "rap" is not well-represented by any factor. Therefore, below we decide to try a five-factor model.

```
# Run a factanal() model for five factors:
factanalResultVarimax <- stats::factanal(dat, factors = 5, rota-
tion="varimax", scores="Bartlett")

factanalResultVarimax
[Output:]
   Call:
   factanal(x = dat, factors = 5, scores = "Bartlett", rotation =
   "varimax")

   Uniquenesses:
    bigband     blues blugrass classicl  country hvymetal
      0.465     0.506    0.248    0.142    0.732    0.862
       jazz  musicals    opera      rap
      0.331     0.416    0.554    0.005

   Loadings:
            Factor1 Factor2 Factor3 Factor4 Factor5
   bigband   0.525   0.254  -0.119   0.316   0.284
   blues     0.162   0.661           0.148
   blugrass  0.175   0.156           0.823  -0.137
   classicl  0.867   0.143                  -0.285
   country                           0.502   0.114
   hvymetal                  0.310          -0.179
```

```
jazz       0.210   0.782
musicals   0.676   0.138                   0.311
opera      0.645   0.155
rap                        0.981           0.146

               Factor1 Factor2 Factor3 Factor4 Factor5
SS loadings      2.010   1.214   1.096   1.073   0.345
Proportion Var   0.201   0.121   0.110   0.107   0.034
Cumulative Var   0.201   0.322   0.432   0.539   0.574

Test of the hypothesis that 5 factors are sufficient.
The chi square statistic is 11.69 on 5 degrees of freedom.
The p-value is 0.0393
```

Above, the test of model appropriateness still shows a significant departure from sufficient fit, though at the much better level of $p = 0.0393$. In fact, if the alpha significance level is 0.01 rather than 0.05, then five factors would be deemed sufficient. We judge five factors to be sufficient on an exploratory basis. This judgment is confirmed by the loadings tale approaching having a simple factor structure, with each variable primarily loaded on one factor with the exception of hvymetal.

Loadings

The portion of model summary output above pertaining to the factor loadings is contained in the element $loadings.

```
factanalResultVarimax$loadings
[Output:]
   Loadings:
            Factor1 Factor2 Factor3 Factor4 Factor5
   bigband   0.525   0.254  -0.119   0.316   0.284
   blues     0.162   0.661           0.148
   blugrass  0.175   0.156           0.823  -0.137
   classicl  0.867   0.143                  -0.285
   country                           0.502   0.114
   hvymetal                  0.310          -0.179
   jazz      0.210   0.782
   musicals  0.676   0.138                   0.311
   opera     0.645   0.155
   rap                       0.981           0.146

                  Factor1 Factor2 Factor3 Factor4 Factor5
   SS loadings      2.010   1.214   1.096   1.073   0.345
   Proportion Var   0.201   0.121   0.110   0.107   0.034
   Cumulative Var   0.201   0.322   0.432   0.539   0.574
```

Though output is not shown here, the command below allows the user to set the number of decimal places and the cutoff for suppressing the printing of coefficients in the table above.

```
print(factanalResultVarimax$loadings, digits = 2, cutoff = .2, sort
= TRUE)
```

It is also possible to assign labels to the factors based on imputation from the loadings. An example is given below. However, this process is subjective, researchers may not agree on labels, and some researchers disparage labeling altogether.

```
base::colnames(factanalResultVarimax$loadings) <- base::c("Tradi-
tional","Blues/Jazz","Rap", "Bluegrass/Country", "Factor 5")

factanalResultVarimax$loadings
[Partial output:]
   Loadings:
            Traditional Blues/Jazz Rap    Bluegrass/Country Factor 5
   bigband    0.525       0.254      -0.119  0.316             0.284
   blues      0.162       0.661              0.148
   blugrass   0.175       0.156              0.823            -0.137
   classicl   0.867       0.143                               -0.285
   country                                   0.502             0.114
   hvymetal                          0.310                    -0.179
   jazz       0.210       0.782
   musicals   0.676       0.138                                0.311
   opera      0.645       0.155
   rap                               0.981                     0.146
```

At first glance , the factor loadings look very different from those for the default `fa()` command. However, `factanal()` only implements ML estimation. In Chapter 4, we used the “pa” (principal axis factoring) method of estimation. This gave results close to SPSS and other major packages. Below, however, we run the `fa()` model with “ml” estimation. The output shows the `factanal()` and `fa()` solutions to approach being the same, rounding apart, when `fa()` uses ML. The uniquenesses, communalities, and complexity coefficients also change between the pa and ml estimation methods.

```
faResultML <- psych::fa(r=music2[,s2],nfactors = 5, fm="ml", rotate
="varimax", covar=FALSE)

faResultML
[Partial output:]
  Standardized loadings (pattern matrix) based upon correlation
  matrix
             ML2   ML4   ML1   ML3   ML5   h2   u2 com
  bigband   0.52  0.25 -0.12  0.32  0.28 0.53 0.465 3.0
  blues     0.16  0.66  0.10  0.15  0.01 0.49 0.506 1.3
  blugrass  0.17  0.16 -0.02  0.82 -0.14 0.75 0.249 1.2
  classicl  0.87  0.14 -0.01 -0.07 -0.28 0.86 0.142 1.3
  country  -0.03 -0.05 -0.01  0.50  0.11 0.27 0.732 1.1
  hvymetal -0.07  0.07  0.31 -0.03 -0.18 0.14 0.862 1.8
  jazz      0.21  0.78  0.08 -0.08 -0.01 0.67 0.331 1.2
  musicals  0.68  0.14 -0.07  0.08  0.31 0.58 0.416 1.6
```

```
opera      0.65  0.15  0.05  0.05  0.03 0.45 0.554 1.1
rap        0.06  0.08  0.98  0.00  0.15 1.00 0.005 1.1
```

Uniquenesses and communalities

Uniquenesses reflect the proportion of variance in a given variable which is not explained by the factor model, while communalities reflect the proportion which is explained by the model. Some researchers drop items for which the uniqueness is greater than the communality. Uniquenesses are retained in the "$uniquenesses" element of the `factanal()` result object. Uniquenesses reflect the propotionCommunalities are simply 1 minus the uniqueness for any input variable. Based on the ML estimation method used by the `factanal()` command,

Uniquenesses

```
factanalResultVarimax$uniquenesses
[Output:]
   bigband     blues  blugrass  classicl   country
 0.4650580 0.5062839 0.2484150 0.1423294 0.7321716
  hvymetal      jazz  musicals     opera       rap
 0.8623077 0.3306949 0.4155420 0.5542794 0.0050000
```

Communalities

```
1 - factanalResultVarimax$uniquenesses
[Output:]
   bigband     blues  blugrass  classicl   country
 0.5349420 0.4937161 0.7515850 0.8576706 0.2678284
  hvymetal      jazz  musicals     opera       rap
 0.1376923 0.6693051 0.5844580 0.4457206 0.9950000
```

Ideally, no variables are poorly measured by the factors, indicated by all communalities being > 0.5. The variables which are poorly measured by this criterion are blues, country, hyvmetal, and opera. Those that are adequately measured by the factor model are bigband, blugrass, classicl, jazz, musicals, and rap.

Factor scores

If scores have been requested in the calling command, then scores are retained in the "$scores" element of the `factanal()` result object. There is one set (row) of scores for each observation in the dataset.

```
base::options(max.print = 10000)
factanalResultVarimax$scores
[Partial output:]
              Factor1         Factor2        Factor3        Factor4
Factor5
1    -0.9512761513 -0.3079181638  0.911697321  2.109077762
```

```
1.214231949
2     -1.2263418705 -1.6074692662  0.239950845  0.963491327
0.158836207
...
1497 -1.5469474849  3.4062007318  0.847099592 -1.511654240
-0.193095871
1499 -0.1134569321 -1.9335134141  0.240553286  0.768363806
-0.152518809
```

The commands below save scores for the first five dimensions as columns in dat2, which is a copy of the original dat data object. The columns Factor1 through Factor5 may be used as variables in regression or other forms of analysis. Note that the save is to dat2 in memory. To save to file, the researcher would have to use the `write.csv()` command.

```
dimscores <- factanalResultVarimax$scores[,1:5]
dat2 <- dat
dat2$Factor1 <- dimscores[,1]
dat2$Factor2 <- dimscores[,2]
dat2$Factor3 <- dimscores[,3]
dat2$Factor4 <- dimscores[,4]
dat2$Factor5 <- dimscores[,5]
View(dat2)
```

Residuals

We can compute the residual matrix using the steps below. We are looking to see if the preponderance of residuals in the residual matrix are low or approach zero.

1 Let "factanalLoadings" be the loadings element of the `factanal()` result object. The factanalLoadings object is of data class "loadings".

```
factanalLoadings <- factanalResultVarimax$loadings
```

2 Create a matrix with uniquenesses on the diagonal and 0's everywhere else. Note that the command `"umatrix <- factanalResultVarimax$uniquenesses"` would create a numeric vector of uniquenesses, not a matrix.

```
umatrix <- base::diag(factanalResultVarimax$uniquenesses)
```

3 Let "observedCormatrix" be the observed correlation matrix. It is also of data class "matrix".

```
observedCormatrix <- factanalResultVarimax$correlation
```

4 Let "expectedCormatrix" be the estimated/fitted correlation matrix, which, of course, is of data class "matrix". The "%*%" operator is the matrix multiplication operator. The `t()` function is the transpose function.

```
expectedCormatrix <- factanalLoadings %*% base::
t(factanalLoadings) + umatrix
```

5 Subtract expected Cormatrix from observedCormatrix to get the residual matrix, which we round to five decimal places.

```
residMatrix <- base::round(observedCormatrix - expectedCormatrix, 5)
residMatrix
[Output:]
          bigband    blues blugrass classicl  country
bigband   0.00000  0.00162  0.00035 -0.00141 -0.00483
blues     0.00162  0.00000 -0.00078  0.00259  0.00585
blugrass  0.00035 -0.00078  0.00000 -0.00003  0.00010
classicl -0.00141  0.00259 -0.00003  0.00000 -0.00029
country  -0.00483  0.00585  0.00010 -0.00029  0.00000
hvymetal  0.01802  0.00868  0.00708 -0.00324 -0.06110
jazz     -0.00189  0.00043  0.00018 -0.00113 -0.00098
musicals  0.00341  0.00754  0.00211  0.00004 -0.01835
opera     0.00431 -0.02466 -0.00246  0.00098  0.02591
rap      -0.00003  0.00000 -0.00001  0.00000  0.00007

         hvymetal     jazz musicals    opera    rap
bigband   0.01802 -0.00189  0.00341  0.00431 -3e-05
blues     0.00868  0.00043  0.00754 -0.02466  0e+00
blugrass  0.00708  0.00018  0.00211 -0.00246 -1e-05
classicl -0.00324 -0.00113  0.00004  0.00098  0e+00
country  -0.06110 -0.00098 -0.01835  0.02591  7e-05
hvymetal  0.00000 -0.01094 -0.00702  0.00620  2e-05
jazz     -0.01094  0.00000 -0.00528  0.01408  1e-05
musicals -0.00702 -0.00528  0.00000 -0.00529  2e-05
opera     0.00620  0.01408 -0.00529  0.00000  0e+00
rap       0.00002  0.00001  0.00002  0.00000  0e+00
```

6 Compute the mean value of absolute residuals. We see that the mean value is well under 0.05, which the researcher may consider acceptable. We can also use the residual matrix to spot outliers. The residual for the correlation of hvymetal with rap is 0.00002, for instance.

```
base::mean(base::abs(residMatrix))
[Output:]
   [1] 0.0051864
```

Visualization

We can plot variables in factor space in the usual manner shown below. We plot observations on the first two factors, though any pair of factors might have been selected. The `plot()` command draws the plot and the `text()` command adds the labels. The result is shown in Figure 9.5.

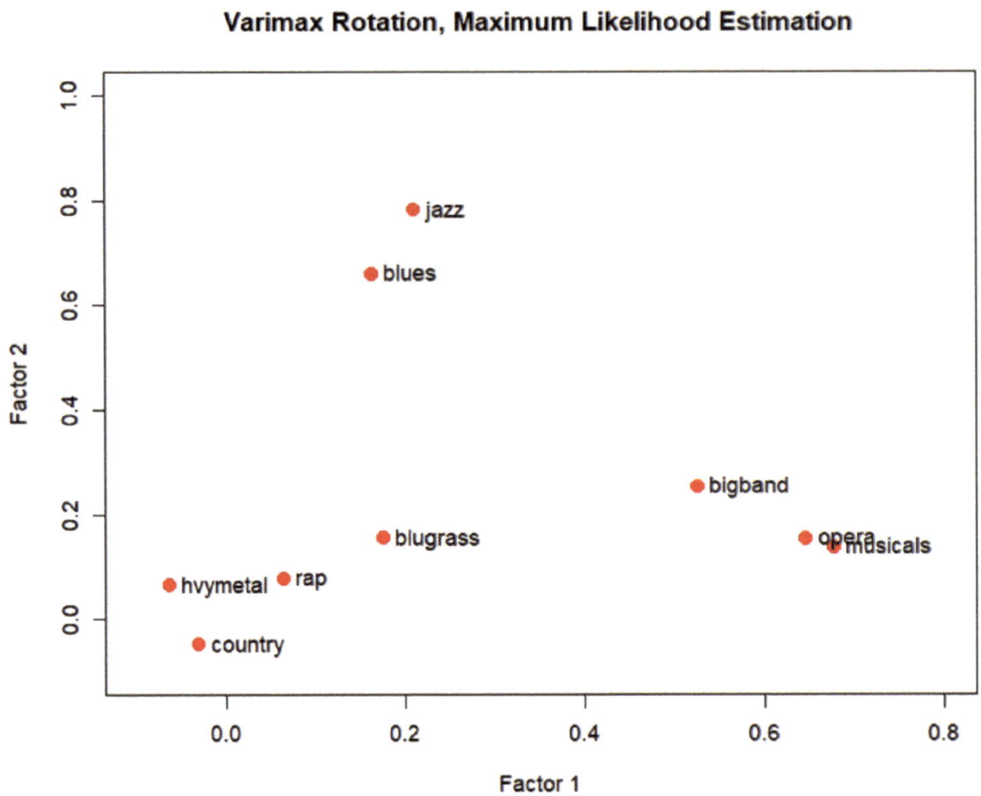

Figure 9.5 Music variables in factor space, first two dimensions, varimax rotation, ML estimation using factanal().

```
base::names(factanalLoadings) <-  base::colnames(dat)
base::plot(x=factanalLoadings[,1], y=factanalLoadings[,2],
    xlab = "Factor 1",
    ylab = "Factor 2",
    ylim = base::c(-.1,1),
    xlim = base::c(-.1,.8),
    pch=16,
    cex=1.5,
    col="red",
    main = "Varimax Rotation, Maximum Likelihood Estimation")
# Add labels
graphics::text(x=factanalLoadings[,1], y=factanalLoadings[,2],
labels = base::colnames(dat), pos = 4)
```

Oblique factor analysis with `fa.promax()`

Introduction

Oblique factor analysis creates correlated factors and therefore is used when the researcher feels able to make the assumption that underlying factors will be correlated.

Promax rotation is the leading oblique alternative to oblimin rotation. Technically, it is an orthoblique method with two stages: computation of an orthogonal solution followed by the transformation of this solution into an oblique one. Promax addresses a number of inherent statistical shortcomings of varimax rotation (see Cureton & Mulaik, 1975). Because promax is computationally faster than oblimin, it is preferred for very large datasets.

Direct oblimin rotation has become the default oblique rotation method in most settings. The "psych" package, for instance, made direct oblimin its default over a decade ago. There are four perspectives pertinent to the choice of rotation methods:

1 Promax solutions are an approximation to oblimin solutions. The oblimin method is seen as giving the exact results, without need for promax's shortcuts now that modern computing has meant that computational speed is rarely an issue.
2 Nonetheless, promax rotation, due to its two-stage computation process, may be seen by the researcher as a desirable compromise between orthogonal and oblique methods.
3 The researcher may run both promax and oblimin solutions on a set of variables, then select the subset which seems stable between the two methods.
4 As discussed in Chapter 3, there is no "correct" rotation method. The best method is the one which gives a solution which is interpretable and advances the researcher's contribution to theory.

The `fa.promax()` program was written by Dirk Enzmann to provide enhancements to promax rotation. This program computes ML factor analysis with promax or varimax rotation. It supports specification of promax power and sorting of loadings. Output includes the correlation matrix of factors and, optionally, matrices of factor scores. Its algorithm differs slightly from that of the `fa()` command, resulting in small differences in coefficients.

Setup

The `fa.promax()` program is not a package but rather is an R function written by Dirk Enzmann. The function is in the file "fa.promax.R", which should be located in the user's working directory. It is a plain text file. The original source is located at http://www.ottersbek.de/software/fa.promax.r. Right-click and select "Save file as" in Chrome (or similar in other browsers).[1] After placing the file "fa.promax.r" in the user's working directory, it is activated by the following commands:

```
base::setwd("C:/Data")
base::source("fa.promax.r")
```

Example

For example data, we use the same "ability.cov" data previously used in Chapters 3, 4, and 7. These data are results of six ability tests (e.g., "reading") and are provided in R's built-in "datasets" package. The contents of ability.cov are shown below.

```
ability.cov$cov
```

```
[Output:]
         general picture  blocks    maze reading   vocab
general   24.641   5.991  33.520   6.023  20.755  29.701
picture    5.991   6.700  18.137   1.782   4.936   7.204
blocks    33.520  18.137 149.831  19.424  31.430  50.753
maze       6.023   1.782  19.424  12.711   4.757   9.075
reading   20.755   4.936  31.430   4.757  52.604  66.762
vocab     29.701   7.204  50.753   9.075  66.762 135.292
```

The fa.promax() model

The purpose here is simply to illustrate the enhanced output made possible by using the `fa.promax()` function rather than `fa()` with `rotation=promax`. Not shown here, but a similar model using `fa()` is obtained by this command: `psych::fa(r=ability.cov$cov, nfactors=2, rotate="promax", fm="ml")`

The `fa.promax()` model and its output is shown below. Coefficients differ slightly between `fa.promax()` and `fa()`. For instance, the correlation of the two requested factors is 0.51 in the former but is 0.54 in the latter. The section of output below labeled "$promax loadings" corresponds to the section of `fa()` output labeled "Standardized loading (pattern matrix)" but again, coefficients are similar but not exactly the same. Also similar but not identical is the section on "$promax.SS" (sum of squares). The "$promax.structure" portion of `fa.promax()` output is similar but not exactly identical to the $Structure element of `fa()` results. The "$uniqueness" portion of `fa.promax()` output is almost identical to the $uniquenesses element of `fa()` results. The primary difference is that `fa.promax()`provides additional sections for "residual variance", "$unrotated.loadings", "varimax.SS", and "varimax.loadings".

```
# The "m" parameter is set to the default power used in the promax
algorithm.
# The default is 3, commonly recommended.
# Higher values of "m" simplify loadings at the cost of greater
inter-factor correlation.
fa.promax(~general+picture+blocks+maze+reading+vocab, factors=2, m=
3, sort = TRUE, covmat = ability.cov)
 [Output:]
  $uniqueness
          residual variance
  reading            0.0524
  blocks             0.2182
  vocab              0.3336
  general            0.4552
  picture            0.5893
  maze               0.7694

  $unrotated.loadings
          Factor1 Factor2
  reading  0.9641 -0.1347
  vocab    0.8154 -0.0391
```

```
blocks   0.4711  0.7483
picture  0.3474  0.5385
general  0.6475  0.3543
maze     0.2530  0.4081

$varimax.SS
               Factor1 Factor2
SS loadings     1.8582  1.7236
Proportion Var  0.3097  0.2873
Cumulative Var  0.3097  0.5970

$varimax.loadings
        Factor1 Factor2
reading  0.9562  0.1821
vocab    0.7848  0.2248
blocks   0.2058  0.8599
picture  0.1561  0.6215
general  0.4994  0.5434
maze     0.1085  0.4678

$promax.SS
               Factor1 Factor2
SS loadings     1.7867  1.7461
Proportion Var  0.2978  0.2910
Cumulative Var  0.2978  0.5888

$promax.loadings
        Factor1 Factor2
reading  1.0027 -0.0601
vocab    0.7975  0.0358
blocks  -0.0606  0.9136
picture -0.0356  0.6583
maze    -0.0368  0.4979
general  0.3727  0.4747

$promax.structure
        Factor1 Factor2
reading  0.9721  0.4513
vocab    0.8158  0.4425
blocks   0.4053  0.8827
general  0.6148  0.6648
picture  0.3001  0.6401
maze     0.2172  0.4791

$corr.factors
        Factor1 Factor2
Factor1    1.00    0.51
Factor2    0.51    1.00
```

```
$n
[1] 112

$chi
objective
    6.107

$df
[1] 4

$p
objective
0.1913263
```

If `fa.promax()` results are sent to a results object (here, "x"), then a variety of output elements are available as named below.

```
x <- fa.promax(~general+picture+blocks+maze+reading+vocab,
factors=2, m= 3, sort = TRUE, covmat = ability.cov)
base::attributes(x)
[Output:]
   $names
    [1] "uniqueness"          "unrotated.loadings" "varimax.SS"
    [4] "varimax.loadings"    "promax.SS"          "promax.loadings"
    [7] "promax.structure"    "corr.factors"       "n"
   [10] "chi"
```

Bayesian factor analysis with BayesFM

Introduction

Bayesian approaches to statistics are increasingly popular. The core idea is that better estimates may be obtained if the researcher incorporates prior knowledge (e.g., results from previous models) into the current model. In this sense, a Bayesian model is a conditional probability model, making estimates conditioned on prior information. Put another way, rather than predicting Y as in regression-based methods, one is predicting Y conditional on priors.

In this section, we illustrate Bayesian factor analysis with the "BayesFM" package authored by Rémi Piatek. There are other R packages for Bayesian factor analysis (links to other packages are at https://www.r-pkg.org/search.html?q=bayesian+factor). The `BayesFM::befa()` command is the primary program within the BayesFM package. The term "`befa()`" stands for "Bayesian Exploratory Factor Analysis" and implements methods set forth by Conti et al. (2014). These methods involve stochastic search on the structure of the factor loading matrix. The number of factors in the solution is not set a priori by the researcher nor is that number a program default. Rather, the number of factors is set during MCMC (Markov Chain Monte Carlo) sampling. However, the researcher may specify the maximum number of factors to be in the solution.

Bayesian methodology is a very large topic, beyond the scope of this book. The reader is referred to such introductions as Crawley (2007) and Robert (2007), or to the tutorials and vignettes that come with the "BayesFM" package.

The steps in the process of building a Bayesian factor model are:

1 Installation and setup
2 Specifying a model
3 Setting initial values
4 Updating the model using some numerical approximation algorithm
5 Summarizing the model
6 Visualizing output
7 Performing posterior predictive checks

Setup

As usual, it is assumed that packages mentioned in `library()` commands have been installed.

```
base::setwd("C:/Data")
base::library(BayesFM)
```

Example

For example data, we use the "fragilestates.csv" data. These data contain 12 measures of state fragility for 179 countries of the world, plus "Country" for the country name in column 1. (This dataset is used later in Chapter 10 with regard to testing for unidimensionality using the `prcomp()` program.)

```
fs <- utils::read.csv("fragilestates.csv", header=TRUE, sep = ",",
stringsAsFactors=TRUE)
View(fs)
# Put Country (column 2 of fs) plus the frailty variables into the
object "frail".
frail <- fs[base::c(2,6:17)]
```

We list variable (column) names in the frail data frame:

```
base::names(frail)
[Output:]
   [1] "Country"     "Security" "Factions"    "Grievance"     "Economy"
   [6] "Inequality" "Flight"    "Legitimacy" "Services"     "Rights"
   [11] "Demog"      "Refugees" "Intervention"
```

The Bayesian model

We now run a Bayesian model on the variables in "frail", omitting the first column (Country) because it is a factor, not a numeric variable. Note that as a random process is used, one must set a random seed to assure the same results on multiple runs. We ask for a model with two latent factors, each with at least three manifest variables. Each manifest variable is allowed to load on at most one latent factor. One way to set the number of factors requested is to apply the Kaiser criterion to the eigenvalues. Below, as only the first two eigenvalues are 1.0 or higher, we call for a two-factor model by setting `Kmax=2`.

```
base::eigen(stats::cor(frail[,2:13]))
[Partial output:]
  eigen() decomposition
  $values
   [1] 8.33144781 1.13197581 0.59216623 0.54133476 0.34056917
   [6] 0.24900193 0.22627393 0.17919393 0.15389509 0.11880107
  [11] 0.07526446 0.06007580
```

```
# Run the MCMC sampler:
base::set.seed(12   3)
model <- BayesFM::befa(frail[,2:13],
    burnin = 1000,
    iter = 10000,
    Nid = 3,
    Kmax = 2)
```

Discussion of `befa()` options used in the model:

- burnin =: The number of initial iterations to skip, excluding them from the model. The default is 1,000.
- iter =: The number of iterations for the model. This must be larger than "burnin" and is typically a very large number. The default is 10,000.
- Nid =: Specifies the minimum number of manifest variables per factor for identification. The default is 3.
- Kmax =: This specifies that maximum number of latent factors. We selected 2 factors based on the number of eigenvalues of 1.0 or higher, as per previous output. Note: Kmax * Nid cannot exceed the number of available manifest variables.

MCMC output must be processed by `post.column.switch()`. Without switching, posterior results for loadings and correlations may not be interpretable. The columns of the factor loading matrix for each MCMC draw, as well as the rows and columns of the correlation matrix of the factors, are reordered by the `post.column.switch()` function to restore the identification of the model a posteriori with respect to the column switching problem. Likewise, the post.sign.switch() function is needed to switch signs on the MCMC draws (sampling) to restore the consistency of the signs of the factor loadings and of the correlations of the latent factors a posteriori.

```
# Post process and summarize:
model <- BayesFM::post.column.switch(model)
model <- BayesFM::post.sign.switch(model)
```

The `summary()` command provides information on the highest posterior probability (HPP) model uncovered by the MCMC sampling process.

```
model.sum <- base::summary(model)
model.sum
[Output:]
   BEFA - Bayesian Exploratory Factor Analysis
   Summary of posterior results

   Maximum number of factors (Kmax) = 2
   Identification restriction (Nid) = 3

   MCMC iterations = 10000
   burn-in period  = 1000

   Metropolis-Hastings acceptance rate = 1

   Posterior frequency of number of latent factors:
     K =  2      100.00%

   ------------------------------------------------------------
   Model parameters

   Factor loadings:

                        dedic    prob    mean      sd    [ 95%    hpd ]
   alpha:Security           1   1.000   5.640   0.314   5.036   6.252
   alpha:Factions           2   0.985   6.918   0.375   6.207   7.673
   alpha:Grievance          2   0.985   5.874   0.335   5.254   6.555
   alpha:Economy            1   0.985   6.096   0.337   5.453   6.771
   alpha:Inequality         1   0.985   5.646   0.308   5.088   6.270
   alpha:Flight             1   0.985   5.526   0.312   4.917   6.134
   alpha:Legitimacy         2   0.985   6.323   0.343   5.665   7.004
   alpha:Services           1   0.985   6.187   0.334   5.540   6.841
   alpha:Rights             2   0.985   5.881   0.325   5.262   6.526
   alpha:Demog              1   0.985   6.236   0.340   5.548   6.877
   alpha:Refugees           1   0.986   5.060   0.292   4.489   5.615
   alpha:Intervention       1   0.985   5.622   0.319   5.022   6.259

   Idiosyncratic variances:

                           mean      sd   [ 95%    hpd ]
   sigma:Security         1.485   0.173   1.171   1.839
   sigma:Factions         1.106   0.194   0.742   1.493
   sigma:Grievance        3.002   0.362   2.312   3.696
   sigma:Economy          1.473   0.174   1.135   1.804
   sigma:Inequality       0.886   0.111   0.675   1.104
   sigma:Flight           1.802   0.208   1.424   2.231
   sigma:Legitimacy       1.106   0.179   0.785   1.473
   sigma:Services         0.672   0.095   0.494   0.862
   sigma:Rights           1.239   0.186   0.910   1.631
```

```
sigma:Demog              0.817    0.109    0.604    1.031
sigma:Refugees           2.332    0.256    1.843    2.845
sigma:Intervention       2.541    0.288    1.989    3.118

Factor correlation matrix:

         mean      sd    [ 95%     hpd ]
R:1:2   0.974   0.005   0.965   0.983
```

Interpreting the summary output above:

1 "Maximum number of factor (kmax)": Our modeling code specified two factors.
2 "Identification restriction (Nid)": Our modeling code set the constraint that each factor should be associated with at least three measured variables.
3 "MCMC iterations": We asked that 10,000 samples be drawn from the data.
4 "burn-in period": We asked that the first 1,000 samples not be used.
5 "Metropolis-Hastings acceptance rate": Low acceptance rates indicate ill-convergence, but this is not a problem here. When we ask for `sum(model$MHacc)` we find all 10,000 samples were accepted.
6 "Posterior frequency of number of latent factors": All samples had two latent factors.
7 "Factor loadings": The "dedic" column shows to which factor the given manifest variable was assigned. For instance, Security is the first variable and it is assigned to Factor 1. Other columns give the probability of the assignment, the mean value across samples, the standard deviation, and the lower and upper 95% confidence limits on the estimated mean. et "model" be a `befa()` results object, then `colMeans(model$alpha)` will return the factor loadings for each variable. Also, `round(colMeans(model$dedic),0)` will return the factor assignments for each variable.
8 "Idiosyncratic variances": Idiosyncratic variance is the difference between total variance and variance associated with the common factors. The lower the idiosyncratic variance of a variable, the more it plays a determining role in defining its assigned factor. Let "model" be a `befa()` results object, the `colMeans(model$sigma)` will return the idiosyncratic variances for each variable. For a discussion of idiosyncratic variances, see Conti et al. (2014).
9 "Factor correlation matrix": Factors created by `befa()` are oblique, not orthogonal. For the two-dimensional solution, there is only one factor correlation, shown in the "mean" column along with its standard deviation and lower and upper 95% confidence bounds.
10 The results object (here, "model", which is of class "befa") will contain the output elements below, all of which are matrices of class "matrix" "array".
   - $alpha: Factor loadings.
   - $sigma: Idiosyncratic variances.
   - $R: Correlation matrix of the latent factors (off-diagonal elements only).
   - $beta: regression coefficients (if any).
   - $dedic: indicators (integers indicating on which factors the manifest variable load).
   - $nfac: Vector of number of 'active' factors across MCMC iterations (i.e., factors loaded by at least Nid manifest variables).
   - $MHacc: Logical vector indicating accepted proposals of Metropolis-Hastings algorithm

Visualization

When applied to a `befa()` results object such as "model", the `plot()` command generates five plots, the most salient of which is the factor loadings plot shown in Figure 9.6. This figure shows that the variables associated with factor 2 are Factions, Grievance, Legitimacy, and Rights. All other variables are associated with factor 1.

```
base::plot(model)
```

Compare Figure 9.6 with Figure 9.7, which is based on allowing as few as two manifest variables for up to six factors. The `befa()` algorithm selects a five-factor model which

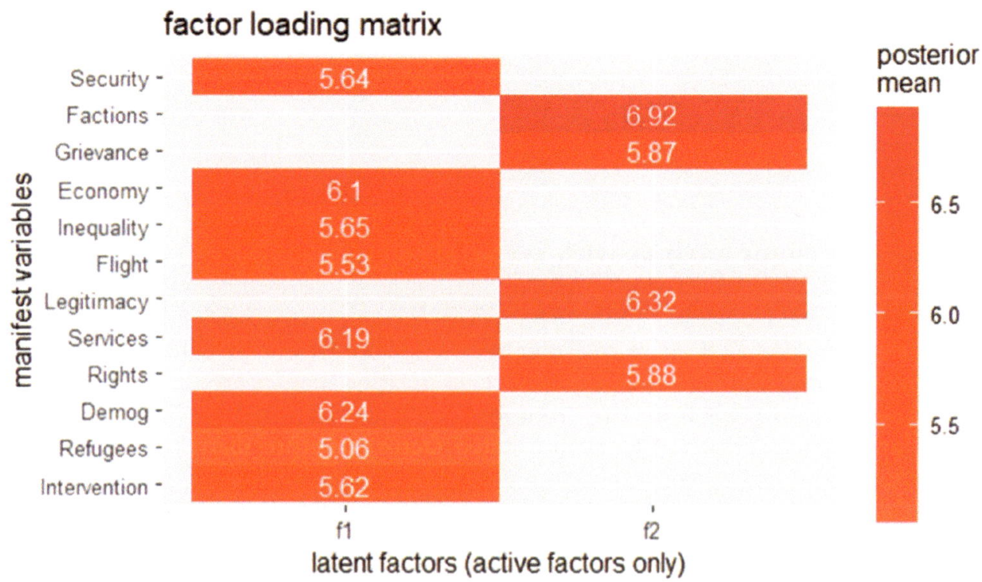

Figure 9.6 Loadings for the two-factor Bayesian model.

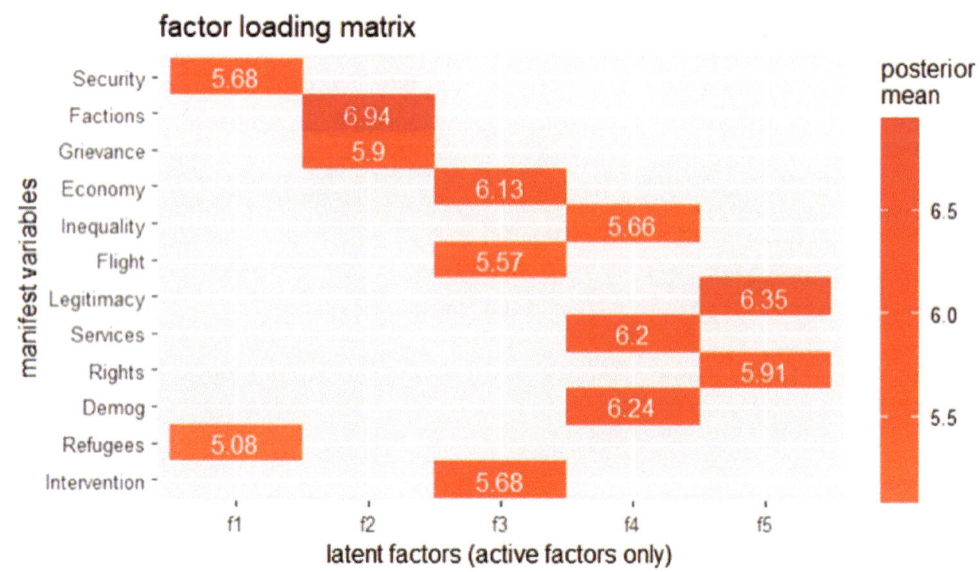

Figure 9.7 Loadings for the five-factor Bayesian model.

differentiates the manifest variables into the corresponding five-factor groupings. For instance, the manifest variables legitimacy and rights are assigned to the same factor (factor 5). Likewise, security and refugees are assigned to the same factor (factor 1).

```
base::set.seed(123)
model2 <- BayesFM::befa(frail[,2:13],
                        burnin = 1000,
                        iter = 10000,
                        Nid = 2,
                        Kmax = 6)
model2 <- BayesFM::post.column.switch(model2)
model2 <- BayesFM::post.sign.switch(model2)
model2.sum <- base::summary(model2)
model2.sum
base::plot(model2)
```

Regularized factor analysis with `fareg()`

Introduction

Regularization is an alternative to factor rotation. Simulations by Scharf and Nestler (2019) have shown that for some purposes, regularization may be preferable to rotation. The `fareg()` command from the "fungible" package, authored by Niels G. Waller (2020), applies regularization to an unrotated factor structure matrix. See Jung and Takane (2008).

Setup and data

As usual, it is assumed that packages mentioned in `library()` commands have been installed.

```
base::setwd("C:/Data")              # Set the local directory
base::library(fungible)             # Has the fareg() command
base::library("psych")              # Has the fa() command

# As in Chapter 4, we use the "ability" data from R's built-in
"datasets" library.
# Invoke the data
utils::data(ability.cov)
# Type help("ability.cov") for data description

# Create the "covmatrix" matrix object from ability.cov, which is a
list.
covmatrix <- ability.cov$cov

# Create the "corrmatrix" matrix object from covmatrix.
corrmatrix <- stats::cov2cor(covmatrix)
```

Loadings and communalities for regularized factor analysis

Regularized PFA may be performed using the `fareg()` command from the "fungible" package, as illustrated below. Let faResultRegularized be the results object of the `fareg()` command. The main output is the $loadings element, which is a matrix of unrotated factor loadings. Below is a list of available result elements:

- $loadings: (Matrix) A matrix of unrotated factor loadings.
- $h2: (Vector) A vector of estimated communality values.
- $L: (Numeric) Value of the estimated penalty parameter.
- $Heywood (Logical) TRUE if a Heywood case is detected (this should never happen).

Type faResultRegularized$loadings to retrieve the loadings matrix.

```
faResultRegularized <- fungible::fareg(corrmatrix, numFactors = 2,
facMethod = "rls")
base::rownames(faResultRegularized$loadings)<- base::rownames(corr-
matrix)
faResultRegularized$loadings
[Output:]
                [,1]        [,2]
   general -0.7662037  0.07407878
   picture -0.5501131  0.36759863
   blocks  -0.6958586  0.40010016
   maze    -0.4204819  0.26139843
   reading -0.7633646 -0.43523986
   vocab   -0.7403272 -0.42556853
```

The following code retrieves the communalities for the six variables in the ability tests data. Recall that communalities are the sum of the squared factor loadings for all factors for a given variable and show the variance in that variable accounted for by all the factors. Below we see that the reading test is best explained by the factors in the regularized factor analysis solution while the maze test is the least well explained.

```
# Get communalities from  fareg() results.
comm1 <- faResultRegularized$h2
# Label them by variable name
comm2 <- stats::setNames(comm1, base::rownames(corrmatrix))
comm2
[Output:]
     general   picture    blocks      maze   reading     vocab
   0.5925557 0.4377531 0.6442993 0.2451342 0.7721593 0.7291930
```

Comparing regularization with unrotated and varimax solutions

In Figures 9.8–9.10, we implement an unrotated PFA solution, a varimax-rotated solution, and a regularized factor solution.

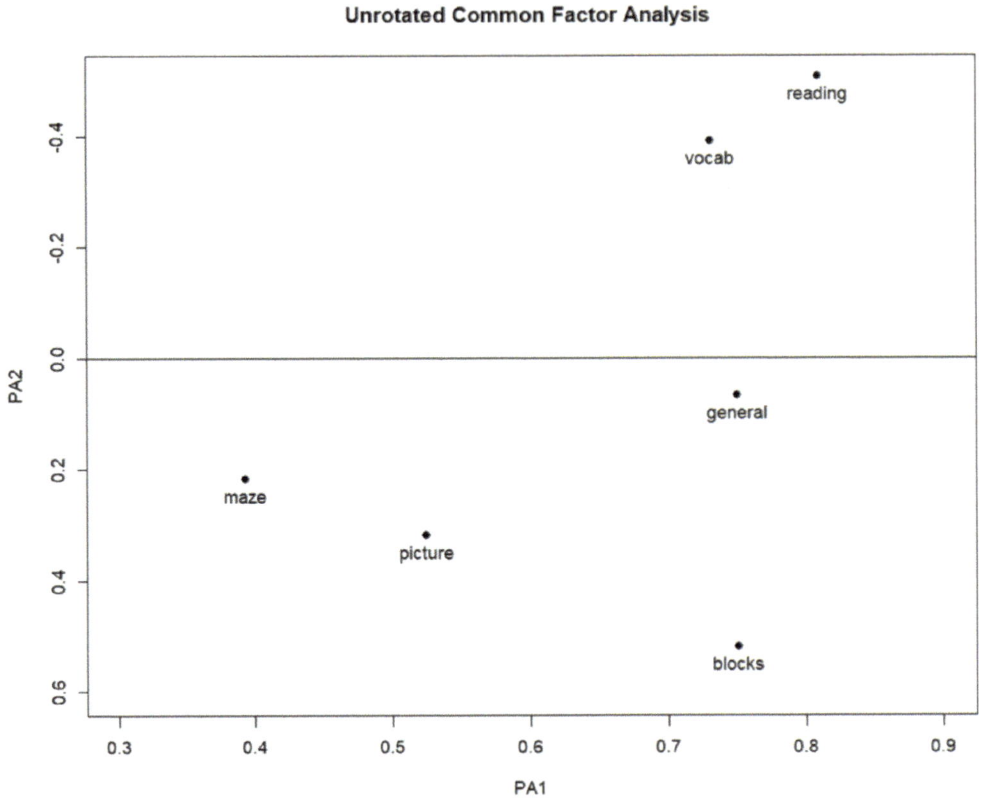

Figure 9.8 Unrotated PFA for ability tests data.

```
# Run the PFA model on the ability test data, with no rotation.
```

```
# Run the PFA model on the ability test data, with no rotation.
```


```
faResultUnrotated <- psych::fa(r=corrmatrix,nfactors = 2, fm="pa",
rotate ="none", covar=FALSE, max.iter=1000)
# Create the vector of test names
testlabels= base::rownames(faResultUnrotated$Structure)
# Visualize the results.
base::plot(faResultUnrotated, labels = testlabels,
xlim=base::c(0.3,0.9), ylim=base::c(0.6,-0.5), main="Unrotated Com-
mon Factor Analysis" )

# Run the PFA model on the ability test data, with varimax rotation.

faResultVarimax <- psych::fa(r=corrmatrix,nfactors = 2, fm="pa", ro-
tate ="varimax", covar=FALSE, max.iter=1000)
# Create the vector of test names
testlabels= base::rownames(faResultVarimax$Structure)
# Visualize the results.
base::plot(faResultVarimax, labels = testlabels,
```

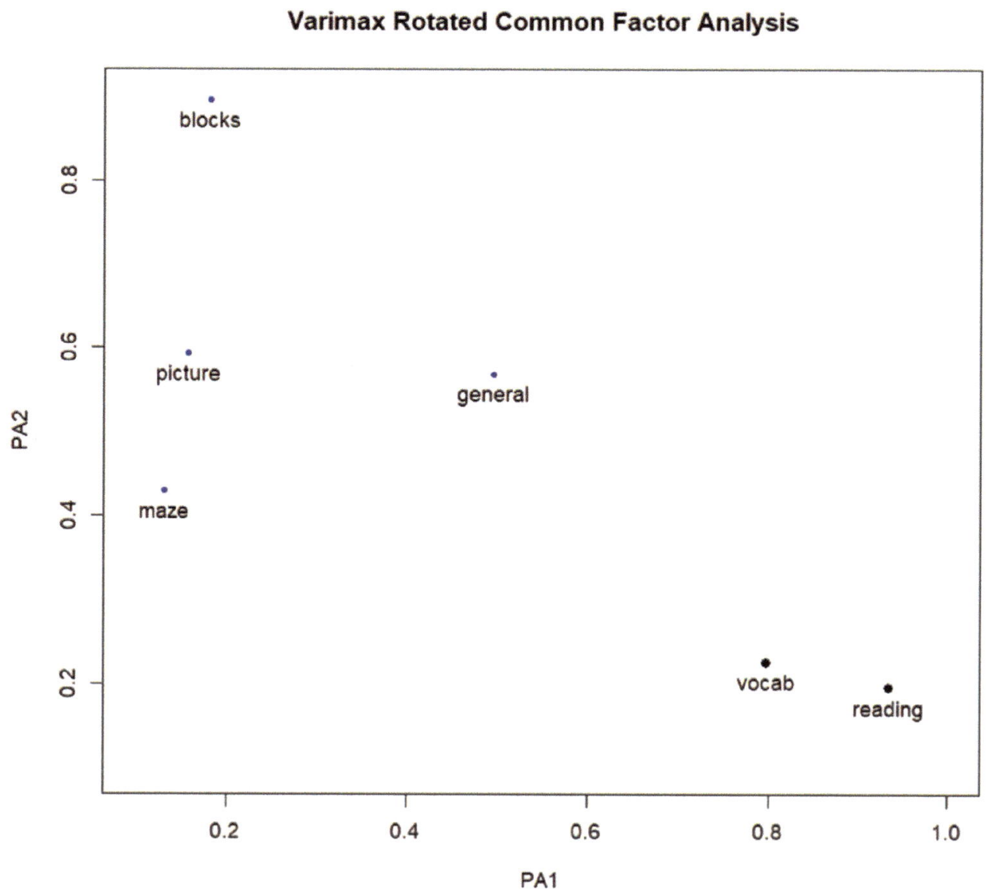

Figure 9.9 Varimax-rotated PFA for ability tests data.

```
xlim=base::c(0.1,1.0), ylim=base::c(0.1,0.9), main="Varimax Rotated
Common Factor Analysis" )

# Run the PFA model on the ability test data, with regularization
# The default facMethod is "rls" for regularized least squares
estimation.
# Change this to "rml" for regularized maximum likelihood
estimation.
faResultRegularized <- fungible::fareg(corrmatrix, numFactors = 2,
facMethod = "rls")

# Visualize the results.
base::plot(faResultRegularized$loadings,pch=16,cex=1.5,
xlim=base::c(-0.80,-0.40),ylim=base::c(-0.5,0.5), xlab="Dimension
1", ylab="Dimension 2", main="Regularized Factor Analysis")
# Add labels
# Create the vector of test names
testlabels= base::rownames(corrmatrix)
```

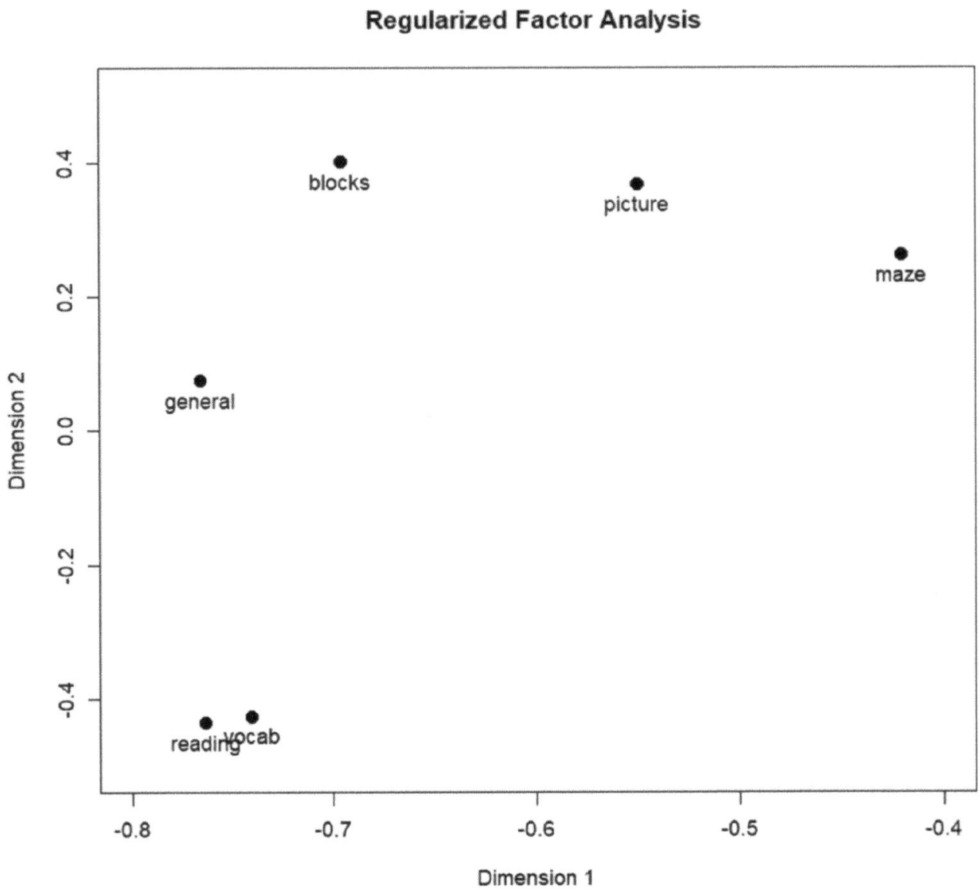

Figure 9.10 Regularized factor analysis for ability tests data.

```
# Place the labels on the plot.
graphics::text(x = faResultRegularized$loadings[,1], y = faResult-
Regularized$loadings[,2], labels = testlabels, pos = 1)
```

In Figures 9.8–9.10, dimension 1 is always anchored by maze-reading and dimension 2 is always anchored by reading-blocks. The difference is that in graphical terms, regularization more clearly shows even more than varimax rotation how the variables are arrayed on the two dimensions.

Note

1 An alternative source for downloading this file is https://github.com/andkov/Steiger_Multivariate/blob/master/functions/fa.promax.R. This method requires cutting-and-pasting. Leave out the example file at the end and be sure to save as plain text.

10 PCA in greater detail

Introduction

In this chapter we discuss the `prcomp()` command from the "stats" package, the `principal()` command from the "psych" package, and the `princomp()` command, also from the "stats" package, in that order. There are three reasons for this seeming redundancy:

Including also the `PCA()` program covered in Chapter 6, all four methods of performing principal components analysis are widely used and therefore are widely found in the literature. Knowing the basics of each method is part of statistical literacy in R.

Each package makes unique contributions (as well as having omissions) and the researcher will want to pick the best tool for the research problem at hand.

We have tried to show different PCA-related tasks for each package, making the coverage more diverse.

In addition to these three programs and the `PCA()` function from the "FactoMineR" package covered in Chapter 6, other common packages for PCA in R include `acp()` from the "amap" package and `dudi.pca()` from the "ade4" package but we do not cover these in this book.

PCA with `prcomp()`

Introduction

In this section, we perform principal component analysis (PCA) with the `prcomp()` command. Though the `PCA()` command discussed in Chapter 6 can yield more detail, `prcomp()` is widely used. Since this command is from R's built-in "stats" base::library, it does not need to be loaded with a `base::library()` command.[1] Some literature suggests that `prcomp()` has higher accuracy as compared to `princomp()` due to its singular value decomposition (SVD) method.[2]

Setup and data

The data used in this section is the music preference data in the "dat" data frame, identical to that used in Chapter 6. This is the complete-cases version of the "music2" data frame on attitudes toward several music genres. Below, we re-create the "dat" data frame. Note that there is a second example of the use of `prcomp()` further below, in the section on "A second biplot example".

DOI: 10.4324/9781003279693-11

```
# Set-up and reading in the data:
base::setwd("C:/Data")
music <- utils::read.csv("musicvars.csv", header=TRUE, sep = ",",
stringsAsFactors=TRUE)

# Create music2 as a complete cases version of music
music2 <- music[stats::complete.cases(music),]
```

Later we will use the sex variable for color-coding observations by gender. As sex was read in as an integer (1 = male, 2 = female). We convert it to a factor for labeling purposes

```
music2$sex <- base::as.factor(music2$sex)

# Create a subset of music2 to analyze and copy it into the object
"dat".
# Column index numbers below are from the names(music2) command.
musicColumns <- base::c(2,3,5,6,8,10,11,13,14,15)
musicsubset <- music2[,musicColumns]
dat <- musicsubset

# Put the correlation matrix in the object datmatrix
datmatrix <- stats::cor(dat,use = "complete.obs")

# musicsubset has 10 variables, so this is the maximum number of
factors
maxfactors <- base::length(musicsubset)
maxfactors
[Output:]
   [1] 10

# View the variables (columns) in the "dat" data frame.
base::names(dat)
[Output:]
   [1] "bigband"  "blues"    "blugrass" "classicl" "country"
   [6] "hvymetal" "jazz"     "musicals" "opera"    "rap"
```

KMO

KMO results for the overall MSA coefficient should be above the threshold of 0.60 to warrant proceeding with PCA with expectation of interpretable results. In Chapter 6, for the `PCA()` command for the same "dat" data frame, it was found that overall MSA was 0.72, above the threshold.

PCA model, component importance, and plotting variables in component space

Below, we run PCA with `prcomp()` on the "dat" data frame of music variables. The "stats::" package prefix is optional but is included here for instructional reasons. The `retx=` option asks that rotated values be returned, which is the default. By setting `center=TRUE` and `scale=TRUE`, variables are standardized (scaled to have unit variance ad mean of 0) prior to analysis. This is not strictly necessary for the dat data frame since all music items were on the same scale.

```
prcompResult <- stats::prcomp(dat, retx=TRUE, center=TRUE, scale =
TRUE)
```

The `attributes()` command reveals the output elements of prcompResult:

```
base::attributes(prcompResult)
[Output:]
    $names
    [1] "sdev"      "rotation" "center"    "scale"     "x"

    $class
    [1] "prcomp"
```

The sdev element, for instance, contains the standard deviations of the principal components. This element may be abbreviated as "sd". Given that the data were centered and scaled, the squared standard deviations are the variances, which are the same as eigenvalues. Standard deviations of 1.0 or greater also have eigenvalues of 1.0 or greater. Note that with each successive component, standard deviations and eigenvalues diminish.

The first four components are above the Kaiser cutoff criterion (above 1.0) to justify their extraction. It is common for the number of extractable components to be approximately the number of variables divided by 3. It is also common for the first few components, or even the first two, to account for the bulk of variance extracted. These, however, are not rules and are not assured results. The coefficients for the eigenvalues (`prcompResult$sdev^2`) are the same as in the "Eigenvalues and the scree plot" section of `PCA()` coverage in Chapter 6 and would be the same whether the option `retx=` is set to TRUE or FALSE.

```
prcompResult$sdev
[Output:]
    [1] 1.7201848 1.2318618 1.1583132 1.0099235 0.8776745
    [6] 0.7816449 0.7312116 0.6797905 0.6409690 0.6106861

prcompResult$sdev^2
[Output:]
    [1] 2.9590359 1.5174835 1.3416894 1.0199456 0.7703126
    [6] 0.6109687 0.5346705 0.4621152 0.4108412 0.3729376
```

The same eigenvalues are given by the `get_eigenvalue()` command from the "factoextra" package, which also returns the variance.percent and cumulative.variance. percent.

```
factoextra::get_eigenvalue(prcompResult)
[Output:]
          eigenvalue variance.percent cumulative.variance.percent
```

```
Dim.1    2.9590359        29.590359                    29.59036
Dim.2    1.5174835        15.174835                    44.76519
Dim.3    1.3416894        13.416894                    58.18209
Dim.4    1.0199456        10.199456                    68.38154
Dim.5    0.7703126         7.703126                    76.08467
Dim.6    0.6109687         6.109687                    82.19436
Dim.7    0.5346705         5.346705                    87.54106
Dim.8    0.4621152         4.621152                    92.16221
Dim.9    0.4108412         4.108412                    96.27062
Dim.10   0.3729376         3.729376                   100.00000
```

Eigenvectors are the "rotation" element. Eigenvectors are used as coordinates to plot variables in component space. Eigenvectors are affected by rotation. Indeed, the point of rotation is to bring variables' coordinates closer to one axis or another in order to improve interpretability.

```
prcompResult$rotation
[Partial output for components 1-4 of 10:]
                 PC1         PC2         PC3         PC4
  bigband  -0.42490265  0.21251002 -0.01050895  0.01800907
  blues    -0.33046965 -0.22873331 -0.33658278 -0.48288160
  blugrass -0.25205551  0.35208430 -0.46320249  0.08011635
  classicl -0.42006262 -0.06973574  0.31635212  0.18070598
  country  -0.06790870  0.47921014 -0.49974623  0.16009928
  hvymetal  0.02360084 -0.46332915 -0.30447423  0.44023152
  jazz     -0.33461849 -0.34458445 -0.12393792 -0.48620860
  musicals -0.42006115  0.09581859  0.25009285  0.15185087
  opera    -0.41403012 -0.02439314  0.20901485  0.30329771
  rap      -0.07247440 -0.44817765 -0.32851373  0.39565793
  …
```

The eigenvectors (rotations) can be used to plot variables in component space using the commands below. The plot is shown in Figure 10.1.

```
# Put variable coordinates (eigenvectors) in the object "coords":
coords <-  prcompResult$rotation
# Create musicLabels as a vector of variable names:
musicLabels <- base::rownames(coords)

# Save the x coordinates to x and the y coordinates to y:
x <- coords[,1]
y <- coords[,2]

# Plot the variables in coordinate space.
# The xlim= and ylim= options expand the plot for better labeling.
# The pch= option sets the symbol to use and cex= sets its size.
base::plot(prcompResult$rotation, pch=16, cex = 1.4, col="blue",
```

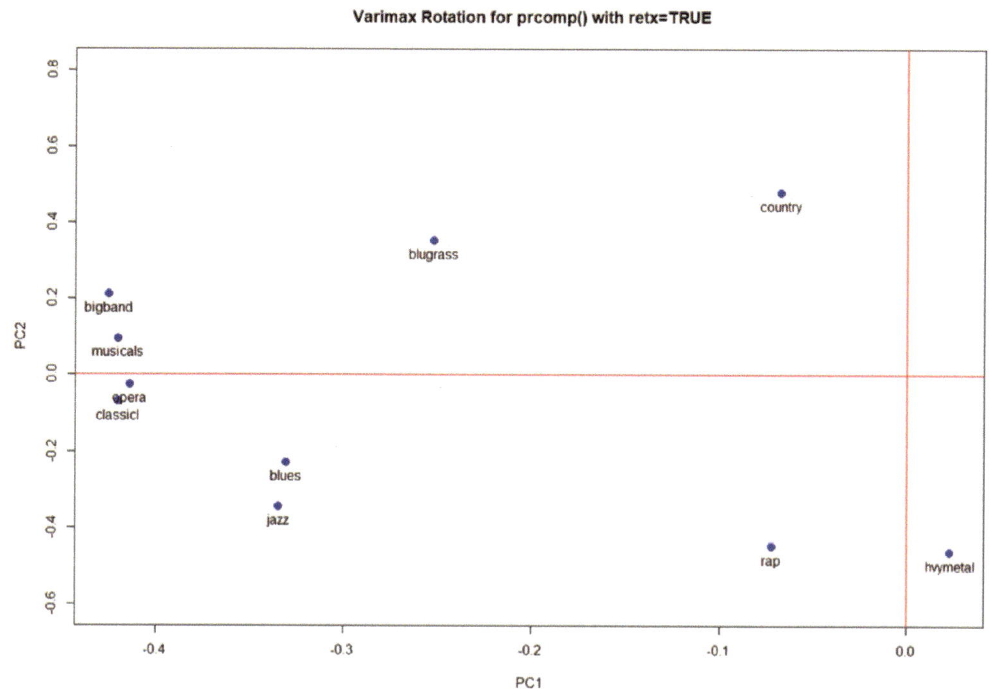

Figure 10.1 Variables plotted in component space using prcomp().

```
ylim = base::c(-0.6, 0.8), main = "Varimax Rotation for prcomp()
with retx=TRUE")

# Add the variable labels and reference lines to create Figure 10.1:
graphics::text(x = x, y = y, labels = musicLabels, pos = 1)
graphics::abline(v=0,col="red")
graphics::abline(h=0,col="red")
```

The `summary()` command returns what may be considered naïve Bayes estimates of the proportion of variance explained by each component. These estimates are the same as the factoextra `PCA()` estimates in Chapter 6 except those were in percentages and the `summary()` output is in decimal format. For these estimates, note that the additional proportion of variance explained drops below 10% after the first four components and below 5% after the first seven components. There are ten components in all since this is the number of variables in the "dat" data frame.

```
base::summary(prcompResult)
[Partial output:]
   Importance of components:
                             PC1    PC2    PC3    PC4     PC5
   Standard deviation     1.7202 1.2319 1.1583 1.0099 0.87767
   Proportion of Variance 0.2959 0.1517 0.1342 0.1020 0.07703
   Cumulative Proportion  0.2959 0.4476 0.5818 0.6838 0.76085
```

```
                          PC6     PC7     PC8     PC9    PC10
Standard deviation     0.7816 0.73121 0.67979 0.64097 0.61069
Proportion of Variance 0.0611 0.05347 0.04621 0.04108 0.03729
Cumulative Proportion  0.8219 0.87541 0.92162 0.96271 1.00000
```

Scree plot and scatterplots

The "factoextra" package supports various multivariate visualizations, including scree plots. The "elbow" of a scree plot is supposed to indicate how many components to extract but this can be a bit subjective. Many prefer to use the Kaiser criterion, which is to extract components with eigenvalues of 1.0 or higher. For these data, that would be four components. In Figure 10.2, note that eigenvalues are the squared standard deviations shown earlier (and also the same as from the `factoextra::get_eigenvalue()` command).

```
base::library(factoextra)
factoextra::fviz_eig(prcompResult,
   choice = "eigenvalue",
   addlabels=TRUE,
   main="Screeplot of eigenvalues, music data")
```

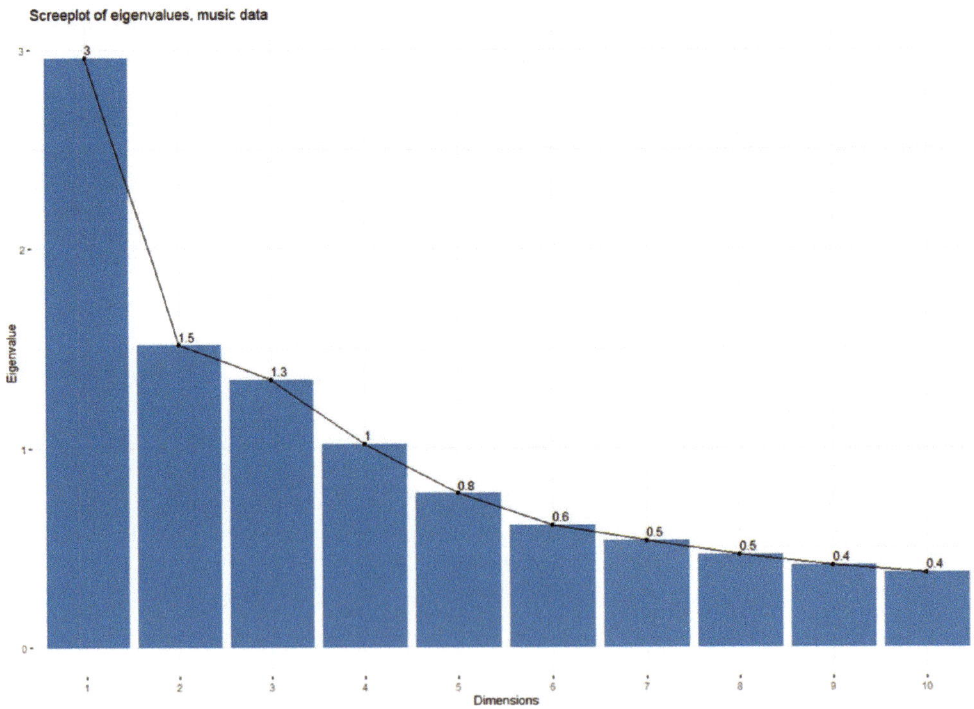

Figure 10.2 Scree plot of eigenvalues, music data.

Cumulative variance plot

The cumulative variance plot in Figure 10.3 shows that in order to have 90% of the variance explained in the example data, we would have to consider eight components. Above, we considered four components, corresponding to almost 70% explained.

```
cumvar <- base::cumsum(prcompResult$sdev^2 /
base::sum(prcompResult$sdev^2))
base::plot(cumvar,
     xlab = "Component",
     ylab = "Amount of  variance explained",
     cex=1.5,
     cex.axis=1.5,
     cex.lab=1.5,
     cex.main =2,
     pch=16,
     col="blue",
     type="b",
     main = "Cumulative variance plot")
graphics::abline(v = 4, col="red", lty=5)
graphics::abline(h = 0.9, col="red", lty=5)
graphics::abline(h = 0.7, col="red", lty=5)
```

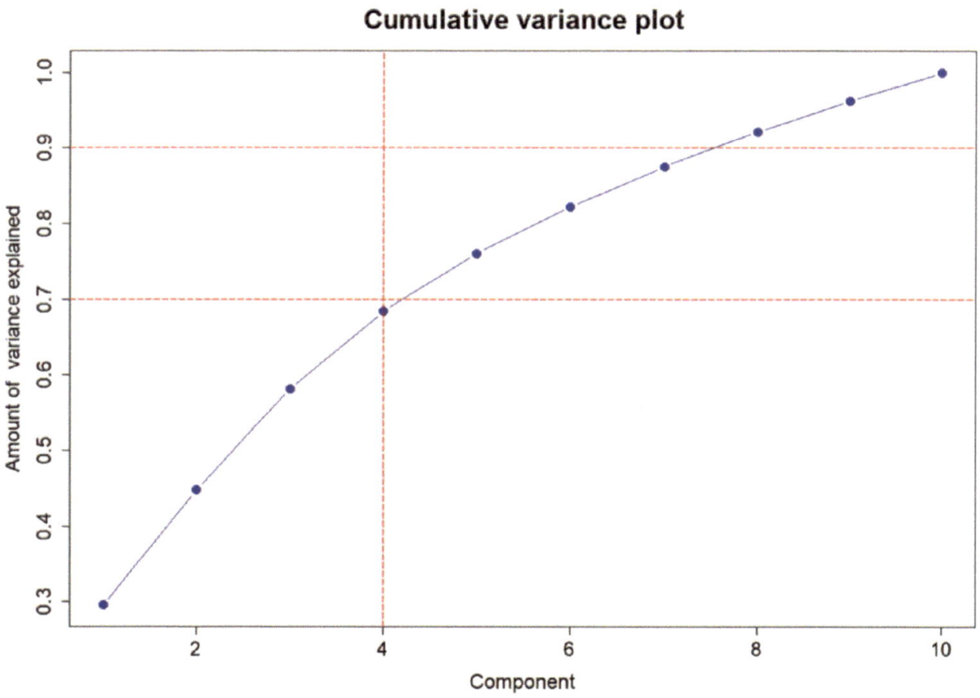

Figure 10.3 Cumulative variance plot, music data.

Scatterplots

Next, we create scatterplots by sex of observations as related to the first four components. To do this we (1) create the data frame dat2 from dat; and (2) add the variable "sex" from the music2 data frame to dat2. As there are ten music preference variables in either dat or dat2, sex gets added to dat2 as column 11.

```
dat2 <- dat
dat2$sex <- music2$sex
```

The panel plot in Figure 10.4 uses the `pairs.panels()` command from the "psych" package. In this figure, the upper triangle contains the correlations. As the components are orthogonal, all correlations are 0. Scatterplots are shown in the lower triangle, with males represented by red dots and females by blue dots. Histograms on the diagonal display the distribution of observations by component, showing no severe departures from a normal distribution.

In the code below, "x" is an element of prcompResult. The term "prcompResult$x[,1:4]" gives the scores of the subjects on the four components. If retx is true (which it is by default), the value of the rotated data (the centered and scaled if requested data multiplied by the rotation matrix) is returned as "x". Hence, cov(x) is the diagonal

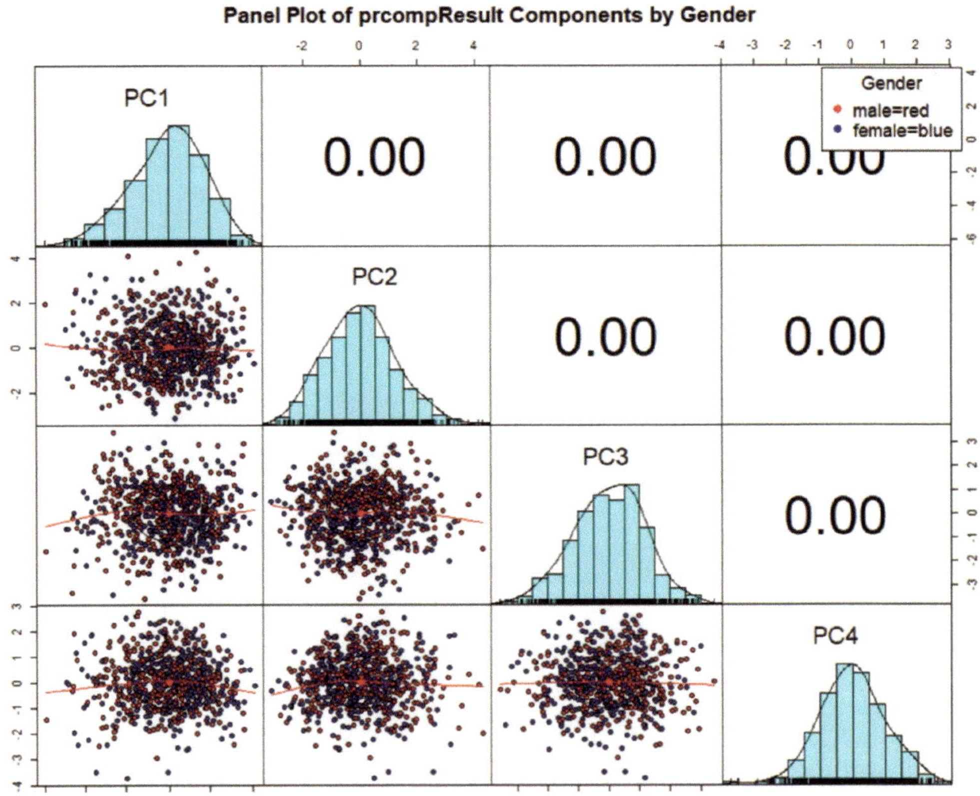

Figure 10.4 Scatterplots, components 1–4, music data.

matrix "diag(prcompResult$sdev^2)" and holds the eigenvalues as shown previously. The pch option sets the symbol shape. The sex variable is coded 1 = male, 2 = female. Setting the bg parameter to red then blue makes red dots the males and blue dots the females.

```
psych::pairs.panels(prcompResult$x[,1:4],
   gap=0, bg = base::c("red", "blue")[dat2$sex], pch=21, main =
   "Panel Plot of prcompResult Components by Gender")
graphics::legend("topright", base::c("male=red","female=blue"),
pch=16, col=base::c("red","blue"),title= "Gender")
```

Observation importance

The "cos2" coefficient, which is the squared cosine of a PCA component with an observation, is a measure of the importance of a given component to a given observation. The cos2 measure is considered a measure of the observation's "quality of representation" in the PCA analysis. In Figure 10.5 we plot the 40 observations with the highest cos2 values for the prcompResult analysis for component 1. If `axes=` were to list two axes, then cos2 in the plot is the sum of cos2 for axis 1 and axis 2. To create this plot, we use the `fviz_cos2()` command from the "factoextra" package.

```
# Create bar chart of top 40 individuals by cos2
base::library(factoextra)
p <- factoextra::fviz_cos2(
     prcompResult,
     # plot individuals rather than variables
     choice ="ind",
     axes = 1,
     fill = "dodgerblue",
     color = "dodgerblue",
     # Sort the top 40 individuals on cos2, descending
     sort.val = "desc",
     # Plot the top 40 individuals with highest cos2 on axis 1; use
top = Inf to plot all
     top = 40,
     xtickslab.rt = 45,
     ggtheme = theme_minimal()
)
p
```

The actual cos2 coefficients are contained in p$data$cos2. Below we list the cos2 value for the first bar in Figure 10.5, which is name = "1334".

```
p$data["1334",]
[Output:]
        name      cos2
   1334 1334 0.9306842
```

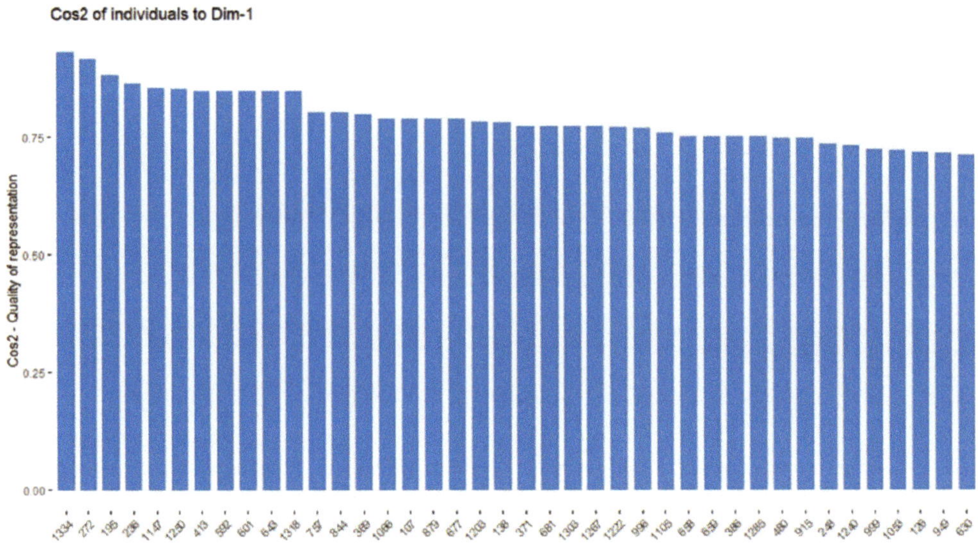

Figure 10.5 Plot of the importance of dimension 1 to the top 40 observations by cos2.

The instruction p$data would list the cos2 for all cases. However, below we list the first several. Note the listings are in name order, not sorted by cos2.

```
utils::head(p$data)
      name          cos2
  1      1 0.034427625
  2      2 0.601656664
  3      3 0.324042426
  7      7 0.451467308
  9      9 0.008089602
  10    10 0.343244212
```

Variable importance

To show variable importance, a variable correlation plot may be displayed using the `fviz_pca_var()` command from the "factoextra" package. This plot is shown in Figure 10.6. We use the result object from the `prcomp()` algorithm.

```
factoextra::fviz_pca_var(prcompResult,
    legend="none",
    labelsize = 5,
    arrowsize = 1.0,
    col.circle = "grey70",
    circlesize = 1.0,
    col.var = grDevices::rainbow(base::ncol(dat)))
```

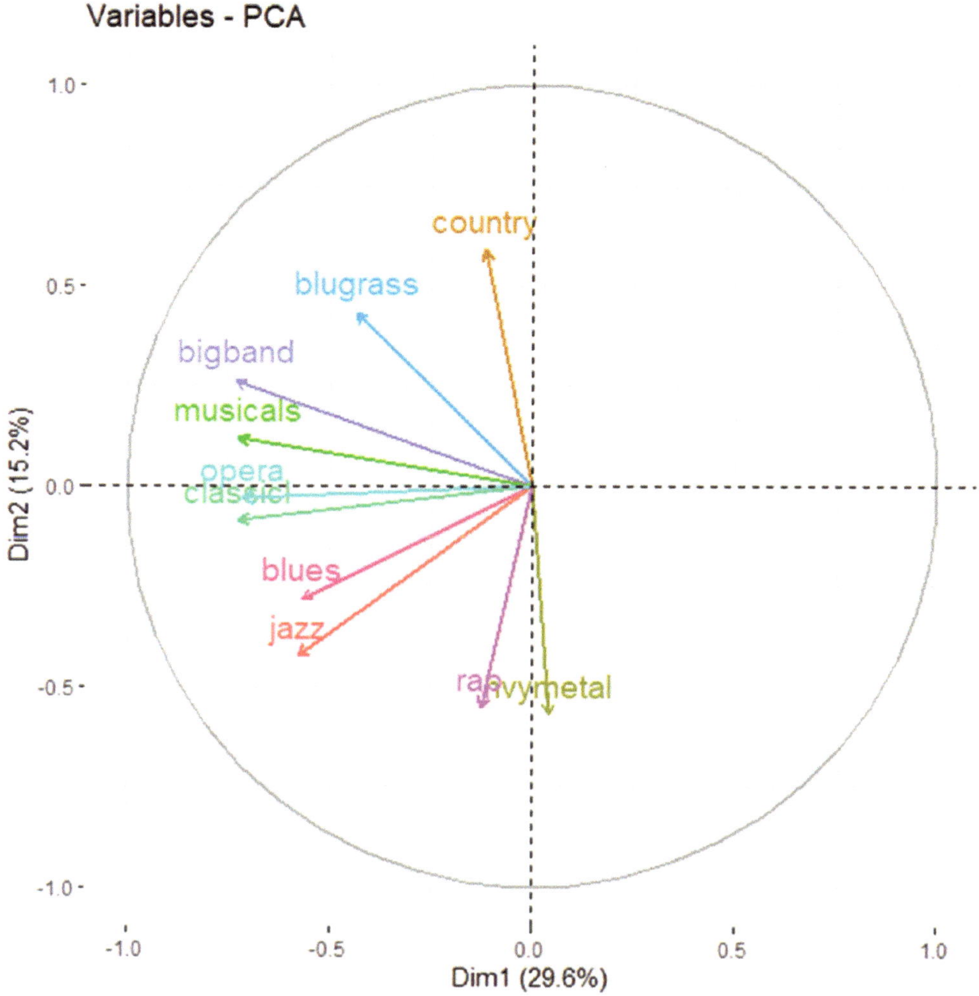

Figure 10.6 Variable correlation plot.

In Figure 10.6, a variable is placed close to the circumference of the variable correlation plot if it is well represented by the principal components. If representation is perfectly based on components 1 and 2, the variable is located on the circumference of the circle. Placement on or near the circumference reflects a high cos2 value. Variables which are not as well represented have lower cos2 coefficients and are located closer to the center of the circle and may be considered less important for the definition of the components.

Variables close to each other have a similar profile (same direction and length) on the components. Variables on opposite sides of the plot, in opposite quadrants, are negatively correlated in projected component space, though they may not be negatively correlated in raw terms.[3] The angle of the variable arrows shows the relation of the variables to the components. The more the arrow is horizontal and further from the origin, the more that variable is associated with PCA 1. The more the arrow is vertical and further from the origin, the more that variable is associated with PCA 2.

It is also possible to obtain a bar chart showing the importance of components 1 and 2 considered jointly to each of the music variables in the PCA analysis using the `fviz_cos2()` command from the "factoextra" package. This is the same as for individuals, but the `choice=` option is set to "var". Though Figure 10.7 is for dimensions 1, note that the `axes=` option may be set to a range of dimensions such as 1:2, or a collection of dimensions (using the `c()` combinations function).

```
factoextra::fviz_cos2(prcompResult,
   choice = "var",
   axes = 1,
   fill = "dodgerblue",
   color = "dodgerblue",
   xtickslab.rt = 45,
   title="Variable importance (contribution) to dimension 1",
   subtitle="Height of bars corresponds to magnitude of the cos2
   coefficient")
```

Loadings

Component loadings are used to infer labels for the components based on which variables load heavily on a given component.

The "rotation" element in a `prcomp()` result contains the eigenvectors. These are sometimes mistakenly called "loadings" since they are a type contribution of the row variable to the column component. Based on eigenvalues or on the `summary()`, output showing components after the fourth accounted for less than 1% of the variance for the example data. Therefore we ask for the eigenvectors only for the first four components.

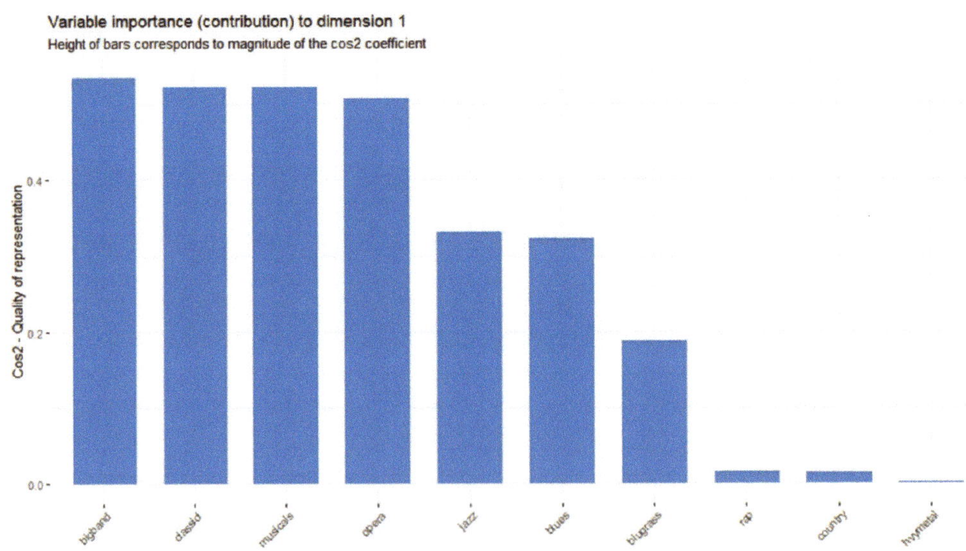

Figure 10.7 Plot of the importance of PCA1 to variables by cos2.

The code below generates the unrotated component loadings for the first four components. Note that `t()` is the transpose matrix command. Loadings are a function of eigenvectors divided by the square root of eigenvalues. These unrotated loadings are almost identical to the unrotated loadings discussed in the "Component loadings" subsection of the previous section on the PCA() function. The results below correspond to the usual factor loadings metric which multiplies the eigenvectors by the square root of the eigenvalues.

```
evectors <- prcompResult$rotation
evalues <- prcompResult$sdev^2
temp <- base::t(evectors)*base::sqrt(evalues)
prcompLoadings <- base::t(temp)
prcompLoadings[,1:4]
[Output:]
                PC1         PC2         PC3         PC4
   bigband  -0.73091109  0.26178298 -0.01217265  0.01818779
   blues    -0.56846888 -0.28176782 -0.38986827 -0.48767349
   blugrass -0.43358206  0.43371920 -0.53653355  0.08091139
   classicl -0.72258534 -0.08590479  0.36643483  0.18249922
   country  -0.11681551  0.59032066 -0.57886264  0.16168803
   hvymetal  0.04059782 -0.57075748 -0.35267652  0.44460017
   jazz     -0.57560565 -0.42448042 -0.14355893 -0.49103351
   musicals -0.72258281  0.11803527  0.28968585  0.15335777
   opera    -0.71220833 -0.03004898  0.24210465  0.30630749
   rap      -0.12466936 -0.55209292 -0.38052178  0.39958425
```

Alternatively, the `eigen.loadings()` function from the "psych" package creates loadings which we may call "component loadings". In output, the columns are PC1 through PC4. The rownames are the ten predictors which were colnames in the "dat" data frame. Note the object within parentheses for `eigen.loadings()` must have been created, like here, from the `eigen()` command or from `princomp()`. This method requires assigning row and column labels. Resulting unrotated loading coefficients are the same.

```
# Create a correlation matrix
dat_matrix <- stats::cor(dat)

#  Compute eigenvalues, used to determine how many components to
retain.
dat_eigen <- base::eigen(dat_matrix)
dat_eigen$values
 [Output:]
    [1] 2.9590359 1.5174835 1.3416894 1.0199456 0.7703126
    [6] 0.6109687 0.5346705 0.4621152 0.4108412 0.3729376

# Compute eigenvectors, used as coordinates to plot variables in
component space.
```

```
dat_eigen$vectors
[Output:]
             [,1]        [,2]        [,3]        [,4]        [,5]
 [1,]  0.42490265 -0.21251002  0.01050895 -0.01800907 -0.05123560
 [2,]  0.33046965  0.22873331  0.33658278  0.48288160 -0.01400821
 [3,]  0.25205551 -0.35208430  0.46320249 -0.08011635 -0.33354877
 [4,]  0.42006262  0.06973574 -0.31635212 -0.18070598 -0.18009170
 [5,]  0.06790870 -0.47921014  0.49974623 -0.16009928  0.25102455
 [6,] -0.02360084  0.46332915  0.30447423 -0.44023152 -0.59629918
 [7,]  0.33461849  0.34458445  0.12393792  0.48620860  0.04035413
 [8,]  0.42006115 -0.09581859 -0.25009285 -0.15185087  0.16395651
 [9,]  0.41403012  0.02439314 -0.20901485 -0.30329771  0.06439548
[10,]  0.07247440  0.44817765  0.32851373 -0.39565793  0.63423014
              [,6]        [,7]        [,8]         [,9]       [,10]
 [1,]  0.566895151 -0.09953625 -0.56554230 -0.114081649  0.32727219
 [2,]  0.005968663  0.02158019  0.33253274 -0.610262527  0.09117204
 [3,] -0.054424380  0.62020429  0.03057592  0.209675008 -0.21854250
 [4,] -0.307108278  0.09920129  0.30989352  0.205435167  0.64118952
 [5,] -0.263712900 -0.55854339  0.13084479  0.072216874  0.15330952
 [6,]  0.130675022 -0.34504850  0.03315338 -0.009641554 -0.07322229
 [7,] -0.172687058 -0.21107196 -0.24760196  0.593384580 -0.16769009
 [8,]  0.436141177 -0.16230255  0.51584756  0.142694057 -0.44593613
 [9,] -0.512900765 -0.03909720 -0.35448341 -0.381301267 -0.39053249
[10,]  0.107652958  0.30278921 -0.03649069  0.072332990  0.13166557

# Compute component loadings for the first four components
temp <- psych::eigen.loadings(dat_eigen)
prcompResultLoadings <- temp[,1:4]

# Assign row and column labels.
base::rownames(prcompResultLoadings) <- base::colnames(dat)
base::colnames(prcompResultLoadings) <- base::c("PC1", "PC2", "PC3",
"PC4")

# Print the unrotated component loadings table.
prcompResultLoadings[,1:4]
[Output:]
                 PC1         PC2         PC3         PC4
bigband   0.73091109 -0.26178298  0.01217265 -0.01818779
blues     0.56846888  0.28176782  0.38986827  0.48767349
blugrass  0.43358206 -0.43371920  0.53653355 -0.08091139
classicl  0.72258534  0.08590479 -0.36643483 -0.18249922
country   0.11681551 -0.59032066  0.57886264 -0.16168803
hvymetal -0.04059782  0.57075748  0.35267652 -0.44460017
jazz      0.57560565  0.42448042  0.14355893  0.49103351
musicals  0.72258281 -0.11803527 -0.28968585 -0.15335777
opera     0.71220833  0.03004898 -0.24210465 -0.30630749
rap       0.12466936  0.55209292  0.38052178 -0.39958425
```

Keeping in mind that the music items were coded in terms of disliking (high values corresponded to disliking), we can say that:

- PC1 was most heavily loaded on not disliking bigband, and not disliking classicl, musicals, or opera, to name the top four. For want of a better label, we call this the "traditional music" component.
- PC2 was most heavily loaded on not disliking rap or hvymetal but disliking country music. These loadings were lower, in the 0.50's. For want of a better label, we call this the "rap & metal" component.
- PC3 was heavily loaded (defined as >= 0.5) on country and blugrass.
- PC4 was most not heavily loaded on any music genre, but had sizable cross loadings for jazz and blues.

The ".dat" data frame reflects relatively weak to moderate loadings on an unrotated basis. which falls short of yielding a simple factor structure, defined as one in which every variable loads strongly (>0.6) on one component and weakly (<0.4) on the rest, with no cross-loadings (no loadings in the 0.4 to 0.6 range).

Normally, however, these judgments would be made on the rotated loadings, which pulls loadings closer to simple factor structure. Some `prcom()` documentation labels the "$rotation" element of a `prcomp()` result as "loadings". The rotation element of prcompResult is shown below for the first four components. These are neither the unrotated nor the rotated component loadings. The `prcomp() command` does not directly output component loadings. Rather the rotations are the eigenvectors, used to plot variables in component space.

```
prcompResult$rotation
[Output:]
                 PC1         PC2         PC3         PC4
  bigband  -0.42490265  0.21251002 -0.01050895  0.01800907
  blues    -0.33046965 -0.22873331 -0.33658278 -0.48288160
  blugrass -0.25205551  0.35208430 -0.46320249  0.08011635
  classicl -0.42006262 -0.06973574  0.31635212  0.18070598
  country  -0.06790870  0.47921014 -0.49974623  0.16009928
  hvymetal  0.02360084 -0.46332915 -0.30447423  0.44023152
  jazz     -0.33461849 -0.34458445 -0.12393792 -0.48620860
  musicals -0.42006115  0.09581859  0.25009285  0.15185087
  opera    -0.41403012 -0.02439314  0.20901485  0.30329771
  rap      -0.07247440 -0.44817765 -0.32851373  0.39565793
```

Next we show the results of the `facto_summarize()` command. The Dim.1 and Dim.2 columns are the unrotated factor loadings as before. The coord and cos2 columns are identical to each other but different from the contrib column.

```
factoextra::facto_summarize(prcompResult,"var",axes=1:2)
[Output:]
             name      Dim.1       Dim.2     coord      cos2    contrib
  bigband  bigband -0.73091109  0.26178298 0.6027613 0.6027613 13.464956
  blues      blues -0.56846888 -0.28176782 0.4025500 0.4025500  8.992477
```

```
blugrass blugrass -0.43358206  0.43371920 0.3761057 0.3761057  8.401745
classicl classicl -0.72258534 -0.08590479 0.5295092 0.5295092 11.828592
country   country -0.11681551  0.59032066 0.3621243 0.3621243  8.089418
hvymetal hvymetal  0.04059782 -0.57075748 0.3274123 0.3274123  7.313992
jazz         jazz -0.57560565 -0.42448042 0.5115055 0.5115055 11.426411
musicals musicals -0.72258281  0.11803527 0.5360582 0.5360582 11.974890
opera       opera -0.71220833 -0.03004898 0.5081436 0.5081436 11.351311
rap           rap -0.12466936 -0.55209292 0.3203490 0.3203490  7.156208
```

The relative importance of each variable on a given component such as PCA1 may be visualized by the `fviz_contrib()` command from the "factoextra" package. Its contrib results are the same as the contrib column above. To visualize for a different dimension, simply change the `axes=` option. The dotted red reference line in Figure 10.8 is the contribution level if all variables made an equal contribution. Five music categories contributed more than average to PCA 1 and five contributed less than average.

Contributions are percentages adding to 100 percent contribution to PCA 1. The first one is bigband, with a contribution of approximately 13.46%. These are calculated as the "cos2" coefficient for a component or combination of components. As discussed earlier, the "cos2" coefficient is the squared cosine of a PCA component. For a given variable and a specified component or set of components, cos2 is a measure of the importance or "contribution" of that variable to the specified component(s). The cos2 measure is considered a measure of the variable's "quality of representation" in the PCA

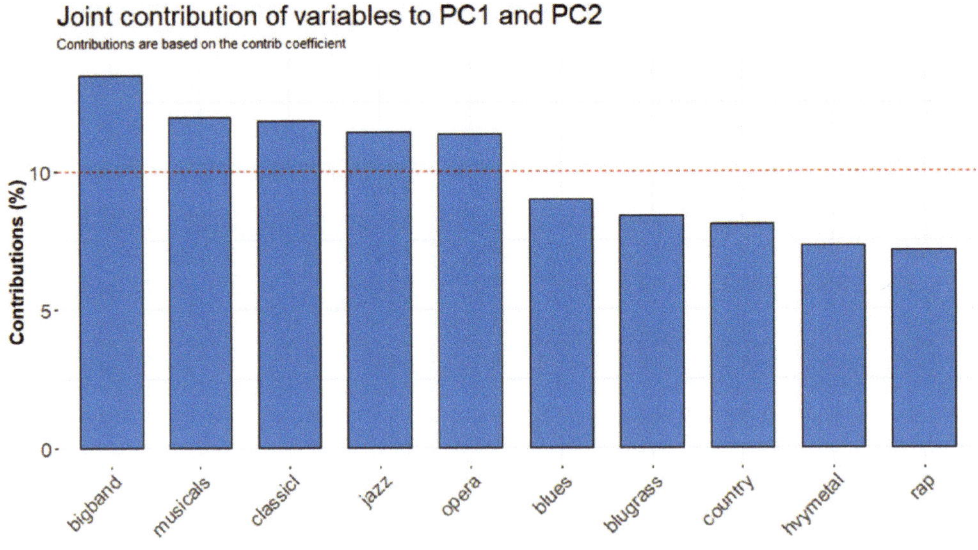

Figure 10.8 Variable contribution to PCA components 1 and 2.

analysis. It is also possible to compute a cos2 coefficient for observations rather than variables, though that Is not done here. The "contribution" is cos2 for a variable (row) divided by the sum of cos2 values representing one or more components. Below we specify components PC1 and PC2, and the contribution is the contribution of a variable to both components jointly.

If desired, the vector of contributions is may be created this way:

```
dd <- factoextra::facto_summarize(prcompResult, element = "var", re-
sult = "contrib", axes = 1:2)

contrib <- dd$contrib
base::names(contrib) <- base::colnames(dat)
contrib
[Output:]
     bigband      blues   blugrass  classicl    country
   13.464956   8.992477   8.401745 11.828592   8.089418
    hvymetal       jazz   musicals     opera        rap
    7.313992  11.426411  11.974890 11.351311   7.156208
```

Finally, we create the plot object, g.

```
g <- factoextra::fviz_contrib(prcompResult,
   choice = "var",
   axes = 1:2,
   color = "black",
   barfill  = "dodgerblue",
   fill = "dodgerblue",
   title="Joint contribution of variables to PC1 and PC2",
   subtitle="Contributions are based on the contrib coefficient",
   xtickslab.rt = 45,
   ggtheme = theme_minimal() +
   ggplot2::theme(plot.title = ggplot2::element_text(size=22)) +
   ggplot2::theme(axis.text=ggplot2::element_text(size=16),
   axis.title=ggplot2::element_text(size=16,face="bold"))
   )
# Plot contributions (Figure 10.8)
g
```

Figure 10.8 and Figure 10.7 have a similar appearance. Figure 10.8 is a contributions bar chart based on the cos2 coefficient expressed as a percentage. Figure 10.7 was a variable importance bar chart based on absolute values of the cos2 coefficient. Also, Figure 10.7 was for variable importance for PCA1 while Figure 10.8 is for joint contributions for PCA1 and PCA2. However, either type of chart could be generated for one or multiple dimensions.

Below is a different way to display the contrib values jointly for components 1 and 2, where g is the graphics object just created:

```
g$data
[Output:]
                  name   contrib
   bigband    bigband 13.464956
   blues        blues  8.992477
   blugrass blugrass  8.401745
   classicl classicl 11.828592
   country    country  8.089418
   hvymetal hvymetal  7.313992
   jazz          jazz 11.426411
   musicals musicals 11.974890
   opera        opera 11.351311
   rap            rap  7.156208
```

Recall that unrotated factor loadings were captured in `prcompResultLoadings[,1:4]`. Judgments about which variables are "heavily loaded" and if simple factor structure exists are usually made with rotated factor loadings. The `prcomp()` program does not output rotated factor loadings. However, we may create them using the `varimax()` command from R's built-in "stats" package.

```
loadingsMatrix <- prcompResultLoadings[,1:4]
rotatedLoadings <- stats::varimax(loadingsMatrix, normalize = TRUE,
eps = 1e-5)
rotatedLoadings
[Partial output:]
   $loadings

   Loadings:
            PC1    PC2    PC3    PC4
   bigband   0.623 -0.137  0.367  0.248
   blues     0.118         0.168  0.862
   blugrass  0.181         0.782  0.162
   classicl  0.817                0.141
   country                 0.842
   hvymetal         0.802
   jazz      0.221                0.843
   musicals  0.785                0.106
   opera     0.801  0.116
   rap              0.780         0.121

                    PC1   PC2   PC3   PC4
   SS loadings    2.418 1.306 1.513 1.601
   Proportion Var 0.242 0.131 0.151 0.160
   Cumulative Var 0.242 0.372 0.524 0.684
```

The output above shows the rotated loadings for prcompResult. By default, loadings less than 0.1 are omitted. We can see the component PC1 is mostly defined by bigband, classicl, musicals, and opera. PC2 is mostly defined by hvymetal and rap. PC3 is mostly defined by blugrass and country. PC4 is mostly defined by blues and jazz. The result comes very close to representing simple factor structure, except that bigband is crossloaded with PC3 at the 0.367 level.

Residual analysis

Residual analysis is useful to assess the goodness of fit of the `prcomp()` model to the data. It was treated in the "Residual analysis" subsection of Chapter 6 on the `PCA()` program. Its use for the `prcomp()` command Is analogous. The `factor.model()` command from the "psych" package serves to reproduce the correlation matrix of the original data from the factor loading matrix. Specifically, the reproduced correlation matrix is the product of the factor loading matrix times its transpose. Residuals are the difference between the observed correlation matrix and the reproduced one. The lower the residuals, the better the fit of the factor model. The higher the residuals, the more likely it is that the model needs additional factors or other respecification.

We compute the reproduced (estimated) correlation matrix based on the unrotated factor loading matrix, here the matrix `prcompLoadings{,1:4}` created earlier. The U2 = FALSE option puts 1's on the diagonal instead of communalities.

```
reproduced <- psych::factor.model(prcompLoadings[,1:4], U2=FALSE)
reproduced
 [Output:]
             bigband      blues   blugrass    classicl     country
   bigband   1.0000000 0.33761422  0.43845287  0.50451600  0.24990469
   blues     0.3376142 1.00000000  0.29398886  0.20311114  0.04690182
   blugrass  0.4384529 0.29398886  1.00000000  0.09420317  0.63034415
   classicl  0.5045160 0.20311114  0.09420317  1.00000000 -0.14890969
   country   0.2499047 0.04690182  0.63034415 -0.14890969  1.00000000
   hvymetal -0.1667087 0.05842017 -0.03995496 -0.02839836 -0.06563460
   jazz      0.3024115 0.74225192  0.10276095  0.31017088 -0.17963324
   musicals  0.5583064 0.18977946  0.22147532  0.64612661  0.01119533
   opera     0.5153187 0.16956814  0.19065443  0.66182911 -0.02516064
   rap      -0.0415068 0.17991939  0.05109471  0.07099906 -0.02647071

             hvymetal        jazz    musicals       opera         rap
   bigband  -0.16670868  0.30241148  0.55830641  0.51531865 -0.04150680
   blues     0.05842017  0.74225192  0.18977946  0.16956814  0.17991939
   blugrass -0.03995496  0.10276095  0.22147532  0.19065443  0.05109471
   classicl -0.02839836  0.31017088  0.64612661  0.66182911  0.07099906
   country  -0.06563460 -0.17963324  0.01119533 -0.02516064 -0.02647071
   hvymetal  1.00000000  0.05122332 -0.13068730  0.03903632  0.62190619
   jazz      0.05122332  1.00000000  0.24892830  0.23754281  0.16453106
   musicals -0.13068730  0.24892830  1.00000000  0.62819159 -0.02403493
   opera     0.03903632  0.23754281  0.62819159  1.00000000  0.13564994
   rap       0.62190619  0.16453106 -0.02403493  0.13564994  1.00000000
```

```
# Compute the observed correlation matrix
observed <- stats::cor(dat)
observed
[Output:]
             bigband      blues      blugrass      classicl      country
bigband   1.00000000 0.29191030  0.3552685022  0.387616359  0.159515495
blues     0.29191030 1.00000000  0.2492435747  0.222994477  0.043999959
blugrass  0.35526850 0.24924357  1.0000000000  0.153399654  0.384974468
classicl  0.38761636 0.22299448  0.1533996542  1.000000000 -0.102384351
country   0.15951549 0.04399996  0.3849744677 -0.102384351  1.000000000
hvymetal -0.09670068 0.06637221  0.0000970178 -0.002071174 -0.101417575
jazz      0.26867198 0.54622704  0.0894267981  0.300304823 -0.087926323
musicals  0.51560288 0.21605862  0.1669519779  0.512220590  0.031025701
opera     0.40006077 0.19440042  0.1721730645  0.570727223  0.026530247
rap      -0.02314970 0.15769581 -0.0200513029  0.011238556 -0.004132701

              hvymetal        jazz    musicals       opera          rap
bigband  -0.0967006771  0.26867198  0.51560288  0.40006077 -0.023149704
blues     0.0663722118  0.54622704  0.21605862  0.19440042  0.157695808
blugrass  0.0000970178  0.08942680  0.16695198  0.17217306 -0.020051303
classicl -0.0020711736  0.30030482  0.51222059  0.57072722  0.011238556
country  -0.1014175748 -0.08792632  0.03102570  0.02653025 -0.004132701
hvymetal  1.0000000000  0.05627548 -0.12171977 -0.01753525  0.278874044
jazz      0.0562754774  1.00000000  0.22997699  0.27009821  0.152292287
musicals -0.1217197749  0.22997699  1.00000000  0.46113771  0.032122487
opera    -0.0175352457  0.27009821  0.46113771  1.00000000  0.102516366
rap       0.2788740438  0.15229229  0.03212249  0.10251637  1.000000000
```

For example, above the reproduced correlation of bigband with blues is approximately 0.338. The observed correlation is approximately 0.292. The residual is observed − expected (which here is reproduced), approximately $0.292 - 0.338 = -0.046$. A rule of thumb is that residuals less than or equal to absolute 0.05 are not large. Thus, this would not be considered a large residual. In a well-fitting factor model, the researcher wants most absolute residuals to be 0.05 or less.

```
# Create the residual matrix
residualMatrix <- observed - reproduced

# Make residuals just the upper triangle of the matrix. This avoids
double-counting.
residuals <- residualMatrix[base::upper.tri(residualMatrix)]
base::mean(residuals)
[Output:]
   [1] -0.03400701

# Get vector for large residuals. This returns TRUE or FALSE for
each correlation.
largeResiduals <- base::abs(residuals) > 0.05
```

```
# Count how many large residuals there are.
# Note that sum() counts, not adds the residual values.
residcount <-base::sum(largeResiduals)
residcount
[Output:]
   [1] 19

# Count how many residuals there are in all.
residtotal <- base::length(residuals)
residtotal
   [Output:]
[1] 45

# Calculate the percentage of residuals which are large.
pctlarge <- residcount/residtotal
pctlarge
[Output:]
   [1] 0.4222222
```

Though a slightly different method was used to arrive at the results above, these are the same values as for residual analysis in the `PCA()` program in Chapter 6. While having 42% of all correlation residuals be absolute 0.05 or greater does show imperfect fit, this is not unusual. A common rule of thumb is to accept PCA results as long as the percent of large residuals does not exceed 50%. Note that the average absolute residual was calculated to be 0.034, below the 0.05 cutoff level.

For space reasons, we do not calculate the mean square residual nor display the histogram of residuals for the `prcomp()` program, but commands and results are the same as shown for the `PCA()` program in Chapter 6.

Biplots and outliers

We now create and interpret a biplot for prcompResult. A biplot is a plot simultaneously showing both variables and observations in relation to the two most important components (the first two) in PCA. Biplots may be produced, among other ways, by the `fviz_pca_biplot()` command from the "factoextra" package. This biplot is Figure 10.9.

```
base::library("factoextra")
factoextra::fviz_pca_biplot(
    prcompResult,
    axes = base::c(1, 2),
    geom = base::c("text"),
    geom.ind = base::c("text"),
    geom.var = base::c("arrow", "text"),
    col.ind = "dodgerblue",
    fill.ind = "white",
    col.var = "darkred",
    fill.var = "white",
    label = "all",
```

```
    addEllipses = TRUE,
    ellipse.level = 0.95,
    title ="PCAresult biplot with outlier ellipse level = 0.95",
     # The remaining options are not used in this example
    # gradient.cols = NULL,
    # invisible = "none",
    # repel = FALSE,
    # habillage = "none",
    # palette = NULL
)
```

Figure 10.9 is flipped compared to the biplot in Figure 6.7 for the `PCA()` method with the same data. However, this is an arbitrary artifact of assigning signs to components and loadings. The interpretation is identical.[4] In this plot, the closer the lines, the higher the correlation of variables and the further apart, the weaker. For instance, the correlation of hvymetal with country or blugrass is very weak and the lines are not close (in fact they go in opposite directions).

To identify outlying observations, an ellipse may be drawn at some level, such as 0.95. Observations outside the ellipse may be considered outliers. The smaller the ellipse level (e.g., 0.80), the smaller the ellipse and the greater the number of outliers.

A second biplot example

For a second biplot example, we use the classic "iris" dataset to show a PCA result for a classification rather than regression problem. The "iris" dataset is one of R's built-in example datasets. It contains five variables: Species (observations are setosa, virginica, or versicolor species) and four predictor variables (sepal width and length, petal width

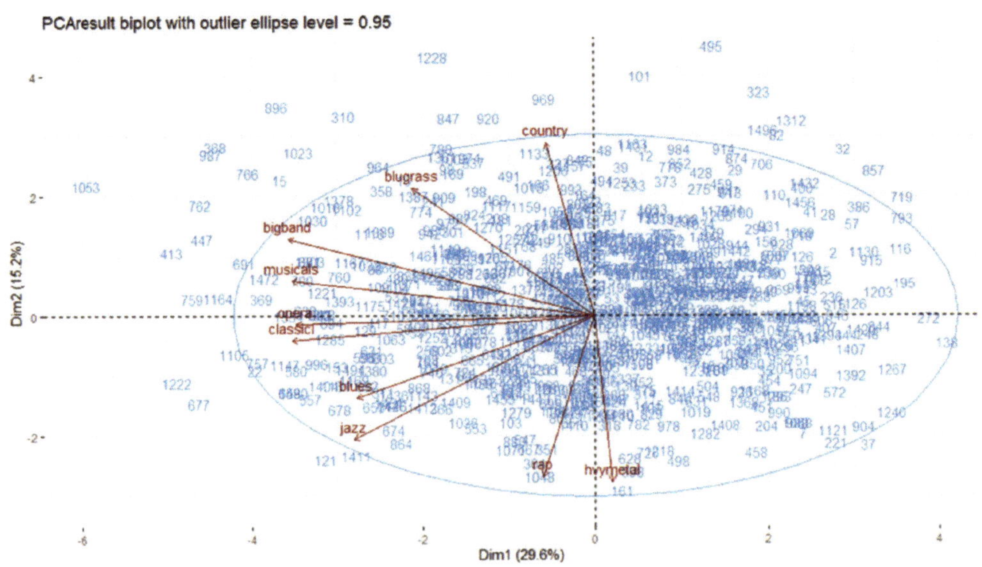

Figure 10.9 Biplot of `prcomp()` solution for music data.

and length). Below we see how well PCA can differentiate the three different flower species.

```
# Setup. The "stats" package is loaded by default as part of R's
System library.
base::library("factoextra")        # For biplot visualization
base::library("ggplot2")           # Used by factoextra
base::library("stats")             # Has the prcomp() command
```

We start by getting the dataset "iris", which is one of the built-in sample datasets in R.

```
utils::data("iris")
```

Next, we list the variables (columns) in iris. We will seek to classify Species by the first four variables.

```
base::names(iris)
[Output:]
   [1] "Sepal.Length" "Sepal.Width"  "Petal.Length"
   [4] "Petal.Width"  "Species"
```

The `summary()` command gives us descriptive statistics. We verify that there are three Species with 50 observations each.

```
base::summary(iris)
[Output:]
      Sepal.Length    Sepal.Width     Petal.Length
      Min.   :4.300   Min.   :2.000   Min.   :1.000
      1st Qu.:5.100   1st Qu.:2.800   1st Qu.:1.600
      Median :5.800   Median :3.000   Median :4.350
      Mean   :5.843   Mean   :3.057   Mean   :3.758
      3rd Qu.:6.400   3rd Qu.:3.300   3rd Qu.:5.100
      Max.   :7.900   Max.   :4.400   Max.   :6.900

      Petal.Width          Species
      Min.   :0.100   setosa    :50
      1st Qu.:0.300   versicolor:50
      Median :1.300   virginica :50
      Mean   :1.199
      3rd Qu.:1.800
      Max.   :2.500
```

The `sapply()` function below provides a way to get a table of predictors by species means. Columns 1–4 in iris are the four predictors and Species is the classification variable. The `sapply()` command loops over the list of four predictors and evaluates the "colMeans" function for "Species". The predictors become rows in the table and the species become columns. We see, for instance, that in comparison with the two other species, the setosa species has wider sepals, narrower petal widths, and shorter petal lengths.

```
# split() divides into groups the data frame or vector in the second
argument (here,
# iris$Species), which must be a factor.
#  sapply() applies colMeans to the groups.
base::sapply(split(iris[1:4], iris$Species), colMeans, na.rm=TRUE)
[Output:]
                setosa versicolor virginica
   Sepal.Length  5.006      5.936     6.588
   Sepal.Width   3.428      2.770     2.974
   Petal.Length  1.462      4.260     5.552
   Petal.Width   0.246      1.326     2.026
```

```
The nrow() function returns the number of data rows in iris. We see
there are 150 observations in the iris data.
base::nrow(iris)
[Output:]
   [1] 150
```

Prior to running PCA, we could divide the iris dataset into a training subset and a validation (test) subset. Then we would develop the PCA model on the training set and cross-validate the solution on the validation set. We do not do so here, but this endnote shows how.[5]

Finally, we perform PCA with `prcomp()` from the "stats" package. In code below, the "-5" leaves out Species, which is column 5 in iris, leaving the four numeric variables.

```
irisResult <- stats::prcomp(iris[,-5],center=TRUE,scale=TRUE)
```

We take the output from `prcomp()`, which is in the object "irisResult", and list its contents. This gives us the component standard deviations and eigenvectors, which are coordinates used for plotting the variables in component space. The eigenvectors are the "$rotation" element in the result. Though these are not "loadings", `prcomp()` documentation notwithstanding, they do show the relationship of the variables to the components (axes).

- Sepal width is clearly associated with PC2.
- Sepal length associates most heavily on PC3 but cross-associates with PC1
- Petal length associates most heavily on PC4 but cross-associates with PC1
- Petal width associates most heavily on PC3 but cross-associates on PC1.

Therefore we expect sepal width to be well differentiated in a biplot further below.

```
irisResult
[Output:]
   Standard deviations (1, .., p=4):
   [1] 1.7083611 0.9560494 0.3830886 0.1439265
```

```
Rotation (n x k) = (4 x 4):
                    PC1         PC2        PC3        PC4
Sepal.Length  0.5210659 -0.37741762  0.7195664  0.2612863
Sepal.Width  -0.2693474 -0.92329566 -0.2443818 -0.1235096
Petal.Length  0.5804131 -0.02449161 -0.1421264 -0.8014492
Petal.Width   0.5648565 -0.06694199 -0.6342727  0.5235971
```

The `summary()` command applied to the principal component result gives us the proportions explained by each component. The first two components explain 95.8% of the variance.

```
base::summary(irisResult)
[Output:]
   Importance of components:
                             PC1    PC2     PC3     PC4
   Standard deviation     1.7084 0.9560 0.38309 0.14393
   Proportion of Variance 0.7296 0.2285 0.03669 0.00518
   Cumulative Proportion  0.7296 0.9581 0.99482 1.00000
```

To visualize the PCA results, we use the `fviz_pca_biplot()` command from the "factoextra" package. The "bi" means the plot visualizes both observations and variables in factor space. By default, it portrays the first two components. The plot is shown in Figure 10.10. The iris biplot shows the components clearly differentiate the species of iris. The larger dot within each ellipse is the species centroid.

```
factoextra::fviz_pca_biplot(irisResult,
    geom.ind = "point",
    pointshape = 21,
    pointsize = 2,
    fill.ind = iris$Species,
    col.ind = "black",
    palette = base::c("dodgerblue","hotpink", "green"),
    addEllipses = TRUE,
    label = "var",
    col.var = "black",
    repel = TRUE,
    legend.title = "Species") +
  ggplot2::ggtitle("PCA biplot, iris dataset") +
  ggplot2::theme(plot.title = ggplot2::element_text(hjust = 0.5))
```

Discussion of Figure 10.10. The y axis shows dimension (component) 2 was associated with sepal width. The x axis shows dimension 1 was associated with the other three predictors. We may say that dimension PC1 is positively correlated with Petal Length, Petal Width, and Sepal Length but is negatively correlated with Sepal Width. By the same token, dimension PC2 is highly correlated with Sepal Width in a negative direction. As seen in `summary()` output above, dimension 1 explains about 73% of the variance and dimension 2 explains approximately another 23%. The closer variable arrows are together, the more highly correlated the variables (e.g., Petal Length and Petal Width).

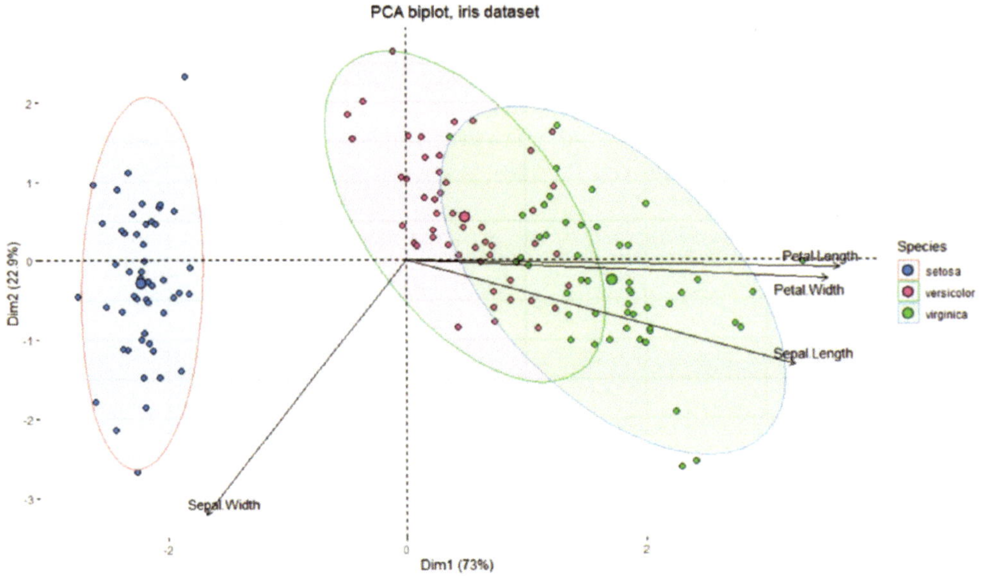

Figure 10.10 Biplot of `prcomp()` solution for iris data, with component ellipses.

Sepal width is only weakly correlated with the other predictors. Observations of the setosa species (blue) are clearly differentiated from other observations, particularly by sepal width. The versicolor and virginica species are not as well differentiated from each other.

Testing for unidimensionality

Sometimes the purpose of factor analysis is not so much to identify multiple dimensions which underlie a set of variables as it is to determine if the set is unidimensional. An example is provided by the "fragilestates.csv" data. This is 2021 data on 179 nations of the world, rated by The Fund for Peace on the "fragility" of their governance. Fragility was an index composed of these 12 variables:

Security	C1: Security Apparatus
Factions	*C2: Factionalized Elites*
Grievance	C3: Group Grievance
Economy	E1: Economy
Inequality	E2: Economic Inequality
Flight	E3: Human Flight and Brain Drain
Legitimacy	P1: State Legitimacy
Services	P2: Public Services
Rights	P3: Human Rights
Demog	S1: Demographic Pressures
Refugees	S2: Refugees and IDPs
Intervention	X1: External Intervention

On each variable, nations were rated from 0 to 10. The 12 indicators were then summed for a "Total" fragility score, which varied in principle from 0 to 120. The observed range was from low fragility of 16.2 for Finland to high fragility of 111.7 for Yemen. The question we examine here is whether these 12 fragility indicators form a single dimension, justifying merging them in the "Total" additive index. We use the `prcomp()` program for PCA to answer this question.

1 First we read in and examine the "fragilestates" data.

```
base::setwd("C:/Data")
fs <- utils::read.csv("fragilestates.csv", header=TRUE, sep =
",", stringsAsFactors=TRUE)

frail <- fs[base::c(2, 6:17)]  # Just the frailty vars plus
Country in col. 1)

base::names(frail)
[Output:]
   [1]  "Country"       "Security"      "Factions"
    [4] "Grievance"     "Economy"       "Inequality"
    [7] "Flight"        "Legitimacy"    "Services"
   [10] "Rights"        "Demog"         "Refugees"
   [13] "Intervention"
```

2 Next we run `prcomp()` without the non-numeric "Country" (name) variable, which is column 1.

```
frailResult <- stats::prcomp(frail[,-1], center=TRUE, scale =
TRUE)
```

3 The "sdev" element of frailResult contains the standard deviations of the principal components. The square is the eigenvalue. Either way, values equal to or above 1.0 meet the Kaiser criterion for retaining the component for extraction and interpretation. We see that the first two components meet this criterion. On this basis, the set of 12 frailty indicators is not unidimensional, but we will explore it further.

```
frailResult$sdev
[Output:]
    [1] 2.8864247 1.0639435 0.7695234 0.7357546
    [5] 0.5835830 0.4990009 0.4756826 0.4233130
    [9] 0.3922947 0.3446753 0.2743437 0.2451037
```

Eigenvectors are found in the "rotation" element of the result. These are coordinates which may be used to plot variables in coordinate space. They are not "loadings" but do show the relationship of the variables to the axes (components). Below we list the eigenvectors for the first two dimensions.

```
frailResult$rotation[,1:2]
```

```
[Output:]
                       PC1         PC2
   Security     -0.3072730 -0.08461090
   Factions     -0.2991090 -0.30313918
   Grievance    -0.2287426 -0.51780473
   Economy      -0.2871035  0.32060010
   Inequality   -0.2994111  0.20887755
   Flight       -0.2671300  0.35385529
   Legitimacy   -0.2948739 -0.32459570
   Services     -0.3149397  0.23761558
   Rights       -0.2873869 -0.33979932
   Demog        -0.3028926  0.21371003
   Refugees     -0.2813413 -0.04955728
   Intervention -0.2841379  0.18916412
```

4 Using the `fviz_pca_var()` command from the "factoextra" package, we visualize the relation of variables to the first two component dimensions. This is shown in Figure 10.11, where we see the "Grievance" indicator is the most deviant from Dimension PC1. Correspondingly, it has the lowest eigenvector (rotation) on PC1.

```
base::library(factoextra)
factoextra::fviz_pca_var(
    frailResult,
    axes = base::c(1, 2),
    geom = base::c("arrow", "text"),
    geom.var = base::c("arrow", "text"),
    addEllipses = TRUE,
    repel = TRUE,
    col.var = "steelblue4",
    fill.var = "white",
    col.circle = "grey40",
    select.var = base::list(name = NULL),
    title = "Variable loading plot, fragilestates data"
)
```

In the code above, repel = TRUE (which is not the default) tries to avoid overplotting variable labels. Under select.var =, the name= option would have restricted the display to the named variable(s). For instance, name = c("Refugees","Economy") would just show these two variables, though ellipses will not be plotted for just two variables. Also, it is possible to define groups of variables but that is not done here. Rather, there is just the one group of all variables and hence only one ellipse. The "geom" options ask for both arrows and text (variable labels).

Figure 10.11 shows variable eigenvector (coordinate) arrows which are the same as one would see in a biplot. The ellipse highlights the fact that the set of variables are associated negatively with dimension PCA1 but are both positive and negative on dimension PCA2. For PCA2, the closer to the horizontal 0 line, the lower the absolute eigenvector on this dimension.

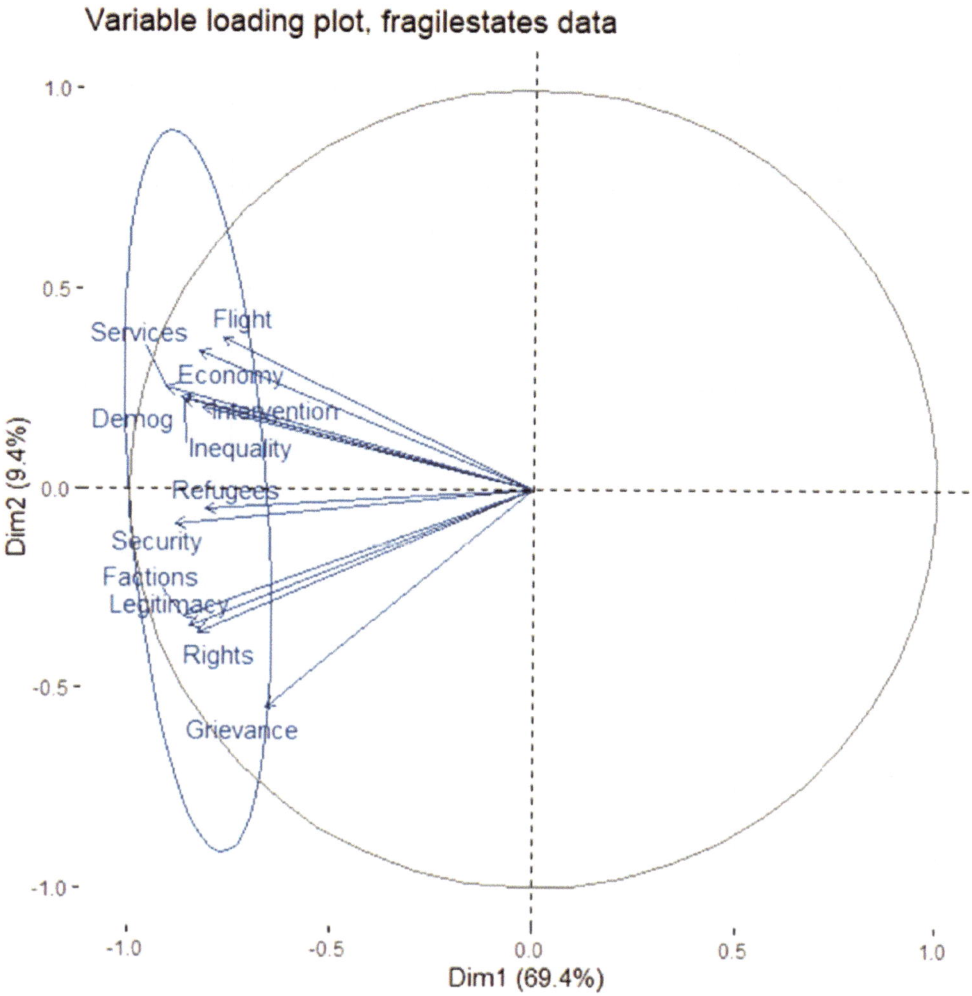

Figure 10.11 Variable loading plot from `prcomp()`, fragilestates data.

5 The usual criterion for a scale being reliable is that Cronbach's alpha is 0.80 or higher (some even allow as low as 0.60). For the frailty data, alpha is 0.96, within confidence limits of 0.95 to 0.97. This illustrates that a scale may be highly reliable by Cronbach's alpha but still not be unidimensional by PCA. In the more stringent view of scaling, a scale should be both.

```
psych::alpha(frail[,-1])
[Partial output:]
   Reliability analysis
   Call: alpha(x = frail[, -1])

    raw_alpha std.alpha G6(smc) average_r
        0.96       0.96    0.97       0.66
   ...
```

```
   lower alpha upper     95% confidence boundaries
  0.95   0.96   0.97
```

6 Next we see if we may obtain a scale which is both reliable and unidimensional by dropping "Grievance" (which is column 4), which the plot showed to be the most deviant on the first component. When we look at the "frailResult2" results using the "sdev" element, now only the first component meets the Kaiser criterion. Most (72%) of the variance may be explained by PCA1 alone, as shown in summary() output below. Moreover, Cronbach's alpha shows that the 11-item scale (without Grievance) is just as highly reliable as the 12-item scale (alpha = 0.96). We conclude that based on the view that a scale should be unidimensional as well as reliable, the "Total" frailty score should be based on the 11 items and "Grievance" frailty should be reported separately.

```
frailResult2 <- stats::prcomp(frail[,-base::c(1,4)], cen-
ter=TRUE, scale = TRUE)

frailResult2$sdev
[Output:]
    [1] 2.8156038 0.9654307 0.7619694 0.6644726
    [5] 0.5060817 0.4985713 0.4234814 0.4211076
    [9] 0.3467480 0.2762939 0.2454885

# Unidimensionality can be confirmed visually:
factoextra::fviz_pca_var(frailResult2, col.var = "black",
repel=TRUE)

[This figure is not shown for space reasons. It is similar to
Figure 10.10 but without the Grievance arrow.]
base::summary(frailResult2)
[Partial output:]
   Importance of components:
                             PC1     PC2     PC3     PC4
   Standard deviation     2.8156 0.96543 0.76197 0.66447
   Proportion of Variance 0.7207 0.08473 0.05278 0.04014
   Cumulative Proportion  0.7207 0.80543 0.85821 0.89835

psych::alpha(frail[,-base::c(1,4)])
[Partial output:]
   Reliability analysis
   Call: psych::alpha(x = frail[, -c(1, 4)])

     raw_alpha std.alpha G6(smc)
         0.96      0.96    0.97
```

Active and supplemental datasets

Sometimes, by research intent or necessity, one develops a PCA model based on a subset of "active" observations. Later, however, it may be useful to plot another subset, which may be new observations, placing these points in factor space as calculated based on the active set. This second subset is called the "supplemental" set.

Though not directly illustrated in this section, a common application would be to display observations for a training set of data, used to create the PCA model, then to plot both training and validation set points on the same plot, where the PCA1 and PCA2 axes were defined by the training set.

To illustrate active and supplemental datasets, we continue to use the frailty data discussed in the previous section. However, we divide it into an active subset, a supplemental subset, and a residual set. For instructional/visualization reasons, we keep the active and supplemental sets small.

1 We read in the frailstates.csv data as before, into the object "fs". We then subset fs into the data frame "frail", which has just the frailty variables plus Country in column 1.

```
base::setwd("C:/Data")
fs <- utils::read.csv("fragilestates.csv", header=TRUE, sep =
",", stringsAsFactors=TRUE)

frail <- fs[base::c(2, 6:17)]
```

2 We then partition the sample into active, supplemental, and residual subsets. Because sampling involves random processes, we set a random seed. A different seed will lead to different samples and different results.

```
base::set.seed(123)
frailsamp <- base::sample(3, base::nrow(frail), replace=TRUE,
prob=base::c(0.12, 0.08, 0.80))

frailsamp
[Output:]
    [1] 3 3 3 1 2 3 3 1 3 3 2 3 3 3 3 1 3 3 3 2 1 3 3 2 3 3 3
   [28] 3 3 3 2 1 3 3 3 3 3 3 3 3 3 3 3 3 3 3 3 3 3 1 3 3 3 3
   [55] 3 3 3 3 1 3 3 3 3 3 1 3 1 1 3 3 3 3 3 3 3 3 3 3 3 3 3
   [82] 3 3 3 3 3 2 1 1 3 3 3 3 3 3 3 3 3 3 3 3 3 3 2 3 1 1 3
  [109] 3 3 2 3 3 2 3 3 3 2 3 3 3 3 3 3 3 2 3 3 3 3 3 1 3 3 3
  [136] 3 1 3 2 3 3 3 3 3 1 3 3 3 3 3 1 3 3 3 3 3 3 3 3 3 3 3
  [163] 3 3 3 3 3 3 3 3 3 3 1 3 3 3 3 3 1

base::sum(frailsamp==1)          # The active set
[Output:]
  [1] 20
```

```
base::sum(frailsamp==2)                    # The supplemental set
[Output:]
   [1] 12

base::sum(frailsamp==3)                    # The residual set
[Output:]
   [1] 147

# Create and name the active and supplemental sets.
frailActive <- frail[frailsamp==1,]
frailSupp <- frail[frailsamp==2,] # Show the first three rows of
the active set.
frailActive[1:3,]
[Output for rows 1 through 3]
          Country Security Factions Grievance Economy
   4  South_Sudan      9.7      9.2       8.8     9.2
   8        Sudan      8.1      9.1       9.5     9.1
   16     Burundi      8.0      8.2       7.3     8.8

      Inequality Flight Legitimacy Services Rights Demog
   4         8.9    6.7        9.5     10.0    8.7   9.5
   8         8.3    7.7        8.8      8.8    8.6   9.1
   16        7.2    5.6        9.3      7.9    9.2   9.2

      Refugees Intervention
   4       9.9          9.3
   8       9.4          8.7
   16      8.2          8.2
```

3 Run `prcomp()` without the Country (name) variable, which is column 1, and also without the "Grievance" variable (column 4, for reasons given in the previous section). Output is for the active set of observations, the location of which is shown in Figure 10.12.

```
fsactiveResult <- stats::prcomp(frailActive[,-base::c(1,4)],
center=TRUE, scale = TRUE)

# Plot of variables against components
factoextra::fviz_pca_var(fsactiveResult, col.var = "black")

[Figure not shown for space reasons. It is similar to Figure
10.11 but without the Grievance arrow.]
```

As with the entire frail data, only PCA 1 has a standard deviation and therefore a variance (eigenvalue) above 1.0, confirming unidimensionality.

```
fsactiveResult$sdev
[Output:]
     [1] 3.06312372 0.88284400 0.74711560 0.59891572 0.55382108
     [6] 0.50596135 0.35449388 0.31330742 0.28201111 0.19802546
    [11] 0.10694101 0.06522968

# Plot of active observations against components
g <- factoextra::fviz_pca_ind(fsactiveResult,
  xlim=base::c(-5.6,6.5),
  ylim=base::c(-1.4,2.5),
  col.ind = "cos2",
  gradient.cols = base::c("blue","green","red"),
  repel = TRUE,
  legend.title = "Importance (cos2)") +
  ggplot2::theme_bw() +
  geom_text(ggplot2::aes(label = frailActive$Country), hjust =
  0.5,  vjust = -1) +
  ggplot2::theme(text = ggplot2::element_text(size=20),
  axis.text.x = ggplot2::element_text(angle=45, hjust=1))

g <- g + ggplot2::labs(title = " Plot of active frailty countries,
prcomp solution", x = "Component PC1", y = "Component PC2")

base::print(g)
```

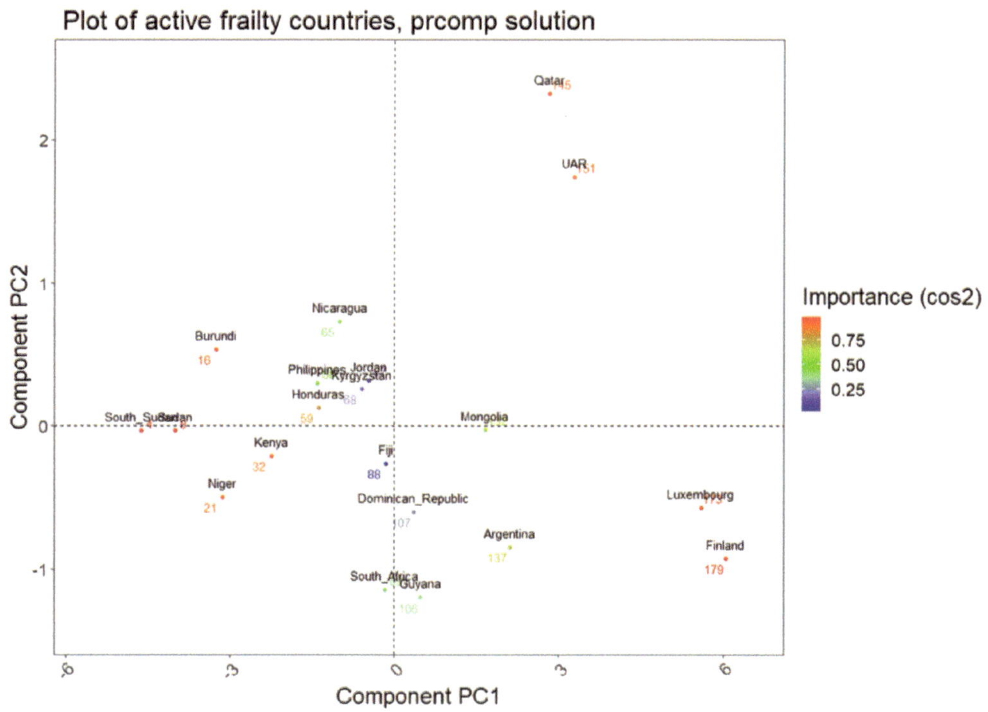

Figure 10.12 Active observations plotted in component space.

4 We now add the supplemental observations in the "frailSupp" object created above. This is a set of 17 "supplemental" cases. Adding is done with the `fviz_add()` command from the "factoextra" package. Figure 10.13 repeats Figure 10.12, but with supplemental points added. Active observations are shown in black and supplemental in blue. The supplemental data was not used in the computation of the PCA1 and PCA2 dimensions, only displayed in the component space defined by PCA1 and PCA2, which were created based on the active set.

```
# Plot of active cases (like Figure 10.12, but without active
points color-coded by cos2 importance.)
g2 <- factoextra::fviz_pca_ind(fsactiveResult,
  xlim=base::c(-5.6,6.5),
  ylim=base::c(-1.4,2.5),
  col.ind = "red",
  repel = TRUE,
  legend.title = "Importance (cos2)") +
  ggplot2::theme_bw() +
  geom_text(ggplot2::aes(label = frailActive$Country),
  col="red",hjust = 0.5,  vjust = -1) +
  ggplot2::theme(text = ggplot2::element_text(size=20),
    axis.text.x = ggplot2::element_text(angle=45, hjust=1))

g2 <- g2 + ggplot2::labs(title = " Fragile states:", subtitle
="Active observations in red, supplemental in blue", x = "Compo-
nent PC1", y = "Component PC2")

# Display the plot so far, but this is not shown here as it is
not the final step.
base::print(g2)

# Get coordinates of supplemental observations, then add them to
the plot.
# Note the use of predict() to apply the PCA result from the ac-
tive set
# to the new data in the supplemental set.  This gives Figure
10.13.
ind.sup.coord <- stats::predict(fsactiveResult,
newdata = frailSupp)

coord <- base::as.data.frame(ind.sup.coord)
base::rownames(coord) <- frailSupp$Country

factoextra::fviz_add(g2,
coord,
    #ind.sup.coord,
    color ="blue",
    addlabel= TRUE
    )
```

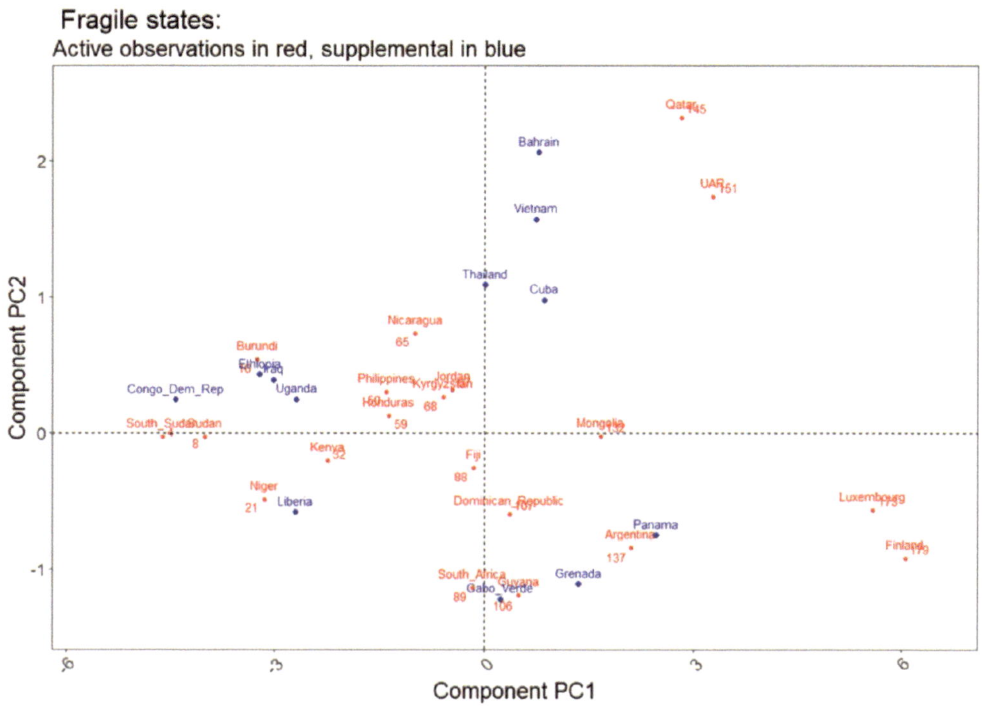

Figure 10.13 Active and supplemental observations plotted in component space.

Prediction with factor scores

Factor analysis is ordinarily thought of as a "non-dependent" procedure, meaning there is no dependent variable. However, factor analysis may be performed on a set of predictors, without the DV, after which the DV may be predicted from the factor scores under a cross-validation design. This can be illustrated based on the "iris" example data, where there were four predictors (sepal and petal width and length) and there was a dependent variable (Species). For this case, it is possible to use factor scores to predict Species in a validation test dataset from component scores calculated based on a training sample.[6] We illustrate the steps below.

1 Create the train and test sets. Recall the iris data is one of the datasets in R's built-in "datasets" package.

```
utils::data(iris)
base::set.seed(123)
```

We divide the population of 150 observations randomly into two groups. The mysamp = 1 group is the training group with 111 observations while the mysamp = 2 group is the test sample with 39 observations.

```
mysamp <- base::sample(2, base::nrow(iris), replace=TRUE,
prob=base::c(0.75,0.25))
mysamp
[Output:]
    [1] 1 2 1 2 2 1 1 2 1 1 2 1 1 1 1 2 1 1 1 2 2 1 1 2 1 1 1 1 1
   [30] 1 2 2 1 2 1 1 2 1 1 1 1 1 1 1 1 1 1 1 1 2 1 1 2 1 1 1 1 2
   [59] 2 1 1 1 1 1 2 1 2 2 2 1 2 1 1 1 1 1 1 1 1 1 1 1 2 1 1 2
   [88] 2 2 1 1 1 1 1 1 1 2 1 1 1 1 1 1 2 1 2 2 1 1 1 2 1 1 2 1 1
  [117] 1 2 1 1 1 1 1 1 1 2 1 1 1 1 1 2 1 1 1 1 2 2 2 1 1 1 1 1 2
  [146] 1 1 1 1 1

base::sum(mysamp==1)
[Output:]
  [1] 111
base::sum(mysamp==2)
[Output:]
  [1] 39
```

We now create the train and test groups. The train data frame has five variables for 111 cases. The test data frame has 39 cases for the same five variables.

```
train <- iris[mysamp==1,]
test <- iris[mysamp==2,]
```

2 Use the train data frame to create the PCA model, in this case using the `prcomp()` command. Note that we remove the fifth variable, which is the DV, "Species". Results are placed in the object "irisPCAtrain", which is of class "prcomp".

```
irisPCAtrain <- stats::prcomp(train[,-5] , center=TRUE,
scale=TRUE)
```

Note that irisPCAtrain contains the principal component loadings for each of the predictor variables used for classification, based on the training data. There are four principal components because that is the maximum number given that there are four input variables.

```
irisPCAtrain
[Output:]
   Standard deviations (1, .., p=4):
   [1] 1.7269869 0.9274809 0.3691816 0.1449146

   Rotation (n x k) = (4 x 4):
                        PC1        PC2        PC3        PC4
   Sepal.Length   0.5235012 0.35586597  0.7280657 -0.2631088
   Sepal.Width   -0.2926943 0.92682966 -0.1950892  0.1313661
   Petal.Length   0.5739480 0.03759012 -0.1397879  0.8059963
   Petal.Width    0.5575484 0.11372413 -0.6421214 -0.5136990
```

3 Use irisPCAtrain to make predictions for the training set, then add Species back into the predictions data frame object called trainpred.

```
trainpred <- stats::predict(irisPCAtrain,train)
trainpred <- data.frame(trainpred, train[5])
utils::head(trainpred)
[Output:]
   1  -2.312079  0.5526473  0.1695292 -0.030009671  setosa
   3  -2.387031 -0.2899027 -0.0339668 -0.043515978  setosa
   6  -2.159484  1.6089421  0.0543102  0.004723069  setosa
   7  -2.485381  0.1295934 -0.3076762  0.030364575  setosa
   9  -2.333102 -1.0851265 -0.1653977  0.003255221  setosa
   10 -2.197358 -0.4371658  0.2569328  0.020500437  setosa
```

4 Repeat for testpred. Note that the test set is used based on the training irisPCAtrain solution.

```
testpred <- stats::predict(irisPCAtrain,test)
testpred <- base::data.frame(testpred, test[5])
utils::head(testpred)
[Output:]
              PC1        PC2         PC3         PC4 Species
   2  -2.086943 -0.6472119  0.22789913 -0.12424610  setosa
   4  -2.314405 -0.5513981 -0.09108714  0.04881811  setosa
   5  -2.445728  0.7323756  0.03460327  0.03337850  setosa
   8  -2.272243  0.2890043  0.12039492  0.01627820  setosa
   11 -2.230081  1.1294067  0.33187698 -0.01626170  setosa
   16 -2.386897  2.8475373  0.09994949 -0.02494784  setosa
```

5 Since Species has three levels, we now use multinomial regression to predict Species from the principal components PC1 and PC2 based on the training object (trainpred). We could use all four principal components, but the first two account for most of the variance. Multinomial regression requires that one of the species be made the reference level. Using the `relevel()` command, we designate setosa as the reference level. The `multinom()` command is from the "nnet" package. Results are placed in the object "mymodel", which is of data classes "multinom" and "nnet".

```
base::library(nnet)
trainpred$Species <- stats::relevel(trainpred$Species,
ref="setosa")
mymodel <- nnet::multinom(Species~PC1+PC2, data=trainpred)
base::summary(mymodel)
[Output:]
   Call:
   nnet::multinom(formula = Species ~ PC1 + PC2, data =
   trainpred)
```

```
Coefficients:
            (Intercept)      PC1       PC2
versicolor   10.351378 14.68261 -3.078157
virginica     2.629596 21.95801 -3.726156
Std. Errors:
            (Intercept)      PC1      PC2
versicolor    234.6882 163.5922 197.4197
virginica     234.6975 163.6031 197.4212
Residual Deviance: 26.95592
AIC: 38.95592
```

6 Use a classification (confusion) table to assess the prediction results for the training sample. In the classification table, rows are predictions and columns are observed values. Correct classifications are those on the diagonal. For this example, there are 7 errors out of 111 cases in the train sample. The results of the prediction were placed in the object "predsTrain", which is of data class "factor" (because multinomial regression is used on a categorical DV). The table in ctableTrain is of data class "table".

```
predsTrain <- stats::predict(mymodel,trainpred)
ctableTrain <- base::table(predsTrain, trainpred$Species)
ctableTrain
[Output:]
   preds        setosa versicolor virginica
     setosa         36          0         0
     versicolor      0         33         3
     virginica       0          4        35

misclassTrain <- 1 - base::sum(base::diag(ctableTrain))/
base::sum(ctableTrain)
misclassTrain
[Output (misclassTrain is a numeric vector of length 1):]
   [1] 0.06306306
```

7 Now compare the 6.3% misclassification rate for the training sample with the misclassification rate for the test sample. It is, as expected, higher, at 15.4%. For this example, there are 6 errors out of 39 cases in the test sample. The misclassification rates will vary depending on the train and test samples, their proportions, and the random seed.

```
predsTest <- stats::predict(mymodel,testpred)
ctableTest <- base::table(predsTest, testpred$Species)
ctableTest
[Output:]
   predsTest    setosa versicolor virginica
     setosa         14          0         0
     versicolor      0          9         2
   virginica       0          4        10
```

```
misclassTest <- 1 - base::sum(diag(ctableTest))/
base::sum(ctableTest)
misclassTest
[Output:]
   [1] 0.1538462
```

Factor scores in `prcomp()`

One possible use of factor analysis is to derive observation-level scores on each component or factor. To illustrate this, we return to the music preference data in the "dat" dataset used to create prcompResultin the "Setup and data" subsection earlier. These scores may be saved and used as predictor variables in other forms of analysis, such as regression. Assuming varimax rotation, by default, these variables will be orthogonal (uncorrelated), with no mathematical possibility of multicollinearity.

```
# Set-up and reading in the data:
base::setwd("C:/Data")
music <- utils::read.csv("musicvars.csv", header=TRUE, sep = ",",
stringsAsFactors=TRUE)

# Create music2 as a complete cases version of music
music2 <- music[stats::complete.cases(music),]

# Create a subset of music2 to analyze and copy it into the object
"dat".
# Column index numbers below are from the names(music2) command.
musicColumns <- base::c(2,3,5,6,8,10,11,13,14,15)
musicsubset <- music2[,musicColumns]
dat <- musicsubset

# Create prcompResult
prcompResult <- stats::prcomp(dat, retx=TRUE, center=TRUE, scale =
TRUE)
```

The `prcompResult$x` element contains the component scores for each observation on each component. Since there were ten variables, there are ten components. The scores for the first two components are in `prcompResult$x[,1:2]`. The command below lists these for the first three observations.

```
utils::head(prcompResult$x[,1:2],3)
[Output:]
        Comp.1      Comp.2
   1 0.5909126 0.578650431
   2 2.7627193 0.810082481
   3 1.8232037 0.009882032
```

To add the component scores for all observations to the "dat" data frame:

```
dat$PCA1 <- prcompResult$x[,1]
dat$PCA2 <- prcompResult$x[,2]
```

To save the "dat" data, with component scores, as a csv file to the default directory, use the command below. Note column names are added to the csv file by default.

```
utils::write.csv(dat, file = "music_dat.csv", row.names = FALSE)
```

Coordinates, contributions, and quality of representation (cos2)

In this summary we give selected commands to list out key statistics associated with a `prcomp()` solution. Let prcompResult be the results from a `prcomp()` command. The `prcomp()` command is in R's built-in "stats" base::library, which does not need explicit loading with a `base::library()` command. Output from the commands summarized here is not shown but rather is left to the reader.

```
# Setup
base::library(factoextra)

# Observation-level results
obsInfo <- factoextra::get_pca_ind(prcompResult)
# For each of the following, observations are rows and dimensions
are columns.
# All are of data classes "matrix" and "array".
# Coordinates are used to plot observations in factor space.
obsInfo$coord
# Contributions reflect an observation's contribution to the defini-
tion of the component.
obsInfo$contrib
# cos2 is "quality of representation"
# Large cos2 means the component is important to explaining the
observation.
obsInfo$cos2

# Variable-level results
varInfo <- factoextra::get_pca_var(prcompResult)
# For varInfo$coord and varInfo$contribt, variables are rows and di-
mensions are columns.
# Both are of data classes "matrix" and "array".
# Coordinates are used to plot variables in factor space.
varInfo$coord
# Contributions reflect a variable's contribution to the definition
of the component.
varInfo$contrib
# cos2 is "quality of representation", shown in Figure 10.14.
# Large cos2 means the component is important to explaining the
variable.
base::plot(varInfo$cos2,
   xlim=base::c(-0.1,0.6),
   pch=16,
   col="red",
   cex=1.5,
```

```
   main="Variable quality of representation (cos2) by first two com-
   ponents")
graphics::text(varInfo$cos2, labels=base::names
musicsubset), adj=-0.3)
```

PCA with principal()

Setup and data

Another package by which PCA is often implemented in R is the `principal()` command from the "psych" package, which is in R's system library. As previous coverage of the `PCA()` command in Chapter 6 and of the `prcomp()` command in this chapter have covered much of the basics of principal components analysis, coverage for the `principal()` command is briefer. However, some new topics such as oblique rotation are introduced.

```
# Set the working directory:
base::setwd("c:/Data")
```

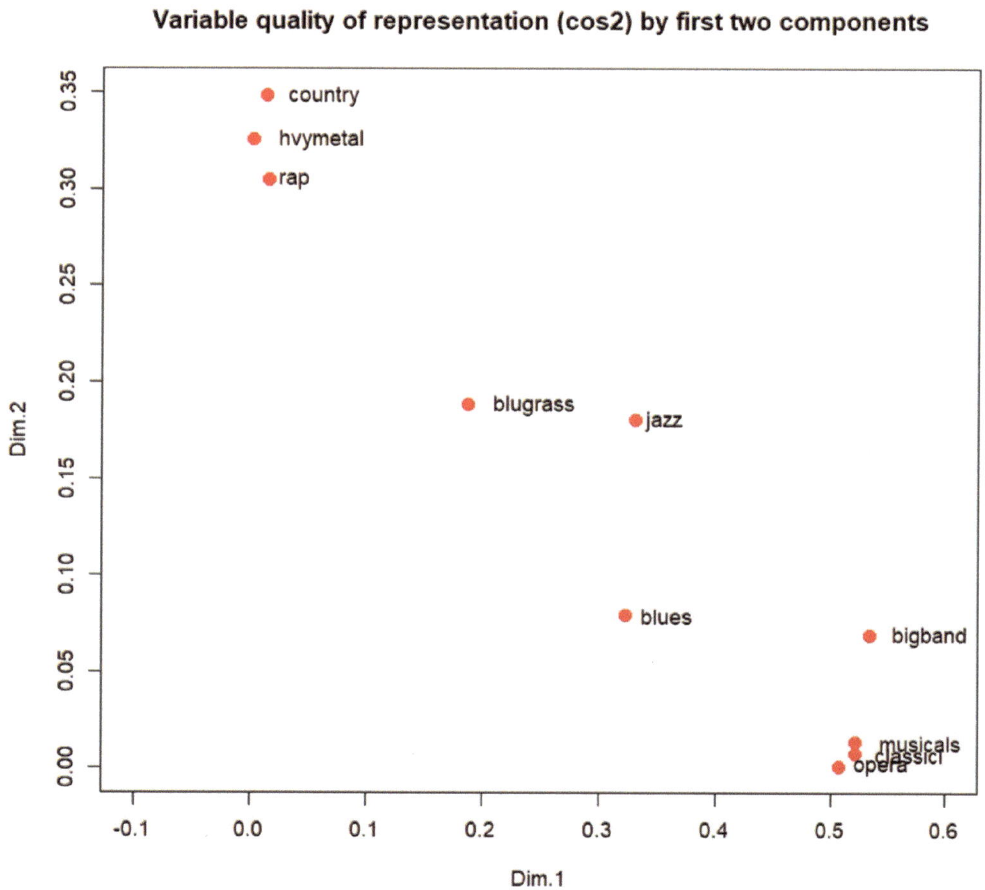

Figure 10.14 Variables by cos2 quality of representation.

```
# Load packages ("stats" is in the R system library but is listed
here for instructional reasons).
base::library(corrplot)          # Used for correlation plots
base::library(GGally)            # Has the ggpairs() command.
base::library(GPArotation)       # Required for oblique rotation
base::library(psych)             # Has the principal() program for PCA
 base::library(stats)            # Has loadings(),cor(), and cov()
functions
```

We continue to use the "dat" data frame based on musicvars.csv, the same as for `PCA()` and `prcomp()`.

```
# Set-up and reading in the data:
music <- utils::read.csv("musicvars.csv", header=TRUE, sep = ",",
stringsAsFactors=TRUE)

# Create music2 as a complete cases version of music
music2 <- music[stats::complete.cases(music),]

# Create a subset of music2 to analyze and copy it into the object
"dat".
# Column index numbers below are from the names(music2) command.
musicColumns <- base::c(2,3,5,6,8,10,11,13,14,15)
dat <- music2[,musicColumns]
```

Correlation plot

We now create a correlation plot with variables arranged by the angular order of eigenvectors (AOE), which reveals correlation clusters. This is Figure 10.15. Note we base it on datmatrix (not on music or music2) which contains Pearson correlations.

```
# Put the correlation matrix in the object datmatrix
datmatrix <- stats::cor(dat,use = "complete.obs")

# Create the correlation plot.
corrplot::corrplot(datmatrix,
   title="Correlation Plot of Music Variables",
   method = "square",
   order="AOE",
   mar=base::c(0,0,1,0))
```

The `ggpairs()` command from the "GGally" package creates paired-variable panel correlation plots, shown in Figure 10.16. Code used to create Figure 10.16 is shown below the figure. We restrict the figure to just five of the ten variables in the "dat" data frame to view a less cluttered chart. We also create "dat2" with these five variables plus the "sex" variable from the music2 data frame. Sex is used in the scatterplot portion of Figure 10.16 to color-code the points. In Figure 10.16, the diagonal contains density plots of the variables. The upper triangle contains the Pearsonian overall, male, and

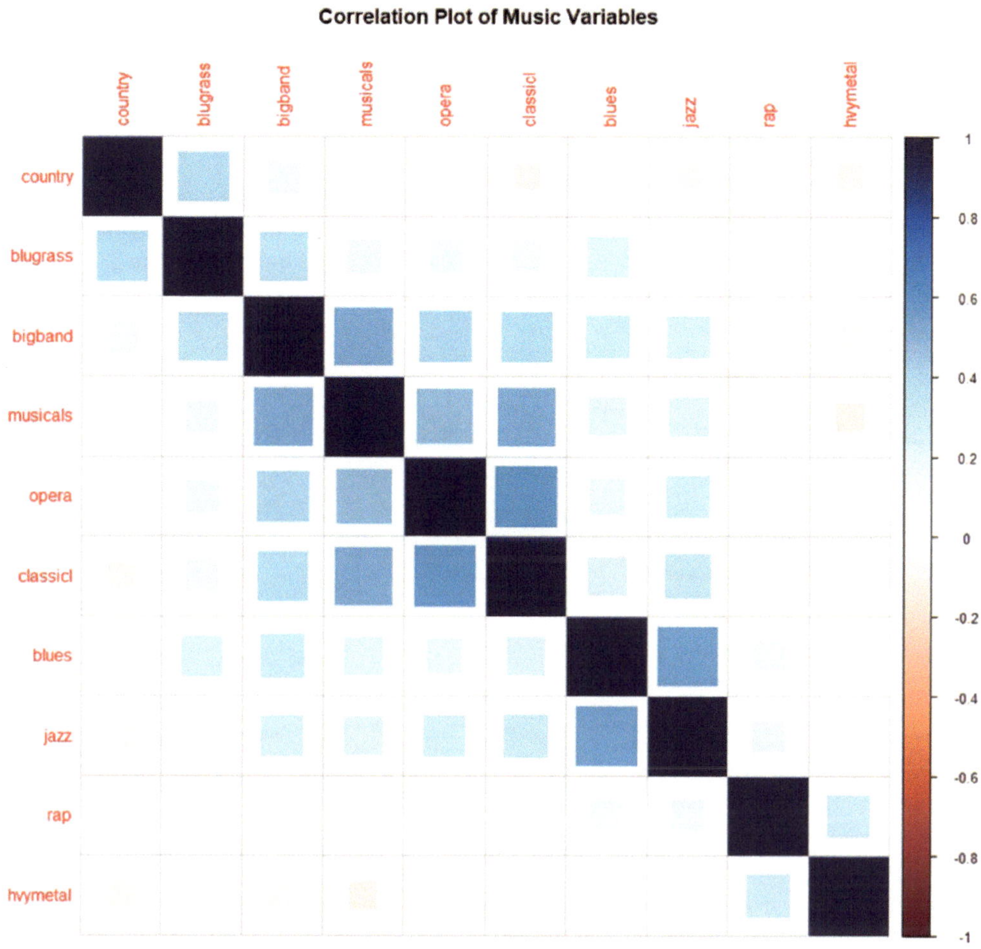

Figure 10.15 Corrplot of music preference variables in the dat matrix.

female correlations for pairs of variables. The lower triangle contains scatter plots of the row variable when paired with the column variable, with loess smoothing lines. Since the music variables were 5-point Likert items, the scatterplot appears as a 5-by-5 matrix of points. All three portions are color-coded by gender, where 1 = red = male and 2 = blue = female. Names of the variables are in tabs at the top and on the right.

The steeper the smoothing line, the more correlated is that pair of variables. Recall that higher values correspond to greater dislike of the genre. Thus, blugrass and country are positively correlated for both genders whereas classicl and country are slightly negatively correlated, indicated by the slightly downward slope of the smoothing line. Blues and country are uncorrelated for women, shown by the almost flat blue smoothing line, but for men, there is a positive association at higher levels (those who strongly dislike country, rating it 5, are more apt to have mixed feelings about blues, rating it 3). The distribution charts on the diagonal show that the largest group of respondents rated each genre 2, corresponding to “like it”.

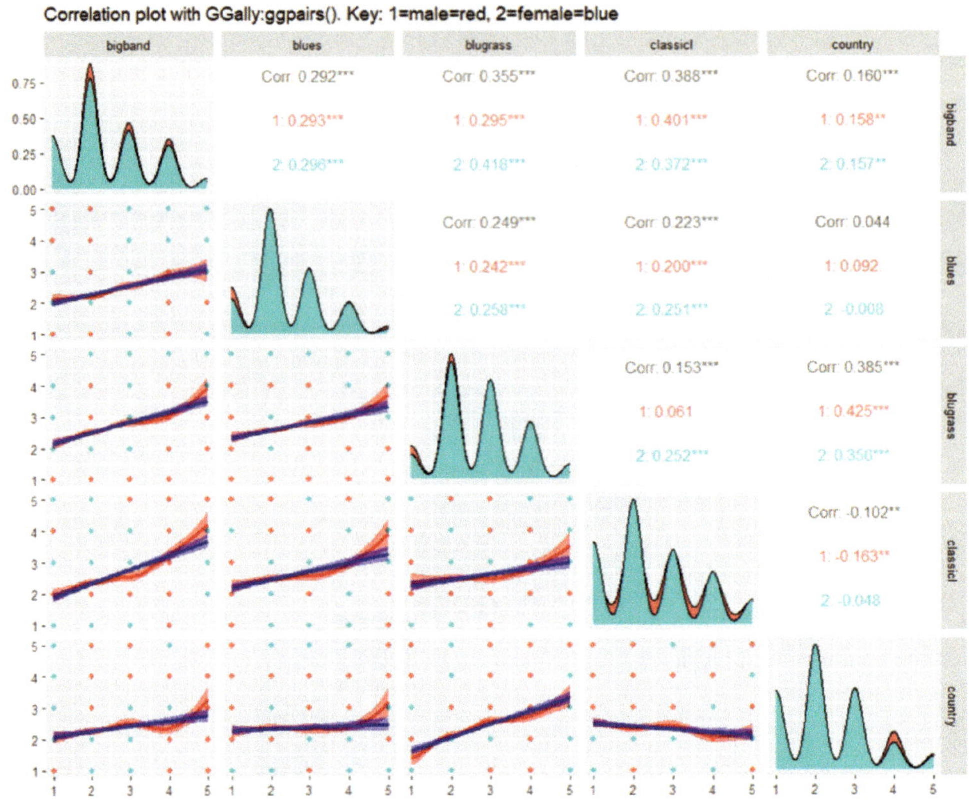

Figure 10.16 Panel of paired correlation plots using `GGallly::ggpairs()`.

The code generating Figure 10.16 is given below.

```
# Copy the music preference variables 1 through 5 into "dat2",
# then add the sex variable from music2.
dat2 <- dat[,1:5]
dat2$sex <- base::as.factor(music2$sex)

# Create the "lowessline_fn" function for the lower triangle.
# Without it, the lower triangle would be scatterplots only without
a line.
loessline_fn <- function(data, mapping, ...){
  p <- ggplot(data = data, mapping = mapping) +
    geom_point() +
    geom_smooth(method=loess, fill="red", color="red", ...) +
    geom_smooth(method=lm, fill="blue", color="blue", ...)
  p
}

# Run ggpairs() to create the panel correlation plot, with title.
# Plots have five peaks because the music variables are 5-point
Likert items.
```

```
# Nonetheless, the correlations are Pearson correlations.
# The plot is color-coded by sex.
g <- GGally::ggpairs(dat2[,1:5],
   title = "Correlation plot with GGally:ggpairs(). Key: 1=male=red,
2=female=blue",
   ggplot2::aes(colour=dat2$sex),
   lower = base::list(continuous = loessline_fn)
)

# Display graph "g". This is Figure 10.16.
g
```

KMO

KMO results for the overall MSA coefficient should be above the threshold of 0.60 to warrant proceeding with PCA with expectation of interpretable results. In Chapter 6, it was found for the same "dat" data frame that overall MSA was 0.72, above the threshold.

Eigenvalues and number of factors to extract

We now compute eigenvalues for the correlation matrix of the "data" data frame in order to determine the number of factors to extract. The `eigen()` command is from R's "base" package. Output below shows the first four factors have eigenvalues of 1.0 or greater, thus meeting the Kaiser criterion for extraction. Therefore we call for four factors in the next run of `principal()`.

```
base::eigen(stats::cor(dat))
[Partial output. Eigenvectors are also output but are not shown
here.]
   eigen() decomposition
   $values
    [1] 2.9590359 1.5174835 1.3416894 1.0199456 0.7703126
    [6] 0.6109687 0.5346705 0.4621152 0.4108412 0.3729376
```

Using the Kaiser criterion (retain components with eigenvalues of 1.0 or greater) we conclude that we should ask for four components.

PCA model using `principal()`

In the model, we use the ten music preference items contained in the "dat" data frame. Based on findings in the previous section, we run the `principal()` command with four factors and varimax rotation. Other rotation choices are "none", "quartimax", "promax", "oblimin", "simplimax", and "cluster". Type `help(principal)` for more information.

By applying the principal() command to "dat" without specifying the number of factors to extract, we get an initial solution with the maximum number of factors, which is 10 since there are ten music variables. The eigenvalues are in the $values element of the result object. They confirm that by the Kaiser criteria, four factors should be extracted.

```
# Load the "psych" base::library
base::library(psych)

# Initial run of principal()
principalResultInitial <- psych::principal(dat)

# Print the eigenvalues
principalResultInitial$values
Output:
   [1] 2.9590359 1.5174835 1.3416894 1.0199456 0.7703126
   [6] 0.6109687 0.5346705 0.4621152 0.4108412 0.3729376
```

We now rerun the analysis requesting four factors and varimax rotation, sending the output to "principalResult".

```
principalResult <- psych::principal(dat, nfactors=4,
rotate="varimax")
# Display component loadings results for principalResult from the
principal() command. # Results are the same as for rotated results
from PCA() and princomp().
principalResult
[Output:]
   Principal Components Analysis
   Call: principal(r = dat, nfactors = 4, rotate = "varimax")
   Standardized loadings (pattern matrix) based upon correlation
   matrix
              RC1   RC4   RC3   RC2   h2   u2 com
   bigband   0.62  0.25  0.37 -0.14 0.60 0.40 2.1
   blues     0.12  0.86  0.17  0.09 0.79 0.21 1.1
   blugrass  0.18  0.16  0.78  0.01 0.67 0.33 1.2
   classicl  0.82  0.14 -0.10  0.02 0.70 0.30 1.1
   country  -0.06 -0.10  0.84 -0.03 0.72 0.28 1.0
   hvymetal -0.06  0.01 -0.05  0.80 0.65 0.35 1.0
   jazz      0.22  0.84 -0.10  0.07 0.77 0.23 1.2
   musicals  0.79  0.11  0.08 -0.10 0.64 0.36 1.1
   opera     0.80  0.07  0.04  0.12 0.66 0.34 1.1
   rap       0.05  0.12  0.02  0.78 0.62 0.38 1.1

                          RC1  RC4  RC3  RC2
   SS loadings           2.42 1.60 1.51 1.31
   Proportion Var        0.24 0.16 0.15 0.13
   Cumulative Var        0.24 0.40 0.55 0.68
   Proportion Explained  0.35 0.23 0.22 0.19
   Cumulative Proportion 0.35 0.59 0.81 1.00

   Mean item complexity =  1.2
   Test of the hypothesis that 4 components are sufficient.
```

```
The root mean square of the residuals (RMSR) is  0.09
 with the empirical chi square  571.42  with prob <  1.8e-115

Fit based upon off diagonal values = 0.87
```

Discussion of above results for the principalResult object

Note that factors are auto-labeled with an "RC" prefix, standing for "rotated component". That mean item complexity is low (1.2) supports the hypothesis that 4 components are sufficient. Near the bottom, note that RMSR = 0.09. Since the model is based on correlations, this is actually standardized RMSR. A value less than 0.05 is widely considered a good fit though some consider below 0.08 to be adequate fit. Here, fit is less than adequate by this criterion. However, another fit measure is "fit-based off-diagonal values" (0.87), which would be interpreted as a good fit.[7]

The h2 column contains the communalities. The higher the communality, the better that variable is represented in the PCA solution. As a rule of thumb, indicators with a communality below 0.50 may be considered for dropping from the factor model. For these data, there are no such variables. The u2 column contains the uniquenesses, which are simply 1 minus the communalities. For further discussion, see the later subsection on "Component loadings".

Scree plot

To consider our decision to extract four factors in more detail, we turn to a graphical method by creating a scree plot of eigenvalues using `plot()` from R's built-in "graphics" base::library. The cex= option sets symbol size. The col= option sets the symbol color. By scree plot criteria, the appropriate number of factors to extract is where there is an "elbow" in the curve, after which adding a factor does not have much effect. However, deciding where the elbow is often a subjective choice. Based on the scree plot in Figure 10.17, different researchers might select the three-, four-, five-, or even six-factor solution. Using the eigenvalue table above with the Kaiser criterion is a less subjective way to decide on the number of factors to select, which was 4 by the K1 rule.

```
base::plot(principalResult$values,
   main = "Scree Plot",
   ylab="Eigenvalue",
   xlab="Number of factors in the solution",
   pch=16,
   cex=1.5,
   cex.axis=1.5,
   cex.lab=1.5,
   cex.main =2,
   col="blue")
graphics::abline(h=1)
graphics::text(x=7, y=1.1, "Kaiser criterion line", cex=1.4)
```

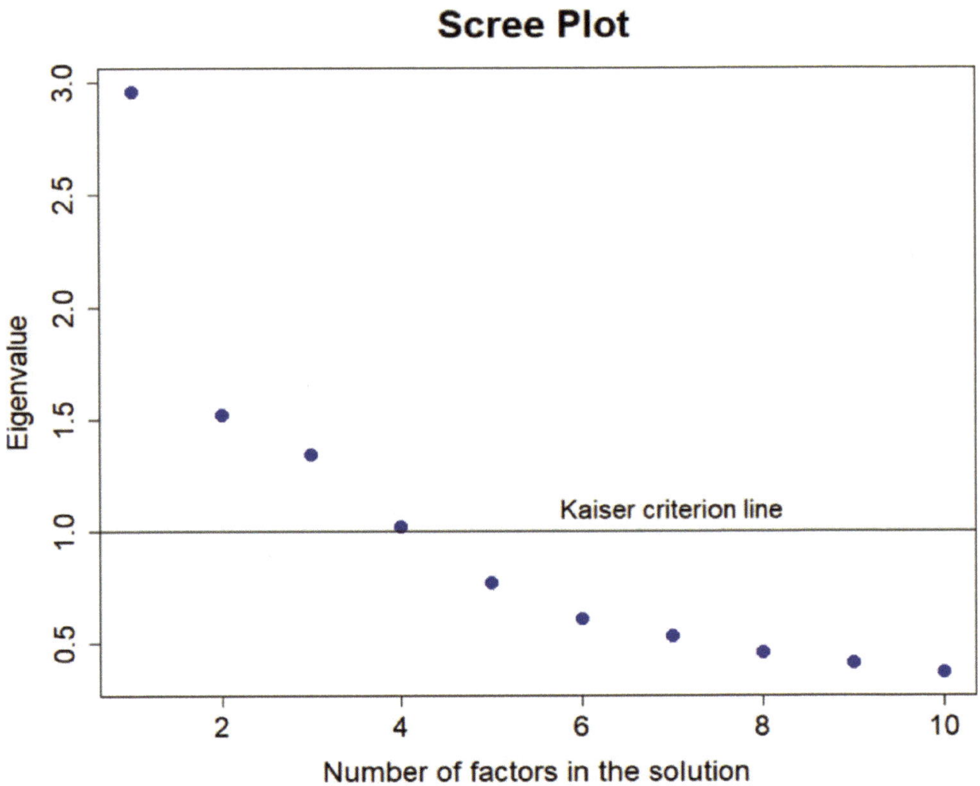

Figure 10.17 Scree plot.

Component loadings

Next, we print the factor loadings. Unrotated and rotated factor loadings are different. Researchers tend to focus on the latter. Researchers often attempt to assign labels to the rotated factors based on which measured variables load heavily on them. Such labels should not be the same as the label for any of the indicator variables. Different researchers might assign different labels and the process is more than a little subjective, leading some researchers to forego labeling factors at all.

Labeling of factors must avoid the "construct identity fallacy" (CIF). The "jingle fallacy" occurs when two factors with the same names refer to different phenomena (Thorndike, 1904). The "jangle fallacy" occurs when the same phenomenon is referenced by two different labels (Kelley, 1927). As noted by Larsen and Bong (2016: 530), who coined CIF as an umbrella term for the jingle-jangle fallacies, Complaints arise from almost every well-researched area in the social and behavioral sciences."

Note that higher scores on the music items in the "dat" data frame used to illustrate the `principal()` command correspond to greater dislike. Thus, a high positive loading on a music item corresponds to disliking that type of music.

A simple printout of component loadings is obtained by using the `loadings()` command from the "stats" package, which is a built-in part of R's system library. Using the loadings() command, by default, loadings less than 0.1 are suppressed. Factors are listed left to right in descending order of variance explained, which displays at the bottom of the output.

```
stats::loadings(principalResult)
[Output not shown. See instead output for print.psych() below.]
```

However, an equivalent but more readable printout is obtained by using the print.psych() command from the "psych" package, also part of R's system library. This command suppresses loadings below a researcher-specified level (e.g., below 0.30), making the table more readable. It also attempts to reorder the variables (rows) so they are sorted high to low in their associated factors. Note: print.psych() works for objects of class "psych", such as those created by the principal() command. It does not work for PCA(), prcomp(), or princomp() objects discussed in other sections.

```
psych::print.psych(principalResult, cut=.3, sort=TRUE)
[Output:]
   Principal Components Analysis
   Call: principal(r = dat, nfactors = 4, rotate = "varimax")
   Standardized loadings (pattern matrix) based upon correlation
   matrix
             item   RC1   RC4   RC3   RC2   h2   u2 com
   classicl     4  0.82                   0.70 0.30 1.1
   opera        9  0.80                   0.66 0.34 1.1
   musicals     8  0.79                   0.64 0.36 1.1
   bigband      1  0.62        0.37       0.60 0.40 2.1
   blues        2        0.86             0.79 0.21 1.1
   jazz         7        0.84             0.77 0.23 1.2
   country      5              0.84       0.72 0.28 1.0
   blugrass     3              0.78       0.67 0.33 1.2
   hvymetal     6                    0.80 0.65 0.35 1.0
   rap         10                    0.78 0.62 0.38 1.1

                          RC1  RC4  RC3  RC2
   SS loadings           2.42 1.60 1.51 1.31
   Proportion Var        0.24 0.16 0.15 0.13
   Cumulative Var        0.24 0.40 0.55 0.68
   Proportion Explained  0.35 0.23 0.22 0.19
   Cumulative Proportion 0.35 0.59 0.81 1.00

   Mean item complexity =  1.2
   Test of the hypothesis that 4 components are sufficient.

   The root mean square of the residuals (RMSR) is  0.09
    with the empirical chi square  571.42  with prob <  1.8e-115

   Fit based upon off diagonal values = 0.87
```

Discussion

- As requested in the generating `principal()` command, output above shows varimax-rotated factor loadings. The eigenvectors are rescaled by the sqrt of the eigenvalues to produce the component loadings more typical in factor analysis.
- Note factor columns are not in numerical order. They are in the order of sum of squared loadings.
- Factor RC1: This is the "Traditional Music" factor (classicl, opera, musicals, bigband), with all four items loading at >0.6.
- Factor RC4: This is the "Blues and Jazz" factor, with both items loading at >0.8
- Factor RC3: This is the "Bluegrass/Country" factor, with both items loading at >0.7. Also, bigband cross-loads moderately on this factor (0.37).
- Factor RC2: This is the "Hvymetal/Rap" factor, with both items loading heavily at >0.7.
- All four components collectively explain some 68% of the variance in the items. This is in the "Cumulative Var" row. The "Proportion Explained" row is the proportion each component explains of the 68% that is explained. The "Cumulative Proportion" row always sums to 1.0 or 100%.
- Root mean square residuals (RMSR), also called root mean square error (RMSE), is 0.09. By rule of thumb, 0.05 or less is good fit (some say 0.06) and <0.08 is adequate fit. Some researchers require good fit for confirmatory research but accept adequate fit for exploratory purposes. RMSE of 0.09 is less than adequate fit by this criterion.
- h2 is the communality. These are the sum of the squared loadings for all factors for a given variable. No variable is below 0.50, which is the rule-of-thumb cutoff flagging the possible need to drop the variable from the model. The u2 column is the uniquenesses, which are simply 1 minus the communalities.

Communalities and uniquenesses

We now display communalities. These are the same as the h2 column in the previous table. Note that $communality is a results element in the principalResult object. As no communality is below 0.5, we do not drop any variables.

```
principalResult$communality
Output:
      bigband     blues  blugrass  classicl   country
    0.6032403 0.7923727 0.6705207 0.6970897 0.7233493
     hvymetal      jazz  musicals     opera       rap
    0.6494623 0.7732286 0.6434947 0.6605826 0.6248134
```

Calculate uniquenesses. The communalities output above captures the same information as uniquenesses.

```
principalResult$uniquenesses
[Output is not shown since uniqueness is simply 1 - communality]
```

Factor scores

We then send factor scores for each of 1,500 observations to the list 'musicscores', and then save them to the default directory as "musicscore.csv". This .csv file will have columns for "RC1", "RC4", "RC3", and "RC2".

```
musicscores <- principalResult$scores
musicscores
[Partial output:]
                RC1          RC4          RC3          RC2
   1    -0.8208442460 -0.230948370  1.6370656181  0.917121811
   2    -1.3933782033 -1.362719071  0.8028653868  0.457830479
   3    -2.1435739760  1.108133997  0.6920017860 -0.192136180. . …
```

To save musicscores to file:

```
utils::write.csv(musicscores, file="musicscores.csv")
```

The command below saves musicscores back to an augmented dat file (in memory, not disk). Four columns, one per component, will appear at the end of the file, with names like "RC1". Rows will continue to be observations (respondents).

```
datWithScores <- base::cbind(dat, musicscores)
```

Residual analysis

Analysis of residuals may be used to assess goodness of fit of the PCA model. The `factor.model()` command from the "psych" package serves to reproduce the correlation matrix of the original data from the factor loading matrix. Residuals are the difference between the observed correlation matrix and the reproduced one. The lower the residuals, the better the fit of the factor model.

```
# Commands for principal() follow those for prcomp()in its "Residual
analysis" section.

# The first few commands are shown below.
reproduced <- psych::factor.model(principalResult$loadings)
observed <- stats::cor(dat)
residualMatrix <- observed-reproduced
residuals <- residualMatrix[base::upper.tri(residualMatrix)]
base::mean(residuals)
[Output:]
   [1] -0.03400701
```

Visualization

Next, we plot variables in rotated component space (recall that the generating `principal()` command called for varimax rotation). As a plot has two dimensions, we select the two most important factors, which are the first two (RC1 and RC4). The plot in Figure 10.18 graphically shows the clustering of variables based on their eigenvector weights (coordinates) on the RC1 and RC4 components.

```
# Set evectors the eigenvector weights of the first two components.
evectors = principalResult$weights[,1:2]

# Plot the variables using evectors. RC1 is the first component,
explaining the most variance.

# RC4 is the second component, shown on the y axis. This is Figure
10.18.
base::plot(evectors,
    pch=16,
    col="blue",
    cex=1.4,
    main = "Music variables in rotated component space",
    xlab="First component (RC1)",
    ylab="Second component (RC4)",
    ylim=base::c(-.2,.4))
# Fill in the plot using variable labels in red
graphics::text(evectors,
```

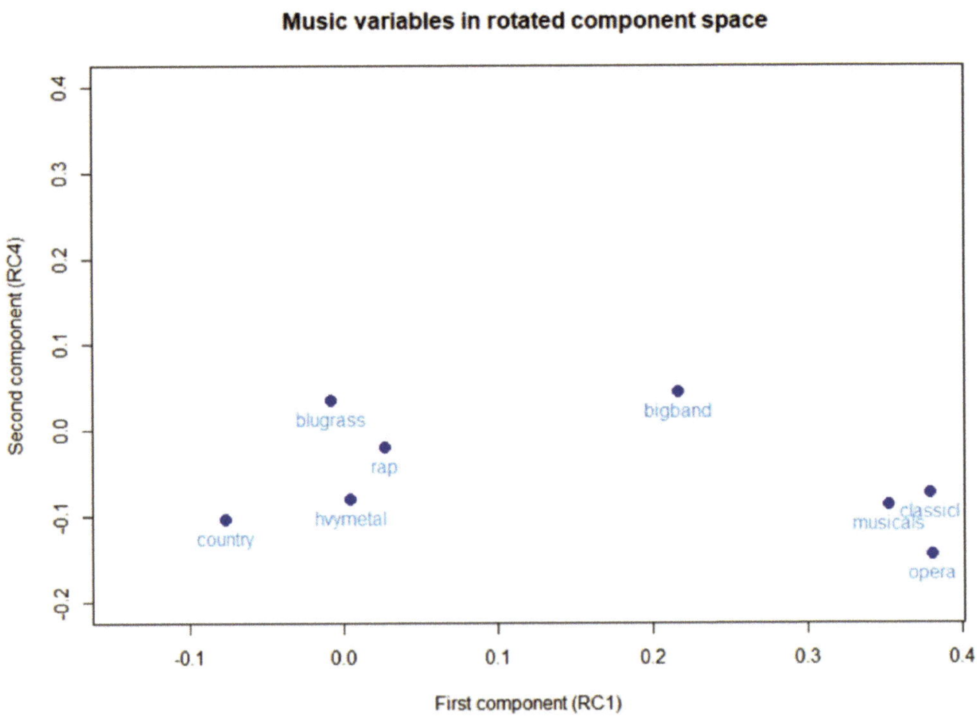

Figure 10.18 Music variables in component space, from `principal()`.

```
    pos=1,
    offset=0.5,
    labels=base::c("bigband",  "blues", "blugrass", "classicl",
    "country", "hvymetal", "jazz","musicals", "opera",  "rap"),
    col="dodgerblue")
```

Oblique rotation

The `principal()` command supports oblique rotation, the leading flavor of which is "oblimin" rotation. All that is needed is to add the option `rotate = "oblimin"`. We do this and then print out its component loadings as we did earlier for orthogonal (varimax) rotation.

```
# This library must be present or principal() results will not be
rotated.
base::library(GPArotation)
principalObliqueResult <- psych::principal(dat, nfactors=4,
rotate="oblimin")
psych::print.psych(principalObliqueResult, cut=0.3, sort=TRUE)
[Output:]
   Principal Components Analysis
   Call: psych::principal(r = dat, nfactors = 4, rotate = "oblimin")
   Standardized loadings (pattern matrix) based upon correlation
   matrix
            item  TC1   TC4   TC3   TC2   h2   u2 com
   classicl    4  0.83                   0.70 0.30 1.1
   opera       9  0.83                   0.66 0.34 1.1
   musicals    8  0.79                   0.64 0.36 1.0
   bigband     1  0.57        0.33       0.60 0.40 1.9
   blues       2        0.89             0.79 0.21 1.0
   jazz        7        0.86             0.77 0.23 1.1
   country     5              0.85       0.72 0.28 1.1
   blugrass    3              0.77       0.67 0.33 1.1
   hvymetal    6                    0.81 0.65 0.35 1.0
   rap        10                    0.78 0.62 0.38 1.0

                          TC1  TC4  TC3  TC2
   SS loadings           2.40 1.63 1.51 1.30
   Proportion Var        0.24 0.16 0.15 0.13
   Cumulative Var        0.24 0.40 0.55 0.68
   Proportion Explained  0.35 0.24 0.22 0.19
   Cumulative Proportion 0.35 0.59 0.81 1.00

    With component correlations of
         TC1  TC4   TC3   TC2
   TC1  1.00 0.34  0.11 -0.03
   TC4  0.34 1.00  0.10  0.12
   TC3  0.11 0.10  1.00 -0.07
   TC2 -0.03 0.12 -0.07  1.00
```

```
Mean item complexity =  1.1
Test of the hypothesis that 4 components are sufficient.

The root mean square of the residuals (RMSR) is  0.09
 with the empirical chi square  571.42  with prob <  1.8e-115

Fit based upon off diagonal values = 0.87
```

Discussion

1 Under orthogonal PCA (e.g., as for varimax rotation illustrated earlier) components are constrained not to correlate. That is, the correlation of any pair of components is 0. Under oblique rotation, as with oblimin rotation, components are unconstrained (they are not constrained to have zero correlation). The output section headed "With component correlations of" shows these correlations. For these data, the correlations are low (the highest is only 0.34 and all others are below 0.2).
2 Because component correlations are low, it is not surprising that oblique rotation does not improve fit over orthogonal rotation. In both cases RMSR is 0.09 and fit based upon off-diagonal values is moderately strong (0.87). Though the exact loading values differ, we would impute the same component meanings and labels in both orthogonal and oblique solutions. Since allowing correlations to be unconstrained does not improve the fit of the component model or change the meaning of the factors, many researchers would revert to orthogonal rotation in order to have components which are independent of rather than overlapping with each other.
3 The upper part of the output, labeled "Standardized loadings (pattern matrix)", may be used to impute meaning to the components based on which items load most heavily on a given component. As mentioned above, in this case the meanings would be the same as discussed for orthogonal rotation.
4 With oblique rotation, there are two matrices: the pattern matrix just discussed and the structure matrix. The structure matrix is not shown in the output above, perhaps because most researchers use the pattern matrix to interpret the components.
5 Some authors, such as Graham, Guthrie, and Thompson (2003) recommend reporting both the pattern and structure matrices. The structure matrix is output by the command below. Where the pattern matrix reflects the unique contributions of items to the components, the structure matrix reflects both common and unique contributions. As mentioned earlier, for oblique rotation the researcher should look at both the structure and pattern coefficients when attributing a label to a factor.
6 Mean item complexity is 1.1, low enough to support the view that four factors are sufficient to explain the data.

To compute the structure matrix:

```
struc <- principalObliqueResult$loadings %*%
principalObliqueResult$Phi
struc
[Output:]
                TC1        TC4          TC3          TC2
   bigband 0.66874373 0.37937460 0.4125534152 -0.158528342
   blues   0.25104814 0.88062849 0.1960037740  0.121697660
```

```
blugrass   0.23805398  0.24359169  0.7942954312 -0.018374115
classicl   0.82329224  0.29567134 -0.0485054499  0.008960831
country   -0.03891517 -0.05866129  0.8344361791 -0.073123897
hvymetal  -0.06573486  0.04733692 -0.0776248594  0.803930846
jazz       0.33816218  0.86517599 -0.0617797700  0.111799599
musicals   0.79519795  0.25853459  0.1309866992 -0.117949445
opera      0.80277893  0.23607199  0.0863290025  0.091327597
rap        0.06354554  0.18403375 -0.0003310031  0.781660574
```

PCA for R with `princomp()`

Introduction

The `princomp()` command from the "stats" library is the last alternative PCA algorithm covered in Chapter 10. However, it is a predecessor to `prcomp()` and other PCA packages described earlier. For most users, these packages have replaced `princomp()`. The `princomp()` program generates an unrotated PCA solution. Rotation is not supported. Nonetheless, `princomp()` is still common in the literature and therefore is covered here though to avoid redundancy it is treated only briefly.

Compared to `prcomp()` In `princomp()` the labels sdev, scale, and center remain the same. However, what is "rotation" in `prcomp()` is "loadings" in `princomp()`. Also what is "x" in `prcomp()` is "scores" in `princomp()`. Visualization using the "factoextra" package is supported for both algorithms.

There are algorithmic differences between `prcomp()` and `princomp()`. In particular the two use different methods of matrix decomposition when computing eigenvalues. Decomposition has to do with the factorization of a matrix into a product of matrices. The `prcomp()` package, like the `PCA()` command, uses SVD whereas the `princomp()` package uses spectral decomposition. The former examines covariances among individual observations whereas the latter examines covariances among variables. Online documentation suggests that the SVD approach has slightly better accuracy, which is one reason `prcomp()` is more widely used than `princomp()`. If Q-mode factor analysis is desired rather than the usual variable-centric R-mode, then `prcomp()`, not `princomp()`, should be used since the former but not the latter can handle data with more columns than rows. Because of algorithmic differences, results such as component scores, will differ.

To provide a brief example of `princomp()` we use the "dat" music preferences object created earlier (for instance, in the `prcomp()` section in this chapter). The "dat" data object has ten music preference variables. As per previous sections, KMO confirms factor adequacy. In code below, if `cor=TRUE`, then the data will be centered and scaled prior to analysis. If `scores=TRUE`, then each component's coordinates are calculated. That is, assuming `scores=TRUE` (the default) in the `princomp()` command, factor scores on the unrotated components for each observation are saved automatically.

Set-up

```
# General setup. It is assumed that package libraries are installed.
base::setwd("C:/Data")
base::library(factoextra)
base::library(psych)
```

```
# Read data into the data frame "dat", same as for earlier PCA
programs.
music <- utils::read.csv("musicvars.csv", header=TRUE, sep = ",",
stringsAsFactors=TRUE)
music2 <- music[stats::complete.cases(music),]
musicColumns <- base::c(2,3,5,6,8,10,11,13,14,15)
dat <- music2[,musicColumns]
```

PCA model using princomp()

The results from `princomp()` will give the same eigenvalues as other PCA packages discussed above only if the cor=TRUE option is set. The causes computation to be done on the correlation matrix of the "dat" data file. If set to FALSE or omitted, computation will be done on the covariance matrix, which is the default for `princomp()`.

```
# Run the princomp() model
princompResult <- stats::princomp(dat, cor = TRUE, scores = TRUE)
```

Results from princomp()

Eigenvalues

Eigenvalues are used to determine how many components to extract. Eigenvalues may be obtained by a call to the $sdev element squared: `princompResult$sdev^2`. Alternatively, below we use the command `get_eigenvalue()` from the "factoextra" package. This returns the eigenvalues and the percentage and cumulative percentage explained by each component. Below, four components are above the 1.0 Kaiser criterion for extraction and retention.

```
factoextra::get_eigenvalue(princompResult)
[Output]
        eigenvalue variance.percent cumulative.variance.percent
   Dim.1   2.9590359        29.590359                    29.59036
   Dim.2   1.5174835        15.174835                    44.76519
   Dim.3   1.3416894        13.416894                    58.18209
   Dim.4   1.0199456        10.199456                    68.38154
   Dim.5   0.7703126         7.703126                    76.08467
   Dim.6   0.6109687         6.109687                    82.19436
   Dim.7   0.5346705         5.346705                    87.54106
   Dim.8   0.4621152         4.621152                    92.16221
   Dim.9   0.4108412         4.108412                    96.27062
   Dim.10  0.3729376         3.729376                   100.00000
```

The following commands will give the same results in a different format:

```
fit <- stats::princomp(dat, cor=TRUE)
base::summary(fit)
[Output not shown]
```

Unrotated eigenvectors

Eigenvectors are used to plot observations in component space. Unrotated eigenvectors for the first four components are contained in the misnamed "loadings" element of the result.

```
eigenvectorsUnrotated <- princompResult$loadings[,1:4]
eigenvectorsUnrotated
 [Partial output]
                 Comp.1      Comp.2      Comp.3      Comp.4
   bigband   0.42490265  0.21251002  0.01050895  0.01800907
   blues     0.33046965 -0.22873331  0.33658278 -0.48288160
   blugrass  0.25205551  0.35208430  0.46320249  0.08011635
   ...
```

Rotation and rotated component loadings

For interpretation of the meaning of components and for display of variables in component space, most social scientists prefer rotated eigenvectors and rotated loadings. The) program, however, does not support rotation. The residual analysis commands used for the `principal()` program earlier in this chapter do not apply to princompResult since this result is for unrotated components.

Component scores

Likewise, component scores for observations (here, survey subjects) cannot be calculated in `princomp()` for rotated components, only for unrotated components. This is normally not what the social scientist wants. Nonetheless, if scores on unrotated components are wanted, they are stored in princompResult$scores. Since there were ten variables, there are ten components. The scores for the first four unrotated components are requested in `princompResult$scores[,1:4]`. The command below lists these for the first three observations.

```
utils::head(princompResult$scores[,1:4],3)
          Comp.1     Comp.2   Comp.3     Comp.4
   1 -0.5912952 0.57902508 2.148103  0.6963492
   2 -2.7645081 0.81060698 1.079507  0.8887319
   3 -1.8243841 0.00988843 1.967090 -1.5442167
   [Output:]
```

Notes

1 Online documentation for `rcomp()` states: "Note: The signs of the columns of the rotation matrix are arbitrary, and so may differ between different programs for PCA, and even between different builds of R." Online documentation further states,

> The calculation is done by a singular value decomposition of the (centered and possibly scaled) data matrix, not by using eigen on the covariance matrix. This is generally the preferred method for numerical accuracy. The print method for these objects prints the results in a nice format and the plot method produces a scree plot.

Unlike princomp(), variances are computed with the usual divisor N.

2 `PCA()` and `prcomp()` both compute principal component analysis. Both use a singular value decomposition approach, which has better numerical accuracy than does the spectral decomposition approach used by `princomp ()`, discussed in a later section uses the spectral decomposition approach.

3 The true variables are in p-dimensional space, but the biplot shows them projected in two-dimensional space.

4 A nice discussion of the flipping issue is found at https://stats.stackexchange.com/questions/88880/does-the-sign-of-scores-or-of-loadings-in-pca-or-fa-have-a-meaning-may-i-revers.

5 The code below creates for cross-validation purposes, the trainIris and testIris subsets for the iris data.

```
set.seed(123)
s <- sample(2, nrow(iris),
         replace = TRUE,
         prob = c(0.7, 0.3))
trainIris <- iris[s==1,]
testIris <- iris[s==2,]
```

6 This section is based on an example using the iris data by Finnstats at https://finnstats.com/index.php/2021/05/07/pca/.

7 Fit based upon off diagonal values is a type of pseudo-R-Squared measure. By one rule of thumb for such measures, 0–3.0 is “weak”, 3.0–6.0 is “moderate”, and above 6.0 is “strong”. Pseudo-R-Squared measures are not to be interpreted as “percent of variance explained”. Some researchers disparage the use of pseudo-R-Squared measures since the interpretation of values is imprecise.

Part II

Additional tools for dimension reduction

11 Sixteen additional methods for dimension reduction (dimRed)

Dimension reduction in the `dimRed()` system

Introduction

Authored by Guido Kraemer, the "dimRed" package contains a variety of dimensionality reduction procedures, mostly drawn from other R packages. It provides a common interface for calling these procedures. Also, it may be used in conjunction with the "coRanking" package as described in Kraemer, Reichstein and Mahecha (2018). The coRanking package, which is discussed in Chapter 12, contains methods and measures for assessing the quality of dimension reduction (DR).

Whereas most functions discussed in this volume are "S3" methods, dimRed uses R's "S4" method. While both S3 and S4 support object-oriented programming, S4 represents a newer generation. Given a results object, such as "pcaResult" for a PCA model, elements are called up using the "@" operator, not the "$" operator used in S3 methods. For instance, `pcaResult@pars` will bring up the listing of parameters for the PCA method. S4 methods organize data into "slots". Thus, "pars" is a slot for the parameters information. Slot elements can have sub-elements. For instance, much of the data of interest to researchers is in the "data" slot, which has two sub-elements: "data" and "meta". Thus `pcaResult@data@data` will hold the PCA results, which are the eigenvector values for each observation for each component. Also, `pcaResult@data@meta` will hold variable information. For the "Iris" data, this is the name of the Species for each observation (setosa, versicolor, virginica). An S4 object may have many slots and elements. The researcher may find it helpful to look at the expanded listing for pcaResult or other S4 objects in the "Environment" tab of RStudio. In R code for dimRed methods described below, slots and their elements and sub-elements are referenced.

Setup

Note that loading the dimRed package masks the `as.data.frame()` function in base R, substituting its own of the same name. It is assumed that packages mentioned in `library()` commands have been installed previously.

```
base::setwd("C:/Data")        # Declare the working directory
base::library(coRanking)      # Performance metrics for dimension
reduction
base::library(dimRed)         # Organizes 19 dimension reduction
procedures
```

DOI: 10.4324/9781003279693-13

The remaining packages, listed in order of their use in this text, are specific for particular dimRed methods or performance quality measures. They are invoked at appropriate places in the text as required.

```
base::library(kernlab)
base::library(vegan)
base::library(lle)
base::library(loe)
base::library(diffusionMap)
base::library(igraph)
base::library("Rtsne")
base::library(fastICA)
base::library(DRR)
base::library(optimx)
base::library(pcaPP)
base::library(energy)

# The Biobase package requires special installation and loading.
utils::install.packages("BiocManager")
BiocManager::install("Biobase")
base::library(Biobase)
# The "NMF" package requires "Biobase".
base::library(NMF)
```

The `embed()` function in dimRed

The `dimred::embed()` function serves as a wrapper around all dimensionality reduction functions supported by the dimRed package. In the example below, the classic "Iris" data is loaded with the `loadDataSet()` command from dimRed. Then the `embed()` command uses the Iris data, renamed "dat", and applies the "PCA" method as supported by dimRed.

Note that dimRed comes with its own `dataSets()` function, which contains the datasets listed below, including "Iris".

```
dimRed::dataSetList()
[Output:]
    [1] "Swiss Roll" "Broken Swiss Roll"  "Helix"    "Twin Peaks"
    [5] "Sphere"     "Ball"    "FishBowl"            "3D S Curve"
    [9] "variable Noise Helix"              "Iris"     "Cube"
```

The "Iris" dataset name starts with an upper case "I" to differentiate it from the usual "iris" dataset which is part of R's "datasets" package. The data are the same, however, external datasets like "iris" (lower case, from R's datasets package) must be converted into dimRed datasets. This has been done for "Iris". To convert the non-dimRed object "iris" manually, one uses R code such as the following:

```
dat <- dimRed::as.dimRedData(Species ~ Sepal.Length + Sepal.Width +
Petal.Length + Petal.Width, data=iris)
```

```
emb <- dimRed::embed(dat, "PCA")
dimRed::plot(emb, type = "2vars")
```

After conversion, the `embed()` and `plot()` commands below will work because "dat" is now a dimRed data object.

Below is an example of minimal code using the `embed()` function for a PCA model of Iris data. The `loadDataSet()` function only works directly with the dimRed datasets listed above. If the "dimRed::" prefix is omitted, `plot()` will still use the dimRed plotting functions in which `type=` may be one of "pairs", "parpl", "2vars", "3vars", or "3varsrgl". The plot (not shown) is a bare-bones representation of observations of Iris species on the first two PCA dimensions.

```
dat <- dimRed::loadDataSet("Iris")
emb <- dimRed::embed(dat, "PCA")
dimRed::plot(emb, type = "2vars")
```

We now present an example of using `embed()` to create a data frame with PCA eigenvector results. This is another way of using the external "iris" data frame (not "Iris") within dimRed.

```
iris_df_pca <- dimRed::as.data.frame(dimRed::embed(base::as.matrix-
(iris[, 1:4]), "PCA"))

# List first five data rows.
utils::head(iris_df_pca, 5)
[Output:]
        PC1        PC2  org.Sepal.  org.Sepal.  org.Petal.      org.
                            Length       Width      Length    Petal.
                                                               Width
1 -2.684126 -0.3193972         5.1         3.5         1.4       0.2
2 -2.714142  0.1770012         4.9         3.0         1.4       0.2
3 -2.888991  0.1449494         4.7         3.2         1.3       0.2
4 -2.745343  0.3182990         4.6         3.1         1.5       0.2
5 -2.728717 -0.3267545         5.0         3.6         1.4       0.2
```

Dimension reduction methods in dimRed

At the current writing, the dimRed package supported over a dozen methods of DR. While most were available in existing R packages, some were uniquely adapted to dimRed. In general, DR methods fall into four categories:

1. Methods based on projections of high-dimensional raw data onto low-dimensional planes representing components or factors (PCA, ICA)
2. Methods using attracting and repelling forces (Force Directed Methods or FDM family, such as the KamadaKawai layout method)
3. Methods based on divergence of statistical distributions (Stochastic Neighborhood Embedding or SNE family)
4. Methods based on the reconstruction of local spaces or points by their neighbors (locally linear embedding or LLE family)

Different methods have different tuning parameters which may affect results greatly. Some require scaled data and some do not. Also, some methods require defining the "distance" between points and there are a variety of distance measures to choose among. For many of its methods of DR, the dimRed package serves as a wrapper. As a wrapper, it allows passing of some parameters, as discussed in the sections below. However, the stand-alone packages which it wraps may support additional parameter settings which cannot be altered in dimRed. In general, dimRed incorporates standard settings.

To see currently supported dimRed methods, type:

```
dimRed::dimRedMethodList()
[Output:]
     [1] "AutoEncoder"            "DiffusionMaps"         "DRR"
     [4] "FastICA"                "KamadaKawai"           "DrL"
     [7] "FruchtermanReingold"    "HLLE"                  "Isomap"
    [10] "kPCA"                 "PCA_L1"               "LaplacianEigenmaps"
    [13] "LLE"                    "MDS"                   "nMDS"
    [16] "NNMF"                   "PCA"                   "tSNE"
    [19] "UMAP"
```

In this chapter, we present the implementation of most of these methods. In Chapter 12, we present metrics for comparatively evaluating their results.

PCA and PCA_L1 methods

The dimRed package functions as a wrapper for the `prcomp()` method of PCA described in Chapter 10. The dimRed documentation notes, "PCA is by far the fastest and simplest method of dimensionality reduction and should probably always be applied as a baseline if other methods are tested." In PCA, one variable may dominate simply because its metric is at a higher scale. Therefore, scaling of variables may be recommended as a pre-processing step. The seminal article is Pearson (1901).

The PCA_L1 method is a variant of PCA often preferred when outliers are thought to be a problem. Standard PCA is sensitive to outliers. The PCA_L2 method is less sensitive. In more technical terms, standard PCA transforms the data so that the L2 reconstruction error is minimized. L2 reconstruction error has to do with reconstructing the original data from the components. Under the L2 method, the variance of the projected data is maximized. In contrast, PCA_L1 minimizes the L1 reconstruction error. This means that it maximizes the sum of the L1 norm of the projected observations.

The dimRed implementation of PCA is a wrapper for the `prcomp()` command from R's built-in "stats" base::library. There are four passable parameters for the PCA model under dimRed. However, if `prcomp()` is run independently, there are many more settable parameters as discussed in Chapter 10.

These are the passable parameters in dimRed for the PCA method:

- "ndim" is the number of requested dimensions (default = 2). The maximum value is the number of predictor variables (4 for the Iris data).
- "center" centers the data in pre-processing if TRUE, which is the default.
- "scale" scales (standardizes) the data if TRUE (FALSE is the default).

Example

Load the data from dimRed's built-in datasets collection. The Iris data was discussed and illustrated in Chapter 4.

```
dat <- dimRed::loadDataSet("Iris")

# Run PCA with needed parameters
# Replicable results will be achieved without need for a set.seed()
command.
pcaResult <- dimRed::embed(
   .data=dat,
   .method="PCA",
   ndim = 2)

# Plot the results using dimRed's plot routine (output not shown).
dimRed::plot(pcaResult,
   type = "2vars",
   main="Principal component analysis (PCA)",
   sub= "Data = Iris")

# A better plot of the same results data. This is Figure 11.1.
plot(pcaResult@data@data,
   col=dat@meta$Species,
   cex = 1.1,
   pch = 16,
   main = "Principal component analysis (PCA)",
   sub= "Data = Iris")

graphics::legend("bottomright",legend=base::levels(dat@meta$Spe-
cies), col=1: base::length(base::unique(dat@meta$Species)), pch=16)
```

In Figure 11.1 the three varieties of iris flower are differentiated on the first two PCA components. Differentiation is primarily on Component 1.

```
# Verify the parameters used
pcaResult@pars
[Output]
   $center
   [1] TRUE

   $scale.
   [1] FALSE
```

As measurement of the iris variables is on similar scales, it was not thought necessary to standardize the data.

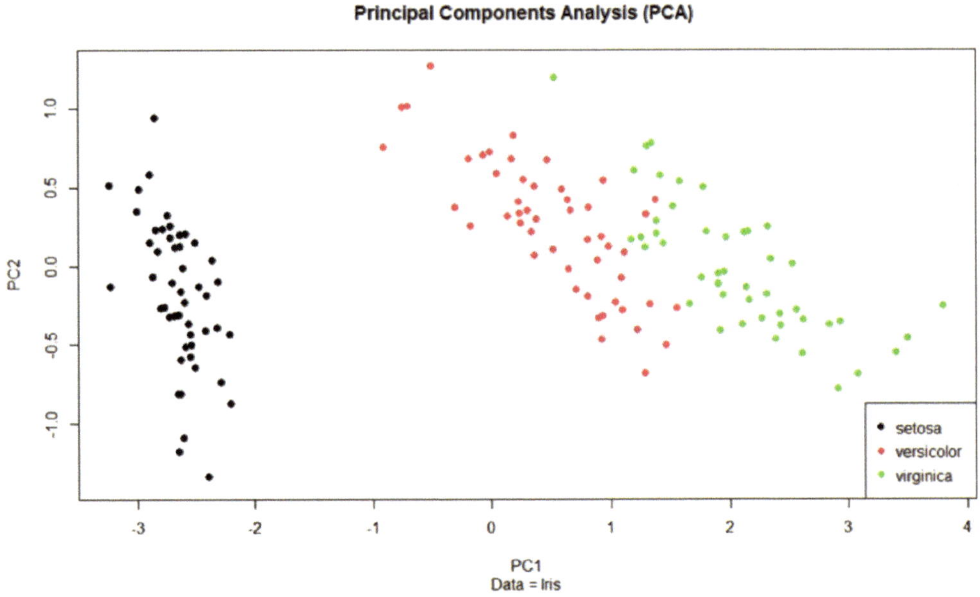

Figure 11.1 PCA on Iris data using method = "PCA" in the dimRed package.

```
# View observation values (eigenvectors) on the two PCA dimensions
pcaResult@data@data
[Partial output:]
                    PC1          PC2
     [1,] -2.684125626 -0.319397247
     [2,] -2.714141687  0.177001225
   ...
    [149,]  1.900941614 -0.116627959
   [150,]  1.390188862  0.282660938
```

Kernel PCA (kPCA)

The kPCA method, which stands for kernel principal component analysis, was also discussed in the section on "kPCA with the recipes package" in Chapter 7. In simple terms, kPCA is a nonlinear version of PCA. In choosing a "kernel" the researcher is selecting which nonlinear function to apply. The most common kerned is the Gaussian kernel, but there are others. Type `help(kpca)` after loading the "kernlab" package to see a full listing. As the method is computer-intensive, kPCA is not recommended for very large datasets. Rather, a subsample should be used. See Schölkopf et al. (1998).

The dimRed package uses the implementation of the kPCA method in the "kernlab" package. The simplified example below is adapted from online dimRed documentation and uses dimRed's built-in version of the familiar "Iris" dataset.

```
# If necessary, load DimRed and the "kernlab" package.
base::library(dimRed)
base::library(kernlab)
```

```
# Load the dimRed version of the Iris data from dimRed's built-in
collection of datasets.
dat <- dimRed::loadDataSet("Iris")

# Run the kPCA method.
kpcaResult1 <- dimRed::embed(dat, "kPCA", ndim=2)

# Plot the results (this figure is not shown).
dimRed::plot(kpcaResult1, type = "2vars", main="kPCA with rbfdot
kernel", sub="Data = Iris")
```

To see the default parameters used in the example above, use the "@" operator on the `embed()` object, for the "pars" slot. The `embed()` object here is kpcaResult1. The sigma parameter sets the inverse kernel width for the radial basis kernel function "rbfdot". For a full listing of parameters to pass and their meanings, type `help(kpca)` after loading the "kernlab" package.

```
kpcaResult1@pars
[Output:]
  $kernel
  [1] "rbfdot"

  $kpar
  $kpar$sigma
  [1] 0.1

  $ndim
  [1] 2
```

To modify the example above to use a kernel other than the default radial basis function ("rbfdot") kernel, use the commands below. In this example, we switch to the "polydot" kernel, which is for polynomial models. The polydot kernel requires the degree and scale parameters be specified. While degree is often set to 2, setting the scale requires a cross-validated grid search. For instructional purposes, however, we set scale as the inverse of the number of features.

```
# Load the data from dimRed's built=in datasets collection
dat <- dimRed::loadDataSet("Iris")

# Calculate a scale value (grid search would be better)
# There are 4 predictor variables.
num_features <- 4
scalefactor <- 1/num_features

# Run kPCA with a polynomial kernel and its needed parameters
# Replicable results will be achieved on each run without need for a
set.seed() command.
kpcaResult2 <- dimRed::embed(.data=dat, .method="kPCA", kernel =
```

```
"polydot", kpar = base::list(degree = 2, scale = scalefactor),
ndim=2)

# Plot the results (this plot is not shown)
dimRed::plot(kpcaResult2, type = "2vars", main="kPCA with polydot
kernel", sub="Data =Iris")

# A better plot of the same results data. This is Figure 11.2.
plot(kpcaResult2@data@data,
    col=dat@meta$Species,
    cex = 1.1,
    pch = 16,
    main = "kPCA Method with polydot kernel",
    sub= "Data = Iris, kernel = polydot, degree = 2, scale = 0.25")

graphics::legend("bottomright",legend=base::levels(dat@meta$Spe-
cies), col=1: base::length(base::unique(dat@meta$Species)), pch=16)
```

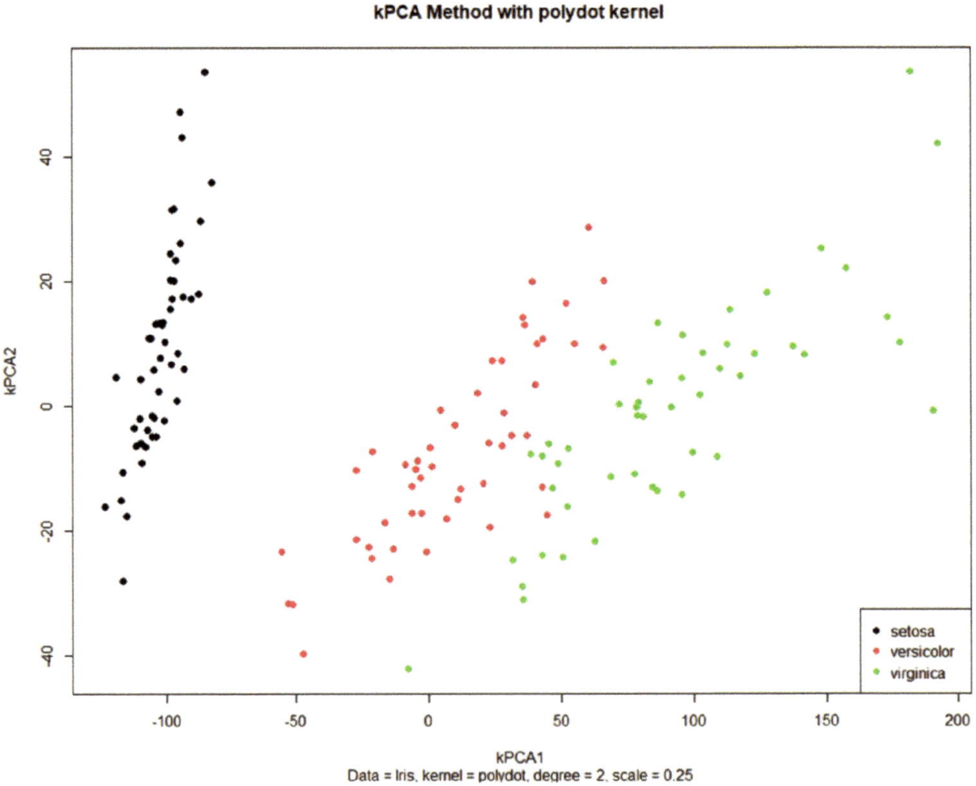

Figure 11.2 kPCA on Iris data using the dimRed package with kernel = polydot.

```
# Verify the parameters used
kpcaResult2@pars
[Output:]
   $kernel
   [1] "polydot"

   $kpar
   $kpar$degree
   [1] 2

   $kpar$scale
   [1] 0.25

   $ndim
   [1] 2

# View observation values on the two kPCA dimensions
kpcaResult2@data@data
[Partial output:]
                  kPCA1        kPCA2
     [1,] -102.3912276  13.20522807
     [2,] -107.0619487  -3.95944822
   …
   [149,]   78.6699734  -0.29579566
   [150,]   46.7283412 -13.25806964
```

Classical multidimensional scaling

Classical multidimensional scaling (CMDS) is a method for visualizing high-dimensional raw data points in low-dimensional Euclidean space, based on a dissimilarity matrix. CMDS is sometimes called just "MDS" or "principal coordinates analysis" (Gower, 1966). This terminology is not to be confused with principal components analysis or PCA. CMDS requires observations be related by a distance measure. While correlation may be the distance measure, there are a number of alternatives. Whatever the distance measure, the clustering of variables or observations is done on a simple distance basis. This contrasts with PCA, which uses partial coefficients which control for other variables in the model when considering how variables are related. In more technical terms, CMDS employs eigenvalue decomposition on a transformed distance matrix, seeking a solution in which distances under DR correspond to observed distances in the input data. MDS in R is discussed in Chapter 5 of Garson (2022). The seminal work on CMDS is Torgerson (1958).

The dimRed package is a wrapper for the `cmdscale()` function from R's "stats" base::library. A distance matrix must be specified. In the dimRed system, the "MDS" method takes the following passable parameters:

- "ndim" is the requested number of dimensions (default = 2).
- "d" is the method for calculating the distance matrix from the input coordinates (default = euclidean).

The following distance measures may be used: "euclidean" (the usual way of measuring distance between two points and the default), "maximum", "manhattan", "canberra" (used with count data), "binary", or "minkowski". Type `help(dist)` for meanings of distance measures.

Example

```
# If not already done, set the working directory and load the DimRed
package.
base::setwd("C:/Data")
base::library(dimRed)

# Load the dimRed version of the Iris data from dimRed's built-in
collection of datasets.
dat <- dimRed::loadDataSet("Iris")

# Load the Iris data from dimRed's built-in datasets collection.
dat <- dimRed::loadDataSet("Iris")
```

We now run CMDS with the needed parameters. The `dist()` function from the "stats" package returns a distance matrix. This function specifies Euclidean distances, to which `dist()` defaults. To use squared Euclidean distances, use `d = function(x) stats::dist(x)^2)`. For manhattan distances, simply set the `method` = option as `d = function(x) stats::dist(x, method="manhattan")`. Replicable results will be achieved on each run without need for a `set.seed()` command.

```
mdsResult <- dimRed::embed(
   .data=dat,
   .method="MDS",
   ndim = 2,
   d = function(x) stats::dist(x, method="euclidean"))

# Plot the results (plot not shown).
dimRed::plot(mdsResult,
   type = "2vars",
   main="Classical MDS",
   sub= "Data = Iris, Distance function = Euclidean")

# A better plot of the same results data. This is Figure 11.3.
plot(mdsResult@data@data,
   col=dat@meta$Species,
   cex = 1.1,
   pch = 16,
   main = "Classical MDS Method",
   sub= "Data = Iris, Distance function = Euclidean")

graphics::legend("bottomright",legend=base::levels(dat@meta$Spe-
cies), col=1: base::length(base::unique(dat@meta$Species)), pch=16)
```

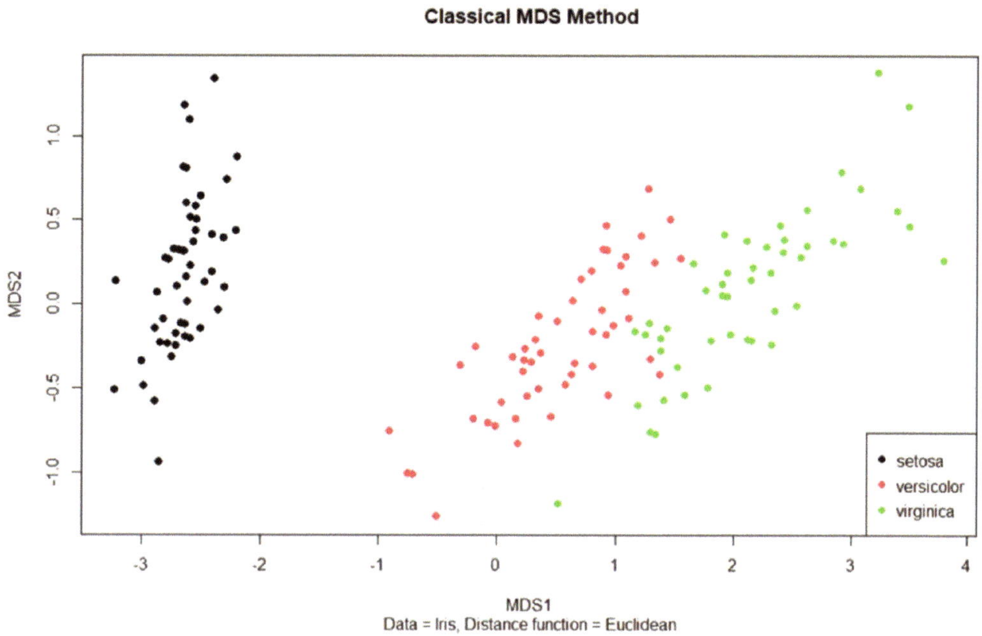

Figure 11.3 Classical multidimensional scaling of Iris data.

```
# Verify the parameters used
mdsResult@pars
[Output:]
    $d
    function(x) stats::dist(x, method="euclidean")

    $ndim
    [1] 2

# View observation values on the two MDS dimensions
mdsResult@data@data
[Partial output:]
                MDS1         MDS2
    1    -2.684125626   0.319397247
    2    -2.714141687  -0.177001225
    …
    149   1.900941614   0.116627959
    150   1.390188862  -0.282660938
```

Non-metric multidimensional scaling

Non-metric multi-dimensional scaling (nMDS) can be seen as a non-linear version of MDS using monotonic regression. Where CMDS, discussed above, seeks a solution using eigen-vector decomposition, nMDS seeks a solution by optimizing a "stress" coefficient to be as low as possible. Stress measures differences between interpoint distances in observed space compared to interpoint distances in component coordinate space, which it calls "S-space". The formula by which stress is computed is given in the online vignette by Kraemer, Reichstein and Mahecha (2019: 9). The seminal work on nMDS is Kruskal (1964).

The dimRed package functions as a wrapper for the `monoMDS()` function in the package "vegan". However, there are other R packages for nMDS, such as `isoMDS()` in the "MASS" package. The nMDS method takes only two parameters which are passable in dimRed:

- "ndim", which is the number of requested dimensions (default is 2).
- "d" is the distance function.

Example

```
# If not already done, set the working directory and load needed
packages.
base::setwd("C:/Data")
base::library(dimRed)
base::library(vegan)

# Load the data from dimRed's built-in data collection.
dat <- dimRed::loadDataSet("Iris")
```

The `embed()` function below specifies Euclidean distances, to which `dist()` defaults. To use squared Euclidean distances, use `d = function(x) stats::dist(x)^2`. For Manhattan distances, simply set `method =` to a supported distance method such as `d = function(x) stats::dist(x, method="manhattan"`.

```
# Run the cMDS model within the dimRed system,
base::set.seed(123)        # Needed for replicable results
nmdsResult <- dimRed::embed(
   .data = dat,
   .method = "nMDS",
   ndim = 2,
   d = function(x) stats::dist(x, method="euclidean"))

# Plot the results using dimRed's native plotting (this plot is not
shown).
dimRed::plot(nmdsResult, type = "2vars")

# A better plot of the same results data. This is Figure 11.4.
plot(nmdsResult@data@data,
    col=dat@meta$Species,
    cex = 1.1,
    pch = 16,
    main = "Non-Metric Multi-dimensional Scaling Method",
    sub = "Data = Iris, d = function(x) stats::dist(x,
    method='euclidean')")
```

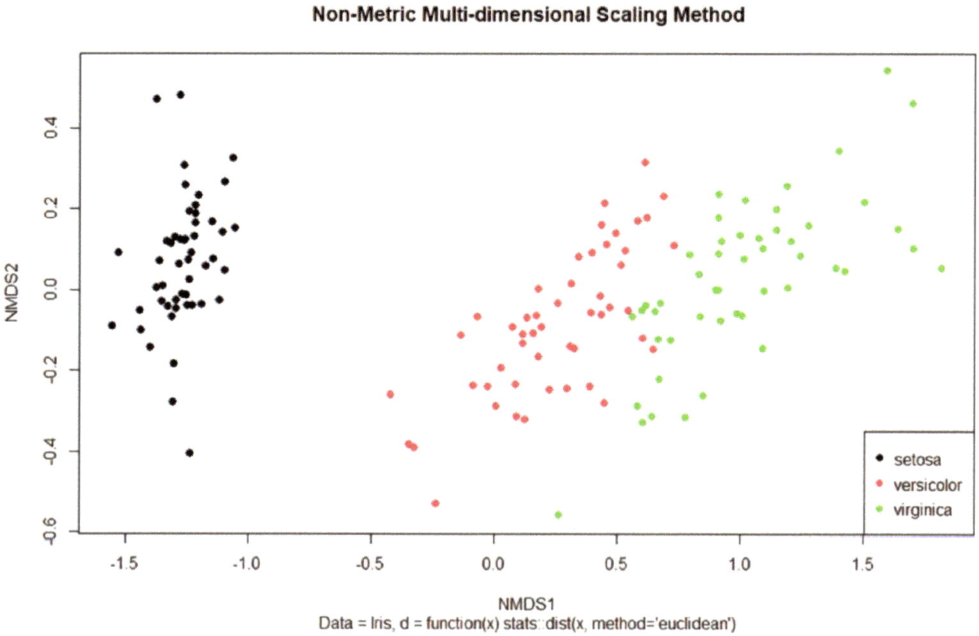

Figure 11.4 Non-metric multidimensional scaling.

```
graphics::legend("bottomright",legend=base::levels(dat@meta$Spe-
cies), col=1: base::length(base::unique(dat@meta$Species)), pch=16)

# View observation values on the two nMDS dimensions
nmdsResult@data@data
[Partial output:]
               NMDS1          NMDS2
   1    -1.277471429   0.1227678799
   2    -1.292705311  -0.0270227922
   ...
   149   0.911589878   0.1781107504
   150   0.665089047  -0.1244003846
```

Locally linear embedding

The basic concept of locally linear embedding (LLE) is that observed data may be reconstructed in component space based on neighborhoods of nearest neighbors in observed space. The "linear" in LLE refers to the assumption that neighborhoods lie in a locally linear pattern. The mathematical solution involves decomposition of eigenvalues of the input matrix. Like Isomap (discussed in a later subsection), LLE is sensitive to selection of the "knn" parameter which sets the number of nearest neighbors used to define neighborhoods. The seminal work on LLE is Roweis and Saul (2000).

Passable parameters in dimRed for LLE models are only two:

- "ndim" is the number of requested dimensions (defaults = 2).
- "knn" is the number of nearest neighbors on which to base the solution (default = 50). The researcher may wish to experiment to determine the threshold at which the solution differentiates the data (for Iris data, differentiates the species) in an optimal manner. Too low a value of knn (e.g., 5 or even 10 for the Iris data) will lead to a result with inadequate differentiation.

The dimRed package is a wrapper for the "lle" package, which must be loaded.

```
# If not already done, set the working directory and load the needed
packages.
base::setwd("C:/Data")
base::library(dimRed)
base::library(lle)

# Load the "Iris" dataset in dimRed format.
dat <- dimRed::loadDataSet("Iris")

# Run the lle model within the dimRed system,
# Replicable results will be achieved on each run without need for a
set.seed() command.
lleResult <- dimRed::embed(dat, "LLE", ndim = 2, knn = 30)

# Plot the results using dimRed's native plotting (this plot is not
shown)
dimRed::plot(lleResult, type = "2vars")

# A better plot of the same results data. This is Figure 11.5.
plot(lleResult@data@data,
    col=dat@meta$Species,
    cex = 1.1,
    pch = 16,
    main = "Local Linear Embedding Method",
    sub = "Data = Iris, knn = 30")

graphics::legend("topleft",legend=base::levels(dat@meta$Species),
col=1: base::length(base::unique(dat@meta$Species)), pch=16)

# View observation values on the two locally linear embedding
dimensions
lleResult@data@data
[Partial output:]
              LLE1        LLE2
  [1,] -0.582625348  1.277719257
  [2,]  1.144646751  1.247143399
 ...
```

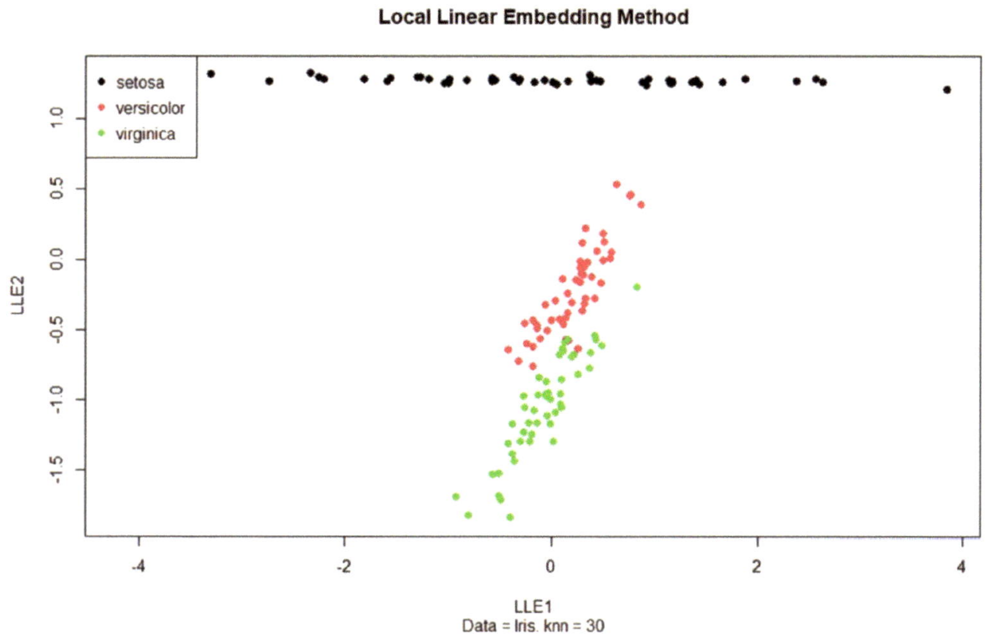

Figure 11.5 Local linear embedding method.

```
[149,] -0.011511722 -0.997995532
[150,]  0.222127082 -0.687886407
```

Yet another approach to plotting LLE results is to create a data frame from lleResult, then plot its eigenvector columns and use Species to color the points. The result is the same as for the foregoing plots but the data frame shows the DV and the component coordinates (eigenvectors) for each observation. The coordinates (eigenvectors) at in the "@data" slot of the lleResult object. We use dimRed's `as.data.frame()` function to convert it to an ordinary data frame.

```
LLEdf <- dimRed::as.data.frame(lleResult@data)
We list a few rows to show the column labels and some contents.
utils::head(LLEdf,3)
[Output:]
     meta.Species        LLE1      LLE2
   1       setosa -0.5826253 1.277719
   2       setosa  1.1446468 1.247143
   3       setosa  1.1379510 1.271129
```

We may now create a basic plot of the observations in coordinate space. This plot is essentially the same as from the earlier command, `dimRed::plot(lleResult, type = "2vars")`. For space reasons, this plot is not shown but is similar to Figure 11.5.

```
dimRed::plot(LLEdf$LLE1, LLEdf$LLE2, col=LLEdf$meta.Species)
```

Hessian locally linear embedding

Hessian locally linear embedding (HLLE) is a DR approach which uses local Hessians in its solution. A Hessian (named after Otto Hess, but sometimes lowercase) is a square matrix of second-order partial derivatives of a scalar-valued function. It describes the curvature or surface of the data. HLLE is a variant of the LLE method and, more generally, may be seen as a modification of the Laplacian Eigenmaps framework. For the dimRed implementation of HLLE, the seminal work is Donoho and Grimes (2003).

The dimRed implementation of HLLE is not a wrapper for a function of an external package. Rather, dimRed's original implementation follows the algorithm in Donoho and Grimes (2003). There are two passable parameters:

- "ndim" is the requested number of dimensions.
- "knn" specifies the number of nearest neighbors used for approximating local coordinate structure.

Example

```
# Set the working directory and load the needed packages (install
first if needed).
base::setwd("C:/Data")
base::library(dimRed)
base::library(RANN)
base::library(RSpectra)

# Load the data from dimRed's built-in data collection.
dat <- dimRed::loadDataSet("Iris")

# Run the HLLE model within the dimRed system with custom
parameters.
hlleResult <- dimRed::embed(
    .data = dat,
    .method = "HLLE",
    ndim = 2,
    knn = 12)

# Plot the results using dimRed's native plotting
dimRed::plot(hlleResult, type = "2vars")

# A better plot of the same results data is shown in Figure 11.6.
# set.seed() assures the same placement of points each time this
block of code is run.
# It does not assure the same placement each time embed() is run.
# set.seed=123
plot(hlleResult@data@data,
    col = dat@meta$Species,
    cex = 1.1,
    pch = 16,
```

```
   main = "Hessian Locally Linear Embedding Method (tSNE)",
   sub = "Data = Iris, knn = 12")

graphics::legend("bottom",legend=base::levels(dat@meta$Species),
col=1: base::length(base::unique(dat@meta$Species)), pch=16)

# View observation values on the two locally linear embedding
dimensions
hlleResult@data@data
[Partial output:]
                  HLLE1          HLLE2
     [1,] -1.414214e-01 -3.998318e-16
     [2,] -1.414214e-01 -1.415919e-16
   …
   [149,] -1.483442e-07 -6.107636e-02
   [150,] -1.483442e-07  3.995098e-04
```

Laplacian eigenmaps

Laplacian eigenmaps, also labeled spectral clustering, are implemented by the dimRed package, which functions as a wrapper for the `spec.emb()` function in the package "loe", which carries out spectral embedding with a normalized or unnormalized Laplacian (Laplacian eigenmaps). See Belkin and Niyogi (2003).

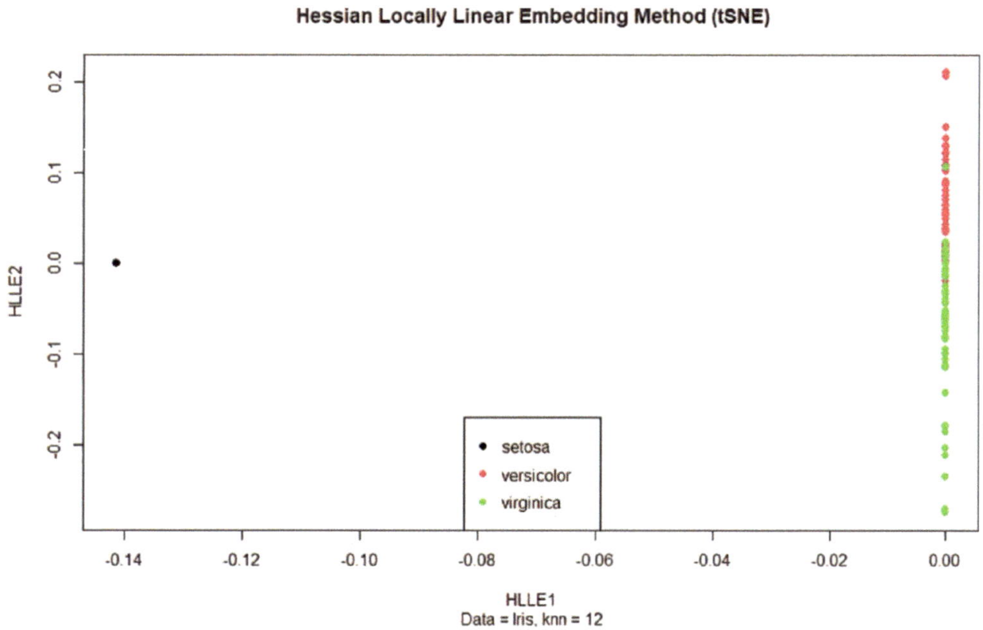

Figure 11.6 Hessian locally linear embedding method.

A "Laplacian" is a spatial second-order derivative which measures how much change there is from one point to another. For any point, the Laplacian tells how close that point is to being a local maximum or local minimum. Because the Laplacian operator can be used to find edges, it is used extensively in image analysis and enhancement. The Laplacian is also widely used in spectral clustering, which identifies communities of graph nodes (observations) based on their edges. It is an alternative, for example, to k-nearest-neighbors clustering, which spectral clustering often outperforms.

Laplacian eigenmaps avoid computational complexity by creating a sparse matrix. This can be done in two ways: (1) restricting analysis to the k nearest neighbors, or (2) specifying an epsilon neighborhood. An epsilon neighborhood is the set of all points whose distance from a given point is less than some specified number epsilon (ε). In dimRed, the default value for epsilon is 0.1. This is the "eps" parameter for the Laplacian Eigenmaps method.

Using a sparse matrix greatly reduces computational complexity and computing time since it is not necessary to estimate distances between all pairs of points. After being rendered sparse, a Gaussian kernel is used to construct a similarity matrix, which in turn is used to compute the Laplacian eigenvalue solution.

Example

```
# If not already done, set the working directory and load the needed
packages.
base::setwd("C:/Data")
base::library(dimRed)
base::library(loe)

# Load data from dimRed's built-in dataset collection
dat <- dimRed::loadDataSet("Iris")

# Initialize an object of class "LaplacianEigenmaps" with its func-
tion in slot "fun".
leim <- dimRed::LaplacianEigenmaps()

# Run the program with default parameters.
leimResultDefault <- leim@fun(dat, leim@stdpars)

# Verify its parameters
leimResultDefault@pars
[Output:]
   $ndim
   [1] 2

   $sparse
   [1] "knn"

   $knn
   [1] 50
```

```
$eps
[1] 0.1

$t
[1] Inf

$norm
[1] TRUE

$d
function (x, method = "euclidean", diag = FALSE, upper = FALSE,
p = 2)
```

```
# Plot the result for the default model. This plot is not shown.
dimRed::plot(leimResultDefault@data@data, col=dat@meta$Species)
```

LaplacianEigenmaps can take the following parameters. The defaults may be overridden to create a custom model. Parameters for the default model are embedded in leim@ stdpars for the example above (Figure 11.7).

- "ndim" is the requested number of dimensions (default = 2)
- "knn" is the number of nearest neighbors for the knn graph (default = 50)
- "eps" is the distance for the epsilon neighborhood graph (default = 0.1).
- "sparse" specifies the method to make the graph sparse (default = knn). If "knn" then a K-nearest neighbor graph is used). If "eps" then an epsilon neighborhood graph is used. Otherwise, a dense graph is used.
- "t" is the method of transforming the distance matrix. To have all distances treated equally, t == Inf treats all distances != 0 equally. The default is Inf. The alternative is to transform the matrix by w = exp(−d2/t), with larger values giving less weight to differences in distance.
- "norm" is a logical (default = TRUE). If TRUE, the normed laplacian is used.

```
# Run the LaplacianEigenmaps function with custom parameters.
# Replicable results will be achieved on each run without need for a
set.seed() command.
leimResultCustom <- dimRed::embed(
   .data=dat,
   .method = "LaplacianEigenmaps",
    ndim = 2,
    sparse="eps",
    eps = 2.0,
    t = Inf,
    norm = TRUE)

# Plot the result for the custom model.
plot(leimResultCustom@data@data,
    col=dat@meta$Species,
    cex = 1.1,
```

```
    pch = 16,
    main = "Laplacian Eigenmap Method",
    sub = "Data = Iris, sparse = eps, eps = 2.0")

graphics::legend("topleft",legend=base::levels(dat@meta$Species),
col=1: base::length(base::unique(dat@meta$Species)), pch=16)
```

The coordinates for the component scores of the cases which are used to plot observations, are retrieved as below. These coordinate scores are similar to eigenvectors.

```
ecoords <- leimResultCustom@data@data
utils::head(ecoords, 3)
[Output:]
                LEIM1         LEIM2
    [1,] -0.1233456 -0.0055532182
    [2,] -0.1225697  0.0002175183
    [3,] -0.1233456 -0.0055532182
```

Diffusion maps

The innovative aspect of the "DiffusionMaps" method of DR is that it takes the input distance matrix and calculates a matrix of transition probabilities, The transition probabilities matrix, labeled P, is mean to simulate a diffusion process between nodes

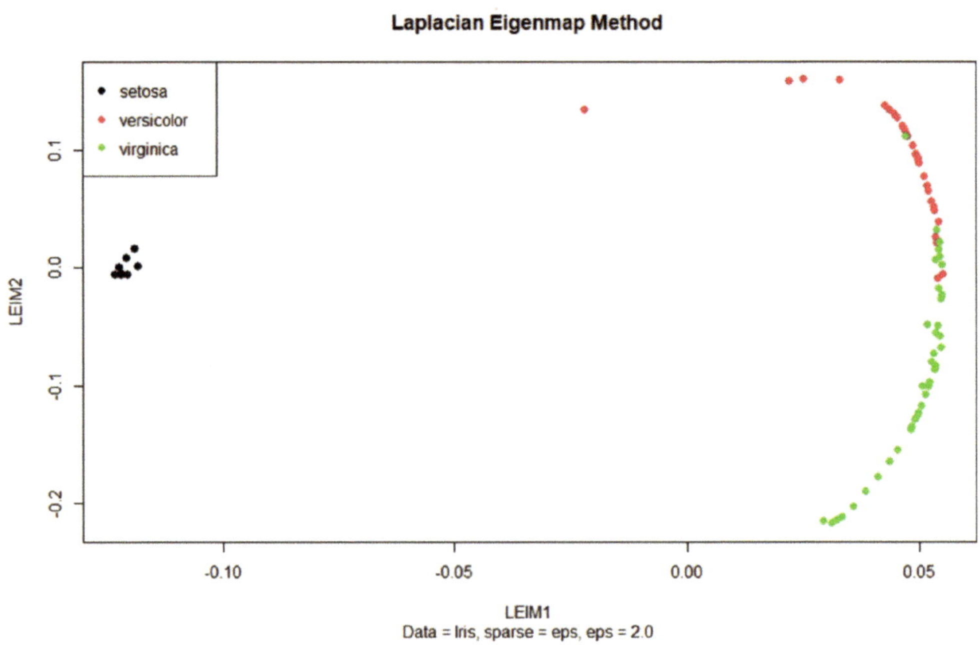

Figure 11.7 Laplacian eigenmap method.

(observations) on the graph based on the distance matrix. Then eigenvalue decomposition of the P matrix is used to calculate coordinates in the solution. The seminal article on diffusion maps is Coifman and Lafon (2006).

The DiffusionMaps method in dimRed is a wrapper for the `diffuse()` function from the package "diffusionMap". The DiffusionMaps model in dimRed supports five parameters:

- "ndim" is the number of requested dimensions (default = 2)
- "d" is a function transforming a matrix row wise into a distance matrix or dist object.
- "eps" is the epsilon parameter value, which is used to control the degree of localness in the diffusion weight matrix and hence in the solution. If set to `eps = "auto"` it will be set to the median distance to the 0.01 * n nearest neighbor.
- "t" is the time-scale parameter. The recommended default value is 0.
- "delta" is the sparsity cut-off for the symmetric graph Laplacian. A higher value results in more sparsity and faster calculation. The default value is 0.00001.

Example

```
# If not already done, set the working directory and load the needed
packages.
base::setwd("C:/Data")
base::library(dimRed)
base::library(diffusionMap)

# Load the data from dimRed's built-in data collection.
dat <- dimRed::loadDataSet("Iris")

# Run the DiffusionMap model within the dimRed system with custom
parameters.
# Due to intrinsic random processes, placement of points will differ
on each run of
# DiffusionMaps though interpretation of clusters will remain the
same.
# This cannot be overridden by use of the set.seed() command.
dmapResult <- dimRed::embed(
    .data = dat,
    .method = "DiffusionMaps",
    eps = 5,
    delta = 0.001,
    ndim = 2)

# Plot the results using dimRed's native plotting
dimRed::plot(dmapResult, type = "2vars")

# A better plot of the same results data:
# set.seed() assures replicable plots for any given run of the em-
bed() command.
```

```
# but point placement will still differ each time embed() is run.
base::set.seed(123)
plot(dmapResult@data@data,
    col = dat@meta$Species,
    cex = 1.1,
    pch = 16,
    main = "Diffusion Map Method",
    sub = "Data = Iris, eps = 5, delta = 0.001")

graphics::legend("bottom",legend=base::levels(dat@meta$Species),
col=1: base::length(base::unique(dat@meta$Species)), pch=16)
```

The coordinates for the component scores of the cases which are used to plot observations, are retrieved as below. These coordinate scores are similar to eigenvectors (Figure 11.8).

```
ecoords <- dmapResult@data@data
utils::head(ecoords, 3)
[Output:]
         diffMap1  diffMap2
    [1,] -9.32115 0.1456493
    [2,] -9.17817 0.1127499
    [3,] -9.58791 0.1733478
```

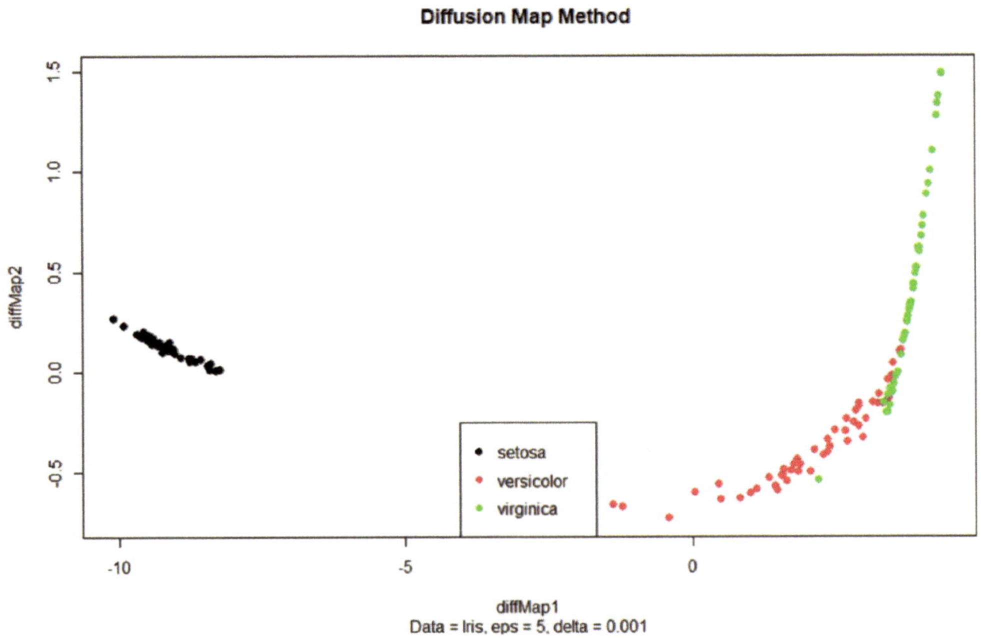

Figure 11.8 Dimension reduction by the diffusion map method.

Force directed methods

Force directed methods are layouts for the placement of nodes(observations) in relation to one another. The strength of a good layout selection is that the structure (clustering) of nodes is visually highlighted. The cost is that the meaningfulness of distances between nodes in the visualization may be diminished, even to the point of not being meaningful at all.

The dimRed package supports three force-directed layout methods: `KamadaKawai()`, `FruchtermanReingold()`, and `DrL()`. These methods are widely employed in network analysis, where they are used in decisions about the most informative way to lay out the nodes (observations) in the network, given their interrelationships (edges, in network terms). These and other layouts for networks in R are discussed further in Garson (2022: ch. 8).

KamadaKawai layout

`KamadaKawai()` is a layout which uses positive and negative connections between nodes as attracting or repelling forces. These forces are seen as “springs” connecting the nodes. The layout seeks to represent nodes in a way which minimizes the total energy of the system of forces. The dimRed package is a wrapper for the function `layout_with_kk()` function from the “igraph” package. In dimRed, it can take the following parameters:

- ndim: (number of dimensions, default = 2, maximum = 3)
- knn: used to make the graph more sparse by limiting it to the knn nearest neighbors, default = 100
- d: a distance function to set weights of graph edges, defaults to Euclidean distance.

The seminal article is Kamada and Kawai (1989).

Example

```
# If not already done, set the working directory and load the needed
packages.
base::setwd("C:/Data")
base::library(dimRed)
base::library(igraph)

# Load the data from dimRed's built-in data collection.
dat <- dimRed::loadDataSet("Iris")

# Run the KamadaKawai model within the dimRed system with custom
parameters.
# Use of set.seed() is not necessary to achieve replicable results.
kkResult <- dimRed::embed(
    .data = dat,
    .method = "KamadaKawai",
    ndim = 2,
    knn = 6,
    d = function(x) stats::dist(x, method="euclidean"))
```

```
# Plot the results using dimRed's native plotting
dimRed::plot(kkResult, type = "2vars")

# A better plot of the same results data:
plot(kkResult@data@data,
    col = dat@meta$Species,
    cex = 1.1,
    pch = 16,
    main = "KamadaKawai Layout Method",
    sub = "Data = Iris, knn = 6, distance = euclidean")

graphics::legend("topleft",legend=base::levels(dat@meta$Species),
col=1: base::length(base::unique(dat@meta$Species)), pch=16)
```

The coordinates for the component scores of the cases which are used to plot observations, are retrieved as below. These coordinate scores are similar to eigenvectors (Figure 11.9).

```
ecoords <- kkResult@data@data
utils::head(ecoords, 3)
[Output:]
    KK1        KK2
    [1,] 3.817382 7.719801
    [2,] 5.145689 7.588586
    [3,] 5.045202 6.841876
```

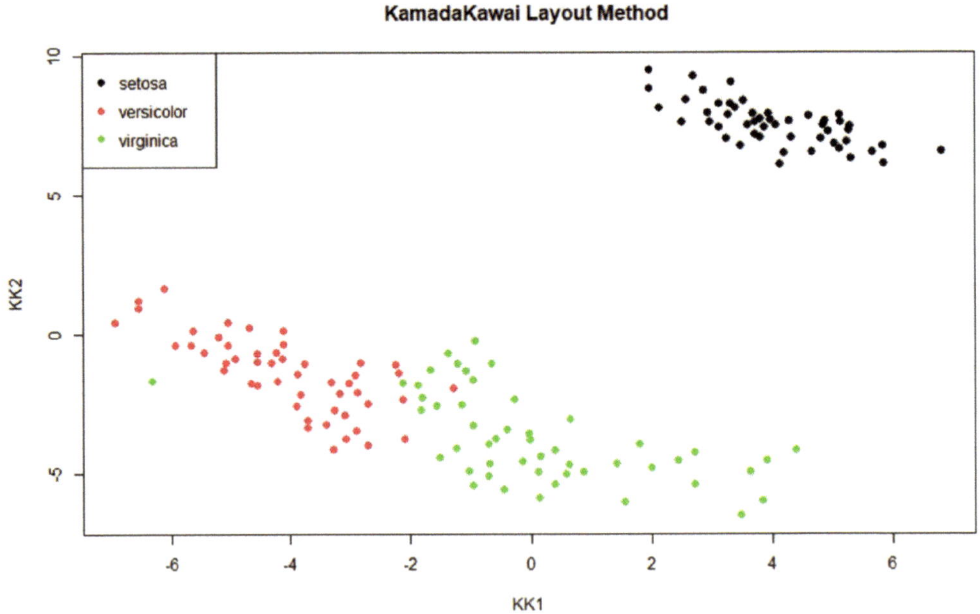

Figure 11.9 Kamada-Kawaii force directed method.

FruchtermanReingold layout

`FruchtermanReingold()`is a layout which puts the data into a circle and places connected points close to each other. The dimRed package is a wrapper for the function `layout_with_fr()` from the "igraph" package. Under dimRed, this layout has the same passable parameters as the `KamadaKawai()` method (ndim, knn, d). The seminal article is Fruchterman and Reingold (1991).

```
# If not already done, set the working directory and load the needed
packages.
base::setwd("C:/Data")
base::library(dimRed)
base::library(igraph)

# Load the data from dimRed's built-in data collection.
dat <- dimRed::loadDataSet("Iris")

# Run the FruchtermanReingold model within the dimRed system with
custom parameters.
base::set.seed(123)                  # Needed to achieve replicable
results.
frResult <- dimRed::embed(
    .data = dat,
    .method = "FruchtermanReingold",
    ndim = 2,
    knn = 6,
    d = function(x) stats::dist(x, method="euclidean"))

# Plot the results using dimRed's native plotting (plot not shown).
dimRed::plot(frResult, type = "2vars")

# A better plot of the same results data. This is Figure 11.10.
plot(frResult@data@data,
    col = dat@meta$Species,
    cex = 1.1,
    pch = 16,
    main = "FruchtermanReingold Layout Method",
    sub = "Data = Iris, knn = 6, distance = euclidean")

graphics::legend("bottomleft",legend=base::levels(dat@meta$Species),
col=1: base::length(base::unique(dat@meta$Species)), pch=16)
```

The coordinates for the component scores of the cases which are used to plot observations, are retrieved as below. These coordinate scores are similar to eigenvectors.

```
ecoords <- frResult@data@data
utils::head(ecoords, 3)
```

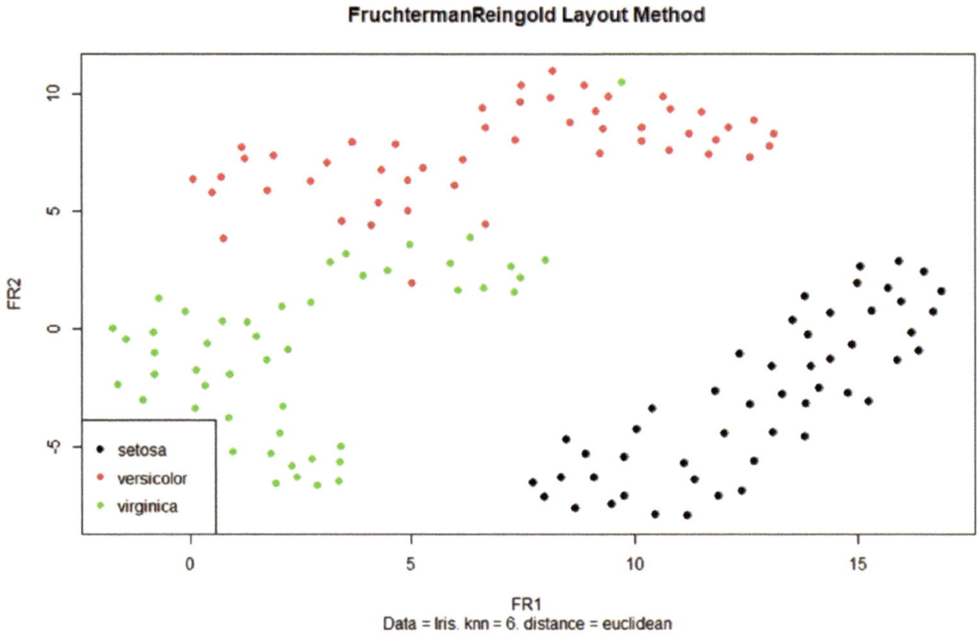

Figure 11.10 Fruchterman-Reingold force directed method.

```
[Output:]
              FR1       FR2
   [1,] 14.387976 -1.300518
   [2,] 10.467132 -7.886805
   [3,]  9.771531 -5.474372
```

DrL layout

The `DrL()` layout uses the force directed algorithm from the DrL toolbox to place nodes. A central objective of the complex algorithm behind this layout is avoiding local minima. The dimRed package is a wrapper for the function `layout_with_drl()` from the "igraph" package. Under dimRed, this layout has the same passable parameters as the previous two layout methods (ndim, knn, d). The seminal article is Martin, Brown and Wylie (2007).

Example

```
# If not already done, set the working directory and load the needed
packages.
base::setwd("C:/Data")
base::library(dimRed)
base::library(igraph)

# Load the data from dimRed's built-in data collection.
dat <- dimRed::loadDataSet("Iris")
```

```
# Run the DrL model within the dimRed system with custom parameters.
base::set.seed(123)
drlResult <- dimRed::embed(
    .data = dat,
    .method = "DrL",
    ndim = 2,
    knn = 6)

# Plot the results using dimRed's native plotting
dimRed::plot(drlResult, type = "2vars")

# A better plot of the same results data is shown in Figure 11.11:
set.seed=123
plot(drlResult@data@data,
    col = dat@meta$Species,
    cex = 1.1,
    pch = 16,
    main = "DrL Layout Method",
    sub = "Data = Iris, knn = 6, distance = euclidean")

graphics::legend("topright",legend=base::levels(dat@meta$Species),
col=1: base::length(base::unique(dat@meta$Species)), pch=16)
```

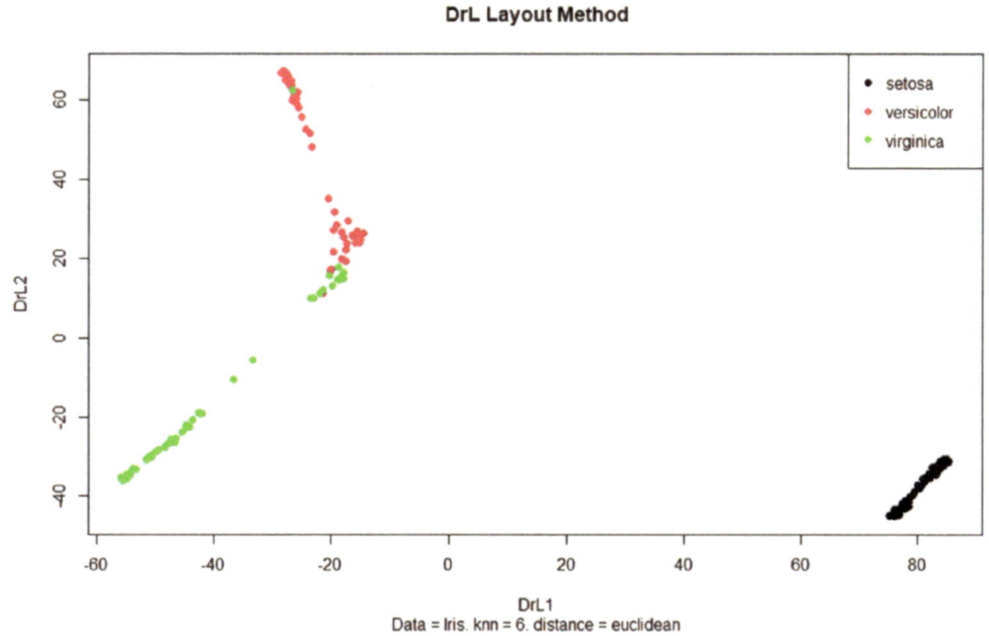

Figure 11.11 DrL force directed method.

The coordinates for the component scores of the cases which are used to plot observations, are retrieved as below. These coordinate scores are similar to eigenvectors.

```
ecoords <- drlResult@data@data
utils::head(ecoords, 3)
[Output:]
            DrL1       DrL2
    [1,] 83.20322 -34.68440
    [2,] 76.13554 -43.61077
    [3,] 77.32829 -42.82084
```

t-Distributed stochastic neighbor embedding (tSNE)

Stochastic neighbor embedding (SNE) uses the t-distribution, hence the dimRed name for this method is tSNE. By using a t-distribution, tSNE preserves local neighborhoods better than standard SNE and penalizes errors in embedding dissimilar points. Standard SNE, in contrast, uses a Gaussian (normal distribution) kernel to compute point similarities. The seminal article is Hinton and Roweis (2003). See also van der Maaten and Hinton (2008).

The dimRed implementation of tSNE is a wrapper for the `Rtsne()` function from the "Rtsne" package, which uses a fast tree-search algorithm. By default, this function involves normalization of data prior to processing and use of an initial PCA step from the `prcomp()` function discussed in Chapter 10.

There are four passable parameters for the tSNE model under dimRed. However, if `Rtsne()` is run independently, there are many more settable parameters. See `help(Rtsne)`.

- "ndim" is the number of requested dimensions (default = 2)
- "d" is a distance function to set weights of graph edges, defaults to Euclidean distance.
- "perplexity" is a parameter which sets neighborhood size (default = 30). The number of observations less 1 must be larger than 3 * perplexity.
  - "theta" sets the approximation for the nearest neighbor search. The larger the value, the faster the search but the less the accuracy. (default = 0.5). Set to 0 for exact accuracy.

Because of the stochastic process involved, the coordinates for the observations and hence their placement in the plot will vary each run, but the identification of clusters of observations will not change in interpretation. This cannot be overridden by a `set.seed()` command when running tSNE inside dimRed. See `help(Rtsne)` for an example of using `set.seed()` to achieve reproducible results when running `Rtsne()` outside the dimRed system. However, placement of points will still differ by seed, with each placement being as "correct" as each other.

Example

```
# If not already done, set the working directory and load the needed
packages.
base::setwd("C:/Data")
base::library(dimRed)
base::library(Rtsne)
```

```
# Load the data from dimRed's built-in data collection.
dat <- dimRed::loadDataSet("Iris")

# Run the tSNE model within the dimRed system with custom
parameters.
tsneResult <- dimRed::embed(
    .data = dat,
    .method = "tSNE",
    ndim = 2,
    perplexity=10,
    theta=0.0)

# Plot the results using dimRed's native plotting
dimRed::plot(tsneResult, type = "2vars")

# A better plot of the same results data is shown in Figure 11.12.
# set.seed() assures replicable plots for any given run of the em-
bed() command.
base::set.seed(123)
plot(tsneResult@data@data,
    col = dat@meta$Species,
    cex = 1.1,
    pch = 16,
    main = "t-Distributed Stochastic Neighbor Embedding Method",
    sub = "Data = Iris, distance=Euclidean, perplexity=10, theta =
    0")

graphics::legend("bottomright",legend=base::levels(dat@meta$Spe-
cies), col=1: base::length(base::unique(dat@meta$Species)), pch=16)
```

The coordinates for the component scores of the cases which are used to plot observations, are retrieved as below. These coordinate scores are similar to eigenvectors.

```
ecoords <- tsneResult@data@data
utils::head(ecoords, 3)
[Output:]
            tSNE1        tSNE2
    [1,] 33.20093 -11.44132694
    [2,] 32.17139   0.29734163
    [3,] 37.27385   0.08656142
```

Independent component analysis (FastICA)

Independent component analysis (ICA) is discussed in Chapter 13, where ICA is implemented in the recipes system. The reader is referred to fuller discussion of ICA there, where a different example (UN votes) is used. Where PCA seeks components which maximize variance explained in the input variables, ICA seeks components which are independent. That is, it assumes the input data reflect a mixture of independent signals.

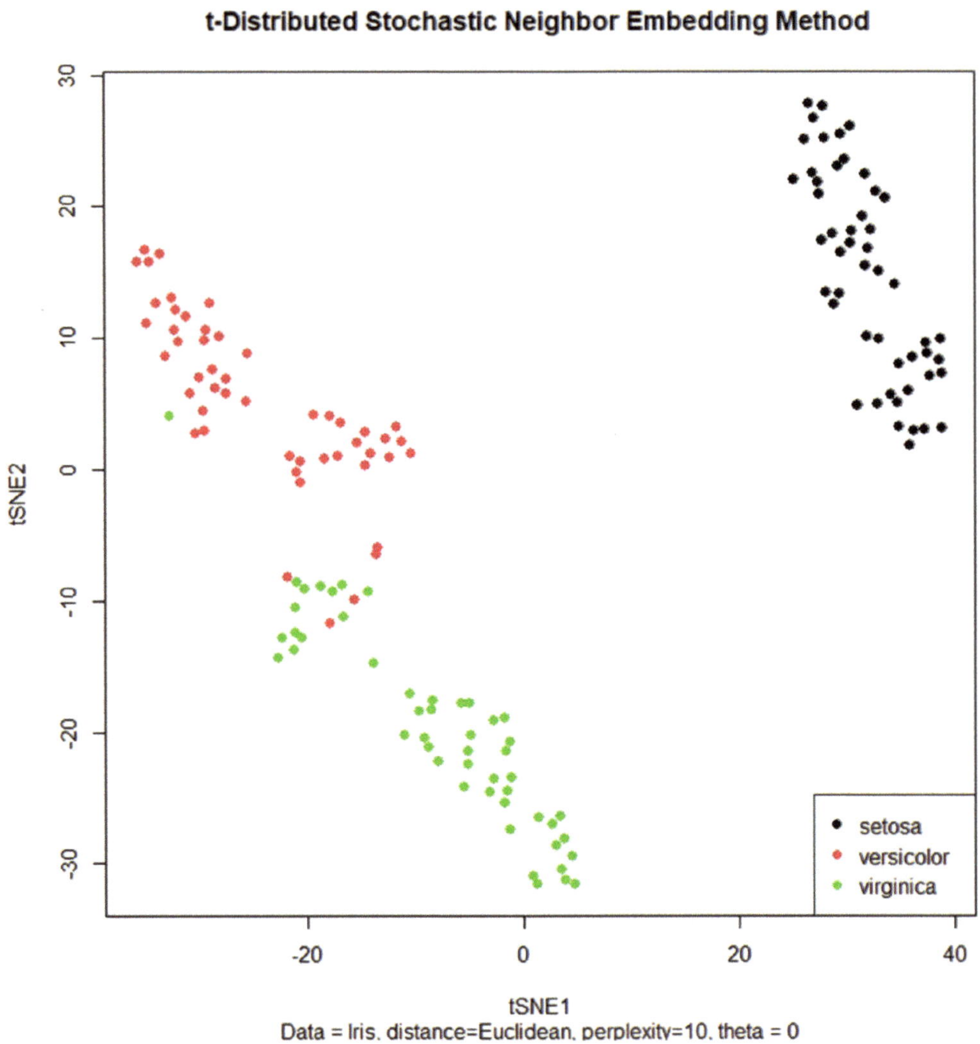

Figure 11.12 Stochastic neighbor embedding method.

The similarity with PCA is that both involve linear rotation of the data. Put another way, both involve linear projection of the data onto the components. The "fast" version of ICA is associated with Hyvarinen (1999).

The dimRed package serves as a wrapper for the `fastICA()` function in the "fastICA" package. The dimRed version is "FastICA". The only passable parameter for FastICA is "ndim", the number of requested dimensions (default = 2).

Example

```
# If not already done, set the working directory and load the needed
packages.
base::setwd("C:/Data")
```

```
base::library(dimRed)
base::library(fastICA)

# Load the "Iris" dataset in dimRed format.
dat <- dimRed::loadDataSet("Iris")

# Run the FastICA model within the dimRed system,
# Replicable results require a set.seed() command.
base::set.seed(123)
ficaResult <- dimRed::embed(
   .data = dat,
   .method = "FastICA",
   ndim = 2)

# Plot the results using dimRed's native plotting (this plot is not
shown).
dimRed::plot(ficaResult, type = "2vars")

# A better plot of the same results data. This is Figure 11.13.
plot(ficaResult@data@data,
   col=dat@meta$Species,
   cex = 1.1,
   pch = 16,
   main = "Independent Components Analysis (FastICA) Method",
   sub = "Data = Iris")

graphics::legend("right",legend=base::levels(dat@meta$Species),
col=1: base::length(base::unique(dat@meta$Species)), pch=16)
```

The coordinates for the component scores of the cases which are used to plot observations, are retrieved as below. These coordinate scores are similar to eigenvectors (Figure 11.13).

```
ecoords <- ficaResult@data@data
utils::head(ecoords, 3)
[Output:]
           ICA1     ICA2
   [1,]  0.5040262 1.372772
   [2,] -0.5026081 1.277214
   [3,] -0.4470063 1.369134
```

Dimensionality reduction via regression (DRR)

Dimensionality Reduction via Regression (DRR) is a nonlinear extension to PCA. It seeks to explain more of the variance than standard PCA. Its purpose is to ensure the

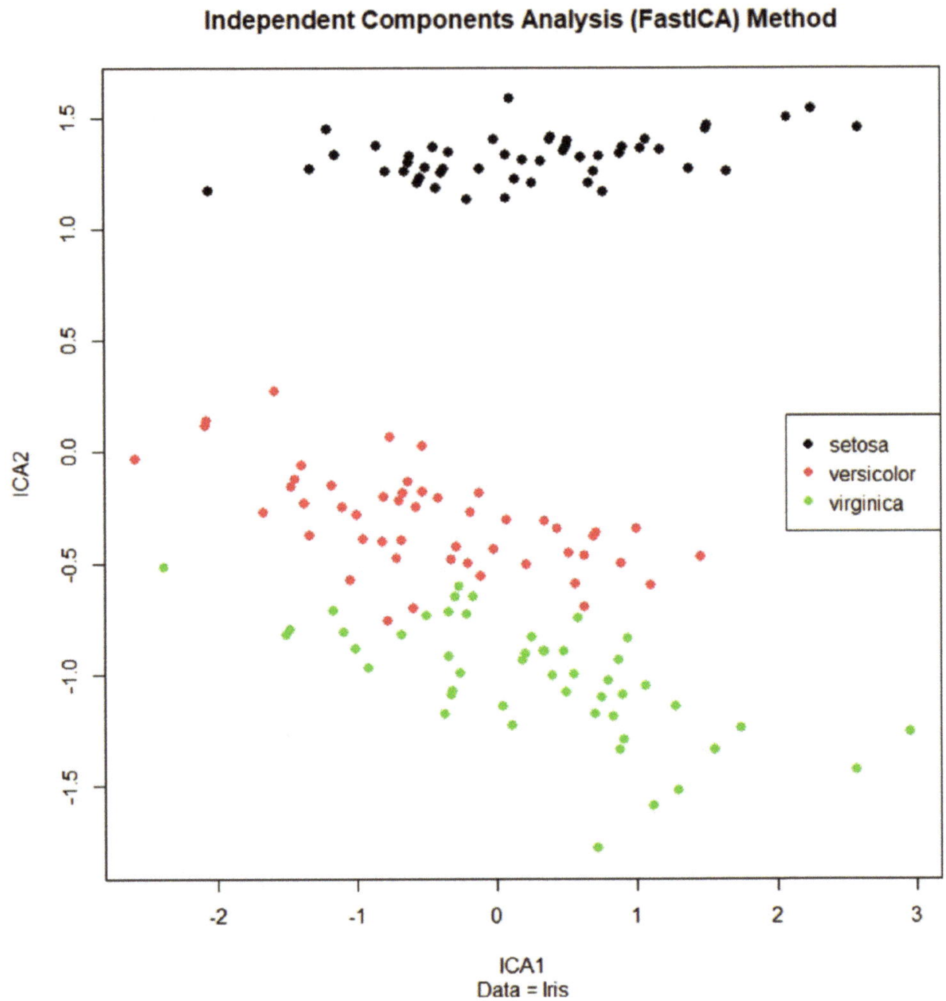

Figure 11.13 Independent component analysis method.

reduction in redundancy between the PCA coefficients, the reduction of the variance of the scores, and the reduction in the reconstruction error by which components are used to reconstruct the input matrix. However, if the data are too complex, DRR may reduce to standard PCA. The less nonlinearity in the data, the less the difference between DRR and PCA results. DRR uses kernel ridge regression (Saunders et al., 1998), a nonlinear form of regression, though other forms of regression might be substituted. For any given rotated component, DRR predicts redundant information based on the remaining components using observation loadings on the given component. The order of components in the initial rotation has a strong effect on DRR results. The seminal paper is Laparra et al. (2015).

The dimRed implementation of DRR is a wrapper for the `drr()` function in the "DRR" package. The DRR method has a large number of passable parameters, though in the example below we accept the defaults except for requesting ndim = 2 to be parallel with other DR procedures treated in this section.

- "ndim" is the number of requested dimensions (default = number of columns in data).
- "lambda" is ridge regression's regularization parameter set. The default is c(0, 10(−3:2)), which is "0.0 0.001 0.01 0.10 1.0 10.00)".
- "kernel" is the kernel to use in ridge regression (default = "rbfdot").
- "kernel.pars" is the list of kernel parameters. The "rbfdot" kernel uses "sigma". The default set is c(0, 10(−3:2)), which is "0 0.001 0.01 0.1 1 10 100".
- "pca", if TRUE (the default), an initial PCA step is performed.
- "pca.center, if TRUE (the default), data a centered prior to the PCA step.
- "pca.scale, if TRUE (FALSE is the default), data a scaled prior to the PCA step.
- "fastcv" , if TRUE (FALSE is the default), `fastCV()` from the CVST package is used instead of standard cross-validation.
- "fastcv.test": If fastcv = TRUE, separate test dataset for fastcv. This method is fast but may miss the global maximum.
- "cv.folds": if fastcv = FALSE, specifies the number of folds for cross-validation (default = 5).
- "fastkrr.nblocks" is an integer. Higher values sacrifice accuracy for speed and less memory required. (Default = 4).
- "verbose", if TRUE (the default), causes cross-validation results to be printed.

Example

```
# If not already done, set the working directory and load the needed
packages.
base::setwd("C:/Data")
base::library(dimRed)
base::library(DRR)

# Load the "Iris" dataset in dimRed format.
dat <- dimRed::loadDataSet("Iris")

# Run the DRR model within the dimRed system,
# Replicable results require a set.seed() command.
base::set.seed(123)
drrResult <- dimRed::embed(
    .data = dat,
    .method = "DRR",
    ndim = 2)

# Plot the results using dimRed's native plotting
dimRed::plot(drrResult, type = "2vars")

# A better plot of the same results data is shown in Figure 11.14.
plot(drrResult@data@data,
    col=dat@meta$Species,
```

```
    cex = 1.1,
    pch = 16,
    main = "Dimensional Reduction via Regression (DRR) Method",
    sub = "Data = Iris")

graphics::legend("bottom",legend=base::levels(dat@meta$Species),
col=1: base::length(base::unique(dat@meta$Species)), pch=16)
```

The coordinates for the component scores of the cases which are used to plot observations, are retrieved as below. These coordinate scores are similar to eigenvectors.

```
ecoords <- drrResult@data@data
utils::head(ecoords, 3)
```

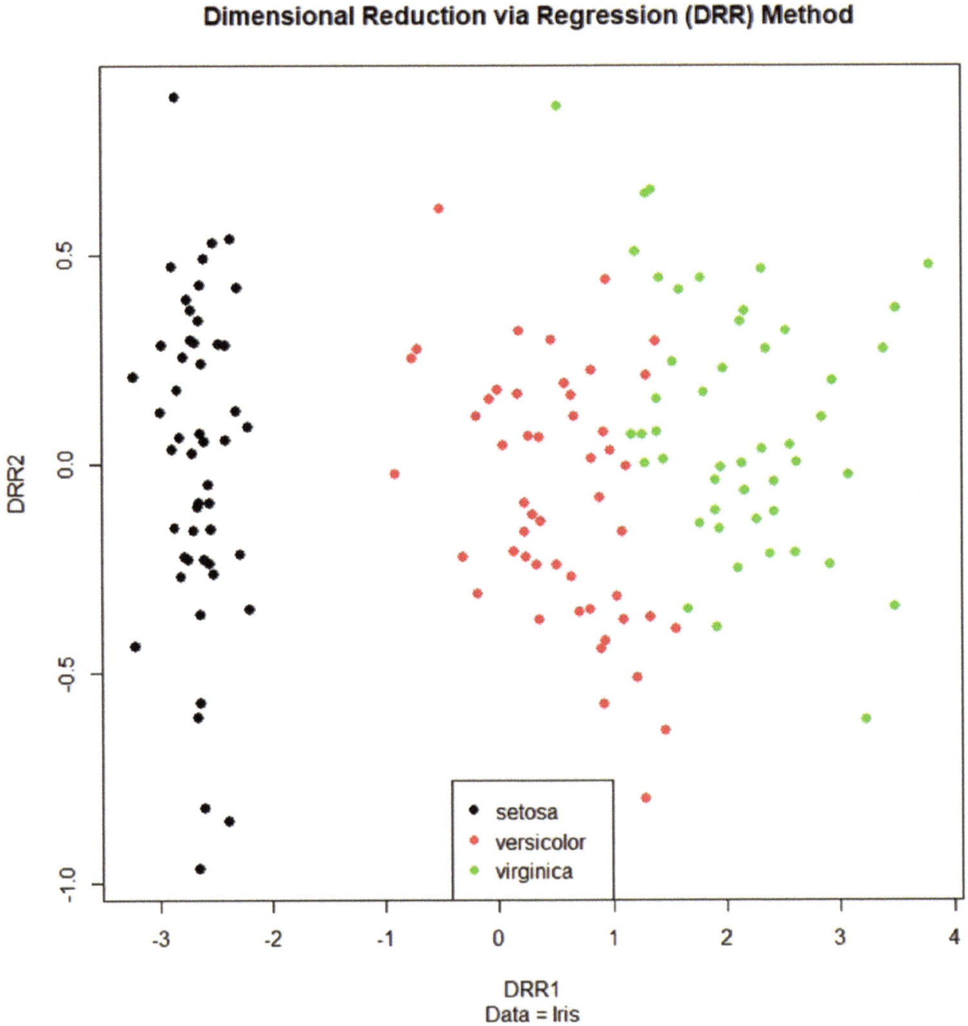

Figure 11.14 Dimensionality reduction via regression method.

```
[Output:]
              DRR1        DRR2
    [1,] -2.684126 -0.16117486
    [2,] -2.714142  0.29381946
    [3,] -2.888991  0.03335853
```

Non-negative matrix factorization (NNMF)

Non-Negative Matrix Factorization (NNMF in dimRed, but NMF elsewhere) decomposes the input matrix into a smaller number of dimensions, constraining the data and the projection to be non-negative. The seminal article is Lee and Seung (1999). A comparative study by Kim and Park (2008: 1) found, "NMF does not simply provide an alternative to K-means [clustering], but rather gives much better and consistent solutions to the clustering problem". See also the extensive article on non-negative matrix factorization at https://en.wikipedia.org/wiki/Non-negative_matrix_factorization.

The dimRed package if a wrapper for the `nmf()` function in the package "NMF". NMF has the following passable parameters:

- "ndim" is the number of requested dimensions.
- "method" specifies the algorithm (default = "brunet"). Type `NMF::nmfAlgorithm()` to see the full list of available methods.

```
NMF::nmfAlgorithm()
[Output:]
    [1] "brunet"     "KL"         "lee"        "Frobenius"
    [5] "offset"     "nsNMF"      "ls-nmf"     "pe-nmf"
    [9] "siNMF"      "snmf/r"     "snmf/l"
```

To see information on a particular method, such as "brunet":

```
NMF::getNMFMethod("brunet")
[Output:]
   <object of class: NMFStrategyIterative>
    name: brunet [NMF]
    objective: 'KL'
    model: NMFstd
    <Iterative schema>
     onInit: none
     Update: function (i, v, x, copy = FALSE, eps = .Machine$dou-
   ble.eps, ...)
     Stop: 'connectivity'
     onReturn: none
```

For further information, see the "NMF::NMF-vignette: An introduction to the package NMF" online documentation for package "NMF".

- "nrun" is the number of rounds of computation.
- "seed" sets the random seed needed for reproducible results.
- "options" is a named list of other options passed to the algorithm.

Example

```
# If necessary, install the "NMF" base::library, which requires the
"Biobase" package.
# Below, the install commands are commented out, assuming both are
already installed.
# For NMF information, type browseVignettes("NMF")
# For Biobase information, type browseVignettes("Biobase")
# utils::install.packages("NMF")
# utils::install.packages("BiocManager")
# BiocManager::install("Biobase")

# If not already done, set the working directory and load the needed
packages.
base::setwd("C:/Data")
base::library(dimRed)
base::library(Biobase)
base::library(NMF)

# Load the "Iris" dataset in dimRed format.
dat <- dimRed::loadDataSet("Iris")

# Run the NMF model within the dimRed system
# .method is the method within dimRed
# method is the method within nmf() in the NMF package
# The "lee" method uses Euclidean distance.
# Replicability is assured by passing the seed via embed()
nnmfResult <- dimRed::embed(
    .data = dat,
    .method = "NNMF",
    ndim = 2,
    seed = 123,
    method = "lee")

# Plot the results using dimRed's native plotting. This plot is not
shown.
dimRed::plot(nnmfResult, type = "2vars")

# A better plot of the same results data, shown in Figure 11.15:
plot(nnmfResult@data@data,
    col=dat@meta$Species,
    cex = 1.1,
```

```
    pch = 16,
    main = "Non-Negative Matrix Factorization (NNMF) Method",
    sub = "Data = Iris, method = lee")

graphics::legend("topright",legend=base::levels(dat@meta$Species),
col=1: base::length(base::unique(dat@meta$Species)), pch=16)
```

The coordinates for the component scores of the cases which are used to plot observations, are retrieved as below. These coordinate scores are similar to eigenvectors (Figure 11.15).

```
ecoords <- nnmfResult@data@data
utils::head(ecoords, 3)
[Output:]
             NNMF1     NNMF2
    [1,] 9.245068 0.9598049
    [2,] 8.344407 1.3169941
    [3,] 8.469283 0.9373583
```

Isomap

Isomap (isometric mapping) may be seen as a variant on CMDS in which geodesic distances are used rather than Euclidean distances. A graph of nearest neighbors is created, then multidimensional scaling is used on this distance matrix. As a generalization, Isomap may be preferred when preservation of local distances is a priority for the researcher while PCA may be preferred when preserving global Euclidean distances is a priority. Isomap

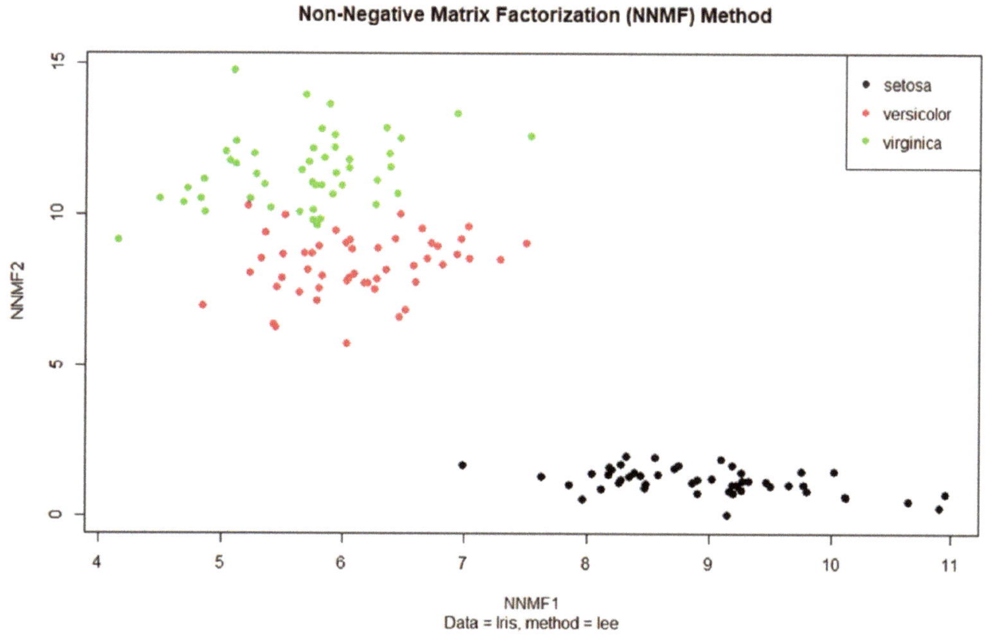

Figure 11.15 Non-negative factorization method.

does not scale well for large datasets and the researcher may wish to use a subsample. The seminal works on isomap are Tenenbaum et al. (2000) and de Silva and Tenenbaum (2004).

Note that though not discussed in this volume, the recipes package supports a `step_isomap()` to implement DR using isomap.

In R, Isomap DR is commonly implemented using the "vegan" package, with the `vegan::isomapdist()` function calculates the geodesic distance matrix and the `vegan::isomap()` function calculates the solution. However, the dimRed package uses its own implementation, which is faster.

In the dimRed system, the "Isomap" method takes the following parameters:

- "ndim" is the requested number of dimensions (default = 2).
- "knn" is the number nearest neighbors to use (default = 50). The Isomap method is sensitive to the knn value. Searching for an optimal knn value is discussed in Kraemer, Reichstein and Mahecha (2019: 19).
- "get_geod" is a logical which, if TRUE, causes the geodesic distance matrix to be kept. The default is FALSE. Later it may be accessed using the `getOtherData(x-)$geod`, where x is the results object.

```
# If not already done, set the working directory and load the needed
packages.
base::setwd("C:/Data")
base::library(dimRed)
base::library(igraph)        # Generates graphs
base::library(RANN)          # Used by dimRed for fast nearest neigh-
bor search

# Load the "Iris" dataset in dimRed format.
dat <- dimRed::loadDataSet("Iris")

# Run Isomap using the embed() function.
isomapResult <- dimRed::embed(dat,
    .method = "Isomap",
    .mute = NULL,
    ndim=2,
    knn = 40)

# Plot the results using dimRed's native plotting. This plot is not
shown.
plot(isomapResult, type = "2vars")

# A better plot of the same results data, shown in Figure 11.16:
plot(isomapResult@data@data,
  col=dat@meta$Species,
  cex = 1.1,
  pch = 16,
  main = "Iris Data, Isomap Method",
  sub = "Data = Iris, method = Isomap, knn = 40")
```

```
graphics::legend("topright",legend=base::levels(dat@meta$Species),
col=1: base::length(base::unique(dat@meta$Species)), pch=16)
```

Another example of the use of Isomap within dimRed is found in Kraemer, Reichstein and Mahecha (2019: 17–21).

The coordinates for the component scores of the cases which are used to plot observations, are retrieved as below. These coordinate scores are similar to eigenvectors.

```
ecoords <- isomapResult@data@data
utils::head(ecoords, 3)
[Output:]
             iso 1       iso 2
    [1,] 3.004500  0.07251985
    [2,] 2.843575 -0.02580613
    [3,] 3.006156 -0.01892208
```

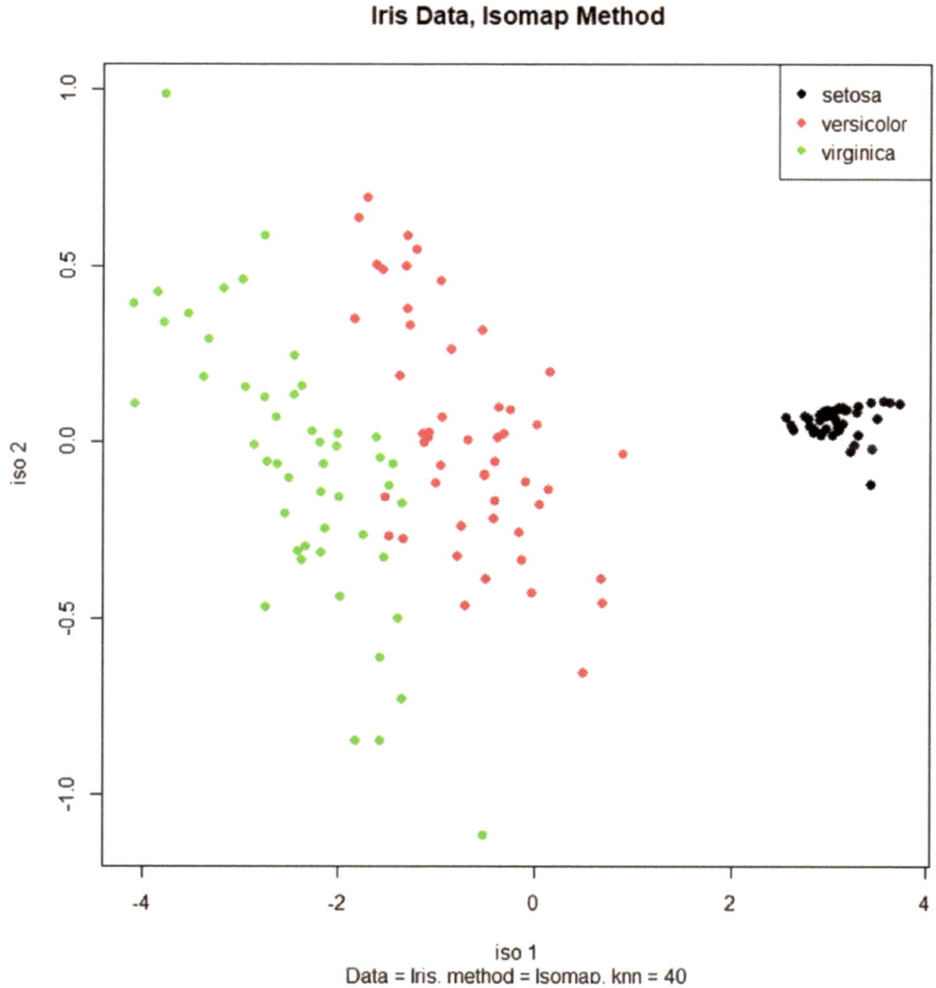

Figure 11.16 Iris data, isomap method.

Autoencoder

Autoencoding, a type of neural network approach to DR, is supported by the "dimRed" package, which uses tensorflow as a backend (for details and issues regarding tensorflow, see https://tensorflow.rstudio.com). A different type of neural network approach is discussed in Chapter 14.

In the dimRed system, the "AutoEncoder" method takes the following parameters:

- ndim: The requested number of dimensions (default = 2).
- n_hidden: The number of neurons in the hidden layers, the length specifies the number of layers, the middle number must be the same as ndim.
- activation: The activation functions for the layers, one of "tanh", "sigmoid", "relu", "elu", everything else will silently be ignored and there will be no activation function for the layer.
- weight_decay: The coefficient for weight decay, set to 0 if no weight decay desired.
- batchsize: If NA, all data will be used for training, else only a random subset of size batchsize will be used
- n_steps: The number of training steps.
- Optional: A list of keras layers that define the encoder and decoder, specifying this, will ignore all other topology-related variables, see "dimRed" documentation.

The R code below creates an autoencoder model for the Iris data:

```
# If not already done, set the working directory and load the needed
packages.
setwd("C:/Data")
library(dimRed)
library(igraph)        # Generates graphs
library(RANN)          # Used by dimRed for fast nearest neighbor search

# Load the "Iris" dataset in dimRed format.
dat <- dimRed::loadDataSet("Iris")

# Run AutoEncoder using the embed() function.
# ndim asks for a two dimension solution.
# n_hidden asks for 100 neurons in the hidden layer, 2 outcome neu-
rons (same as ndim), 10
# levels. Default was c(10,2,10). A large number of models is
possible.
set.seed(123)
aeResult <- embed(dat,
.method = "AutoEncoder",
.mute = NULL,
ndim=2,
n_hidden = c(100,2,10)
)

# Verify model parameters (shows both user-set and default
parameters).
aeResult@pars
```

```
[Partial output:]
$ndim
   [1] 2

$n_hidden
   [1] 100   2  10

$activation
   [1] "tanh" "lin"  "tanh"

$weight_decay
   [1] 0.001

$learning_rate
   [1] 0.15

…
$batchsize
   [1] NA

$n_steps
   [1] 500

# Plot the results using dimRed's native plotting. This plot is not
shown.
plot(aeResult, type = "2vars")

# A better plot of the same results data, shown in Figure 11.17:
plot(aeResult@data@data,
col=dat@meta$Species,
cex = 1.1,
pch = 16,
main = "Iris Data, AutoEncoder Method",
sub = "Data = Iris, method = autoEncoder, ndim = 2, n_hidden =
c(100,2,10)")

legend("topright",legend=levels(dat@meta$Species), col=1:
length(unique(dat@meta$Species)), pch=16, cex=1.5)
```

Note that the scaling of axes and the orientation and location of points will vary depending on the random seed. However, the clustering and separation of points will support the same interpretation (e.g., setosa will be seen to be well differentiated from versicolor and virginica). The coordinates for the component scores of the cases which are used to plot observations, are retrieved as below. These coordinate scores are similar to eigenvectors.

```
ecoords <- aeResult@data@data
head(ecoords, 3)
```

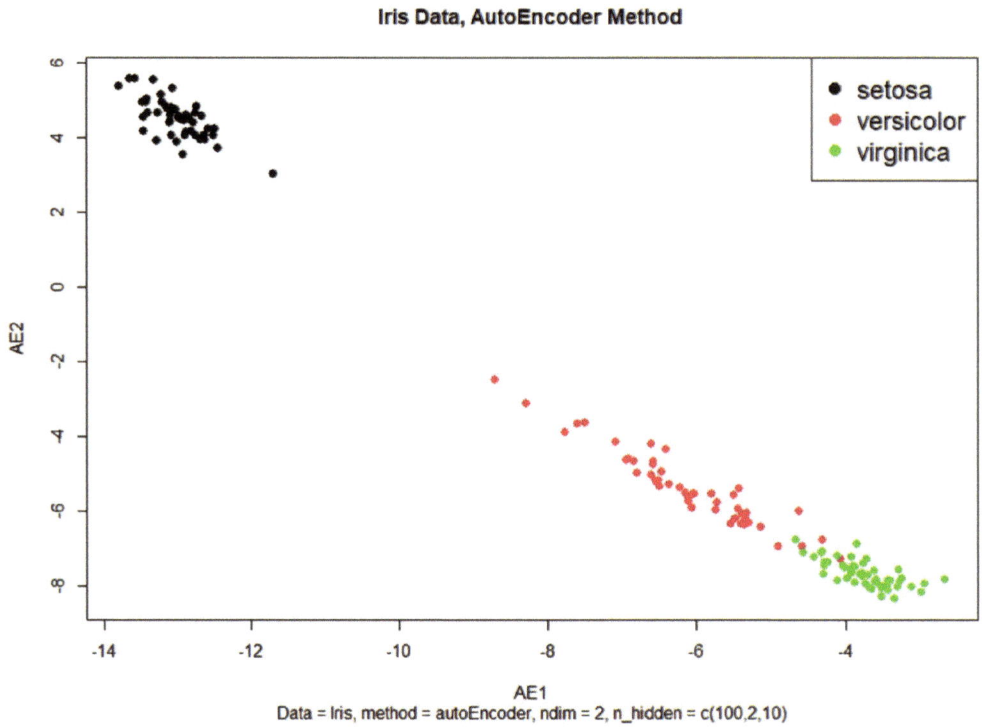

Figure 11.17 Iris data, autoencoder method.

```
[Output:]
            AE1      AE2
[1,] -13.03554 4.744034
[2,] -12.50854 4.035209
[3,] -12.88419 4.580023
```

12 Metrics for comparing and evaluating dimension reduction models

Performance quality metrics for dimRed

Introduction

Given the large number of dimension reduction procedures, "eyeballing" the results is not an adequate assessment method when judging relative model performance. Fortunately, the "dimRed" package provides eight metrics for assessing the model quality of alternative dimension reduction procedures. By using dimRed's performance metrics, the researcher can assess which procedure is most appropriate for the research task at hand. The selected procedure may be implemented within dimRed, or the researcher may choose to use directly the package for which dimRed is a wrapper since in most cases more settings and options are available directly than within dimRed itself.

Setup

```
# Set the working directory.
# It is assumed that packages listed in library() statements have
been installed previously.
base::setwd("C:/Data")
base::library(dimRed)
```

Available model performance metrics may be listed using the following command:

```
dimRed::dimRedQualityList()
[Output:]
    [1] "Q_local"                   "Q_global"
    [3] "mean_R_NX"                 "AUC_lnK_R_NX"
    [5] "total_correlation"         "cophenetic_correlation"
    [7] "distance_correlation"      "reconstruction_rmse"
```

The meaning of each of these performance metrics is given below.

1 "reconstruction_rmse"

Reconstruction error is the difference between the input data matrix and the same data matrix as reconstructed based on the observations' scores on the components. This

DOI: 10.4324/9781003279693-14

measure may only be applied, however, if the dimension reduction procedure being utilized provides for inverse mapping. The PCA and DRR methods do support inverse mapping but most methods supported by dimRed do not. If supported, then root mean square error of reconstruction is in the words of Kraemer, Reichstein and Mahecha (2018: 14), "The fairest and most common way to assess the quality of a dimensionality". The dimRed package provides the `reconstruction_rmse()` and `reconstruction_error()` functions.

For instance, to compare PCA and DRR model performance on the Iris data, we run both procedures for the maximum number of dimensions, which may be computed by the expression `ndims(dat)`, which for the Iris data is 4 – the same as the number of input variables. A `set.seed()` command is needed for replicable results.

```
dat <- dimRed::loadDataSet("Iris")
base::set.seed(123)

drrResultAll <- dimRed::embed(
    .data = dat,
    .method = "DRR",
    ndim = dimRed::ndims(dat))

pcaResultAll <- dimRed::embed(
    .data=dat,
    .method="PCA",
    ndim = dimRed::ndims(dat))

rmse_metrics <- base::data.frame(
rmse_drr = dimRed::reconstruction_error(drrResultAll,
error_fun="rmse"),
rmse_pca = dimRed::reconstruction_error(pcaResultAll,
error_fun="rmse"))

#  Reconstruction RMSE for each of four components for the DRR and
PCA methods.
rmse_metrics
[Output]:
         rmse_drr     rmse_pca
    1 2.311902e-01 2.925821e-01
    2 1.297184e-01 1.591888e-01
    3 6.483437e-02 7.693535e-02
    4 4.821235e-16 4.113526e-16
```

We may compute the column averages for rmse_metrics.

```
averages <- base::colMeans(rmse_metrics)
averages
[Output:]
      rmse_drr  rmse_pca
    0.1064357 0.1321766
```

Since RMSE is a measure of error, lower is better. By the averages above, DRR model performance is better on average than PCA performance, based on the RMSE metric. However, this averaging weights all four components equally, whereas in terms of percent of variance explained, component 1 is the most important and component 4 the least. Therefore, the researcher should inspect the entire rmse_metrics object. Inspection of the output in the rmse_metrics object (above) shows RMSE to be lower for DRR for the first three components but not the fourth, where PCA is the lowest. Note also that model performance coefficients will differ if error_fun = "mae" rather than "rmse". MAE is mean absolute error, an alternative performance measure.

2-3. "Q_local" and "Q_global"

Q_local and Q_global are quality criteria based on the co-ranking matrix, which is discussed in a subsequent subsection. Each coefficient assesses the quality of the overall embedding. The Q_local model performance metric calculates the Q_local score as a metric to assess the quality of a dimensionality reduction. Specifically, it seeks to measure quality with respect to the local variables (the input variables). Q_local varies from 0 to 1. Perfect dimension reduction as it pertains to local variables would be Q_local = 1.0.

The Q_global score is the same thing but as applied to the global variables, which are the component dimensions. Q_global varies from 0 to 1. Perfect dimension reduction as it pertains to global variables would be Q_global = 1.0.

The seminal article is Lee and Verleysen (2010), wherein the authors also define an additional quality assessment criterion, Q_avg, which is the simple average of Q_local and Q_global. For a perfect dimension reduction, Q_avg would be 1.0. For further reference, see also Fanaee-T and Thoreson (2019).

Usage takes the form Q_local(dimRed result object, ndim = number of dimensions) or Q_global(dimRed result object).

```
QlocalPCA <- dimRed::Q_local(pcaResultAll, ndim
=dimRed::getNDim(pcaResultAll))

QglobalPCA <- dimRed::Q_global(pcaResultAll)

QavgPCA <- (QlocalPCA + QglobalPCA)/2

Qmetrics <- (base::c(QlocalPCA, QglobalPCA, QavgPCA))
base::names(Qmetrics) <- base::c("Q_local", "Q_global", "Q_avg")

Qmetrics
[Output:]
      Q_local  Q_global   Q_avg
    0.9733333 0.9905961 0.981964
```

The Q-local coefficient shows dimension reduction for the PCA method is strong at the local level, meaning that the local neighborhood is preserved in component space. That Q_global is also strong means that at the global level that PCA embedding of local data in component space is only highly effective.

4.5. "mean_R_NX" and "AUC_lnK_R_NX"

R_NX is a measure used to assess the quality of a dimensionality reduction. It varies from 0 to 1, with 1 being the perfect embedding of the observed data in component space. As it is a normalized measure, values originally measured on different scales may still be compared. R_NX may be thought of as indicating the relative improvement over a random embedding of observed data in component space. See Lee, Renard, Bernard, Dupont and Verleysen (2013).

Mean_R_NX is the average agreement rate, as per Lee et al. (2013). This coefficient also varies from 0 to 1. In the graphical output, R_NX is shown on the y axis, plotted against K representing the size of the neighborhood. As customary, K is shown on a logarithmic scale, from 100 to 101 to 102 (1 to 10 to 100). R_NX varies from 0 to 1 for all magnitudes of K (that is, R_NX is calculated for all neighborhoods). For formulas and computation see Lee, Peluffo-Ordóñez and Verleysenl (2014).

The AUC measure is typically reported to quantify model performance of the dimension reduction procedure. AUC_lnK_R_NX is the area under the curve in log plots of R_NX. The higher the curve, the larger the area under the curve. Therefore, the larger the AUC, the better the model performance. The AUC value assesses model performance at all scales of the input data.

Example

In this example, we use the object "pcaResult" but results of other methods could also be employed. We compare PCA with DRR and LLE methods.

```
# Load the data from dimRed's built=in datasets collection
dat <- dimRed::loadDataSet("Iris")

# Run each method (PCA, DRR, and LLE) with needed parameters
# Replicable results will be achieved without need for a set.seed()
command.
pcaResult <- dimRed::embed(
   .data=dat,
   .method="PCA",
   ndim = 2)

base::set.seed(123)
drrResult <- dimRed::embed(
   .data = dat,
   .method = "DRR",
   ndim = 2)

# The LLE method requires installation of the "lle" package
utils::install.packages("lle")
base::library(lle)

lleResult <- dimRed::embed(dat, "LLE", ndim = 2, knn = 30)
```

We now compute mean_R_NX for each method, finding PCA best, DRR second, and LLE third in performance by this criterion:

```
dimRed::mean_R_NX(pcaResult)
[Output:]
   [1] 0.92203

dimRed::mean_R_NX(drrResult)
[Output:]
   [1] 0.8768867

dimRed::mean_R_NX(lleResult)
[Output:]
   [1] 0.6697965
```

We now use the dimRed::plot_R_NX()function to plot results for all three methods. The higher the curve, the better the performance of that model. The legend also prints the AUC values for each method, again with higher being better performance. By the R_NX quality criterion, PCA emerges as better performing than the other two methods. In part, this reflects the fact that nonlinearity is not a major aspect of the Iris data.

```
# Specify which methods to apply and then create an object
containing the embedding results:
embed_methods <- base::as.list(base::c("PCA", "DRR", "LLE"))
data_emb <- base::c(pcaResult, drrResult, lleResult)
base::names(data_emb) <- embed_methods

# Plot the R_NX curves, shown in Figure 12.1.
# The package "ggplot2" must be installed.
utils::install.packages("ggplot2")
base::library(ggplot2)
# The package "tidyr" must be installed.
utils::install.packages("tidyr")
library(tidyr)

# ndim=NA uses the original number of embedding dimensions.
# weight= specifies weighting used for calculating the AUC value
(default = "inv").
# weight may be one of "inv", "log", or "log10".  This only affects
AUC values, not the curves.
dimRed::plot_R_NX(data_emb, ndim=NA, weight="log10") +
    ggplot2::theme(legend.title = ggplot2::element_blank(),
              legend.position = base::c(0.5, 0.1),
              legend.justification = base::c(0.5, 0.1))
```

Below is a different approach which uses the specified method with all parameters being the defaults. This method species the methods by dimRed name and then uses embed()

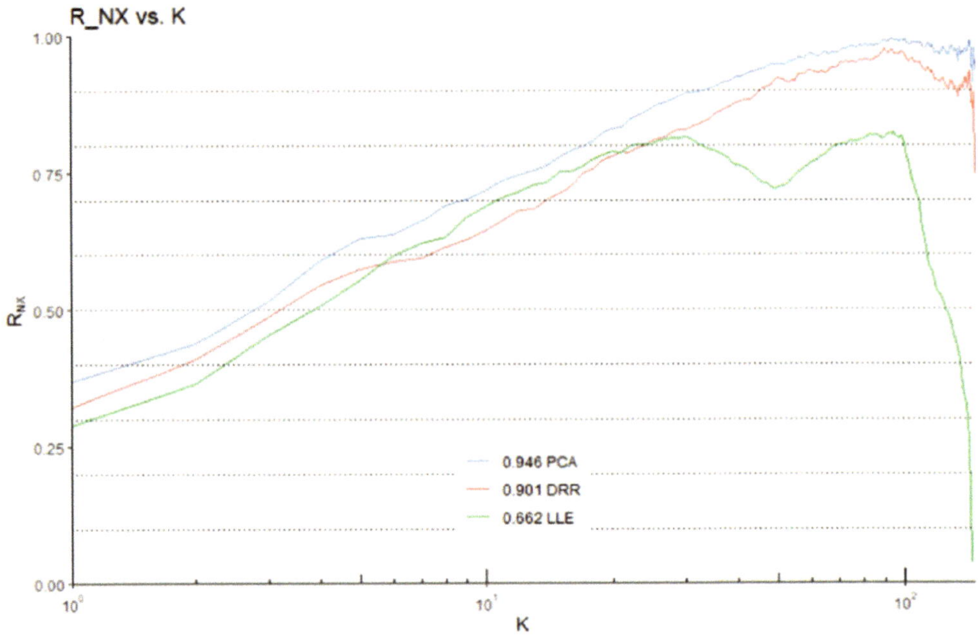

Figure 12.1 PCA, DRR, and LLE model performance by the R_NX criterion.

with all parameters set to defaults. Comparing the AUC values for custom embedding such as that above with default embedding is one way to assess the effect of setting custom parameters.

```
## Specify which methods to apply
embed_methods <- base::c("PCA", "DRR", "LLE")

## Apply dimensionality reduction for each named method.
data_emb <- base::lapply(embed_methods, function(x) dimRed::embed(-
dat, x))
base::names(data_emb) <- embed_methods

## Plot the R_NX curves for each method. This figure is not shown.
dimRed::plot_R_NX(data_emb) +
    ggplot2::theme(legend.title = ggplot2::element_blank(),
                   legend.position = base::c(0.5, 0.1),
                   legend.justification = base::c(0.5, 0.1))
```

```
[Output figure not shown]
```

If the researcher wishes to capture AUC values individually, the `AUC_lnK_R_NX()` function may be used. For example:

```
AUCpca <- dimRed::AUC_lnK_R_NX(pcaResult, weight = "log10")
AUCpca
[Output:]
   [1] 0.9458781
```

4-6. "total_correlation"

The `total_correlation()` function gives the mean squared correlations of the original axes for the input data with the axes in reduced dimensions (component space). Rotation may maximize total_correlation, which may have a value greater than 1.0. Total_correlation is a measure of performance quality of the dimensionality reduction. The passable parameters for this function are these:

- "naxes" is the number of axes to use for optimization.
- "cor_method" is the type of correlation to use. May be one of "pearson", "kendall", or "spearman" but at this writing, only "pearson" was operative.
- "is.rotated": If FALSE the object is rotated.

```
# Load needed libraries (install first if necessary):
base::library(dimRed)
base::library(optimx)
base::library(pcaPP)

# For convenience, re-create pcaResult.
pcaResult <- dimRed::embed(
   .data=dat,
   .method="PCA",
   ndim = 2)

# Calculate the total correlation for pcaResult
dimRed::total_correlation(pcaResult, naxes = ndims(pcaResult), cor_
method = "pearson", is.rotated = FALSE)
[Output:]
   [1] 0.9331478
```

4.7. "cophenetic_correlation"

Associated with the work of Sokal and Rohlf (1962), cophenetic correlation is the correlation of the lower triangle of the distance matrix in high-dimensional space with that in low-dimensional space. High-dimensional space has many dimensions and corresponds to the input data. Low-dimensional space has few dimensions and corresponds to component space in a factor analysis context.

Passable parameters for the `cophenetic_correlation()` method are listed below:

- "d" is the distance function to use.
- "cor_method" is the type of correlation to use. May be one of "pearson", "kendall", or "spearman". This requires installing the library "pcaPP".

Example

```
# Load needed libraries if not already loaded:
base::library(dimRed)
base::library(pcaPP)

# Compute cophenetic correlation for pcaResult.
# This was created in the "total_correlation" section above.
dimRed::cophenetic_correlation(pcaResult, d = stats::dist,   cor_
method = "pearson")
[Output:]
   [1] 0.9983836
```

4.8. "distance_correlation"

This quality measure calculates the distance correlation between the distance matrices in high-dimensional (observed data) and low-dimensional (component) space. Distance correlation takes nonlinear as well as linear association into account when comparing matrices. It does not have any passable parameters other than the name of the dimRed result object.

Example

```
# Load the needed libraries (install first if necessary
base::library(dimRed)
base::library(energy)

# Compute the distance correlation for pcaResult.
dimRed::distance_correlation(pcaResult)
[Output:]
   [1] 0.9993446

# Compute the distance correlation for lleResult (created above).
dimRed::distance_correlation(lleResult)
[Output:]
   [1] 0.9340794
```

By the distance correlation criterion, the PCA result outperforms the LLE result.

Multi-method multi-measure comparison

DimRed's `quality()` command outputs quality measures either for methods with default parameters or result objects that have custom parameters. This section describes the former. It is adapted from Kraemer, Reichstein and Mahecha (2019: 29–21). In this example, we compare the PCA, nMDS, DRR, LLE, and HLLE methods, but any dimRed method might be assessed. The quality measures we consider are reconstruction_rmse, AUC_lnK_R_NX, Q_local, Q_global, and cophenetic_correlation.

Warning: When interpreting the tabular or graphic quality results (see Figure 12.2), keep in mind that higher is better performance for all quality methods *except* reconstruction error, where lower is better performance. Also, for reconstruction error, there are quality measures only for PCA and DRR as these are the only two which support reconstruction of the input matrix from the components.

```
# Declare working directory and load needed libraries (install.pack-
ages first if needed).
base::setwd("C:/Data")
base::library(dimRed)
base::library(vegan)
base::library(RANN)          # Needed for HLLE
base::library(RSpectra)      # Needed for HLLE

# Set a random seed
base::set.seed(123)

# Load the dimRed version of the Iris data.
dat <- dimRed::loadDataSet("Iris")

# The next line supports labeling of the methods to be embedded.
# Alternatively, uncomment the line below and use it for all
methods:
# embed_methods <- dimRedMethodList()
embed_methods <- base::c("PCA","nMDS","DRR","LLE", "HLLE")

# Create a vector of quality methods names.
quality_methods <- base::c("reconstruction_rmse","AUC_lnK_R_NX","Q_
local", "Q_global", "cophenetic_correlation")

# Next, initialize the quality_results matrix.
quality_results <- base::matrix(
NA, base::length(embed_methods), base::length(quality_methods),
dimnames = base::list(embed_methods, quality_methods))

# Then fill in the quality_results matrix and plot it in four steps.
# 1. Initialize "embedded"_data as an object of class list.
embedded_data <- base::list()

# 2. Loop through embed_methods and quality_methods
for (e in embed_methods) {
   embedded_data[[e]] <- dimRed::embed(dat, e)
   for (q in quality_methods)
      base::try(quality_results[e, q]<-
      dimRed::quality(embedded_data[[e]], q))
}
```

```
# 3. Display the results
quality_results
[Output:]
   reconstruction_rmse AUC_lnK_R_NX  Q_local  Q_global  cophenetic_
                                                        correlation
PCA          0.1591888   0.68989109 0.7399126 0.9724874  0.9983836
nMDS                NA   0.67893933 0.6816779 0.9598853  0.9971662
DRR          0.1591888   0.64208369 0.6600112 0.9539861  0.9962904
LLE                 NA   0.52280103 0.6060402 0.8504117  0.7763836
HLLE                NA   0.08756103 0.4318102 0.8668339  0.2685545

# 4.  Plot the results. The plot comparing results is Figure 12.2.
mycolors <- grDevices::rainbow(base::length(embed_methods))

graphics::barplot(quality_results,
main = "Performance measures for dimension reduction methods",
   sub = "Methods with default parameters",
   xlab = "Quality Measures", ylab = "Quality",
   col = mycolors,
   beside = TRUE) # Grouped bars
graphics::legend("topleft", legend=base::rownames(quality_results),
   col= mycolors, pch=15, cex=1.0, pt.cex=1)
```

Figure 12.2 shows that the PCA method of dimension reduction is better or equal to other methods when using each method's default parameter settings, for the Iris data. The DRR method performs equally well by the reconstruction mse and cophenetic correlation criteria, The MDS method is equal to PCA on the cophenetic correlation criterion and almost as good as PCA on the other criteria. The HLLE method performs markedly worse than other methods on all four quality measures.

The "coRanking" package

Introduction

Co-ranking compares ranks in the high-dimensional raw data with ranks in the low-dimensional component space. Co-rankings serve to help the researcher understand how the embedding process is working. Embedding, of course, is the mapping of high-dimensional raw data to low-dimensional component space.

A coRanking matrix is thus another way of evaluating the performance of a dimension reduction procedure such as PCA. The "coRanking" package in R, authored by Guido Kraemer, creates a co-ranking matrix and provides tools for visualizing it. The matrix is created by the function `coRanking::coranking()`. It is visualized using the function `coRanking::imageplot()`. It is also possible to create a plot of the R_NX model performance quality metric using the `R_NX()` function discussed above. The package also contains other functionality not discussed here. A critical article on co-ranking is Lee and Verleysen (2009).

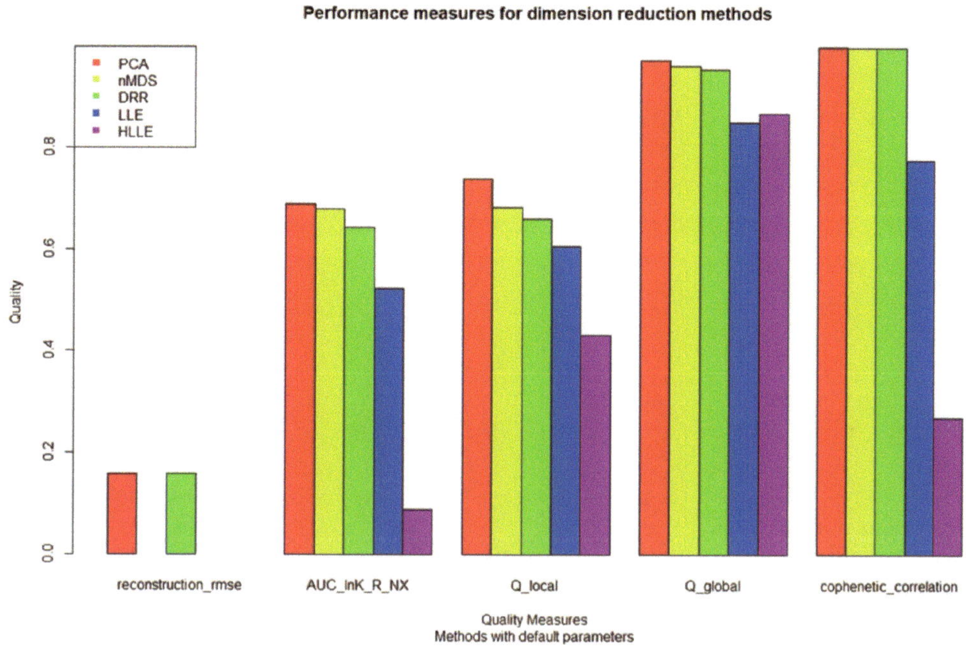

Figure 12.2 Model performance comparison by six quality measures.

Co-ranking plots

Example

```
#Setup: Load needed libraries (install first if needed):
base::library(dimRed)
base::library(coRanking)
base::library(psych)
base::setwd("C:/Data")

# Make sure the Iris dataset is loaded.
Iris <- dimRed::loadDataSet("Iris")

# Put the dimRed data object Iris into a data frame called "df".
df <- dimRed::as.data.frame(Iris)

# Create "dat" as an object with only the numeric columns of df.
dat <- df[,2:5]

# Run princomp() on dat to get a PCA results object.
princompResult <- stats::princomp(dat)

# Take the scores from the PCA result and create a co-ranking
matrix.
```

```
# The co-ranking matrix is a square n-by-n matrix, where n is the
number of observations.
Q.pca <- coRanking::coranking(dat, princompResult$scores[, 1:2])

# Optionally, list a small portion of the coranking matrix
Q.pca[1:3,1:4]
[Output:]
      R
   Ro   1  2  3  4
     1 56 26 25 21
     2 20 32 15 23
   3 14 21 28 21

# Use the imageplot() function to visualize the Q.pca coranking
matrix.
# This is Figure 12.3.
coRanking::imageplot(Q.pca, main = "PCA", sub="Values are log scaled
for better contrast.")
```

In the co-ranking plot in Figure 12.3, dots on the diagonal lines are pairs of points whose output distances in component space exactly match their distances in raw input space. Figure 12.3 is discussed further in the following "Metrics" section.

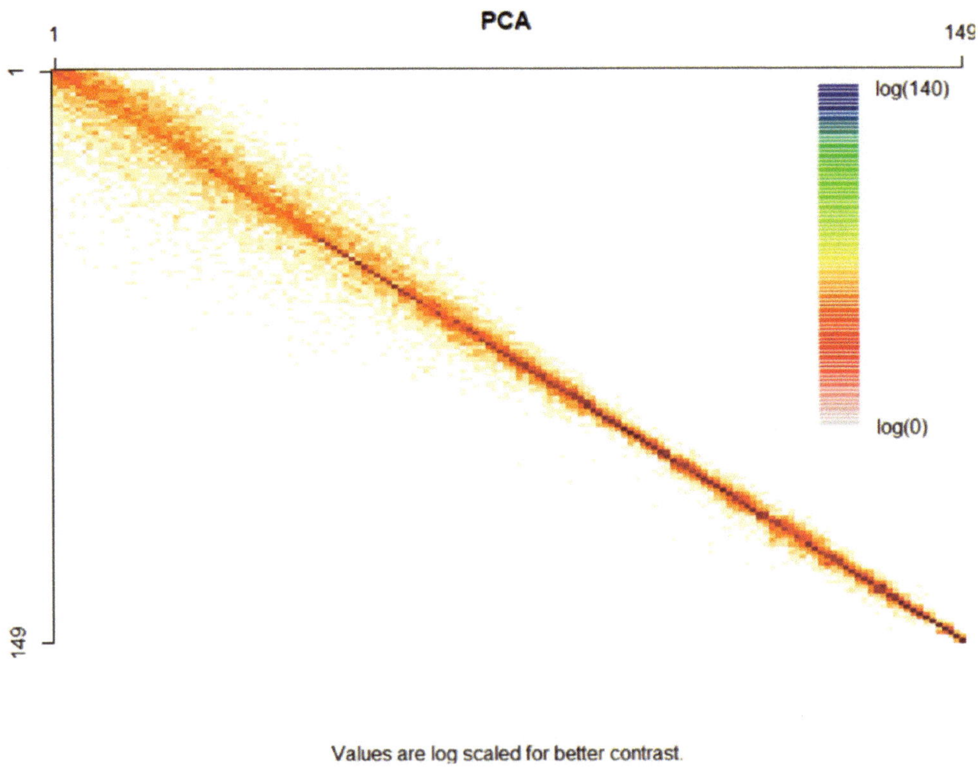

Figure 12.3 coRanking plot.

Metrics for co-ranking: T&C and LCMC

The "trustworthy" coefficient is a measure of false-neighbors, with lower being better. In the co-ranking plot, points in the lower triangle represent pairs of points which are visualized closer in component space than they are in the input data and thus are false neighbors. See Venna and Kaski (2001 and 2006), for a discussion of trustworthy and continuity.

The "continuity" coefficient is a measure of tear-ups, which occur when points are put far apart in (2D) component space compared to raw input space. Lower continuity is fewer tear-ups and is better. In the co-ranking plot, dots in the upper triangle are pairs whose distances are greater in component space than in raw input space and thus are tear-ups.

All methods of embedding raw input data in component space involve some degree of creation of false neighbors and tear-ups. For a given dataset, some methods may perform better than others.

In the coRanking plot in Figure 12.3, the x axis is "R" in the co-ranking matrix (Q.pca). The y axis is "Ro" in the co-ranking matrix. A well-fitting embedding should cluster the co-ranking values around the diagonal of the matrix, close to the diagonal throughout the range. If the values are predominantly in the lower triangle of the plot, then the embedding collapses (compresses) the original high-dimensional structure of the input data in comparison with the low-dimensional component structure, causing distant points to be closer (false neighbors). If the values are predominantly in the upper triangle, then close points in the original input structure are more distant in component space (tear-ups). In Figure 12.3 there are somewhat more points in the lower triangle than the upper and therefore somewhat more false neighbors than tear-ups.

The trustworthy and continuity (T&C) metrics attempt to detect and report where a given embedding goes wrong. However, the dimRed and coRanking packages instead use the local continuity meta-criterion (LCMC). Unlike T&C, LCMC is a single metric measuring goodness of embedding. While LCMC does not distinguish the two sorts of undesired events (false neighbors and tear-ups), it does give a single measure of overall performance of the dimension reduction for any given neighborhood size. There is an `LCMC()` function in both the dimRed and in the coRanking package but the latter is dominant when `base::library(coRanking)` is invoked more recently than `base::library(dimRed)`. Alternatively, use the package prefix as below to make sure the LCMC() function is that of the desired package. Only the coRanking version of LCMC works with objects of class "coranking". For LCMC, higher values indicate better performance of the dimensionality reduction. There is an LCMC value for each observation but we can take the mean as the overall value of LCMC for the embedding of the Iris data using PCA.

Recall that the "Q.pca" object created above was the co-ranking matrix for the default PCA solution using the `princomp()` program. Q.pca is an object of class "coranking". The <DS>LCMA() </DS>function calculates an LCMC value for neighborhoods of a given size, K. Alternatively, K may be a vector with multiple sizes. We assume that the Q, pca object created earlier is still in the R environment.

```
# Determine which neighborhood size returns the highest LCMC value:
base::which.max(LCMC_PCA)
[Output:]
   25

# Use neighborhood size of 25 to display the corresponding LCMC
value
```

```
# The "L" assures K is an integer.
LCMC_PCA <- coRanking::LCMC(Q.pca, K=25L)

LCMC_PCA
[Output:]
       [1] 0.7226148
```

The performance of embedding using a `princomp()` PCA solution, the LCMC metric optimized at neighborhood size = 25, is 0.72.

For comparison, we may compute LCMC for a factor analysis FA solution from the "psych" package.

```
faResult <- psych::fa(r=dat,nfactors = 2, rotate ="varimax")
Q.fa <- coRanking::coranking(dat, faResult$scores[, 1:2])
LCMC_FA <- coRanking::LCMC(Q.fa, K=1:base::nrow(Q.fa))
base::which.max(LCMC_FA)
[Output:]
    [1] 32

# Below we ask for LCMC values for neighborhood size optimized at 25
and 32.
LCMC_FA <- coRanking::LCMC(Q.fa, K=base::c(25L,32L))

LCMC_FA
[Output:]
       [1] 0.4479481 0.4641932
```

Whether optimized at a neighborhood size of 25 or 32, It is seen that LCMC for the `fa()` solution is much lower than for the default PCA solution from `princomp()`. Therefore, by the LCMC criterion, we judge the embedding by the PCA solution to be superior to that for the FA solution.

Using the R_NX curve to assess improvement of embedding

We can confirm that the PCA model is better performing than the FA model by using the "coRanking" package in conjunction with dimRed to plot the R_NX curve for both solutions for the Iris data as implemented by `princomp()` and `fa()` using default parameters. R_NX measures improvement of a given embedding over a random embedding for the neighborhood size K. Values range from 0 for a random embedding to 1 for a perfect embedding. We start by computing the matrix of R_NX values. This is a numeric vector of length 150 for the Iris data (150 is the sample size, which is the maximum neighborhood size).

```
R_NX_vector_pca <- coRanking::R_NX(Q.pca)

# Plot the R_NX curve for PCA.
```

```
dimRed::plot(R_NX_vector_pca, col="blue", xlab = "Neighborhood Size,
K", ylab = "R_NX Quality Metric", main = "Model Performance by
Neighborhood Size", sub="PCA model in blue, FA model in red")

# Add to the plot the R_NX curve for the FA solution. This is Figure
12.4,
R_NX_vector_fa <- coRanking::R_NX(Q.fa)
graphics::points(R_NX_vector_fa, col="red")
```

Recall that R_NX is a measure of improvement of the embedding of high-dimensional data (here, the Iris data) in low dimensional (component) space compared to random embedding. The x axis is an index of neighborhood size K, going from 0 to the number of observations (150 for the Iris data). The formula for R_NX (type `help(R_NX)`) is applied to the K-by-K upper left block of the co-ranking matrix, for successive values of K. The y axis reflects model quality and ranges from 0 (the solution is no better than random embedding) to 1 (perfect embedding). When comparing two methods, such as PCA and FA, the higher curve is the better performing model. By this criterion, the PCA model has superior embedding compared to FA at all neighborhood sizes.

Cross-validation with dimRed

Cross-validation, of course, is training the model with one subset of the data and testing it on another subset. These are often labeled the "train" and "test" sets respectively, though other labels are used also. For instance, the test set is sometimes called the

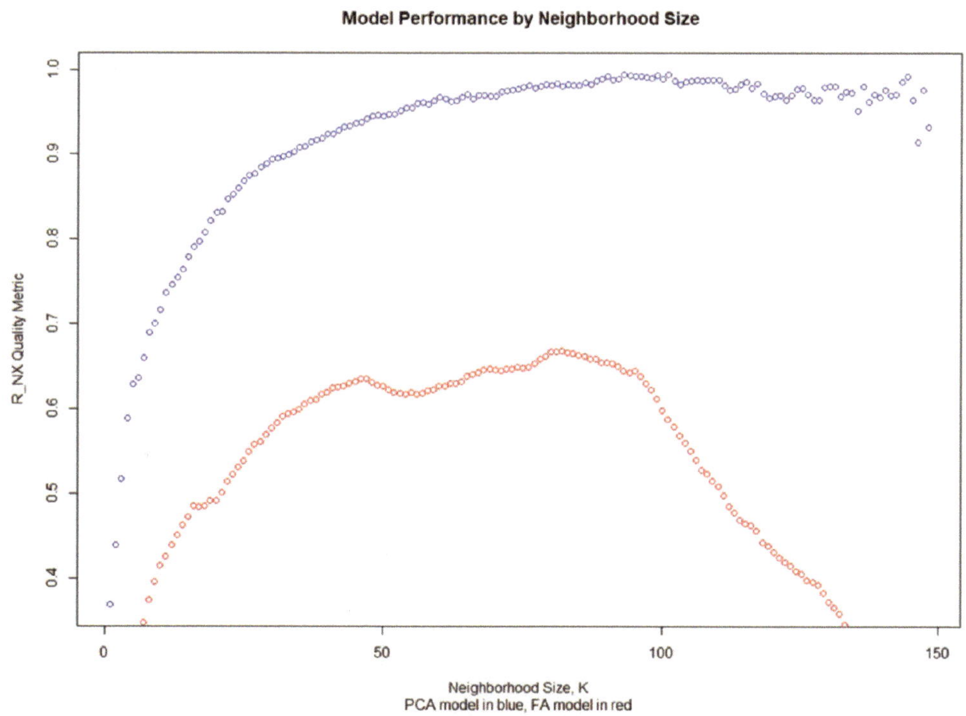

Figure 12.4 R_NX Curves for the PCA and FA models.

validation set. The simplest form of cross-validation is the one-fold train-test method illustrated below.

```
# Example
# Put the dimRed data object Iris into a dimRed data object called
"irisdat".
irisdat <- dimRed::loadDataSet("Iris")

# Define the size of the training sample, here 70% of the sample.
sampleSize <- dimRed::nrow(irisdat)
trainSize <- base::round(.70*sampleSize)

# Create a vector of random observation index numbers for the train-
ing sample.
# sample() is from R's "base" package.
# To get replicable results, set.seed() must be used, otherwise the
sample differs each run.
base::set.seed(123)

trainSample <- base::sample(dimRed::nrow(irisdat),  size=trainSize,
replace=FALSE)

# Create two PCA result objects, one for the train set and one for
the test set
pcaTrainResult <- dimRed::embed(irisdat[trainSample,], "PCA", .mute
= NULL)

pcaTestResult <- dimRed::embed(irisdat[-trainSample,], "PCA", .mute
= NULL)

# Plot both results. This is Figure 12.5.
# par() sets display parameters to one row, two columns (side by
side plots)
graphics::par(mfrow=base::c(1,2))

dimRed::plot(pcaTrainResult, type = "2vars", main="Iris Training
Sample, PCA Result")

dimRed::plot(pcaTestResult, type = "2vars", main="Iris Test Sample,
PCA Result")
# Reset the display parameters for single plots
graphics::par(mfrow=base::c(1,1))
```

Package dimRED multi-method multi-measure comparison with custom parameters

The R code in this section creates a multi-method, multi-measure comparison bar graph similar to that in the "Multi-method multi-measure comparison" section, but with

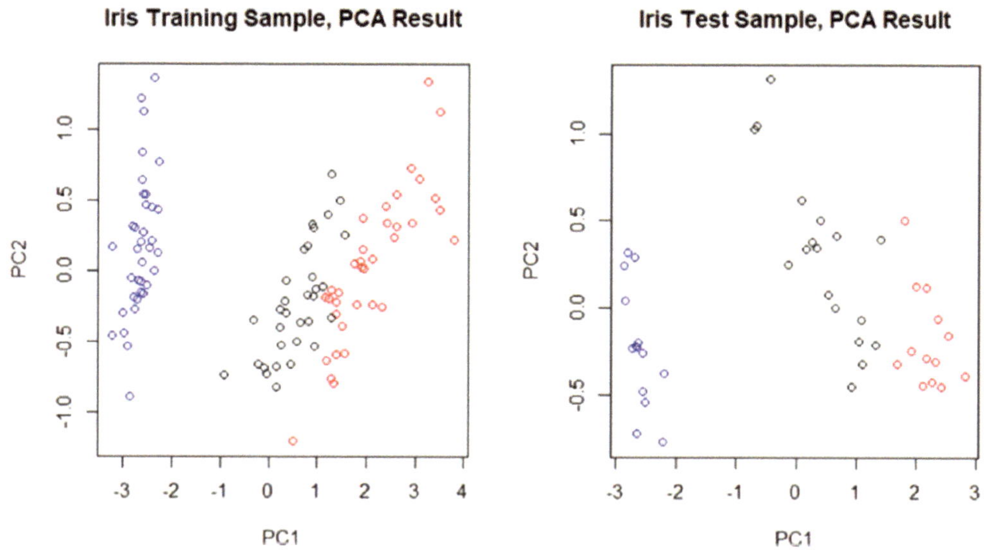

Figure 12.5 Training and test results, iris data.

custom rather than default parameters. For code in this section to work, all dimRed result objects must be in the environment (here, pcaResult, nmdsResult, drrResult, lleResult, and hlleResult). The custom parameters were the ones treated for each result (e.g., pcaResult) in the text, but actually a wide range of other custom settings might be specified. Note that two-dimensional solutions are requested. The result is shown in Figure 12.6.

```
# Setup; Load required packages (install first if needed):
base::setwd("C:/Data")
base::library(dimRed)
base::library(optimx)
base::library(pcaPP)
base::library(psych)

# Load data
dat <- dimRed::loadDataSet("Iris")
# Re-create results objects
pcaResult <- dimRed::embed(
   .data=dat,
   .method="PCA",
   ndim = 2)

base::set.seed(123)          # Needed for replicable results
nmdsResult <- dimRed::embed(
   .data = dat,
   .method = "nMDS",
```

```
    ndim = 2,
    d = function(x) stats::dist(x, method="euclidean"))

base::set.seed(123)
drrResult <- dimRed::embed(
    .data = dat,
    .method = "DRR",
    ndim = 2)

lleResult <- dimRed::embed(dat, "LLE", ndim = 2, knn = 30)


hlleResult <- dimRed::embed(
    .data = dat,
    .method = "HLLE",
    ndim = 2,
    knn = 12)


# The next line is for labels for the methods
methodlabels <- base::c("PCA","nMDS","DRR","LLE", "HLLE")

# List the custom result objects. All must be active in the R
environment.
embed_methods <- base::c("pcaResult", "nmdsResult", "drrResult",
"lleResult", "hlleResult")

# List the quality measures
quality_methods <- base::c("reconstruction_rmse","AUC_lnK_R_NX","Q_
local", "Q_global", "cophenetic_correlation")

# The part below initializes the quality_results matrix
quality_results <- base::matrix(
NA, base::length(embed_methods), base::length(quality_methods),
dimnames = base::list(methodlabels, quality_methods)
)

# The part below fills in the quality_results matrix
# Initialize embedded_data as an object of class list.
embedded_data <- base::list()

# Loop through embed_methods and quality_methods
numMethods <- base::length(embed_methods)
numMeasures <- base::length(quality_methods)
```

```
for (e in 1:numMethods) {
for (q in 1:numMeasures){
   x <- embed_methods[e]
   x2<- eval(parse(text=x))
   y <- quality_methods[q]
   base::try(quality_results[e, q]<- quality(x2,y))
  }
}


# Display the results
quality_results
[Output:]
  reconstruction_rmse AUC_lnK_R_NX   Q_local  Q_global   cophenetic_
                                                          correlation
PCA            0.1591888    0.6898911 0.6526642 0.3899371    0.9983836
nMDS                  NA    0.6789393 0.6179195 0.4003979    0.9971662
DRR            0.1591888    0.6424127 0.5871881 0.3832801    0.9961054
LLE                   NA    0.5761980 0.5931848 0.3225239    0.7401983
HLLE                  NA    0.3799067 0.4687204 0.2010937    0.7141958

# Plot the results. This is Figure 12.6.
mycolors <- grDevices::rainbow(base::length(embed_
methods))

graphics::barplot(quality_results,
```

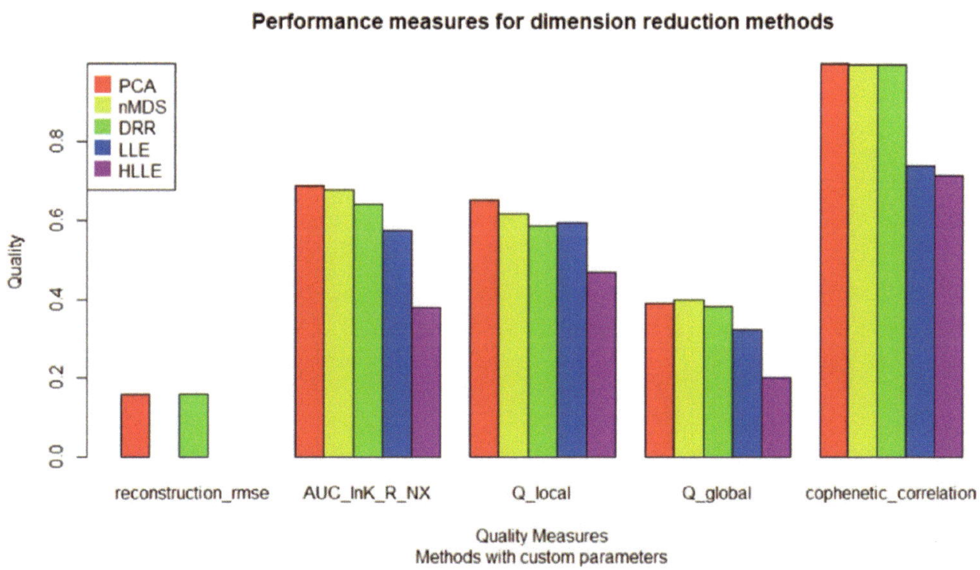

Figure 12.6 Performance measures for five-dimension reduction methods.

```
main = "Performance measures for dimension reduction methods",
sub = "Methods with custom parameters",
  xlab = "Quality Measures", ylab = "Quality",
  col = mycolors,
  beside = TRUE) # Grouped bars
graphics::legend("topleft", legend=base::rownames(quality_results),
  col= mycolors, pch=15, cex=1.0, pt.cex=2)
```

13 Recipes

An alternative system for dimension reduction

Introduction

This chapter illustrates dimension reduction using the `recipe()` command from the package of the same name. The "recipes" package for computing and pre-processing design matrices is not just for a PCA design but also for three other types of dimension reduction: ICA (independent components analysis), KPCA (kernel principal component analysis), and UMAP (uniform manifold approximation and projection).

The recipes design framework

The approach using PCA under "recipes" replaces R's usual one-step formula method with three steps which define how variables are treated. These three main steps are reflected in the `recipe()`, `prep()`, and `bake()` functions in that order, as described below.

1 `recipe()`: The `recipe()` step defines the operations on the data and the associated roles for variables. There is an extensible set of possible subordinate steps, which may be chained using the piping operator. There are three types of principal components steps:
   - `step_pca()`: Define a principal component analysis (PCA) operation. Normalizing of variables is recommended prior to running PCA.
   - `step_ica()`: Define an independent components analysis (ICA) model, which, like PCA, estimates statistically independent components. As with PCA, normalizing of variables is recommended prior to running ICA.
   - `step_kpca()`: Define a kernel principal component analysis (KPCA) model. The kernel function involves a broader dimensionality than PCA. For example, under a quadratic kernel function, each variable is represented by its original values as well as its square. This nonlinear mapping can potentially help uncover better representations of the original data. This step requires previous installation of the "dimRed" and "kernlab" packages. The kernel function and other parameters are specified in the kernel and kpar slots of the options argument to `step_kpca()`. Like PCA, normalizing of variables is recommended prior to running KPCA. Note, however, that in future versions, documentation at this writing states that this step is slated to be phased out in favor of the steps below
     - `step_kpca_rbf()` for radial basis function PCA.
     - `step_kpca_poly()` for polynomial kernel PCA.

DOI: 10.4324/9781003279693-15

- `step_umap()`: Define a uniform manifold approximation and projection (UMAP) model for dimension reduction. This step is available only after installing and loading the "embed" package. This method is derived from the field of topology, where it Is used to project a set of features onto a smaller space. Models are usually supervised (training data are provided) but may be unsupervised as well. Like PCA, normalizing of variables is recommended prior to running UMAP.

Other steps include:

- `step_center()`: Center variables to have mean = 0.
- `step_dummy()`: Create dummy variables from categorical variables.
- `step_impute()`: A series of commands for imputing by different methods. Currently available are `step_bagimpute()`, `step_knnimpute()`, `step_lowerimpute()`, `step_meanimpute()`, `step_medianimpute()`, `step_modeimpute()`, and `step_rollimpute()`.
- `step_normalize()`: Standardize (center and scale) variables so sd = 1 and mean = 0.
- `step_nzv()`: Remove redundant variables that have the same or approximately the same values for every data point.
- `step_range()`: Require numeric data to be within a pre-defined range of values.
- `step_scale()`: Normalize so standard deviation = 1 (does not center).

A full list of available steps is under the "S" listings when in RStudio one clicks the "Packages" tab and goes to the "recipes" package. The "recipes" reference page is at https://recipes.tidymodels.org/reference/index.html.

Also, instructions for creating one's own custom steps are here: https://www.tidymodels.org/learn/develop/recipes/.

2. `prep()`: In this step, any parameters defined by the `recipe()` step are estimated, though results are not yet output. For instance, if `step_pca()` is in the recipe, then principal components and related values are estimated.
3. `bake()`: The `bake()` step applies the operations defined and estimated for the model, in this example a PCA model. Output is presented in this step. If using cross-validation, we would have a train set and a test set, and we would `prep()` on the train set and `bake()` on the test set. Type `help(step_ica)` and view the "Examples" section for an illustration of cross-validation.
4. `juice()` is another function in the "recipes" package which is used to present results of the prep step. As of "recipes" version 0.2.0, `juice()` is superseded in favor of `bake(object, new_data = NULL)`.

Libraries and setup for this section

Prior to trying this section "hands-on", the following libraries should be loaded. Some may well require the reader to install them first with the `install.packages()` command.

```
base::library(dimRed)      # Required for KPCA models.
base::library(dplyr)       # Supports "%>%" piping and more
                           # Starting with R 4.1, "|>" is a built-in
                             piping operator
base::library(embed)       # Required for UMAP models.
```

```
base::library(fastICA)       # Required for ICA models.
base::library(ggplot2)       # Leading visualization package in R
base::library(kernlab)       # Required for KPCA models.
base::library(lubridate)     # Utility for date variables
base::library(recipes)       # Has recipe(), prep(), and  bake()
commands
base::library(tidyr)         # Has pivot_wider()
base::library(tidyverse)     # Suite of tidy tools discussed below
base::library(unvotes)       # Contains the example data

# Set the working directory, using your own folder address.
base::setwd("C:/Data")
```

Though we have loaded all needed libraries explicitly above, this is redundant since `base::library(tidyverse)` will load the core packages in tidyverse. These are:

- `ggplot2`, for data visualization.
- `dplyr`, for data manipulation.
- `tidyr`, for data tidying.
- `readr`, for data import.
- `purrr`, for functional programming.
- `tibble`, for tibbles, a superset of data frames.
- `stringr`, for strings.
- `forcats`, for factors.

For additional information on tidyverse use, see https://tidyverse.tidyverse.org/.

The unvotes example data

The example data are from the package "unvotes" (Voeten, 2013) as found in the package "unvotes" compiled by David Robinson.[1] The data cover UN roll call votes from 1946 to 2019. These data were also used by Julia Silger to illustrate PCA.[2] The unvotes package contains three tibble-type data frames:

1 un_votes: This data frame, which we later re-create as the "votes" data frame, contains a history of each country's UN votes, with four columns and 869,937 rows. The columns are rcid, country, country_code, and vote. For instance, the first row is:

```
         rcid country          country_code vote
        <dbl> <chr>            <chr>        <fct>
  1         3 United States    US           yes
```

The row number (1) is the vote. The rcid variable is the roll call id code (3) for the issue group for that vote. Country and country_code are long and short country labels. The vote variable is coded "yes", "abstain", or "no". All columns are character type except rcid, which is numeric.

2 un_roll_call_issues: This data frame, which we later re-create as the "issues" data frame, shows in which of six possible issue areas the vote was. There are 3 columns and 5,745 rows. The columns are rcid, short_name, and issue. For instance, for rcid = 985 the issue is "Nuclear weapons and nuclear material" and its short name is "nu".

Other rcid's are also in the same issue group. To illustrate further, the first row is this (me stands for Middle East, the short_name for the issue):

```
     rcid short_name issue
    <int> <chr>      <fct>
  1    77 me         Palestinian conflict
```

3 un_roll_calls: This data frame, which we later re-create as the "rollcalls" data frame, contains the date, description, and other information on each vote, in nine columns and 6,202 rows. The columns (variables) are rcid, session, importantvote, date, unres, amend, para, short, and desc. For instance, the first row is shown below:

```
rcid session importantvote date  unres  amend  para short  descr
<int> <dbl> <int> <date>     <chr>  <int> <int> <chr>     <chr>
1  3  1     0       1946-01-01 R/1/66 1      0     AMENDM..."TO ADOPT ...
```

Note that the "rcid" variable (roll call ID) is a link variable across all three data frames, allowing joining of their variables for purposes such as making figures and tables.

After installing and loading the "unvotes" package, load the data frames, copying them into the more simply named data frame objects "votes", "issues", and "rollcalls":

```
base::library(unvotes)
votes <- base::data.frame(un_votes)
issues <- base::data.frame(un_roll_call_issues)
rollcalls <- base::data.frame(un_roll_calls)
```

Data levels: a note of caution

It is known that PCA based on binary variables is problematic for reasons discussed in Chapter 8. Here dimension reduction is not based on binary variables but rather is based on UN roll call votes, all of which are measured as 1 = "yes", 2 = "abstain", and 3 = "no". The vote variables could be conceived as nominal in data level if no order is assumed to exist across the levels. This would require correspondence analysis or some other procedure compatible with nominal data. Second, such variables may be considered to be as ordinal in data level, with three levels, for which CATPCA would be appropriate if the three levels are seen as "true" or "natural" ordered ranks. Third, such variables may be conceived as three-level binned version of an underlying continuous construct which ranges from positivity to negativity. By using PCA, we are making this third data assumption. Binning a continuous variable, however, loses information compared to the original. That is, it is a form of measurement error. Measurement error, in turn, increases standard error and attenuates effect sizes. This makes the result of PCA with binned variables a "floor": variance explained is, on average, at least as high as PCA results report and insofar as some variables are dismissed due to low effect size in relation to standard errors, this may reflect the inflation of Type II error when using binned variables.

Illustration of use of the unvotes data

Before illustrating PCA with the "recipes" package, below we present a brief illustration of the use of the unvotes data. The illustration shows how to create a table of the percent of "yes" votes by year and within year, by each country.

In the code below, we join votes (four variables, including rcid) with rollcalls (nine variables including rcid) to get the data frame "joined" (12 variables, including rcid). An inner join is one which joins all variables in each of the two input data frames. The `by="rcid"` parameter names the link variable. The votes data frame is piped to the `inner_join()` command, which joins it with the rollcalls data frame to create "joined". Note that for piping we use the "|>" piping operator built into R 4.1 and higher (otherwise we would have used the %>% operator from the dplyr package).

```
base::library(dplyr)
base::library(lubridate)

joined <- votes |>
  dplyr::inner_join(rollcalls, by = "rcid")

base::names(joined)
[Output:]
     [1] "rcid"            "country"          "country_code"
     [4] "vote"            "session"          "importantvote"
     [7] "date"            "unres"            "amend"
    [10] "para"            "short"            "descr"
```

In the next block of code, we create the desired table, placing it in the object "yes_by_year_country". First we pass the "joined" data frame to the `group_by()` function, which groups by year and within year by country (because it is listed second). Then the grouped result is piped to the `summarize()` command, which counts the votes using the `n()` function, calculates the percent yes (the `mean()` function is from R's base package), and thus makes the table shown below.

```
yes_by_year_country <- joined |>
  dplyr::group_by(year = lubridate::year(date), country) |>
  dplyr::summarize(votes = dplyr::n(),
             percent_yes = base::mean(vote == "yes"))

yes_by_year_country
[Output:]
       year country       votes percent_yes
      <dbl> <chr>         <int>       <dbl>
    1  1946 Afghanistan      17       0.412
    2  1946 Argentina        43       0.698
    3  1946 Australia        43       0.558
    4  1946 Belarus          43       0.442
    5  1946 Belgium          43       0.605
    6  1946 Bolivia          43       0.698
```

```
 7  1946 Brazil          43         0.605
 8  1946 Canada          42         0.643
 9  1946 Chile           43         0.605
10  1946 Colombia        42         0.310
# ... with 10,451 more rows
```

As a further illustration, drawn from documentation for the "unvotes" package, we may graph the voting pattern (in terms of percent of yes votes) of the United States in each of the six UN roll call issue areas. This is shown in Figure 13.1.

In the code below, the previously created (above) "joined" object is piped to a series of steps using the "|>" piping operator:

1. `filter()`: Only the "United States" rows from the joined data frame are used.
2. `inner_join()`: The variables from the issues data frame are joined using "rcid" as the link variable.
3. `group_by()`: Data are grouped by year and then within year by issue.
4. `summarize()`: Votes are counted and the percent_yes calculated.
5. `filter()`: We re-filter for more than 5 votes.
6. `ggplot()`: The specification of the plot is started, asking for year on the x axis and percent_yes on the y axis. Lines 7–9 are not the result of piping but are further specifications of the `ggplot()` figure connected by plus signs.
7. `geom_point()`: A scatterplot of points is requested with points represented by the letter "v" (for vote) in dark green, size 3.
8. `geom_smooth()`: A loess smoothing line is requested.
9. `facet_wrap()`: A panel plot by issue is requested. There are six issue areas, so there will be six plots in the panel.
10. `theme()`: Panel titles are made larger than default.
11. `labs()`: Sets labels for the figure.

```
joined |>
  dplyr::filter(country == "United States") |>
  dplyr::inner_join(issues, by = "rcid") |>
  dplyr::group_by(year = lubridate::year(date), issue) |>
  dplyr::summarize(votes = n(), percent_yes = base::mean(vote ==
"yes")) |>
  dplyr::filter(votes > 5) |>
  ggplot2::ggplot(aes(year, percent_yes)) +
  ggplot2::geom_point(color="darkgreen", pch="v", cex=3) +
  ggplot2::geom_smooth(se = FALSE) +
  ggplot2::facet_wrap(~ issue) +
  ggplot2::theme(strip.text=ggplot2::element_text(size=14))   +
labs(
    x = "Year",
    y = "Percent Yes", fill = NULL,
    title = "US Voting on UN Issues",
    subtitle = "UN roll call data, 1946 - 2019"
  )
```

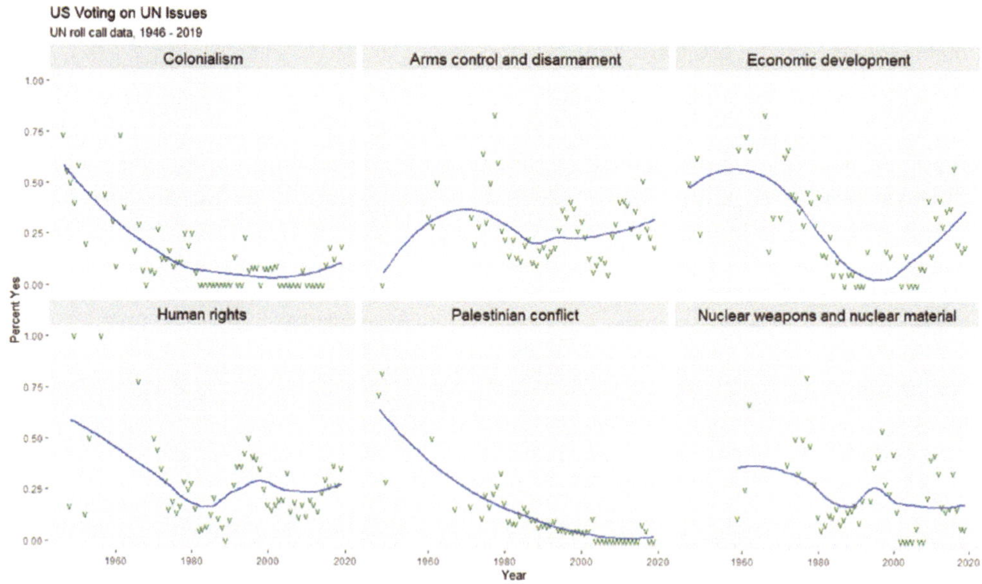

Figure 13.1 Panel plot of UN voting by the United States in six issue areas.

PCA: standard deviations, variances, eigenvectors, eigenvalues, contributions, and loadings

Before turning in the next section to PCA and other dimension reduction procedures using the "recipes" package, in this section we use a small dataset (the classic "iris" dataset) to illustrate how it compares to the `prcomp()` command from R's built-in "stats" package discussed in an earlier section. Specifically, we want to highlight how the two packages approach standard deviations, variances, eigenvectors, eigenvalues, contributions, and component loadings. This is important both to better understand these terms and to understand differences in labeling them.

The "iris" dataset is part of R's built-in "datasets" package. Type `help("iris")` for more information on these data.

```
# Setup for this subsection
# Libraries "base" and "stats" should load automatically as part of
R's System library.
base::library(base)          # Has the eigen() command
base::library(dplyr)         # Has the filter() command
base::library(psych)         # Has the eigen.loadings() command
base::library(recipes)       # Has the recipe() and prep() commands
base::library(stats)         # Has the cor() and prcomp() commands

# Invoke the iris dataset
utils::data(iris)
```

The prcomp approach

Running the `prcomp()` program

This step runs the `prcomp()` program and lists out the eigenvector matrix, labeled "rotations". Eigenvectors have to do with the direction of the location of a variable in component (factor) space. Eigenvalues have to do with the length along the vector for a given variable.

```
# The "-5" removes the 5th column, which Is labels, leaving only the
four numeric variables.
irisResult <- stats::prcomp(iris[,-5],center=TRUE, scale=TRUE)

irisResult
[Output:]
   Standard deviations (1, .., p=4):
   [1] 1.7083611 0.9560494 0.3830886 0.1439265

   Rotation (n x k) = (4 x 4):
                       PC1         PC2        PC3        PC4
   Sepal.Length  0.5210659 -0.37741762  0.7195664  0.2612863
   Sepal.Width  -0.2693474 -0.92329566 -0.2443818 -0.1235096
   Petal.Length  0.5804131 -0.02449161 -0.1421264 -0.8014492
   Petal.Width   0.5648565 -0.06694199 -0.6342727  0.5235971
```

Eigenvectors

The "rotation" from step 1 is the eigenvectors matrix, with variables as rows and components as columns. Output is not shown as it duplicates the "Rotation" portion of output in step 1 above.

```
irisResult$rotationp
```

Eigenvectors are the "coordinates" used for plotting variables in component space. The command `plot(irisResult$rotation)` will give a bare-bones version of such a plot.

Standard deviations and variances

The standard deviations are in the "sdev" element of "irisResult".

```
irisResult$sdev
[Output:]
 [1] 1.7083611 0.9560494 0.3830886 0.1439265
```

However, we are more interested in variances, which are the square of standard deviations. As can be seen in step 5, the variances are the eigenvalues.

```
irisVariances <- irisResult$sdev^2
```

```
irisVariances
[Output:]
   [1] 2.91849782 0.91403047 0.14675688 0.02071484
```

Percent of variance explained

The percent of variance explained by each component is simply its variance divided by the sum of variances for all variables. The first component always explains the most and the last component the least.

```
pctExplained = irisVariances/base::sum(irisVariances)
pctExplained
[Output:]
    [1] 0.729624454 0.228507618 0.036689219 0.005178709
```

Eigenvalues and eigenvectors

It can be seen below that the variances are the same as the eigenvalues. By the Kaiser criterion, one would retain components with eigenvalues of 1.0 or greater, which here is the first one. There are other criteria, however, for selecting the number of components of interest.

For `prcomp()`, variances are the square of the "sdev" element, and these are the eigenvalues. The eigenvalues below are for PCA1 through PCA4.

```
iris_eigen <- irisResult$sdev^2

iris_eigen
[Output:]
   [1] 2.91849782 0.91403047 0.14675688 0.02071484
```

The same result is obtained from the `eigen()` command. However, there is a difference. The object iris_eigen is a simple numeric vector. The object iris_eigen2 is a list with both "values" and "vectors" as attributes. Thus the commands below `iris_eigen2$values` and `iris_eigen2$vectors` only work for iris_eigen2, not iris_eigen.

First, a correlation matrix is created, needed to calculate eigenvalues using the `eigen()` command from R's "base" package.

```
iris_matrix <- stats::cor(iris[,1:4])
[Output not shown, but a correlation matrix is created.]
```

Second, eigenvalues are calculated using `eigen()`.

```
iris_eigen2 <- base::eigen(iris_matrix)

iris_eigen2$values
[Output:]
   [1] 2.91849782 0.91403047 0.14675688 0.02071484
```

Eigenvectors are in the "rotation" element of the irisResult object created earlier by the `prcomp()` command. Eigenvectors are the coordinates used for plotting variables in component space:

```
irisResult$rotation
                    PC1         PC2        PC3        PC4
   Sepal.Length  0.5210659 -0.37741762  0.7195664  0.2612863
   Sepal.Width  -0.2693474 -0.92329566 -0.2443818 -0.1235096
   Petal.Length  0.5804131 -0.02449161 -0.1421264 -0.8014492
   Petal.Width   0.5648565 -0.06694199 -0.6342727  0.5235971
```

Alternatively, we can derive eigenvectors from the iris_eigen2 object created above using the `eigen()` command from the "base" package. For this method, we must first add column and row names to the matrix. This labeling will carry forward to other output further below in this example.

```
base::colnames(iris_eigen2$vectors) <- base::c("PCA1", "PCA2",
"PCA3", "PCA4")
base::rownames(iris_eigen2$vectors) <- base::c("Sepal.Length", "Se-
pal.Width",  "Petal.Length", "Petal.Width")
iris_eigen2$vectors
[Output:]
                    PCA1        PCA2       PCA3       PCA4
   Sepal.Length  0.5210659 -0.37741762  0.7195664  0.2612863
   Sepal.Width  -0.2693474 -0.92329566 -0.2443818 -0.1235096
   Petal.Length  0.5804131 -0.02449161 -0.1421264 -0.8014492
   Petal.Width   0.5648565 -0.06694199 -0.6342727  0.5235971
```

Component contributions to variance explained by a component

The sum of squared eigenvectors across all variables for a given component is always 1.

```
# Create the iris correlation matrix
irisMat <- stats::cor(iris[,1:4])

# Compute eigenvalues and eigenvectors using eigen() from R's "base"
package.
irisEigen <- base::eigen(irisMat)

# Contributions to the importance of each component are the square
of the eigenvectors.
contributions <- irisEigen$vectors^2
base::rownames(contributions) <- base::c("Sepal.Length", "Sepal.
Width",  "Petal.Length", "Petal.Width")
contributions
[Output:]
                      [,1]         [,2]       [,3]       [,4]
   Sepal.Length 0.27150969 0.1424440565 0.51777574 0.06827052
   Sepal.Width  0.07254804 0.8524748749 0.05972245 0.01525463
```

```
   Petal.Length 0.33687936 0.0005998389 0.02019990 0.64232089
   Petal.Width  0.31906291 0.0044812296 0.40230191 0.27415396

# The contributions from the first component sum to 1.0
contributions[,1]
[Output:]
 [1] 0.27150969 0.07254804 0.33687936 0.31906291

base::sum(contributions[,1])
[Output:]
 [1] 1

# This is true for all components. Thus the contributions for four
components sum to 4.
base::sum(contributions[,])
[Output:]
 [1] 4
```

Therefore, it is also true that the contributions are percentages. We may say, for instance, that Sepal.Length contributes about 27% to the variance explained by the first component. The order of contributions for a given component is the order of importance of the variables in explaining the variance explained by the first component. Put another way, since we infer the meaning of components from the contributions, we may say that Sepal.Length contributes about 27% to the meaning of PCA1.

Component loadings

Technically, factor loadings come from a factor analysis such as principal factor analysis (PFA) rather than from PCA. For this reason, `prcomp()` and `princomp()` do not directly output component loadings. If `principal()` is used for PCA, then the procedure below is not needed. The `eigen.loadings()` function from the "psych" package creates pseudo-factor loadings which we may call "component loadings". These correspond to the usual social science definition of the factor loadings metric: loadings = eigenvectors times the square root of the eigenvalues.

In the social science sense of "loadings", contributions are not loadings, though documentation in R not infrequently labels them as such. Below we compute loadings defined as eigenvectors times the square root of eigenvalues. The coefficients for loadings so defined differ from contribution coefficients above. The similarity is that both contributions and loadings give the same order of variable importance. For PCA1 this order (descending) is Petal_Length, Petal_Width, Sepal_Length, and Sepal_Width. The difference is that loadings are appropriate for assessing such common rules of thumb as the simple factor structure criterion. This criterion states that ideally, all variables would load "heavily" (often > 0.6) on one and only one factor and "weakly" (often < 0.4) on others, in order to attain the simple factor structure which makes interpretation of components much easier. Put another way, the metric for a variable being "heavily loaded" on a component is different for loadings and contributions. Social scientists customarily use loadings, not contributions, when assessing the strength of a variable for a given component.

Below we list the component loadings for the iris data. In output, the columns are PCA1 through PCA4. The rows are the four predictors. Note the object within parentheses for eigen.loadings() must have been created, like here, from the eigen() command or from princomp(). This method, unlike the alternative below it, requires assigning column labels again. The loadings in iris_loadings are the same as from the principal() command's "$loadings" element from the command output <- principal(iris[-5], nfactors=4, rotate="none").

```
iris_eigen2 <- base::eigen(iris_matrix)
iris_loadings <- psych::eigen.loadings(iris_eigen2)
base::colnames(iris_loadings) <- base::c("PCA1", "PCA2", "PCA3",
"PCA4")

iris_loadings
[Output:]
                   PCA1        PCA2        PCA3        PCA4
  Sepal.Length  0.8901688 -0.36082989  0.27565767  0.03760602
  Sepal.Width  -0.4601427 -0.88271627 -0.09361987 -0.01777631
  Petal.Length  0.9915552 -0.02341519 -0.05444699 -0.11534978
  Petal.Width   0.9649790 -0.06399985 -0.24298265  0.07535950
```

Below, alternative R code also generates the same component loadings. Note that t() is the transpose matrix command. Loadings are a function of eigenvectors divided by the square root of eigenvalues.

```
evectors <- iris_eigen2$vectors
evalues <- iris_eigen2$values
temp <- base::t(evectors)*base::sqrt(evalues)
loadings <- base::t(temp)

loadings
[Output:]
                   PCA1        PCA2        PCA3        PCA4
  Sepal.Length  0.8901688 -0.36082989  0.27565767  0.03760602
  Sepal.Width  -0.4601427 -0.88271627 -0.09361987 -0.01777631
  Petal.Length  0.9915552 -0.02341519 -0.05444699 -0.11534978
  Petal.Width   0.9649790 -0.06399985 -0.24298265  0.07535950
```

Under either method, we can see from the absolute values of the loadings that Sepal.width is most associated with PCA2 and the other variables are most associated with PCA1.

Below we demonstrate that loadings are the same as the correlations of variables with the components. We use the PC1 component as an example.

```
# Create a PCA results object using principal().
principalOut <- psych::principal(iris[-5], nfactors=4,
rotate="none")

#  Note that the loadings for PC1 are the same as shown above.
principalOut$loadings
```

```
[Output:]
   Loadings:
                PC1    PC2    PC3    PC4
   Sepal.Length  0.890  0.361 -0.276
   Sepal.Width  -0.460  0.883
   Petal.Length  0.992                0.115
   Petal.Width   0.965         0.243

                   PC1   PC2   PC3   PC4
   SS loadings    2.918 0.914 0.147 0.021
   Proportion Var 0.730 0.229 0.037 0.005
   Cumulative Var 0.730 0.958 0.995 1.000

# We save the component scores to "scores".
scores <- base::as.data.frame(principalOut$scores)

# We make a copy of iris.
iris2 <- iris

# We add the scores for PC1 to iris2.
iris2$PC1 <- scores$PC1

# We get the correlation of the variables with the scores for PC1.
# We note the correlations are the same as the loadings.
stats::cor(iris2[1:4],iris2$PC1)

[Output:]
                    [,1]
   Sepal.Length  0.8901688
   Sepal.Width  -0.4601427
   Petal.Length  0.9915552
   Petal.Width   0.9649790
```

The "step_pca" approach under recipes

Running step_pca under "recipes"

The first two steps under "recipes" are the recipe step and the prep step. For the recipe step, note that there are two subsidiary steps: step_normalize and step_pca. Output from step_pca has an index number of 2 as it is second to be invoked.

```
# The recipe step
iris_recipe <- recipes::recipe(Species ~., data = iris) |>
   recipes::step_normalize(all_predictors()) |>
   recipes::step_pca(all_predictors()
   )

# The prep step
iris_prep <-  recipes::prep(iris_recipe)
```

Eigenvectors

Eigenvector coefficients are in step_pca, which is step 2. Below we reference step 2, the "res" results element, then the "rotation" sub-element.

```
# Compute iris eigenvectors
eigenvectors <- iris_prep$steps[[2]]$res$rotation

eigenvectors
[Output:]
                      PC1          PC2        PC3        PC4
   Sepal.Length  0.5210659 -0.37741762  0.7195664  0.2612863
   Sepal.Width  -0.2693474 -0.92329566 -0.2443818 -0.1235096
   Petal.Length  0.5804131 -0.02449161 -0.1421264 -0.8014492
   Petal.Width   0.5648565 -0.06694199 -0.6342727  0.5235971
```

An alternative method of getting the eigenvector table is to use the `tidy()` command. The output is in a long format but the coefficients are the same as above. For the `prcomp()` method in this alternative, the eigenvector label is "value", not "rotation".

```
irisEigenvectorsLong <- embed::tidy(iris_prep, 2, type = "coef")

irisEigenvectorsLong
[Output:]
   # A tibble: 16 x 4
      terms           value component id
      <chr>           <dbl> <chr>     <chr>
    1 Sepal.Length  0.521    PC1       pca_MVqBD
    2 Sepal.Width  -0.269    PC1       pca_MVqBD
    3 Petal.Length  0.580    PC1       pca_MVqBD
    4 Petal.Width   0.565    PC1       pca_MVqBD
    5 Sepal.Length -0.377    PC2       pca_MVqBD
    6 Sepal.Width  -0.923    PC2       pca_MVqBD
    7 Petal.Length -0.0245   PC2       pca_MVqBD
    8 Petal.Width  -0.0669   PC2       pca_MVqBD
    9 Sepal.Length  0.720    PC3       pca_MVqBD
   10 Sepal.Width  -0.244    PC3       pca_MVqBD
   11 Petal.Length -0.142    PC3       pca_MVqBD
   12 Petal.Width  -0.634    PC3       pca_MVqBD
   13 Sepal.Length  0.261    PC4       pca_MVqBD
   14 Sepal.Width  -0.124    PC4       pca_MVqBD
   15 Petal.Length -0.801    PC4       pca_MVqBD
   16 Petal.Width   0.524    PC4       pca_MVqBD
```

We now convert to wide format using `pivot_wider()` from the "tidyr" package. The result is an eigenvectors table with the same values as for the previous `prcomp()` example. The "[, 1:3]" part drops the fourth column, which was the id column in the wide format above.

```
irisEigenvectors <- irisEigenvectorsLong[,1:3] |>
    tidyr::pivot_wider(names_from = "component", values_from =
"value")

irisEigenvectors
[Output:]
  # A tibble: 4 x 5
    terms             PC1     PC2      PC3     PC4
    <chr>           <dbl>   <dbl>    <dbl>   <dbl>
  1 Sepal.Length    0.521  -0.377    0.720   0.261
  2 Sepal.Width    -0.269  -0.923   -0.244  -0.124
  3 Petal.Length    0.580 -0.0245   -0.142  -0.801
  4 Petal.Width     0.565 -0.0669   -0.634   0.524
```

Standard deviations and variances

The `step_pca()` step, which is step 2 in this example, generates variances, cumulative variances, percent variance (explained), and cumulative percent variance (explained).

```
irisVarianceinfo <- embed::tidy(iris_prep, 2, type = "variance")

irisVarianceinfo
[Output:]
  # A tibble: 16 x 4
     terms                         value component id
     <chr>                         <dbl>     <int> <chr>
   1 variance                     2.92           1 pca_dN8U9
   2 variance                     0.914          2 pca_dN8U9
   3 variance                     0.147          3 pca_dN8U9
   4 variance                     0.0207         4 pca_dN8U9
   5 cumulative variance          2.92           1 pca_dN8U9
   6 cumulative variance          3.83           2 pca_dN8U9
   7 cumulative variance          3.98           3 pca_dN8U9
   8 cumulative variance          4              4 pca_dN8U9
   9 percent variance            73.0            1 pca_dN8U9
  10 percent variance            22.9            2 pca_dN8U9
  11 percent variance             3.67           3 pca_dN8U9
  12 percent variance             0.518          4 pca_dN8U9
  13 cumulative percent variance 73.0            1 pca_dN8U9
  14 cumulative percent variance 95.8            2 pca_dN8U9
  15 cumulative percent variance 99.5            3 pca_dN8U9
  16 cumulative percent variance 100             4 pca_dN8U9
```

The irisVarianceinfo object is a data frame in which the variance information is in rows 1–4 (for these data, which have four components) and columns 1–3 (terms, value, component). Below, we put this into the object "irisVariances2" to avoid overwriting "irisVariances" from the `prcomp()` method above. The coefficients are the same.

```
irisVariances2 <- irisVarianceinfo[1:4, 1:3]

irisVariances2
[Output:]
   # A tibble: 4 x 3
     terms     value component
     <chr>     <dbl>     <int>
   1 variance 2.92          1
   2 variance 0.914         2
   3 variance 0.147         3
   4 variance 0.0207        4
```

One may also list out standard deviations, which are the square root of the variances

```
Iris_SD <- base::sqrt(irisVariances2$value)

Iris_SD
[Output:]
   [1] 1.7083611 0.9560494 0.3830886 0.1439265
```

Eigenvalues

The variances from irisVariances2 above are the eigenvalues. Therefore, if we want a separate object for it, this is simply a matter of copying irisVariances2. By the Kaiser rule, eigenvalues of 1.0 or greater are retained in the model, but there are other rules of thumb.

```
irisEigenvalues <- irisVariances2

irisEigenvalues
[Output not given as it is the same as for irisVariances2 above.]
```

An alternative method is to use the `tidy()` command as described below:

```
# Get iris_prep results for type=variance, step 2 (step_pca)
iris_prep_variance <- embed::tidy(iris_prep, 2, type = "variance")

# Strip out the rows for terms="variance". These are the
eigenvalues.
irisEigenvalues2 <- dplyr::filter(iris_prep_variance, terms ==
"variance")
# List only the first the columns, omitting the fourth id column.
irisEigenvalues2[,1:3]
# A tibble: 4 x 3
      terms     value component
      <chr>     <dbl>     <int>
```

```
1 variance 2.92         1
2 variance 0.914        2
3 variance 0.147        3
4 variance 0.0207       4
```

Component loadings

As explained above in the `prcomp()` section, component loadings are pseudo-factor loadings. They are computed based on the usual factor loadings metric which multiplies the eigenvectors by the square root of the eigenvalues.

```
# Create evectors2, leaving out column 1, which is the "terms"
(variable) label
evectors2 <- irisEigenvectors[,-1]
# Use only the "value" column of irisEigenvalues when creating
evalues2
evalues2 <- irisEigenvalues$value
# Calculate the loading in the next two lines
temp2 <- base::t(evectors2)*base::sqrt(evalues2)
loadings2 <- base::t(temp2)
# Add rownames back in
base::rownames(loadings2) <- base::c("Sepal.Length", "Sepal.Width",
"Petal.Length", "Petal.Width")
# Print the loadings table
loadings2
[Output:]
                    PC1         PC2         PC3         PC4
Sepal.Length  0.8901688 -0.36082989  0.27565767  0.03760602
Sepal.Width  -0.4601427 -0.88271627 -0.09361987 -0.01777631
Petal.Length  0.9915552 -0.02341519 -0.05444699 -0.11534978
Petal.Width   0.9649790 -0.06399985 -0.24298265  0.07535950
```

We see that Sepal.Width is heavily loaded on PC2 and all others are heavily loaded on PC1.

PCA with the recipes package

In this section, we return to analysis of the UN voting data. We assume the needed packages and data have been loaded, as per previous sections. Of the three data frames, we will use only "votes" and "issues", which should be showing in the RStudio "Environment" tab.

We start by creating a wide-format version of the votes data frame. The wide version has rows as countries and votes as columns. The wide version is given the title "votes_wide". To do this we use the `pivot_wider()` command from the "tidyr" package.

```
# Re-create the votes and votes_wide data frames.
base::library(unvotes)
votes <- base::data.frame(un_votes)
```

```
# The factor(), as.numeric(), and  paste() functions are all in R's
base package.
votes_wide <- votes |>
  dplyr::select(country, rcid, vote) |>
  dplyr::mutate(
     vote = base::factor(vote, levels = base::c("no", "abstain",
"yes")),
     vote = base::as.numeric(vote),
     rcid = base::paste0("rcid_", rcid)) |>
tidyr::pivot_wider(names_from = "rcid", values_from = "vote", val-
ues_fill = 2)
```

The resulting data frame has 200 country observations for 6,203 variables, which are the votes (except column 1 is "country"). Thus, in wide format, the columns are the votes (the first is rcid_3) and the rows are the 200 country names. There are 31 unique rcid's for the votes. Overall, unvotes_df has 200 observations for 6,203 variables.

```
base::names(votes_wide)
[Partial output:]
       [1] "country"    "rcid_3"     "rcid_4"     "rcid_5"
       [5] "rcid_6"     "rcid_7"     "rcid_8"     "rcid_9"
     . . .
     [993] "rcid_954"   "rcid_955"   "rcid_956"   "rcid_957"
     [997] "rcid_958"   "rcid_959"   "rcid_960"   "rcid_962"
     [ reached getOption("max.print") -- omitted 5203 entries ]
```

There are no missing values:

```
base::sum(base::is.na(votes_wide))
[Output:]
    [1] 0
```

The recipe step

The `recipe()` step defines the model, which here includes step_pca. All variables in votes_wide are used. The tilde (~) instructs the algorithm to predict using all variables. The "country" column is made the new id variable. That is, `update_role()` retains "country" as an id variable but does not include it in calculating principal components. Variables have a "predictor" role unless reassigned. Therefore, the `all_predictors()` specification refers to all variables except "country". It is possible to have more than one id variable. If there were a "region" variable, one could write: `update_role(country, region, new role = "id")`, for instance. All predictors are normalized to have a mean of 0 and a standard deviation of 1. The PCA model here calls for a 5-component solution.

```
base::library(recipes)

pca_recipe <- recipes::recipe(~., data = votes_wide) |>
```

```
  recipes::update_role(country, new_role = "id") |>
  recipes::step_normalize(all_predictors()) |>
  recipes::step_pca(all_predictors(), id ="pca", num_comp = 5)
```

Above there are two steps (step_normalize and step_pca). Each step can be given a step id. The step id for `step_pca()` above is "pca". Later, when we refer to id = pca it will be known this refers to `step_pca()`, without the need to refer to the step sequence number in the recipe (2, for `step_pca()` in this recipe).

The prep step

The `prep()` step for a recipe with at least one pre-processing operation, estimates the required parameters from a training set that can be later applied to other datasets. Here, however, all data are used. Output shows there is one id variable and 6,202 predictor variables (each of the 6,202 UN rcid_ votes is a predictor).

```
pca_prep <- recipes::prep(pca_recipe)

pca_prep
[Output:]
   Data Recipe
   Inputs:
         role #variables
           id          1
    predictor       6202

   Training data contained 200 data points and no missing data.
   Operations:
   Centering and scaling for rcid_3, rcid_4, rcid_5, ... [trained]
   PCA extraction with rcid_3, rcid_4, rcid_5, ... [trained]
```

Variable factor loadings and means

In this section, we explain how results are stored in and retrieved from the recipes prep step. Notice that the recipe above had two steps – normalize and pca. Various values from the recipe and prep steps are stored under the "steps" element of pca_prep, as described below. Alternatively, the `tidy()` command can retrieve variable means for the first step and variable factor loadings for the second step, as noted below.

Step 1: step_normalize

Step 2: step_pca

Means for 6,202 rcid's from step 1. Each rcid is a UN vote. The same information in different format is given by the command `tidy(pca_prep, 1)`

```
varMeans <- pca_prep$steps[[1]]$means
```

```
utils::head(varMeans,10)
[Output:]
   rcid_3   rcid_4   rcid_5   rcid_6   rcid_7
    2.055    1.875    2.030    1.925    2.035
    rcid_8   rcid_9  rcid_10  rcid_11  rcid_12
   2.185    2.225    2.220    2.205    1.995
```

Eigenvectors for variables. Starting with the first (rcid_3) from recipe step 2, eigenvalues are computed below. Rows are rcid_ codes (these are the variables), columns are PC1 through PC200. The number of observations (countries) is 200, so there cannot be more than 200 components even though there are 6,204 variables.

```
pcaVectors <- pca_prep$steps[[2]]$res$rotation
```

Below we show a partial listing of eigenvectors for the first rcid vote, which was rcid_3. This format groups loadings by variable (the rcid_ votes in this case).

```
utils::head(pcaVectors, 1)
[Output:]
               PC1           PC2          PC3        PC4        PC5
rcid_3 0.005855396 -0.0006301897 -0.001046442 0.01367928 0.01850898
              PC6         PC7         PC8         PC9         PC10
rcid_3 0.02998734 0.008124332 -0.01252836 -0.02470439 -0.007249137
              PC11          PC12        PC13        PC14        PC15
. . .
              PC196         PC197       PC198       PC199         PC200
rcid_3 0.0008610137 -0.001340806 0.01300814 -0.02980315 -0.005964077
```

In this two-step recipe, the standard deviations are in step 2, which was `step_pca()`. The same information in different format is given by the command below, which groups eigenvectors by component. Negative values are shown in red.

```
pcaVectors2 <- embed::tidy(pca_prep,2, type="coef")
```

```
pcaVectors2
[Partial output:]
      terms       value component id
      <chr>       <dbl>   <chr>   <chr>
    1 rcid_3    0.00586     PC1     pca
    2 rcid_4   -0.00262     PC1     pca
    . . .
    9 rcid_11   0.00218     PC1     pca
   10 rcid_12  -0.00183     PC1     pca
   # ... with 1,240,390 more rows
```

Next we display variances.

```
pcaVarianceinfo <- embed::tidy(pca_prep, id = "pca", type =
"variance")
```

```
pcaVarianceinfo
[Output:]
   # A tibble: 800 x 4
      terms     value component id
      <chr>     <dbl>     <int> <chr>
    1 variance 1690.          1 pca
    2 variance  706.          2 pca
    3 variance  495.          3 pca
    4 variance  387.          4 pca
    5 variance  207.          5 pca
    6 variance  151.          6 pca
    7 variance  126.          7 pca
    8 variance   90.6         8 pca
    9 variance   78.6         9 pca
   10 variance   74.2        10 pca
   # ... with 790 more rows
```

However, the "terms" column of pca_varianceinfo has four distinct types of rows: "variance", "cumulative variance", "percent variance", and "cumulative percent variance". There are 200 rows for each of the four types of coefficients, one per country in the UN example data.

```
pcaVarianceinfo |>
dplyr::count(terms)
[Output:]
   # A tibble: 4 x 2
     terms                           n
     <chr>                       <int>
   1 cumulative percent variance   200
   2 cumulative variance           200
   3 percent variance              200
   4 variance                      200
```

Below we create a data frame with just the "percent variance" rows. Percents are not decimals. For instance, ten percent is 10, not 0.10.

```
pcaPctvariance <- dplyr::filter(pcaVarianceinfo, terms == "percent
variance")
```

Similarly, we can create a data frame with just the "variance" rows, which are the eigenvalues.

```
pcaEigenvalues <- dplyr::filter(pcaVarianceinfo, terms ==
"variance")
```

Next, we calculate how many rows account for more than 5% of the variance in all the input variables. We use this to draw a vertical cutoff reference line at this number (four for these data). That is, there are four principal components accounting for 5% or more of the total variance.

```
cutoff <- 5
vlinePosition <- base::sum(pcaPctvariance$value > cutoff)

vlinePosition
[Output:]
   [1] 4

# Make sure the "ggplot2" graphics library is still loaded:
base::library(ggplot2)

# Create the plot for Figure 13.2.
pcaPctvariance |>
   ggplot2::ggplot(aes(component, value)) +
   ggplot2::geom_col(fill="dodgerblue") +
   ggplot2::geom_vline(xintercept= vlinePosition, color="red") +
   ggplot2::theme(axis.text.x=ggplot2::element_text
   (size=6,angle=90,hjust=1)) +
   ggplot2::theme(axis.title=ggplot2::element_text
   (size=16,face="bold")) +
   ggplot2::theme(plot.title = ggplot2::element_text
   (color = "darkblue", size = 18, face = "bold"), plot.subtitle =
   ggplot2::element_text(color = "darkblue", size=16) ) +
   ggplot2::labs(
      x = "Principal Components",
      y="Percent Variance Explained",
      title = "Variance explained contributions by principal
      component",
      subtitle = base::paste("Red vertical line at ", base::as.
      character(cutoff), "% explained"))
```

Now we plot again for cumulative variance explained. The vlinePosition object is reused. This plot is Figure 13.3.

```
pcaVarianceinfo|>
   dplyr::filter(terms == "cumulative percent variance") |>
   ggplot2::ggplot(aes(component, value)) +
   ggplot2::geom_col(fill="dodgerblue") +
   ggplot2::geom_vline(xintercept= vlinePosition, color="red") +
   ggplot2::theme(axis.text.x=ggplot2::element_text(size=6,an-
   gle=90,hjust=1)) +
   ggplot2::theme(axis.title=ggplot2::element_text(size=16,-
   face="bold")) +
   ggplot2::theme(plot.title = ggplot2::element_text(color
   = "darkblue", size = 18, face = "bold"), plot.subtitle =
```

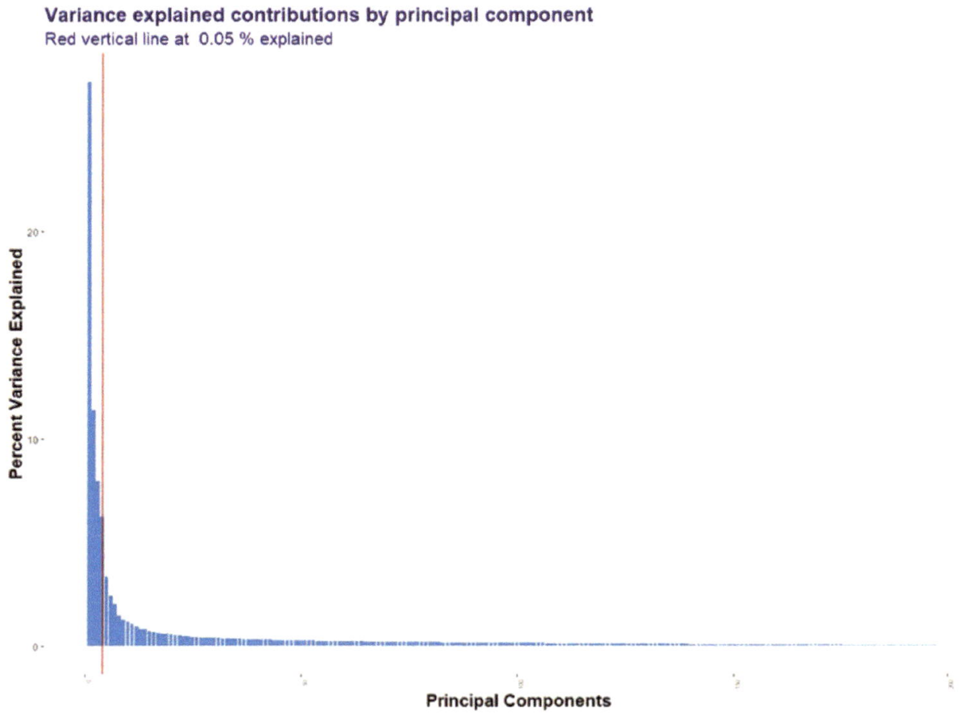

Figure 13.2 Percent variance explained, PCA components, UN data.

```
ggplot2::element_text(color = "darkblue", size=16) ) +
ggplot2::labs(
   x = "Principal Components",
   y = "Cumulative Percent Variance Explained",
   title = "Cumulative Variance explained by principal
   component",
   subtitle = base::paste("Red vertical line at ", base::as.
   character(cutoff), "% explained"))
```

Eigenvalues

Eigenvalues reflect how well a given factor (or component) correlates with the variables. The first factor is always the largest and explains the highest percentage of variance in the data. Remaining factors are successively small and explain successively less of the variance. In a model with as many factors as variables, the cumulative percent explained by the last factor is always 100%. However, it is common for the first 1–4 factors to explain the bulk of the variances. It is common to follow the Kaiser rule, which is to extract all factors with eigenvalues of 1.0 or greater.

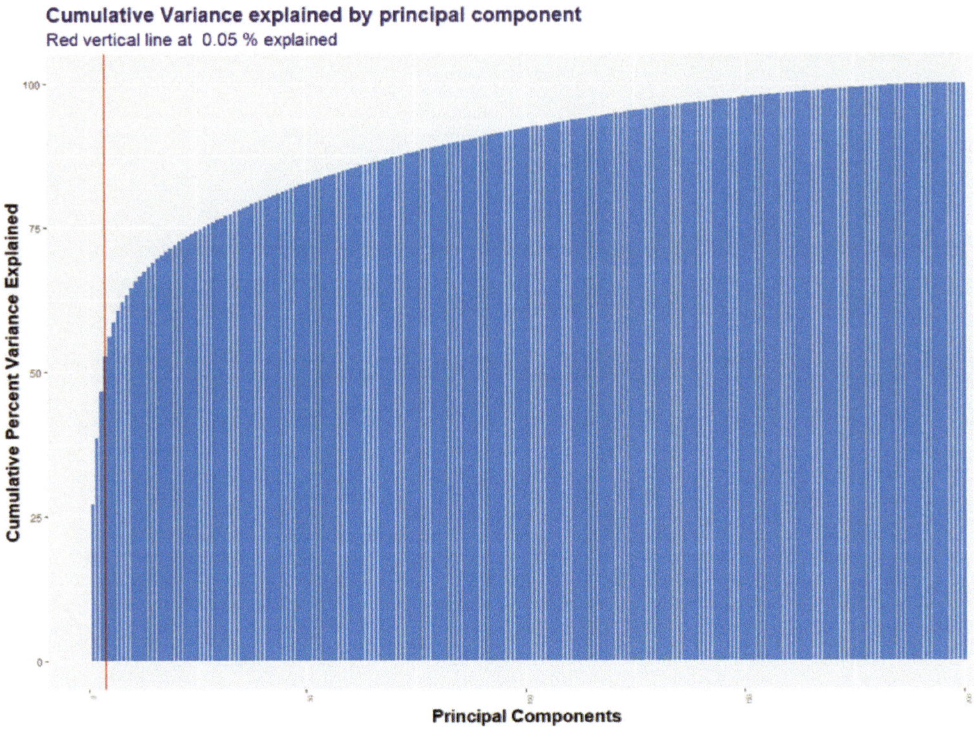

Figure 13.3 Cumulative variance explained, PCA components, UN data.

Earlier we created the tibble data frame object "pcaEigenvalues", which are the eigenvalues, which in turn are the variances (the squared standard deviations).

```
pcaEigenvalues <- dplyr::filter(pcaVarianceinfo, terms ==
"variance")

utils::head(pcaEigenvalues)
[Output:]
     terms    value component id
     <chr>    <dbl>     <int> <chr>
   1 variance 1690.         1 pca
   2 variance  706.         2 pca
   3 variance  495.         3 pca
   4 variance  387.         4 pca
   5 variance  207.         5 pca
   6 variance  151.         6 pca
```

Above we also created the tibble data frame object "pcaPctvariance". For convenience, this command is reproduced below. These are the percent of variance explained, which are the eigenvalues expressed as a percentage of the sum of eigenvalues for a given component. There is one eigenvalue and one percent of variance explained per component.

In the current example, there are 200 components. We may look at the first six percents of variance explained using the `head()` function:

```
pcaPctvariance <- dplyr::filter(pcaVarianceinfo, terms == "percent
variance")

utils::head(pcaPctvariance)
[Output:]
   # A tibble: 6 x 4
     terms            value component id
     <chr>            <dbl>     <int> <chr>
   1 percent variance 27.2          1 pca
   2 percent variance 11.4          2 pca
   3 percent variance  7.98         3 pca
   4 percent variance  6.24         4 pca
   5 percent variance  3.33         5 pca
   6 percent variance  2.44         6 pca
```

Above, note that the actual eigenvalues are in the "value" columns. If we sum the value for all 200 components, we get 100 percent. The percent of variance explained is the squared standard deviation (the variance) divided by the sum of all variances, for all variables in a column, for each column (component).

```
base::sum(pcaPctvariance$value)
[Output:]
   [1] 100
```

Component scores

The component scores of each country on each component are contained in the "template" list of the "pca_prep" object above. The pca_prep object is of data class "recipe". Below we convert this list to a tibble-type data frame (pca_scores) with 200 rows (one per country) for all columns, and list out its first six rows. A tibble is simply a more versatile version of a data frame. One could add variables to this data frame and then use these country-level component scores as variables in other forms of analysis, such as regression. Scores are also used to locate observations in factor space, as in Figure 13.4.

```
# Compute scores:
pca_scores <-base::as.data.frame(pca_prep$template[1:200,])
# Display first six rows:
utils::head(pca_scores)
[Output:]
        country        PC1       PC2        PC3       PC4        PC5
1 United States -131.73250 -23.08938 -88.846122 30.779376   4.431943
2 Canada         -48.19139 -67.75602 -26.296447  1.483281 -10.155615
3 Cuba            41.94006  19.78626  -1.246994 -6.355840  35.911770
```

```
4 Haiti            20.97030 -14.25018  -6.703383 18.978925  14.131150
5 Dominican
  Republic         20.43326 -24.68591  -7.637462 29.233792   8.062470
6 Mexico           42.95596 -23.21231  -4.588424  6.000366  20.353717
```

The bake step

The `bake()` step implements the model contained in the object pca_prep, which is an object of class "recipe". For this example, `ggplot()` is used to visualize the results for the PCA model for the first two components, shown in Figure 13.4. Note that estimates for components PC1 and PC2 were done in the previous `prep()` step and then piped to the `bake()` step.

Observe also that In `geom_text()`, overlap is checked and point labels are pulled from the edges toward the center. Also, the larger the size of the text, the more points will be unlabeled due to their overlapping. If `check_overlap=` is set to FALSE, then all labels will print but many will overlap.

```
# In the data, the United States is row 1 and Russia is row 32, out
of 200 countries.
# Create a color palette with the United States in blue, Russia in
red, all others in dark grey.
palette <- base::rep("grey20", 200)
palette[1] <- "blue"
palette[32] <- "red"

# Create a point size vector with the United States and Russia in
size 6, all others in size 4.
pointsize <- base::rep(4, 200)
pointsize[1] <- 6
pointsize[32] <- 6

# The bake step
# Here we color points by palette, but if there were another charac-
ter variable such as
# "continent", we could use:
# geom_point(aes(color = continent), alpha = 0.7, size = 2) +

recipes::bake(pca_prep, new_data = NULL) |>
    ggplot2::ggplot(aes(PC1, PC2, label = country)) +
    ggplot2::geom_point(color = palette, alpha = 0.6, size = poin-
tsize, pch=16) +
    ggplot2::geom_text(check_overlap = TRUE, hjust="inward",
size=pointsize, color=palette) +
    ggplot2::theme(axis.text=ggplot2::element_text
(size=16, hjust=1)) +
    ggplot2::theme(axis.title=ggplot2::element_text
```

```
(size=16, face="bold")) +
    ggplot2::theme(plot.title = ggplot2::element_text(color = "dark-
blue", size = 18, face = "bold"), plot.subtitle = ggplot2::element_
text(color = "darkblue", size=16) ) +
    ggplot2::labs(
        title = "PCA Model",
        subtitle = "UN roll call data, 1946 - 2019"
  )
```

In Figure 13.4, for instance, we can see that UN voting by the United States has been closer in factor space to Israel than to any other country. It is closer on PCA component PC2 than PC1. Russia is in the opposite quadrant on both dimensions compared to the United States. One can make out certain clusters, such as one for Scandinavia (Denmark is one of the unlabeled points due to overlap, but is close to Sweden). However, in issue terms, what do these dimensions mean? That is addressed in the next section.

Joining data from another data frame

Up to now, we have only used the votes_wide data frame in the three-step recipes process. Now we illustrate how to link to data from another data frame, namely the

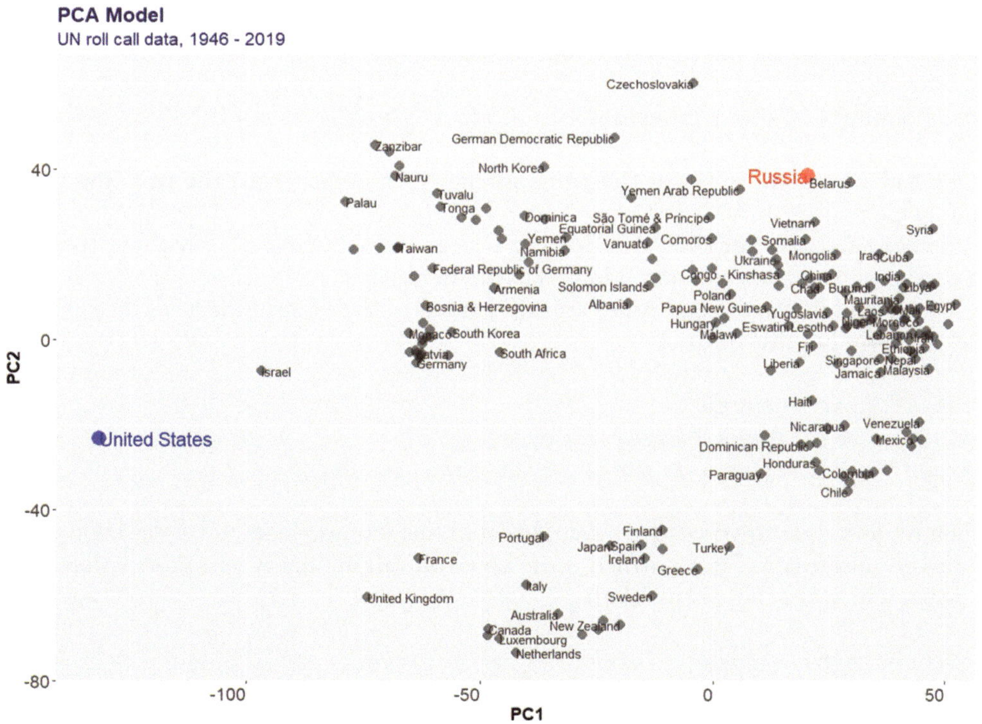

Figure 13.4 PCA output for UN voting data.

issues data frame. The link field is "rcid", which is a column in both data frames. The rcid number is the index number of a UN vote in the votes_wide data frame. However, multiple rcid's may be in a given issue, defined in the issues data frame, such as "Nuclear weapons and materials". The code below creates a bar chart revealing the most prominent issues in UN voting, grouped by principal component from the previous PCA analysis. This enables us to assess which issues contributed to each of the principal components. As such, it provides information enabling the researcher to impute an issue-related label to each component.

Step A: Use `tidy()` on the pca_prep object created earlier in order to join UN votes with issues. In Step A, again several operations are chained together by the "|>" piping operator, which forwards information from a previous step to the next one.

1. `tidy()`: Returns a data frame based on pca_prep, looking at the first two components. This tibble-type data frame as 4 columns (terms, value, component, and id) for 1,240.500 rows. The term variable has contents such as rcid_3, rcid_4, etc. The value variable has factor loadings for the component named in the "component" variable, which has the contents "PC1" or "PC2" for each row. The id variable is used internally by the algorithm
2. `filter()`: Selects the component variable if one of PC1, PC2, PC3, or PC4.
3. `left_join()`: Joins the variables in the issues data frame.
4. `mutate()`: Creates "rcid_" as a new variable.
5. `filter()`: Drops rows in issues with missing values.
6. `group_by()`: Groups rows by component.
7. `top_n()`: Selects the top 8 rows for each group based on value, which is the factor loading.
8. `ungroup()`: Cancels group_by.

If we had set `id = "pca"` as the name for step_pca earlier, then the first line below could be:

`components_by_issues <- tidy(pca_prep, id="pca") |>`, without the need to recall that step_pca was step 2 in this particular recipe.

We assume that votes and issues are still loaded in the environment, as they were at the beginning of this tutorial:

```
base::library(unvotes)
votes <- base::data.frame(un_votes)
issues <- base::data.frame(un_roll_call_issues)
```

Then we join issue to votes to get a single dataframe (components_by_issues) which has both vote information (rcid, which is the vote roll call id) and issues information (such as issue, which is a short issue label).

```
components_by_issues <- embed::tidy(pca_prep, 2) |>
  dplyr::filter(component %in% paste0("PC", 1:4)) |>
  dplyr::left_join(issues |>
  dplyr::mutate(terms=paste0("rcid_", rcid))) |>
  dplyr::filter(!base::is.na(issue)) |>
  dplyr::group_by(component) |>
```

```
  dplyr::top_n(8, base::abs(value)) |>
  dplyr::ungroup()
```

Though not required, examination of the components_by_issues object shows it is a tibble-type data frame with the seven variables (columns) listed below. The first three rows of data in components_by_issues are also listed below.

```
base::names(components_by_issues)
[Output:]
   [1] "terms"       "value"       "component"   "id"
   [5] "rcid"        "short_name"  "issue"

utils::head(components_by_issues,3)
[Output:]
     terms      value component id     rcid short_name issue
     <chr>      <dbl> <chr>     <chr> <int> <chr>      <fct>
   1 rcid_2447 0.0214 PC1       pca    2447 ec         Economic
   development
   2 rcid_2864 0.0214 PC1       pca    2864 hr         Human rights
   3 rcid_2887 0.0216 PC1       pca    2887 di         Arms control &
   disarmament
```

Step B creates a bar chart of the results of Step A. This is shown in Figure 13.5. Only two steps are connected by a components_by_issues to `mutate()`, and then `mutate()` to `ggplot()`. However, `ggplot()` has multiple steps connected by plus signs.

1 `mutate`: Create value as a new variable. It contains absolute values of the old value variable, which is overwritten.
2 `ggplot()`and `aes()`: Start specifying the esthetics (aes) of the figure. The value variable, which is the factor loading, is to be plotted. The terms variable is the rcid code. Votes are reordered by terms, then by value within terms. A unique fill color is to be associated with each issue.
3 `geom_col()`: This asks for column charts, which are bar charts, The position-"dodge" setting By default, multiple bars occupying the same x position will be stacked atop one another but here we ask that bars be "dodged", which means placed side-to-side, which is what position_dodge()calls for. In Figure 7.5 note that some rows contain more than one color/issue bar as a result.
4 `facet_wrap()` calls for a panel of bar lots, one per component. Scales are allowed to be free in the y axis direction. Freeing scales makes it easier to see patterns within each panel (here, each component), but by the same token, it may be more difficult to compare across panels.
5 `labs()`: This adds labels to the figure.

```
# Assure that the "forcats" library is still loaded, needed for the
fct_reorder() command.
base::library(forcats)

# Create Figure 13.5.
```

```
components_by_issues |>
  dplyr::mutate(value = base::abs(value)) |>
  ggplot2::ggplot(aes(value, forcats::fct_reorder(terms, value),fill
= issue)) +
  ggplot2::geom_col(position = "dodge") +
  ggplot2::facet_wrap(~component, scales = "free_y") +
  ggplot2::labs(
    x = "Factor loadings (contributions) of top UN roll call votes",
    y = "Roll Call ID", fill = NULL,
    title = "Top issues for each PCA component",
    subtitle = "UN roll call data, 1946 - 2019"
  )
```

Can we impute issue-based labels to the four principal components (PC1 through PC4) in the model? Based on the top and secondary issues associated with each component, one possible labeling is this:

- PC1: The human rights/economic development component
- PC2: The colonialism/human rights component
- PC3: The economic development/colonialism component
- PC4: The nuclear weapons /arms control component

However, it is not necessary to attach labels at all. We may simply say that when the variables (here, the votes) are reduced to four underlying factors, most of the variance and hence most of the meaning is accounted for by economic development, colonialism, and human rights, except for dimension 4, which has to do with arms control and nuclear weapons. The dimension accounting for most variance is PC1, which is mostly

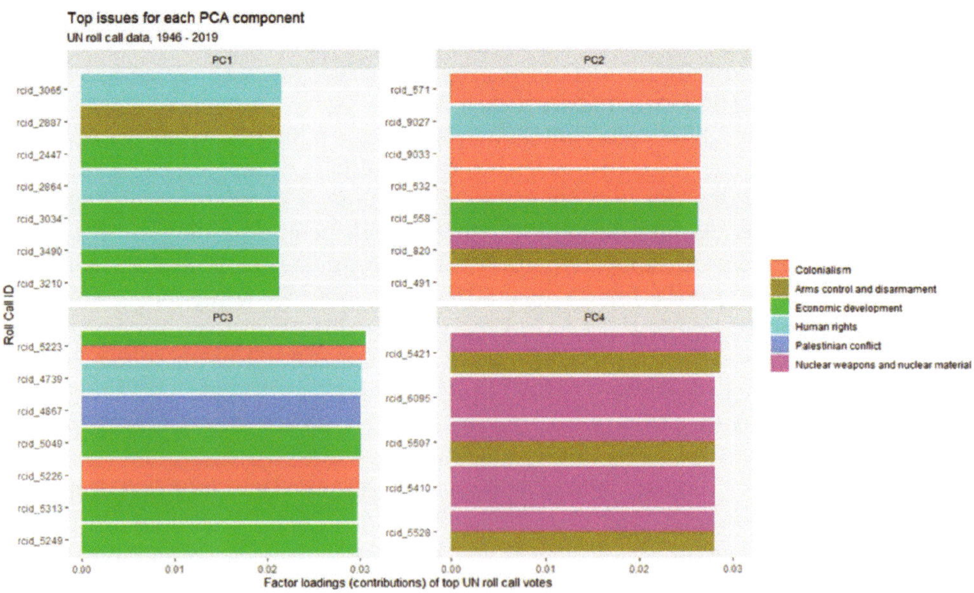

Figure 13.5 Top issues associated with each PCA component, UN voting data.

associated with human rights and economic development. The dimension accounting for the least variance is PC4.

ICA with the recipes package

ICA is an independent components analysis. Like PCA, it is a dimension reduction procedure which estimates components which are statistically independent. Note that in ICA, these are called "independent components", not "principal components". ICA is known for its use in signal processing, where it is employed to separate a multivariate signal into components which are additive. This requires that the components be statistically independent. An oft-cited teaching example is the "cocktail party problem", where the listener must separate signals to understand the speech of one person among many in a noisy room (Hyvärinen, 1983).

Hyvärinen (1983: 2) observed, "What distinguishes ICA from PCA and factor analysis is that it uses the non-Gaussian structure of the data, which is crucial for recovering the underlying components that created the data" and consequently "ICA is able to find the underlying components and sources mixed in the observed data in many cases where the classical methods fail." The term "non-Gaussian" means that ICA does not assume normally distributed continuous data as does PCA. That is, ICA can exploit data which do not have normal distributions. Hyvärinen (1983: 15), however, also noted that "It has been realized that non-Gaussianity is in fact quite widespread in any applications dealing with scientific measurement devices (as opposed to, for example, data in the social and human sciences)." As social scientists live in a largely Gaussian data environment, the advantages of ICA may be diminished. In fact, it may be reasoned that for data that are Gaussian, a procedure like PCA which takes advantage of the normal distribution assumption will have greater statistical power if its assumptions are met.

ICA is an unsupervised method, meaning it is not necessary to supply examples of "correct" outcomes or even to designate an outcome variable. Rather, it is a data-driven method suitable for the exploratory research purpose of uncovering the structure of the data without reference to causal hypotheses. For more information, see the tutorial by Aapo Hyvärinen and Erkki Oja, which may be found at http://cis.legacy.ics.tkk.fi/aapo/papers/IJCNN99_tutorialweb/.

Below, we run an ICA model for the same UN voting data as above for PCA. We ask for the same number of components also. For purposes of running the ICA model below, it is assumed that libraries listed in the earlier "Libraries and setup" section have been loaded. Examination of the R code for the ICA model reveals how the "recipes" method makes it easy to implement new dimension reduction models such as ICA by re-using code, in this case from the PCA model.

```
# Set the working directory and load needed packages.
base::setwd("C:/Data")
base::library(broom)               # Supports tidy() for data
management
base::library(forcats)             # Has fct_reorder() needed for
Figure 13.5
base::library(ggplot2)             # The leading R visualization
package
```

```
base::library(recipes)              # Supports step_ica()
base::library(tidyr)          # Supports pivot_wider() for wide data
format
base::library(unvotes)        # Has the UN voting data

# Re-create the votes data frame.
votes <- data.frame(un_votes)

# Re-convert to wide format, in which variables are columns and
votes are rows.
# The factor(), as.numeric(), and paste0() functions are from R's
"base" package.
votes_wide <- votes |>
  dplyr::select(country, rcid, vote) |>
  dplyr::mutate(
     vote = factor(vote, levels = base::c("no", "abstain", "yes")),
     vote = as.numeric(vote),
     rcid = paste0("rcid_", rcid)) |>
  tidyr::pivot_wider(names_from = "rcid", values_from =
  "vote", values_fill = 2)

# Copy votes_wide into the "data_ica" object.
data_ica <- votes_wide
```

The recipe step for ICA

Setting the random seed in step_ica is needed to assure replicable results.

```
ica_recipe <- recipes::recipe(~., data = data_ica) |>
  recipes::update_role(country, new_role = "id") |>
  recipes::step_normalize(all_predictors()) |>
  recipes::step_ica(all_predictors(), seed=123,num_comp = 5)
```

The prep step

For the votes_wide data, the "prep" step can take a while (about 10 minutes on the author's machine).

```
ica_prep <- recipes::prep(ica_recipe)
ica_prep
 [Output:]
   Data Recipe
   Inputs:
         role #variables
           id          1
    predictor       6202
```

```
Training data contained 200 data points and no missing data.
Operations:
Centering and scaling for rcid_3, rcid_4, rcid_5, ... [trained]
ICA extraction with rcid_3, rcid_4, rcid_5, ... [trained]
```

In recipes, for the tidy method, `step_ica` creates a tibble-type dataframe with the columns "terms" (the selected variables), "value" (the loadings), and "component" (the labels for the independent components. Though documentation calls the values "the loadings", these are not factor loadings in the PCA meaning. They are not based on the usual factor loadings metric which multiplies the eigenvectors by the square root of the eigenvalues. Indeed, ICA does not produce eigenvalues or eigenvectors in the PCA sense. Therefore, we use the label "icaVectors" rather than "loadings". The ICA vectors are in the data frame column icaVectors$value below. The value will vary somewhat depending on the random seed.

```
icaVectors <- broom::tidy(ica_prep, 2)
utils::head(icaVectors)
[Output:]
   # A tibble: 6 x 4
     terms    component      value id
     <chr>    <chr>          <dbl> <chr>
   1 rcid_10  IC1       -0.00110   ica_EoYnc
   2 rcid_10  IC2       -0.000303  ica_EoYnc
   3 rcid_10  IC3        0.000428  ica_EoYnc
   4 rcid_10  IC4        0.0000402 ica_EoYnc
   5 rcid_10  IC5       -0.00183   ica_EoYnc
   6 rcid_100 IC1        0.000737  ica_EoYnc
```

The icaVectors object may be converted to a wide-format matrix (icaVectormatrix below) with the independent components as columns and the variables (here the rcid roll call votes) as rows.

```
icaVectormatrix <- icaVectors |>
tidyr::pivot_wider(names_from = "component", values_from = "value")

utils::head(icaVectormatrix,3)
[Output:]
# A tibble: 3 x 7
  terms     id              IC1       IC2        IC3        IC4       IC5
  <chr>     <chr>         <dbl>     <dbl>      <dbl>      <dbl>     <dbl>
1 rcid_10   ica_EoYnc -0.00110  -0.000303  0.000428   0.0000402 -0.00183
2 rcid_100  ica_EoYnc  0.000737  0.000196  0.0000389 -0.00124    0.00112
3 rcid_1000 ica_EoYnc -0.00107  -0.000326 -0.000217  -0.000432  -0.000737
```

What is more directly useful, however, is that the "template" list of ica_prep takes the icaVectormatrix above and organizes it by observation (here, country), such that each country has a coordinate on each of the requested independent components (here, 5). Below we convert the template listing to a data frame with 200 rows (one per country) for all columns, and list out its first six rows. Though some documentation calls these "loadings", again they are not factor loadings in the PCA sense. Rather they are coordinates useful for mapping countries in ICA component space.

```
icaCoordinates <- ica_prep$template

utils::head(icaCoordinates)
[Output:]
   # A tibble: 6 x 6
     country                IC1     IC2     IC3     IC4     IC5
     <fct>                <dbl>   <dbl>   <dbl>   <dbl>   <dbl>
   1 United States       0.0576 -4.73    -1.93    1.47   -1.17
   2 Canada             -0.563  -0.804   -2.90    0.509 -0.446
   3 Cuba               -1.21    0.141    2.03   -0.352 -1.49
   4 Haiti              -0.929  -0.0339   0.304   1.09   -0.627
   5 Dominican Republic -0.930  -0.0137 -0.133   1.67   -0.329
   6 Mexico             -1.59    0.426    0.340   0.555 -0.937
```

The same icaCoordinates may be obtained by the command below. The coordinates could be considered country "scores" on the five ICA dimensions.

```
icaCoordinates2 <- recipes::juice(ica_prep)
[Output not shown:]
```

The bake step

We now create a color palette with the United States in blue, Russia in red, all others in dark grey. Note that in the data, the United States is row 1 and Russia is row 32, out of 200 countries. This is Figure 13.6.

```
# The rep() repeat command if from R's "base" base::library.
palette <- base::rep("grey20", 200)
palette[1] <- "blue"
palette[32] <- "red"

# Create a point size vector with the United States and Russia in
size 6, all others in size 2.6.
pointsize <- base::rep(2.6, 200)
pointsize[1] <- 6
pointsize[32] <- 6
```

The bake step below passes the country coordinates from the template discussed above to visualization commands using the "ggplot2" package. Replacing the first line with icaCoordinates |> will create the identical map as the present line below, recipes::bake(ica_prep, new_data = NULL) |>.

```
# Note the first two components are now labeled IC1 and IC2.
recipes::bake(ica_prep, new_data = NULL) |>
    ggplot2::ggplot(aes(IC1, IC2, label = country)) +
    ggplot2::geom_point(color = palette, alpha = 0.6, size = poin-
tsize, pch=16) +
    ggplot2::geom_text(check_overlap = TRUE, hjust="inward",
size=pointsize, color=palette) +
    ggplot2::theme(axis.text=ggplot2::element_text(size=16,
hjust=1)) +
    ggplot2::theme(axis.title=ggplot2::element_text(size=16,
face="bold")) +
    ggplot2::theme(plot.title = ggplot2::element_text(color = "dark-
blue", size = 18, face = "bold"), plot.subtitle = ggplot2::element_
text(color = "darkblue", size=16) ) +
    ggplot2::labs(
        title = "ICA Model",
        subtitle = "UN roll call data, 1946 - 2019. USA = Blue dot;
Russia = Red dot")
```

Compare Figure 13.6 to PCA output in Figure 13.4. Some aspects remain the same, though perhaps are easier to discern in the ICA map: the United States is far from Russia on IC2; the United States is close to Israel on both dimensions; Russia is close to Poland

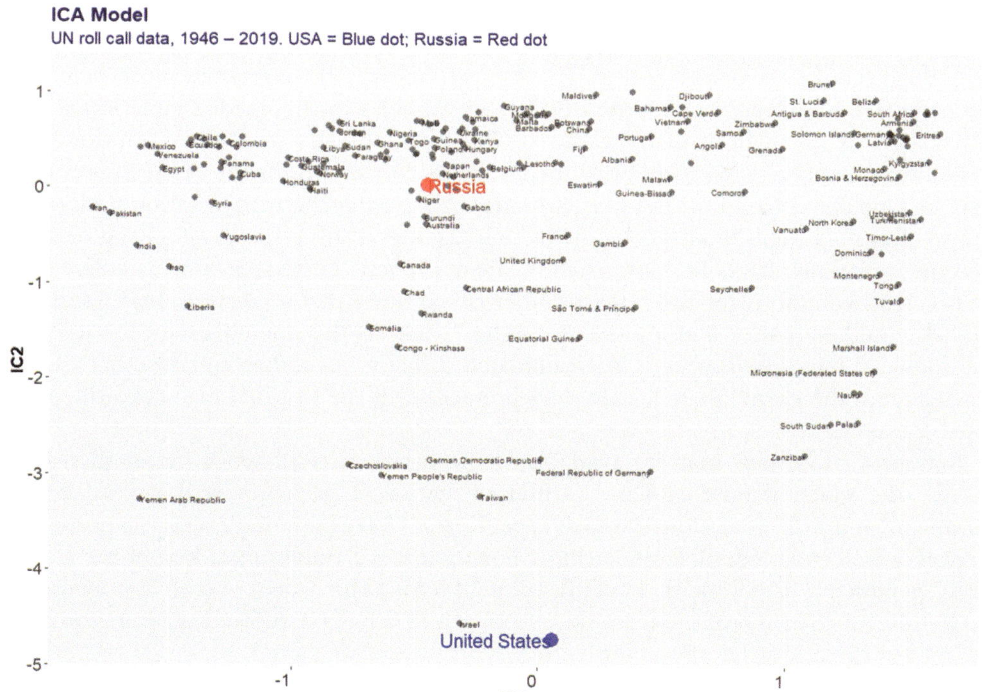

Figure 13.6 ICA output for UN voting data.

on both dimensions but is close to Cuba only on IC2. (Set check_overlap to FALSE to see all labels, but with many overwritten.)

It is also possible to create a plot similar to Figure 13.6, but just for selected countries, as per this endnote.[3] The range of the IC1 and IC2 axes, however, will be adjusted to the actual range for the countries which are selected

KPCA with the recipes package

The kPCA method is the kernel principal component analysis (kernel PCA) method associated with the work of Bernhard Schölkopf and others (Schölkopf, 1998; Schölkopf, Smola, & Mūller, 1998). PCA attempts to find a low-dimensional linear solution which accounts for a high percentage of the variance of input variables. However, PCA is a linear solution. KPCA attempts to find a low-dimensional linear or nonlinear solution. In this sense, KPCA is a generalization of PCA to handle nonlinear dimensionality reduction. While KPCA does allow the possibility of using a linear kernel, the main focus is on selection of one of various nonlinear kernels (e.g., quadratic) listed further below. This nonlinear mapping can potentially help find a better fit of the model to the original data.

In general, KPCA involves projecting the original data to a nonlinear feature space. An analogy from mapping is projecting Cartesian coordinates (linear) to polar coordinates (nonlinear). Then eigenvalues for principal components are extracted based on the nonlinear feature space (the higher dimension). The resulting solution is non-linear principal components analysis.

The analysis process using kernels involves projecting input data to a higher data plane, applying the "kernel trick", and arriving at a nonlinear PCA solution. If one wanted a conventional quadratic equation for two variables, y and z, one would need to expand them to five variables: y, z, y ⋆ z, y2, and z2. If there were more than two variables in a nonlinear model, the number of variables in the needed equation would expand greatly, perhaps unmanageably so. The "kernel trick" provides a solution. The kernel matrix serves a role similar to the covariance matrix in linear PCA. Eigenvalues may be calculated based on the kernel matrix, becoming the principal components in the smaller m-dimensional space onto which the kernel function projects the original n-dimensional raw data. This approach is called a "trick" because it eliminates the need to calculate equations for the large number of variables that a conventional nonlinear model would engender. Put another way, the "trick" is that the covariance matrix for the higher dimensional space is not calculated directly but rather eigenvectors are extracted from the kernel matrix, allowing projection of the original data onto the principal components.

A variety of kernels may be available, some of which will work better than others for the given dataset at hand. Different statistical packages will have different command-word abbreviations for each type. Kernels have parameters and the researcher may choose to override default settings. For instance, a polynomial kernel has a "degree" parameter, usually with a default value of 3 for a third-degree (cubic) polynomial pattern for an equation of the form f(x) = ax3 + bx2 + cx + d, plotted as a line with two bends, but the researcher could set degree to a value other than 3.

Implementation of KPCA under step_kpca with the "recipes" package relies on the "kernlab" package and its `kpca()` command. The `options=` specification in `step_kpca` sets the kernel and its parameters, following the kernlab framework. This is seen in the default options for `step_kpca` below. Notice that the `options=` specification provides both for naming the desired kernel and naming its parameters.

```
step_kpca(
  recipe,
  ...,
  role = "predictor",
  trained = FALSE,
  num_comp = 5,
  res = NULL,
  options = list(kernel = "rbfdot", kpar = list(sigma = 0.2)),
  prefix = "kPC",
  skip = FALSE,
  id = rand_id("kpca")
)
```

Choice of a kernel and its associated parameters is critical to a well-fitting KPCA model. Often exploration of different kernels and different parameters is needed to optimize a KPCA solution. Type `help("kpca"0)` to see kernlab's `kernel=` types listed below and also the `kpar=` options (not listed here).

vanilladot:
: Linear kernel function: Emulates linear PCA.

rbfdot:
: Radial Basis Function (RBF) kernel ("Gaussian"). This is a general purpose kernel recommended when there is no prior knowledge of the patterns in the data. Useful for spiral, crescent, or "half-moon" patterns.

anovadot
: ANOVA radial basis function kernel: An alternative RBF kernel. Useful for multi-dimensional regression problems.

Laplacedot:
: Laplacian kernel: An alternative RBF kernel. Also a general purpose kernel recommended when there is no prior knowledge of the patterns in the data. Largely equivalent to an exponential kernel.

polydot:
: Polynomial kernel: Useful when the data pattern is polynomial in shape. Often used in image recognition problems.

tanhdot:
: Hyperbolic tangent kernel: Useful as a proxy for neural network analysis.

besseldot
: Bessel kernel: Also a general purpose kernel recommended when there is no prior knowledge of the patterns in the data.

splinedot
  Spline kernel: A piece-wise cubic polynomial kernel.

Below, we run a KPCA model for the same UN voting data as for PCA and ICA and asking for the same number of components.

Note that KPCA requires previous installation of the "dimRed" and "kernlab" packages. The kernel function and other parameters are specified in the kernel and kpar slots of the options argument to step_kpca(). The code below is similar to that for the PCA and ICA models under "recipes", whose approach makes it easy to switch to the KPCA model.

```
# Setup
base::setwd("C:/Data")
base::library(broom)         # Supports tidy() for data management
base::library(dplyr)         # Has select(), mutate(), and other
                               functions.
base::library(ggplot2)       # The leading R visualization package
base::library(kernlab)       # Supports kernel PCA.
base::library(recipes)       # Supports step_kpca()
base::library(tidyr)         # Supports pivot_wider() wide data format
base::library(unvotes)       # Has the UN voting data

# Re-create the votes and votes_wide data frames.
votes <- base::data.frame(un_votes)

# Re-convert to wide format, in which variables are columns and
votes are rows.
# The factor(), as.numeric(), and paste0() functions are from R's
"base" package.
votes_wide <- votes |>
  dplyr::select(country, rcid, vote) |>
  dplyr::mutate(
     vote = base::factor(vote, levels = base::c("no", "abstain",
"yes")),
     vote = base::as.numeric(vote),
     rcid = base::paste0("rcid_", rcid)) |>
  tidyr::pivot_wider(names_from = "rcid", values_from =
"vote", values_fill = 2)

# Copy votes_wide into the "data_ica" object.
data_kpca <- votes_wide
```

The recipe step

To create the KPCA recipe, we use anovadot as the kernel, with two degrees. This is because the default rbfdot kernel concentrated all but two countries in the lower right of the figure.

```
kpca_recipe <- recipes::recipe(~., data = data_kpca) |>
    recipes::update_role(country, new_role = "id") |>
    recipes::step_normalize(all_predictors()) |>
    recipes::step_kpca(all_predictors(), num_comp = 5, options =
base::list(kernel = "anovadot", kpar = base::list(sigma = 0.2,
degree=2)))
```

The prep step

```
kpca_prep <- recipes::prep(kpca_recipe)

kpca_prep
 [Output:]
   Data Recipe
   Inputs:
         role #variables
           id          1
    predictor       6202
   Training data contained 200 data points and no missing data.
   Operations:
   Centering and scaling for rcid_3, rcid_4, rcid_5,... [trained]
   Kernel PCA (anovadot) extraction with rcid_3, rcid_4,
   ... [trained]
```

For PCA, the command `tidy(pca_prep, 2, type="variance")` produced a tibble-type data frame with columns labeled terms, value, component, and id. However, for KPCA the corresponding command `tidy(kpca_prep, 2, type="variance")` yields only columns labeled terms and id. The same is true for `type="coef"`. Moreover, the "terms" for PCA included variance, cumulative variance, percent variance, and cumulative percent variance. The "terms" for KPCA are only the variable names (the votes, such as rcid_3). The upshot of this is that we cannot compute eigenvectors and eigenvalues for `step_kpca()` as we did earlier above in `step_pca()`, nor is it intended that we should do so under KPCA.

Means for 6,202 rcid's from step 1 (step_normalize) are shown below. Each rcid is a UN vote. The same information in a different format is given by the command `broom::tidy(kpca_prep, 1)`.

```
kpca_means <- kpca_prep$steps[[1]]$means
utils::head(kpca_means,10)
[Output:]
   rcid_3  rcid_4  rcid_5  rcid_6  rcid_7
    2.055   1.875   2.030   1.925   2.035
    rcid_8  rcid_9 rcid_10 rcid_11 rcid_12
    2.185   2.225   2.220   2.205   1.995
```

The coordinates of each country on each component are contained in the "template" list of the "kpca_prep" object above. The kpca_prep object is of data class "recipe". Below we convert this list to a data frame with 200 rows (one per country) for all columns, and list out its first six rows. Though some documentation calls the resulting values "loadings", these are not factor loadings in the PCA meaning. They are not based on the usual factor loadings metric which multiplies the eigenvectors by the square root of the eigenvalues. However, they can be used as coordinates to map the observations (here, countries) in higher-dimensional KPCA space (here, specifically on dimensions kPC1 and kPC2).

```
# The kpcaCoordinates object is a traditional data.frame object.
# The as.data.frame() function is from R's "base" base::library.
kpcaCoordinates <- base::as.data.frame(kpca_prep$template[1:200,])

utils::head(kpcaCoordinates)
[Output:]
         country      kPC1       kPC2       kPC3       kPC4       kPC5
1 United States -29098.06 -33004.789 -24139.111 -20975.297   1742.086
2 Canada        -16785.15 -48832.709  -2390.528 -10925.806  10748.253
3 Cuba           25920.94   2879.849  -9596.456   2610.915 -28344.048
4 Haiti          15179.14  -8125.036   1314.271 -16173.692  -8704.308
5 Dominican
  Republic       16514.47 -12197.090   1687.432 -22010.564   -6105.919
6 Mexico         30436.82 -13878.422   3728.001  -9825.137  -14912.658
```

The same kpcaCoordinates may be obtained by the commands below. The coordinates could be considered country "scores" on the five kPCA dimensions.

```
kpcaCoordinates2 <- recipes::juice(kpca_prep)
[Output not shown: The object is a tibble-type data frame object.]

kpcaCoordinates3 <- base::as.data.frame(kpca_prep$template)
[Output not shown. The object is a traditional data.frame object]
```

The bake step

As a prelude to the bake step, we create a color palette with the United States in blue, Russia in red, all others in dark grey. Note that in the data, the United States is row 1 and Russia is row 32, out of 200 countries.

```
# Create a color palette with United States in blue, Russia in red,
all others dark grey .
palette <- base::rep("grey20", 200)
palette[1] <- "blue"
palette[32] <- "red"

# Create a point size vector with United States and Russia in size
6, all others size 2.6 .
pointsize <- base::rep(2.6, 200)
pointsize[1] <- 6
pointsize[32] <- 6
```

```
# The bake step. Note the first two components are now labeled kPC1
and kPC2.
recipes::bake(kpca_prep, new_data = NULL) |>
    ggplot2::ggplot(aes(kPC1, kPC2, label = country)) +
    ggplot2::geom_point(color = palette, alpha = 0.6, size = poin-
tsize, pch=16) +
    ggplot2::geom_text(check_overlap = TRUE, hjust="inward",
size=pointsize, color=palette) +
    ggplot2::theme(axis.text=ggplot2::element_text
(size=16, hjust=1)) +
    ggplot2::theme(axis.title=ggplot2::element_text
(size=16, face="bold")) +
    ggplot2::theme(plot.title = ggplot2::element_text(color = "dark-
blue", size = 18, face = "bold"), plot.subtitle = ggplot2::element_
text(color = "darkblue", size=16) ) +
    ggplot2::labs(
        title = "KPCA model, kernel = anovadot",
        subtitle = "UN roll call data, 1946 - 2019")
```

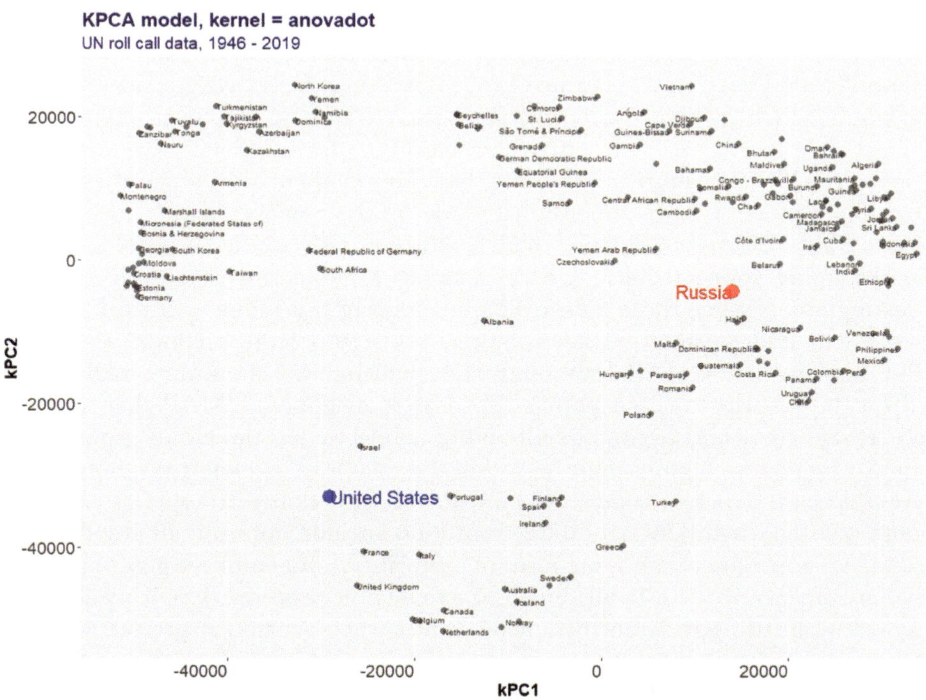

Figure 13.7 Countries by KPCA coordinates for UN voting data.

How does Figure 13.7 (KPCA) compare with Figure 13.4 (PCA) and Figure 13.6 (ICA)? Under all three approaches, the United States and Russia are distant. Also in all three approaches, a Scandinavian and an East European cluster roughly appear, though overlapping suppresses some country labels. All three could be construed to divide nations into three major clusters, with some outliers, but these clusters differ somewhat. The United States is closest to Israel in KPCA as it was in PCA and ICA. It may be noted that when run with default settings (kernel = rbfdot), the resulting figure (not shown) for KPCA is strikingly different, highlighting Cyprus and Israel as outliers with all other countries placed in a tight cluster in the lower right quadrant.

UMAP with the recipes package

A fourth dimensionality reduction procedure supported by the "recipes" package is UMAP. UMAP stands for "uniform manifold approximation and projection" and is associated with the work of Leland McInnes et al. (2018). Unlike PCA, ICA, and KPCA, UMAP is not supplied with the "recipes" package but rather with the "embed" package. UMAP derives from the study of topology, where the challenge is to reduce complex topological data by mapping observed data to a simpler, higher-level but lower-dimensionality model. However, UMAP is applicable outside topology and below we illustrate its use as a step under the "recipes" package for the same UN voting data as for PCA, ICA, and KPCA above (Voeten, 2013).

UMAP structures data in terms of simplices. A 0-simplex is a dot. A 1-simplex is a line connecting two points. A 2-simplex is a triangle connecting three points. A 3-simplex is a pyramid connecting four points. Mathematically, higher-level simplices are possible. Simplices may be "glued" together along their faces to form "simplicial complexes", which in turn can be further abstracted into "simplicial sets" or "fuzzy simplicial complexes". Each successive abstraction is a form of data reduction. In general, what the UMAP algorithm is doing, based on mathematical graph theory, is first attempting to create a high-dimensional model of the observed data, and secondly to optimize a low-dimensional graph which is as structurally similar as possible.

To determine connectedness, UMAP extends a radius outward from each point, connecting points when those radii overlap. Choosing this radius is critical. Too small a choice will lead to small, isolated clusters while too large a choice will connect everything together. UMAP overcomes this challenge by choosing a radius locally, based on the distance to each point's nth nearest neighbor. UMAP then makes the graph "fuzzy" by decreasing the likelihood of connection as the radius grows. Finally, by stipulating that each point must be connected to at least its closest neighbor, UMAP ensures that local structure is preserved in balance with global structure.

Some ways in which UMAP differs from PCA include not normalizing data, using nearest neighbors rather than some form of correlation, and employing symmetrization of higher dimensions. While all dimension reduction procedures will yield different mappings when using different parameter values, the variation is greater for UMAP than, say, PCA. That is, there is more incentive to use UMAP in an exploratory manner, exploring the effect of varying parameter settings. For more information on UMAP, go to https://umap-learn.readthedocs.io/en/latest/how_umap_works.html.

```
# UMAP setup
base::setwd("C:/Data")      # Declare the working directory
```

```
base::library(embed)         # Needed by recipes
base::library(ggplot2)       # Leading R visualization package
base::library(recipes)       # Supports UMAP within recipes framework
base::library(unvotes)       # Has the UN voting data

# Read in the UN voting data.
votes <- base::data.frame(un_votes)

# Create a wide format version of the UN voting data.
# The factor(), as.numeric(), and  paste() functions are all in R's
base package.
votes_wide <- votes |>
  dplyr::select(country, rcid, vote) |>
  dplyr::mutate(
     vote = factor(vote, levels = base::c("no", "abstain", "yes")),
     vote = as.numeric(vote),
     rcid = paste0("rcid_", rcid)) |>
  tidyr::pivot_wider(names_from = "rcid", values_from =
"vote", values_fill = 2)

# Copy the wide version into the tibble-type data frame object
"umap_data".
umap_data <- votes_wide
```

In the recipe step which follows, step_umap() calls for the research to input three critical parameters. Widely differing results may flow from different settings.

The num_comp= parameter is the number of UMAP components to be calculated. As the map in Figure 13.8 displays only the first two components, this is all we ask for. Usually, the first two components explain the bulk of the variance, though this varies.

The neighbors= parameter is the number of nearest neighbors used to construct the target simplicial set. This is the most important parameter because it controls the balance between local versus global structure. The lower the number of neighbors, the more the focus on local structure, but at the expense of seeing the "big picture" of the global topological structure. Conversely, the larger the number of neighbors which UMAP connects when constructing the graphic representation of high-dimensional data, the better the representation of global structure, but at the expense of the fine detail which a focus on local structure provides. This parameter varies from 1 to n, where n is sample size.

The min_dist= parameter is the effective minimum distance between embedded points in low-dimensional space. The min_dist parameter controls how tightly points are clustered together. Smaller min_dist focuses on preservation of the global topological structure while large min_dist focuses on local structure at the expense of the "big picture" of global structure. This parameter varies from approaching 0 to approaching 1.0.

To focus on global structure, we set the number of neighbors high (100) and the minimum distance low (0.05).

The recipe step

The recipe step uses the earlier-created "votes" data to implement step_umap. In turn, it is composed of two steps: (1) step_normalize and (2) step_umap. Note that step_umap requires setting two random seeds in order to obtain replicable results.

```
umap_recipe <- recipes::recipe(~., data = umap_data) |>
  recipes::update_role(country, new_role = "id") |>
  recipes::step_normalize(all_predictors()) |>
embed::step_umap(all_predictors(),neighbors=100,
num_comp=2, min_dist=0.05, seed = base::c(123,465) )

# Default settings for step_umap are these: step_umap(all_
predictors(),outcome = NULL, neighbors=15, num_comp=2, min_
dist=0.01, learn_rate=1,  seed = sample(10^5, 2) )
```

The prep step

The prep step executes the recipe but does not, of itself, yet display results.

```
umap_prep <- recipes::prep(umap_recipe)
umap_prep
[Output:]
    Data Recipe
    Inputs:
          role #variables
            id          1
     predictor       6202
    Training data contained 200 data points and no missing data.
    Operations:
    Centering and scaling for rcid_3, rcid_4, rcid_5, rcid_6,
    rcid_7, rcid_8, ... [trained]
    UMAP embedding for rcid_3, rcid_4, rcid_5, rcid_6, rcid_7,
    rcid_8, ... [trained]
```

For PCA, the command `broom::tidy(pca_prep, 2, type="variance")` produced a tibble-type data frame with the columns terms, value, component, and id. However, for UMAP the corresponding command `broom::tidy(umap_prep, 2, type="variance")` yields only the columns labeled terms and id. The same is true for `type="coef"`. Moreover, the "terms" for PCA included variance, cumulative variance, percent variance, and cumulative percent variance. The "terms" for UMAP are only the variable names (the votes, such as rcid_3). The result is that we cannot compute eigenvectors and eigenvalues for `step_upmap()` as we did earlier above in `step_pca()`, nor is it intended that we should do so under UMAP.

Means for 6,202 rcid's from step 1 (step_normalize). Each rcid is a UN vote. The same information in different format is given by the command `tidy(umap_prep, 1)`.

```
umap_means <- umap_prep$steps[[1]]$means
utils::head(umap_means,10)
```

```
[Output:]
    rcid_3   rcid_4   rcid_5   rcid_6   rcid_7
     2.055    1.875    2.030    1.925    2.035
    rcid_8   rcid_9  rcid_10  rcid_11  rcid_12
     2.185    2.225    2.220    2.205    1.995
```

The coordinates of each country on each umap dimension are contained in the "template" list of the "umap" object above. The umap_prep object is of data class "recipe". Below we convert this list to a data frame with 200 rows (one per country) for all columns, and list out its first six rows. Though some documentation calls the resulting values "loadings", these are not factor loadings in the PCA meaning. They do not add to 1.0 and are not based on the usual factor loadings metric which multiplies the eigenvectors by the square root of the eigenvalues. However, they can be used to map the observations (here, countries) in higher-dimensional UMAP space (here, specifically on dimensions umap_1 and umap_2).

```
base::set.seed(123)
umapCoordinates <- base::as.data.frame(umap_prep$template[1:200,])

utils::head(umapCoordinates)
[Output:]
             country     UMAP1      UMAP2
  1    United States -2.230635 -0.6010791
  2           Canada -2.695099 -2.5119908
  3             Cuba  1.976070 -0.3187036
  4            Haiti  2.209742 -1.1871446
  5 Dominican Republic  1.781401 -1.3210992
  6           Mexico  2.352238 -1.4956225
```

We put the two sets of coordinates into the objects umap_1 and umap_2;

```
umap_1 <- umapCoordinates$UMAP1
umap_2 <- umapCoordinates$UMAP2
```

The same umapCoordinates may be obtained by the commands below. The coordinates could be considered country "scores" on the five umap dimensions.

```
umapCoordinates2 <- recipes::juice(umap_prep)
[Output not shown: The object is a tibble-type data frame object.]

umapCoordinates3 <- base::as.data.frame(umap_prep$template)
[Output not shown. The object is a traditional data.frame object]

# In the data, the United States is row 1, Russia row 32, Israel row
59,  our of 200 countries.
# Create a color palette with United States blue, Russia red, Israel
in green, others dark grey .
```

```
palette <- base::rep("grey40", 200)
palette[1] <- "blue"
palette[32] <- "red"
palette[59] <- "orange"


# Create a point size vector with United States and Russia in size
6, all others size 3 .
pointsize <- base::rep(3, 200)
pointsize[1] <- 6
pointsize[32] <- 6
pointsize[59] <- 6


# The bake step, which produces Figure 13.8.
# Note that the components are now labeled umap_1 and umap_2.
recipes::bake(umap_prep, new_data = NULL) |>
  ggplot2::ggplot(aes(umap_1, umap_2, label = country)) +
    ggplot2::geom_point(color = palette, alpha = 0.6, size = poin-
tsize, pch=16) +
    ggplot2::geom_text(check_overlap = TRUE, hjust="inward",
size=pointsize, color=palette) +
    ggplot2::theme(axis.text=ggplot2::element_text(size=16,
hjust=1)) +
    ggplot2::theme(axis.title=ggplot2::element_text(size=16,
face="bold")) +
    ggplot2::theme(plot.title = ggplot2::element_text(color = "dark-
blue", size = 18, face = "bold"), plot.subtitle = ggplot2::element_
text(color = "darkblue", size=16) ) +
    ggplot2::labs(
        title = "UMAP Model",
        subtitle = "UN roll call data, 1946 - 2019")
```

While UMAP does display a noticeably different pattern of countries in its higher-dimension space, it is similar to PCA, ICA, and KPCA solutions in locating the United States far from Russia. The Scandinavian cluster is still observable near the bottom of the graph, but the East European cluster is no longer discernible The United States is also very close to Israel, but due to label overlap only the orange dot for Israel appears, not the country label. Keep in mind that cluster sizes and distances between clusters are highly sensitive to parameter settings in UMAP and therefore are difficult or impossible to interpret. This is due to using local distances when constructing a graph such as in Figure 13.8.

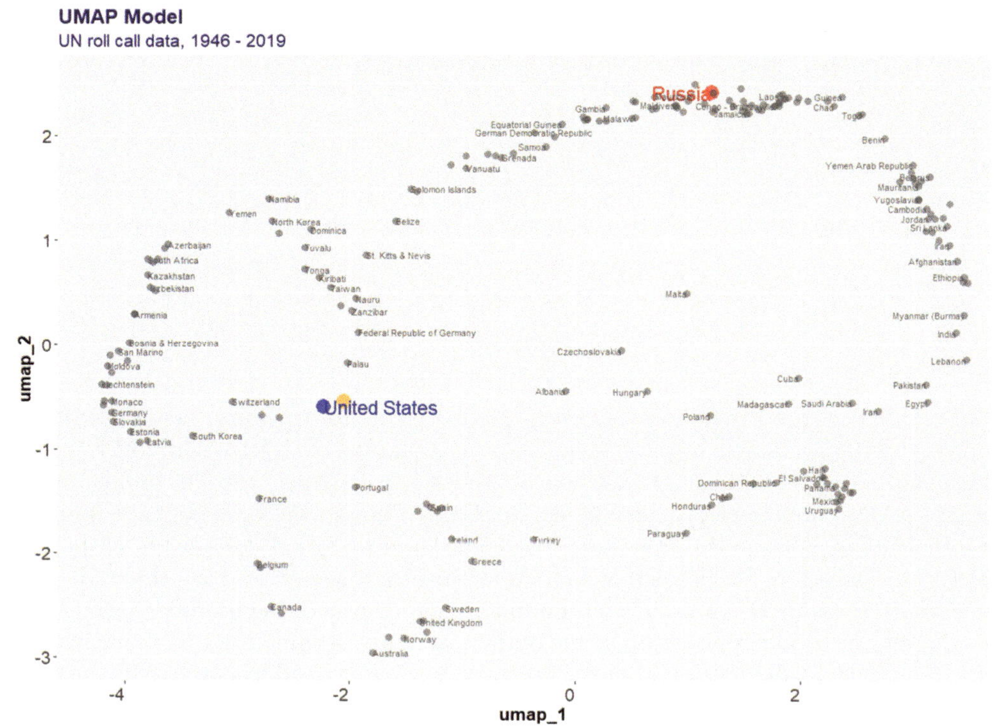

Figure 13.8 UMAP output for UN voting data.

Notes

1 https://cran.rstudio.com/web/packages/unvotes/vignettes/unvotes.html.
2 See https://juliasilge.com/blog/un-voting/. R code in this section was adapted from this article.
3 Note that the axes will shift based on the particular countries selected.

```
selectedCountries <- data_ica[c(1,32,35,36,52,59,72),]
bake(ica_prep, new_data = selectedCountries) |>
 ggplot2::ggplot(aes(IC1, IC2, label = country)) +
 ggplot2::geom_point(color = “black”, fill=”red”, alpha = 0.6,
size = 2, pch=23) +
 ggplot2::geom_text(check_overlap = FALSE, hjust=”inward”,
size=4) +
 labs(
 title = “ICA Model, Selected Countries”,
 subtitle = “UN roll call data, 1946 - 2019”
 )
```

14 Factor analysis for neural models

Introduction

R's "caret" package provides a modeling environment for training, cross-validating, and comparing predictive models, including neural network models.[1] Moreover, it provides automatic 10-fold (or greater) cross-validation of models and common metrics for comparing model performance across methods. Though the caret package supports hundreds of models,[2] its authors have not considered principal component or principal factor analysis to be predictive models within its scope. However, there are two exceptions. First, there is a "pca" method in its data pre-processing module. Second, the "pcaNNet" method of prediction is one of the caret-supported models. This means that a PCA model may be developed in the pre-processing stage and its components may be used for prediction in the neural network analysis stage. See Kuhn (2008).

Example data and setup

As sample data, we use the classic "iris" dataset. This dataset is located in R's built-in "datasets" package and does not need to be loaded. Type `help(iris)` for more information, such as this note:

> This famous (Fisher's or Anderson's) iris data set gives the measurements in centimeters of the variables sepal length and width and petal length and width, respectively, for 50 flowers from each of 3 species of iris. The species are iris setosa, versicolor, and virginica.

```
utils::data(iris)    # Invokes the classic "iris" data frame.
View(iris)           # This activates the data frame and puts it in
the RStudio viewer.
```

As a setup for neural network analysis based on PCA factor analysis, we declare the working directory and load the needed R packages. When caret loads, it automatically loads additional needed packages including "lattice" and "ggplot2". This assumes the three packages below have been installed on the user's machine with the `install.packages()` command.

Prior to trying this section "hands-on", the following libraries should be loaded. Some may well require the reader to install them first with the `install.packages()` command.

DOI: 10.4324/9781003279693-16

```
base::library(caret)        # Supports preProcess() function.
# Set the working directory, using your own folder address.
base::setwd("C:/Data")
```

PCA in caret pre-processing

PCA-based neural network analysis in caret has two steps: pre-processing and then neural network analysis itself. In the `preProcess()` step of caret we name the dataset, specify the method as "pca", and ask to extract however many components are needed to explain 95% of the variance. If just the PCA solution is desired, this step may be all that the researcher needs. Had we wished to specify a specific number of components to extract, say 3, we would replace `"thresh = 0.95"` with `"pcaComp = 3"` in the R code below.

The `method=pca` option in the `preProcess()` procedure has caveats of which the researcher should be aware:

- The pca method in `preProcess()` automatically centers and scales data. If scaling is not wanted, similar results may be obtained from `prcomp()`, which centers but does not scale. Scaling, however, is typically recommended for PCA.
- The pca method in `preProcess()` does not support factor variables as PCA inputs. While some documentation suggests converting categorical variables to dummy variables and then proceeding, this method is inferior to polychoric or heterogeneous PCA. The `preProcess()` method is best used when all input variables are numeric, as they are in the iris example. This method can still use `predict()` to predict a categorical outcome variable, as done for "Species" in the iris example.

The actual `preProcess()` command takes the form below. Results are placed in the object we label "preProc".

```
preProc <- caret::preProcess(iris, method="pca", thresh = 0.95)

preProc
[Output:]
   Created from 150 samples and 5 variables

   Pre-processing:
     - centered (4)
     - ignored (1)
     - principal component signal extraction (4)
     - scaled (4)

PCA needed 2 components to capture 95 percent of the variance
```

The preProc object contains many information elements, listed below.

```
base::attributes(preProc)
[Output:]
   $names
    [1] "dim"                  "bc"                   "yj"
```

```
 [4] "et"                  "invHyperbolicSine" "mean"
 [7] "std"                 "ranges"            "rotation"
[10] "method"              "thresh"            "pcaComp"
[13] "numComp"             "ica"               "wildcards"
[16] "k"                   "knnSummary"        "bagImp"
[19] "median"              "data"              "rangeBounds"

$class
[1] "preProcess"
```

To see how the variables have been processed, display the "method" element. The output below shows that the Species variable was ignored (because it is a factor, which is non-numeric). All numeric variables were centered, scaled, and were inputs to PCA. If only certain variables were meant to have been processed, this would have been specified in the `preProcess()` command (e.g., `preProc2 <- caret::preProcess(iris[, c(1:2,5)], method="pca", thresh = 0.95)` would perform PCA using only the first, second, and fifth columns in iris).

```
preProc$method
[Output:]
$pca
[1] "Sepal.Length" "Sepal.Width"  "Petal.Length" "Petal.Width"

$ignore
[1] "Species"

$center
[1] "Sepal.Length" "Sepal.Width"  "Petal.Length" "Petal.Width"

$scale
[1] "Sepal.Length" "Sepal.Width"  "Petal.Length" "Petal.Width"
```

Computing variable importance on the components

The squared eigenvectors for all variables always sum to 1.0 in PCA. Therefore, the squared eigenvalue for a given component is its percent contribution to the meaning of the component, often called its "contribution" or "importance". Some documentation confusingly calls the squared eigenvalues "loadings", which are also used to assess variable importance to components and which give the same order of variable importance.

However, the common social science definition for a PCA "loading" is this formula: loadings = eigenvectors * sqrt(eigenvalues). However, `method=pca` in `preProcess()` in caret does not directly return eigenvalues, so loadings in this meaning cannot be computed from `preProcess()` output. A workaround is discussed further below.

Eigenvectors are contained in the "rotation" element of the preprocessing result (preProc). These are sometimes labeled "standard deviations". Eigenvectors are also "coordinates" used to map variables in component space.

```
preProc$rotation
[Output:]
             PC1         PC2
Sepal.Length  0.5210659 -0.37741762
Sepal.Width  -0.2693474 -0.92329566
Petal.Length  0.5804131 -0.02449161
Petal.Width   0.5648565 -0.06694199
```

Squared eigenvectors are the "contribution" of each variable to each component. These constitute percentages of variances explained by the column component. They are percentages since the sum of squared eigenvectors for a component is always 1.0.

```
preProc$rotation^2
[Output:]
                    PC1          PC2
Sepal.Length 0.27150969 0.1424440565
Sepal.Width  0.07254804 0.8524748749
Petal.Length 0.33687936 0.0005998389
Petal.Width  0.31906291 0.0044812296
```

For coding convenience, we put the squared eigenvectors in data frame "x". We then set the order of eigenvectors to be decreasing. For readability, we assign rownames (variable names) to the x$PC1 and x$PC2 vectors.

```
x <- base::as.data.frame(preProc$rotation^2)
PC1order <- base::order(x$PC1,decreasing=TRUE)
PC2order <- base::order(x$PC2,decreasing=TRUE)

base::names(PC1order) <- base::rownames(x[PC1order,])
names(PC2order) <- base::rownames(x[PC2order,])
```

List variables by order of importance by contributions on PC1.

```
PC1order
[Output:]
Petal.Length  Petal.Width Sepal.Length  Sepal.Width
           3            4            1            2

# List squared eigenvectors by order of importance by contributions
PC1
x[PC1order,]
[Output:]
                    PC1          PC2
Petal.Length 0.33687936 0.0005998389
Petal.Width  0.31906291 0.0044812296
Sepal.Length 0.27150969 0.1424440565
Sepal.Width  0.07254804 0.8524748749
```

List variables by order of importance of contributions on PC2.

```
PC2order
[Output:]
   Sepal.Width Sepal.Length  Petal.Width Petal.Length
             2            1            4            3

# List squared eigenvectors (contributions) by order of importance
on PC2
x[PC2order,]
[Output:]
                     PC1          PC2
   Sepal.Width  0.07254804 0.8524748749
   Sepal.Length 0.27150969 0.1424440565
   Petal.Width  0.31906291 0.0044812296
   Petal.Length 0.33687936 0.0005998389
```

Component loadings

Contributions are not component loadings in the usual social science definition: loadings = eigenvectors * sqrt(eigenvalues). Though both contributions and loadings give the same order of variable importance, social scientists customarily use loadings, not contributions, to assess whether a variable is "heavily" or "weakly" loaded on a component. This assessment is important when the researcher attempts to use loadings to impute a meaningful label to a component.

Unfortunately, the caret `preProcess()` function returns eigenvectors and not eigenvalues. Without having eigenvalues, component loadings cannot be calculated directly within `preProcess()`. However, it is easy to obtain component loadings using the `eigen()` command from the "base" package in R run on the iris correlation matrix. The `eigen()` command will also return eigenvectors, which are identical to those returned by `preProcess()` in its rotation element.

```
# Calculate component loadings using the eigen() command from R's
"base" package.
iris_cor_matrix <- stats::cor(iris[1:4])
x <- base::eigen(iris_cor_matrix)
vctrs <- x$vectors
vals <- x$values
load <- vctrs*base::sqrt(vals)
base::rownames(load) <- base::colnames(iris[1:4])

# Print component loadings
load
                    [,1]         [,2]        [,3]       [,4]
   Sepal.Length  0.89016876 -0.644765592  1.22927920  0.4463713
   Sepal.Width  -0.25750946 -0.882716269 -0.23364106 -0.1180813
   Petal.Length  0.22234964 -0.009382456 -0.05444699 -0.3070261
   Petal.Width   0.08129782 -0.009634726 -0.09128865  0.0753595
```

Plotting variables and observations in component space

The rotation element also may be used to plot variables in component space. A bare-bones plot is obtained with the commands below.

```
# coordinates <- preProc$rotation
# plot(preProc$rotation, pch=16, cex=2, col="red")
```

However, a better plot can be had with a little more coding effort, shown in Figure 14.1.

```
coordinates <- preProc$rotation
base::plot(coordinates,
   pch=16,
   col="red",
   cex.axis=1.4,
   cex.lab = 1.2,
   cex.main = 1.5,
   xlab = "PC1", ylab = "PC2",
   # ylim and xlim set manually to best display data
   ylim = base::c(-1.0, 0.1),
   xlim = base::c(-.4, 0.6),
   cex=1.8,
   main="PCA with method=pca in caret's preProcess() ")
# Draw reference lines at 0
graphics::abline(h=0, v=0, col="blue")
# Add variable labels
x <- coordinates[,1]
y <- coordinates[,2]
graphics::text(x,y,labels=base::rownames(coordinates), cex=1.4,
pos=2, offset=0.5)
```

We now turn to plotting observations in component space. First, we use `predict()` to obtain the observation scores. The `predict()` command is from R's automatically-loaded "stats" package, which in turn automatically uses caret's predict.preProcess method.

```
irisScores <- stats::predict(preProc,iris[,-5])
```

It is helpful to know the names and order of levels of Species as the DV. There are three levels, in the order below.

```
base::unique(iris$Species)
[Output:]
   [1] setosa     versicolor virginica
   Levels: setosa versicolor virginica
```

Finally, we plot observations in component space. This is Figure 14.2.

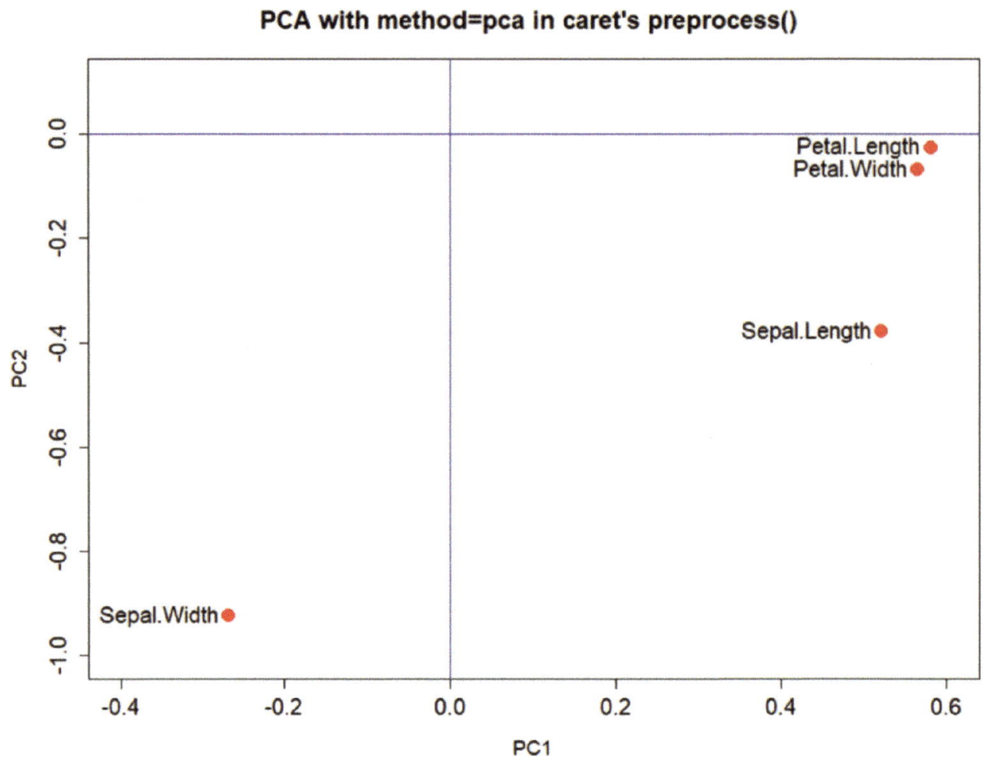

Figure 14.1 Variable plot: PCA with `method=pca`, iris data.

```
base::plot(irisScores,
   col=iris$Species,
   pch=16,
   main="Iris Observations in Component Space")

graphics::legend("topright",legend=base::levels(iris$Species),col=1:
base::length(base::unique(iris$Species)), pch=16)
```

Use of PCA in the pcaNNET modeling method

While `method=pca` in the `preProcess()` command in caret does implement a standard PCA solution, as illustrated above, it largely duplicates what other PCA programs can do. What is unique in using PCA under caret is the integrated second `train()` step in which principal components from pre-processing are passed to the "pcaNNet" method. The `pcaNNET()` function first runs a PCA solution and then the resulting principal components are used in a neural network model. See Ripley (1996).

While there is a stand-alone `nnet()` function (in the "nnet" package), this approach is deprecated in favor of the `train()` function in the caret package using `method = "nnet"` and `preProc = "pca"`.

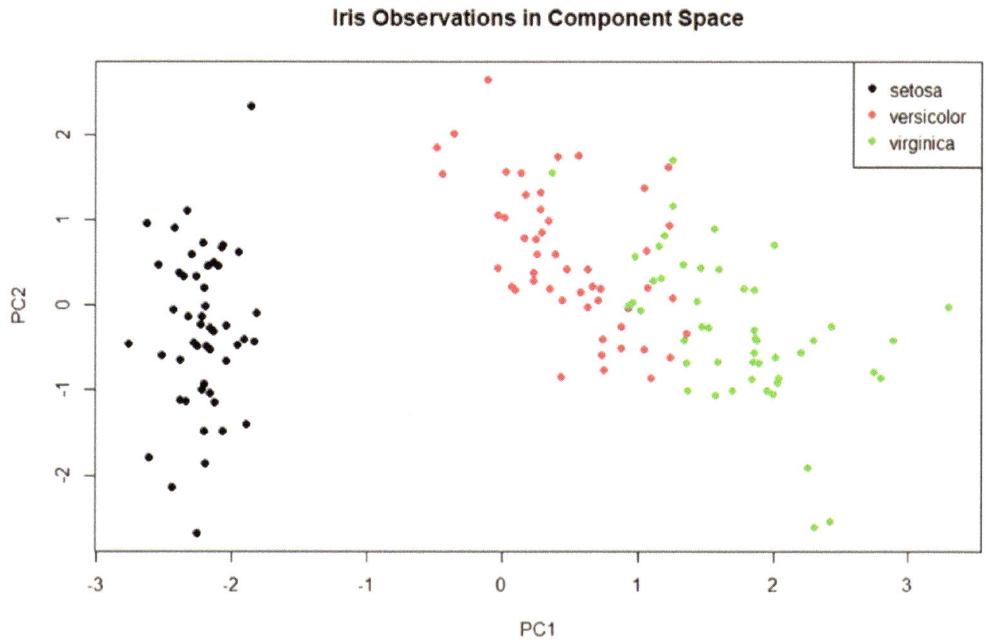

Figure 14.2 Plot of observations for PCA with method=pca, iris data.

```
# Setup
base::setwd("C:/Data")
base::library(caret)          # Environment for training and testing
models
base::library(nnet)           # The package used by caret for neural
network modeling
```

One of the great advantages of modeling with the caret package is that cross-validation of the model is automated. We do this by creating the "ctrl" control object. Defining "ctrl" may be thought of as the first step in the two-step caret model training process. The second step, below, is the `train()` command itself, which uses the ctrl object.

In creating the "ctrl" object below, we set the threshold for the number of components at 95% explained. The `method=` parameter is set to repeated cross-validation. Enter `method="none"` if no cross-validation is wanted. The `number=` parameter sets the number of iterations in a fold of cross-validations Here we set it at 20 but 10 is the default. The `repeats=` parameter sets the number of times that folds are repeated. Here we set repetitions at two, but one is the default. The `p= parameter` gives the proportion of observations to be used for the training sample in the cross-validation process, here 70%. The default summary function is used.

```
ctrl <- caret::trainControl(
  preProcOptions = base::list(thresh = 0.95),
  method = "repeatedcv",
  number = 20,
```

```
  repeats = 2,
  p = .70,
  summaryFunction = defaultSummary
)

# Alternatively, we could specify a two-component model: :
# preProcOptions = list(pcaComp = 2)
```

Next is the caret "train" step for method = "pcaNNet", preprocess = "pca". Since cross-validation involves random sampling, we set the random seed first # so that we may later obtain replicable results.

```
base::set.seed(123)

irisModel <- train(Species ~ .,
    data=iris,
    method = "pcaNNet",
    preprocess = "pca",
    trControl = ctrl)
```

After irisModel converges, it is an object of data class "train" and has many output elements as listed below.

```
base::attributes(irisModel)
[Output:]
     [1] "method"       "modelInfo"    "modelType"    "results"
     [5] "pred"         "bestTune"     "call"         "dots"
     [9] "metric"       "control"      "finalModel"   "preProcess"
    [13] "trainingData" "resample"     "resampledCM"  "perfNames"
    [17] "maximize"     "yLimits"      "times"        "levels"
    [21] "terms"        "coefnames"    "xlevels"
```

Using these output elements, we may show various aspects of the results.

```
# Show whether the model is a classification or a regression one.
irisModel$modelType
[Output:]
    [1] "Classification"

# Show the DV and other variables in the model.
irisModel$terms
[Partial output shows the model, with Species as DV and others as
IVs:]
Species ~ Sepal.Length + Sepal.Width + Petal.Length + Petal.Width
. . .

# For the categorical DV (Species), show its levels, which are nomi-
nal, not ordinal.
irisModel$levels
```

```
[Output:]
    [1] "setosa"     "versicolor" "virginica"
   attr(,"ordered")
   [1] FALSE

# Show just the predictor variables.
irisModel$coefnames
   [Output:]
 [1] "Sepal.Length" "Sepal.Width" "Petal.Length" "Petal.Width"
```

Summary of the neural model

A basic summary of the final model is found in the "$finalModel" element of the results object.

```
irisModel$finalModel
[Output:]
   Neural Network Model with PCA Pre-Processing
   Created from 150 samples and 4 variables
   PCA needed 3 components to capture 99 percent of the variance
   a 3-3-3 network with 24 weights
   options were - decay=0.1
```

A 3-3-3 neural network has three input nodes, three hidden (middle) nodes, and three output nodes.

Tuning parameters for the neural model

The algorithm searched for the best tuning parameters. This lists them. The "6" is just the row number in results (shown below) for the best tuning combination. Best size is 3 (three hidden nodes). Best decay is 0.1. Also called regularization, decay is used for updating neural weights between iterations.

```
irisModel$bestTune
[Output:]
   6    3   0.1
```

Alternative values for the tuning parameters may be plotted as shown in Figure 14.3.

```
base::plot(irisModel,
   lwd=2,
   pch=16,
   cex=1.4,
   main="Best Tuning Parameter Search",
   sub="For data=iris, method = pcaNNet, preprocess = pca" )
```

Figure 14.3 shows graphically the results of the search for best tuning parameters. The best parameters were number of hidden units (size) = 3 and weight decay = 0.1. More details may be output with the commands below.

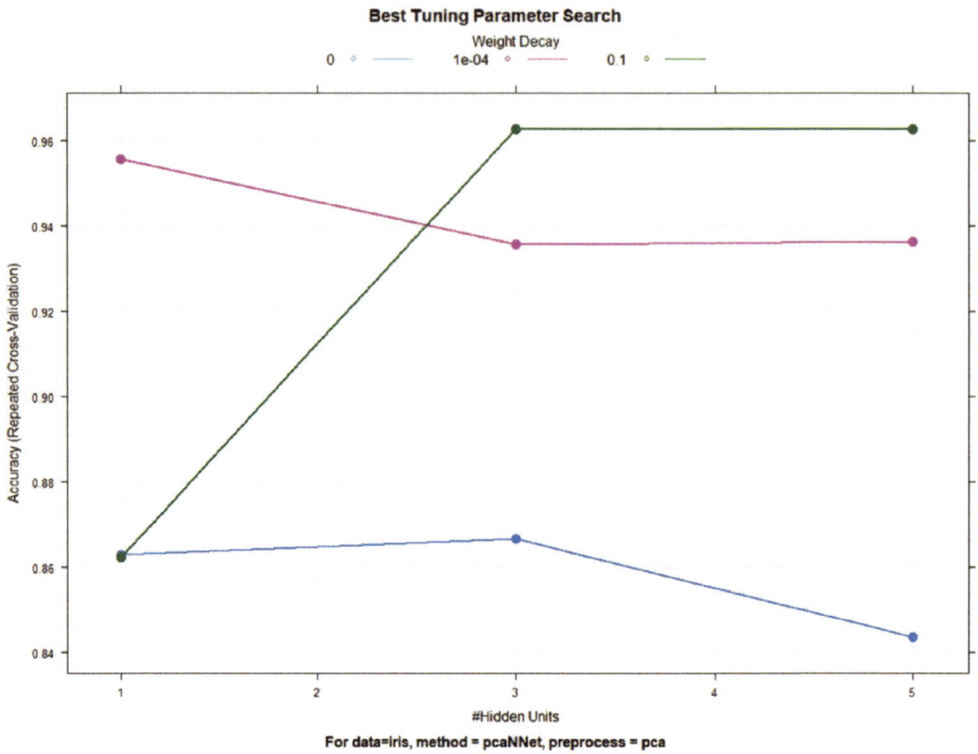

Figure 14.3 irisModel accuracy by number of hidden nodes for three decay settings.

```
# List the primary metric for assessing model performance.
irisModel$metric
[Output:]
 [1] "Accuracy"
```

Above, "accuracy" is used as the metric to assess neural model performance. For a classification model, accuracy is the percent of observations classified correctly,

```
# List all model performance measures.
irisModel$perfNames
[Output - Kappa is a second model performance measure in output.]
   [1] "Accuracy" "Kappa"

# List model performance measures and their standard deviations for
each model tried.
irisModel$results
[Output:]
     size decay  Accuracy     Kappa AccuracySD   KappaSD
   1    1 0e+00 0.8627976 0.7984498 0.16494593 0.2392743
   2    1 1e-04 0.9556548 0.9319521 0.07270119 0.1130928
```

```
3     1 1e-01 0.8623512 0.7926370 0.12094180 0.1804137
4     3 0e+00 0.8665675 0.8073719 0.21915710 0.3111927
5     3 1e-04 0.9357143 0.9033890 0.10619065 0.1584126
6     3 1e-01 0.9627976 0.9437845 0.06156084 0.0929555
7     5 0e+00 0.8434524 0.7665769 0.21818125 0.3227846
8     5 1e-04 0.9363095 0.9025112 0.08565788 0.1318752
9     5 1e-01 0.9627976 0.9437845 0.06156084 0.0929555
```

Above, each row is a model with different size and decay parameters. Convergence on the final solution occurred in iteration 9 (the bottom row), when size had been set to 5 and decay to 0.1. However, iteration 6 achieved identical accuracy with a more parsimonious model (three hidden nodes rather than five) and so was selected. Also displayed are the two performance measures and their standard deviations.

Predictions and accuracy for the neural model

As a baseline, we can list out the training data. For this example, this is the entire iris data frame.

```
irisModel$trainingData
[Output not shown]
```

Below, we use `predict()` to get the actual predicted classifications.

```
# Get neural network predictions of Species (the DV).
# There are 150 predictions, one for each observation
irisNNetPreds <- stats::predict(irisModel)
irisNNetPreds
[Partial Output:]
   [1] setosa     setosa     setosa     setosa
. . .
[148] virginica  virginica  virginica
```

For model performance, we print out the confusion matrix, with a variety of measures of model strength. The confusion matrix compares observed with predicted values of the dependent variable, which is Species. A key metric is accuracy, which is 97.33%. Other coefficients are explained further, for instance, in Garson (2022: Chapter 4).

```
caret::confusionMatrix(iris$Species, irisNNetPreds)
[Output:]
Confusion Matrix and Statistics

            Reference
Prediction   setosa versicolor virginica
  setosa         50          0         0
  versicolor      0         47         3
  virginica       0          1        49
```

```
Overall Statistics

               Accuracy : 0.9733
                 95% CI : (0.9331, 0.9927)
    No Information Rate : 0.3467
    P-Value [Acc > NIR] : < 2.2e-16

                  Kappa : 0.96

 Mcnemar's Test P-Value : NA

Statistics by Class:

                     Class: setosa Class: versicolor Class: virginica
Sensitivity                 1.0000            0.9792           0.9423
Specificity                 1.0000            0.9706           0.9898
Pos Pred Value              1.0000            0.9400           0.9800
Neg Pred Value              1.0000            0.9900           0.9700
Prevalence                  0.3333            0.3200           0.3467
Detection Rate              0.3333            0.3133           0.3267
Detection Prevalence        0.3333            0.3333           0.3333
Balanced Accuracy           1.0000            0.9749           0.9661
```

Summary: Accuracy and other model performance measures are shown in the output of caret's `confusionMatrix()` command. On a cross-validated basis, predicting Species of iris from PCA components, the neural model has a 97.33% accuracy, misclassifying only four observations.

Autoencoder with dimRed

We note that autoencoders are another type of neural network analysis approach. The "dimRed" package, discussed in Chapter 11, gave a simple example of using that package for dimension reduction purposes.

Notes

1 To get a list of all possible caret methods, type: `names(caret::getModelInfo())`.
2 To get a list of all possible caret methods, type: `names(caret::getModelInfo())`.

15 Factor analysis for time series data

Introduction

Functional PCA (FPCA) aids in understanding a time series, such as wages over time. Analysis involves only two variables: the variable of interest (e.g., wages) and the time variable (years of workforce experience in the example in this section). A second variable of interest may be analyzed separately in the same manner. Alternatively, the variable of interest may not be a simple time series such as wages but rather may be the time series of a construct as predicted by a set of covariates (e.g., customer satisfaction). The example, using the wages dataset, provides unemployment rate as a second possible time series of interest but we leave unemployment to the reader to experiment with. In this section, we focus on wages as the time series of interest. An online vignette (tutorial) on `FPCA()` is found at https://cran.r-project.org/web/packages/fdapace/vignettes/fdapace-Vig.html.

With standard PCA one finds the principal components (vectors) which maximally explain variance in the data, with the first component explaining the most, the second component explaining the most remaining variance, and so on. In PCA, the meaning of the components is inferred from the component loadings of measured variables (usually multiple variables). With FPCA we use functions (which define curves) rather than linear vectors. Each component reflects a different function. Basically, the time series for one given variable is decomposed into multiple functions (curves) which explain variation in a single variable over time. The set of functional components are the "modes of variation". For a single time series such as wages as a variable of interest, there will be multiple modes of variation. The first FPCA component is the mode which explains the most variation, with later components explaining successively less variation.

Longitudinal PCA may be implemented as functional principal component analysis using the `FPCA()` function of the "fdapace" package (Dai et al., 2021), maintained by Alvaro Gajardo Cataldo. Also called "principal component analysis through conditional expectation" (PACE), this is a nonparametric procedure. PACE is a flexible alternative to random effects modeling of longitudinal data. FPCA scores may be used for outlier detection, as discussed in the "Outlier detection" section further below.

To be clear, FPCA is a tool to analyze a single time series such as wages. Unlike the usual forms of PCA, FPCA does not create components based on multiple predictor variables. Rather, the components in FPCA reflect explanatory dimensions for the single longitudinal variable of interest (wages) based on patterns across the time variable. FPCA component scores may be used as variables in regression or other subsequent analyses. When component scores are used this way in PCA, each component represents the

DOI: 10.4324/9781003279693-17

cluster of variables which most heavily load on it. When used in FPCA, each component represents a mode (function) explaining how a single time series variable varies over time.

A seminal article on functional PCA is Yao et al. (2005). For recent applications of the method, see Han et al. (2018) and Johnson et al. (2021). For a summary of the mathematics involved, see https://en.wikipedia.org/wiki/Functional_principal_component_analysis. Code in this section was adapted from the Rickert (2021), which uses the same dataset ("wages", which is contained in the "brolgar" package).

While the `FPCA()` function was designed for sparse data, it may also be used on dense data. Longitudinal data may be considered sparse if there are 20 or fewer time periods per observation (subject), and may be considered dense if there are 21 or more time periods. This, of course, is an arbitrary cutoff. The `dataType=` option in the `FPCA()` function may be set to "Sparse" or "Dense". Data declared "Sparse" are subject to smoothing, regardless of the actual number of time periods. For this reason, the researcher may prefer to declare dense data as sparse if smoothing is desired. For sparse designs, observation trajectories over time are fitted, with confidence bands, even for observations with very few longitudinal observations.

The `FPCA()` function supports a large number of options and companion functions. For instance, time profiles (curves) of observations may be visualized using the `CreatePathPlot()` function of the "fdapace" package. Warning: Paths created with this command may model the path of a given subject for time periods for which there are no actual observations. For instance, a subject may have observations only for the first quarter of the time series but a path will be extrapolated for that subject for the entire range of the time variable.

Setup

The following commands and packages are needed for various parts of this tutorial on longitudinal functional PCA. The reader will need to run `install.packages()` first for any package not installed locally.

```
base::setwd("C:/Data")      # Set the working directory (yours may
differ)

base::library(brolgar)      # Graphically browse longitudinal data;
                            # Also has "wages" dataset
                            # (type help(wages) )
base::library(dplyr)        # Utility with pull() and Data mgt.
functions.
base::library(fdapace)      # Contains the FPCA() function
base::library(gghighlight) # Adds highlights to selected data.
base::library(ggplot2)      # Leading visualization package for R
base::library(ks)           # Needed for outlier detection using
kernedensity
base::library(Rcpp)         # Needed by FPCA()

utils::data(wages)          # Invoke the example data, from
"brolgarpackage.
```

Example data

The "wages" dataset is supplied with the "brolgar" package.[1] It consists of wages data from the National Longitudinal Survey of Youth (NLSY). The sample covered male high-school dropouts, ages 14–17 inclusive. The data source is Singer & Willett (2003). The "wages" dataset has the following variables (columns):

[1] "id": A subject ID variable for 888 subjects. IDs are not sequential and range from 31 to 12543. Each subject has a varying number of rows based on the number of measurement times, averaging about seven rows.
[2] "ln_wages": The natural log of hourly wages, adjusted for inflation, to 1,990 dollars.
[3] "xp": Experience, measured as years in the workforce. This is the "time" index variable, with t0 for each subject starting on their first day at work. The number of time points and values of time points for each subject can differ.
[4] "ged": Coded 1 if the subject had a GED certificate, otherwise coded 0.
[5] "xp_since_ged": For GED holders, this is change in experience since getting a GED. Those without a GED are coded 0.
[6] "black": Coded 1 if the subject is black, otherwise coded 0.
[7] "hispanic": Coded 1 if the subject is hispanic, otherwise coded 0.
[8] "high_grade": Highest grade completed (an integer value).
[9] "unemploy_rate": The unemployment rate in the local geographic region at each measurement time.

The format of the "wages" object is a tsibble, which is an R format for a time series tibble, where a tibble is like a data frame but with more functionality. Each subject has a unique id and has multiple data rows for that id, one for each time of measurement. Different subjects may have different numbers of data rows since some are measured more often than others. There is no "year" variable but the experience variable ("xp") serves as the time variable, as explained in the subsection which follows.

```
base::library(brolgar)
utils::data(wages)
base::class(wages)
[Output:]
       [1] "tbl_ts" "tbl_df" "tbl" "data.frame"
```

Visualizing longitudinal data with the "ggplot2" package

In this brief section we explore the example data by means of a plot seeking to establish whether wages tend to increase over time for the high school dropouts in the study. Time is measured by the "xp" variable. This variable reflects experience, as measured by years in the workforce. The xp measure is a decimal value which increases with each successive measurement time. For instance, the first subject is id = 31, whose xp value is 0.015 at time 1 and is 6.984 at time 8.

```
# Setup: Making sure the "ggplot2" package is still active in the R
environment.
base::library(ggplot2)
```

```
# See if ln_wages go up with experience (xp)
ggplot2::ggplot(
   data = wages,
   aes(x = xp, y = ln_wages, color = id)) +
   geom_line(aes(group = id)) +
   geom_point() +
   geom_smooth(se=FALSE, method= "loess", col="red") +
   ggtitle ("Log wages by experience, all subjects, with loess
smoothing line") +
   labs (y= "Natural log of wages", x = "Years in the workforce")
```

The upward slant of the loess smoothing line in Figure 15.1 (similar to a regression line, but nonlinear) shows that wages do increase over time as high school dropouts gain experience in the workforce.

Visualizing longitudinal data with the "brolgar" package

The "brolgar" package provides tools for exploring multivariate longitudinal data. In this sub-section, we show only one of the large number of possible brolgar tools. In the example below we use brolgar's `key_slope()` function in combination with ggplot2 commands to plot exceptions to the general trend of wages going up over time. For these exceptions, wages went down as years in the workforce went up.

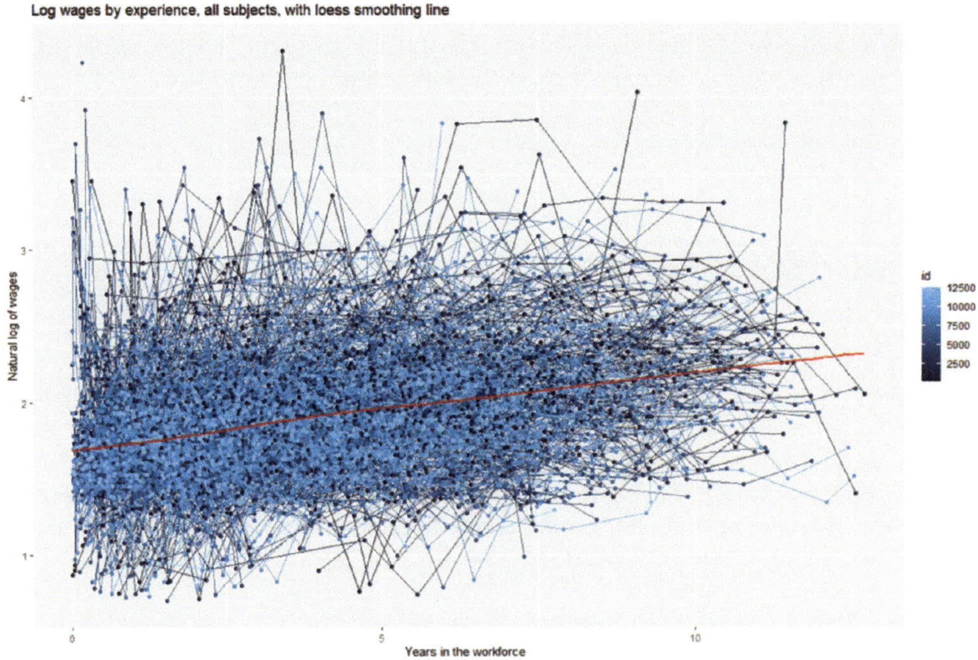

Figure 15.1 Longitudinal plot of wages by years experience.

```
# Setup
base::library(brolgar)
base::library(gghighlight)

# Select out the first 499 data rows in wages. A smaller sample will
plot more clearly
# but 499 is needed to get all rows for the included 13 subjects
(id's).
dat <- wages[1:499,]

# Use brolgar's key_slope() function to create the object
wages_slope.
# wages_slope is a tibble-type data frame with all the variables
from wages
# but adding new columns for the intercept and slope when ln_wages
is predicted from xp.
wages_slope <- dat |>
  brolgar::key_slope(ln_wages ~ xp) |>
  dplyr::left_join(dat, by = "id")

# Use ggplot() to create the line plot.
# The plot is put in the object negSlopesPlot but is not yet
displayed.
negSlopesPlot <- ggplot2::ggplot(wages_slope,
  ggplot2::aes(x = xp,
  y = ln_wages,
  col = "darkred",
  group = id)) +
  ggplot2::geom_line() +
  ggplot2::labs(title = "Log wages by experience, selected subjects,
highlighting negative slopes",
    subtitle = "Red = negative slope. Green = positive slope. La-
beled by subject id",
    caption = "Data source: National Longitudinal Survey of Youth",
    y= "Natural log of wages",
    x = "Years in the workforce")

# Use gghighlight() to differentiate the positive and negative
slopes.
# This also displays the plot in Figure 15.2.
negSlopesPlot +
gghighlight::gghighlight(.slope_xp < 0,
  unhighlighted_params = base::list(color = "lightgreen"))
```

Data preparation for FPCA for longitudinal data

FPCA requires data to be in a particular format described in this section. The data must be reshaped so that the time variable (here, xp) and the time series variable (here, wages)

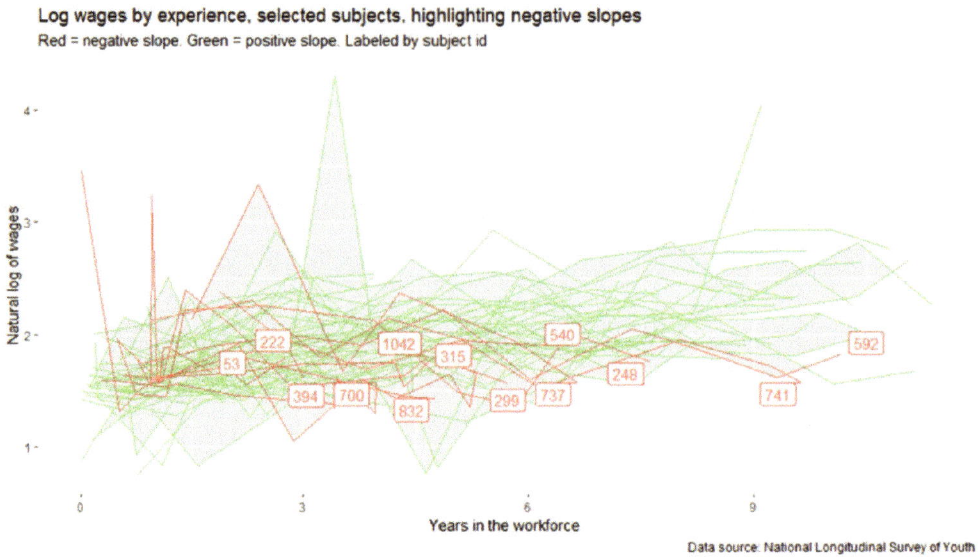

Figure 15.2 Trajectories of wages over time, negative slopes highlighted.

are separate columns of lists. Each row is single subject with a unique id. In this example, rather than one value being in a cell, the cell may hold a list of many values. For instance, for subject 31 in the column "Wages", the cell entry might have eight values, one for each measurement time:

```
c(1.491, 1.433, 1.469, 1.749, 1.931, 1.709, 2.086, 2.129)
```

The data preparation steps are given below.

Step 1. Make the object df as a copy of wages.

```
# The df object is a time series tibble (a "tsibble", tbl_ts), as
was wages.
# However, there are still multiple data rows per subject id.
df <- wages
```

Step 2. Create the object df2 as a tibble (not a tsibble) from df, with selected variables.

```
# df2 still has multiple data rows per subject id.
df2 <- df |> select(id, xp, ln_wages, unemploy_rate) |> as_tibble()
```

Step 3. Initialize some values

```
# uid is an integer vector of unique id's for the 888 subjects
uid <- base::unique(df2$id)
# N (upper case) is 888, the number of subjects.
N <- base::length(uid)
```

```
# The next lines initialize to 0 the variables to be in df3 below.
# Thus, for instance, wages is a numeric vector of 888 values of 0.
Wages <- base::rep(0,N)
Experience <- base::rep(0,N)
UnemployRate <- base::rep(0,N)
```

Step 4. Loop through the 888 subjects to populate the predictors. This step will involve a small delay.

```
# Wages and other predictors become objects of class "list".
for (k in 1:N){
   Wages[k] <-  df2 |> dplyr::filter(id == uid[k]) |> dplyr::se-
lect(ln_wages) |> dplyr::pull() |> base::list()
   Experience[k]  <-   df2 |> dplyr::filter(id == uid[k]) |>
dplyr::select(xp)  |> dplyr::pull() |> base::list()
   UnemployRate[k] <-  df2 |> dplyr::filter(id == uid[k]) |>
dplyr::select(unemploy_rate) |> dplyr::pull()   |> base::list()
}
```

Step 5. We now create df3 as a tibble consisting of lists, one for each predictor. For instance, the first subject in df3 is uid = 31. This subject was measured at eight time periods. The cell entry in the Wages column thus has eight values, listed previously above. Subject 31 has all data for all predictor variables in a single data row of similarly-formatted cells. The df3 object is a tibble.

```
df3 <- dplyr::tibble(uid, Wages, Experience, UnemployRate)
We can also view some of the data in df3 with the glimpse() command.
dplyr::glimpse(df3)
[Partial output:]
   Rows: 888
   Columns: 4
   $ uid     <int> 31, 36, 53, 122, 134, 145, 155, 173, 206, 207,
   222, 223, 22~
   $ Wages   <list> <1.491, 1.433, 1.469, 1.749, 1.931, 1.709,
   2.086, 2.129>, ~
   $ Experience     <list> <0.015, 0.715, 1.734,
   2.773,3.927,4.946,5.965,6.984>~
   $ UnemployRate    <list> <3.21, 3.21, 3.21,3.30,2.89,2.49,2.60,4.
   80>,<4.89,.7
   ...
```

The data in df3 are now in the format required by the `FCPA()` command, which implements longitudinal PCA. Specifically, the `FPCA()` function computes the functional principal components with these parameters:

- `datatype` = Specifies if the data are "Dense" or "Sparse". The wages data are sparse.
- `error` = `FALSE` calls for using a model that does not adjust for unobserved error. The default is TRUE, which assumes the data do contain measurement error.

- `kernel = epan` calls for using the Epanechnikov for smoothing the pooled data when calculating means and covariances (the default is "gauss", which is a Gaussian kernel). The kernel may be one of "rect", "gauss", "epan", "gausvar", or "quar". Note that data declared "Dense" are assumed noiseless so no smoothing is performed.
- `maxK` is the maximum number of principal components to select. Default is min(20, N − 2, nRegGrid − 2), where N is the number of subject curves, and nRegGrid is the number of support points in each direction of covariance surface. Note this is not the number of components returned, which may be fewer if the FVEthreshold is lower than the default 0.99. (See discussion below in the "Number of components" section.)
- `verbose = TRUE` causes diagnostic messages to be displayed (default = FALSE).

Step 6. Next we create the FPCA result object for Wages. In the command below, the first argument is the outcome variable (df3$Wages) and the second argument (df3$-Experience) is and must be the time variable. Warning: Processing time can take several minutes. Be patient.

```
wagesResult <- fdapace::FPCA(df3$Wages, df3$Experience, base::list(-
dataType="Sparse", error=FALSE, kernel="epan", maxK = 10,
verbose=TRUE))
```

Step 7. We also create unemployResult. This can be switched with wagesResult to illustrate using a second time series. Processing time can take several minutes.

```
unemployResult <- fdapace::FPCA(df3$UnemployRate, df3$Experience,
base::list(dataType= "Sparse", error=FALSE, kernel= "epan", maxK =
10, verbose=TRUE))
```

Number of components

By default, the number of components returned is the smallest number of components required to explain 99% of the variance (here, the variance in Wages, where the default returns nine components). Using the function `select()` one can determine the number of relevant components according to AIC, BIC, or an FVE (Fraction-of-Variance-Explained) threshold lower than the 99% default.

The `SelectK()` function will return the number of components needed to achieve any specified percent of variance explained. For instance, the command below shows seven components are needed to attain 95% explained.

```
fdapace::SelectK(wagesResult, criterion = 'FVE', FVEthreshold =
0.95)
[Output:]
    $K
    [1] 7

    $criterion
    [1] 0.9672186
```

To change the `FPCA()` command to specify the FVE threshold, add FVEthreshold= to the parameter list. The command below will return 7 components for the wages data.

```
wagesResult7 <- fdapace::FPCA(df3$Wages,df3$Experience, base::list(-
dataType="Sparse", error=FALSE, kernel="epan", maxK = 10, ver-
bose=TRUE, FVEthreshold=0.95))

wagesResult7
[Output for threshold = 0.95):]
   Functional Principal Components Object for sparse data.

   The optimal number of components selected is: 7, and
   the first K (<=3) eigenvalues are:
    0.959
    0.243
    0.113
```

Component trends over time

The `CreateModeOfVarPlot()` function from the "fdapace" package shows the mode of variation of eigenvectors over time (here, Experience is the time variable) on the x axis. In Figure 15.3, the red line is the functional mean and corresponds to Q = 0. The dark grey area shows variation above and below the mean and for the first component (the first mode of variation). The light grey area shows variation above and below the mean and for the second component (the second mode of variation).

```
# If rainbowPlot = FALSE (the default), the plot will be in shades
of grey.
fdapace::CreateModeOfVarPlot(wagesResult, rainbowPlot = FALSE, main
= "Variation Around the Mean for Eigenvectors over Time", xlab="-
Years Experience in the Workforce (xp)", ylab = "ln_wages")
```

Figure 15.3 shows that the high school dropout subjects had relatively similar wages at the start of observation but that the variation around increasing wages (the red line) increased through seven years of work experience but began to converge a little for the highest year (12).

Diagnostic plots

The plot method in FPCA summarizes the results. The `plot.FPCA()` function is used "behind the scenes" when the `plot()` command is issued. It prints a matrix of plots for the design, mean function, scree, and the first three eigenfunctions (a.k.a. components, dimensions). Note that the default label for the time access in the "Mean Function" and "First 3 Eigenvalues" plots is "s" for support.

```
# Obtain a diagnostic plot for wagesResult. This is Figure 15.4.
base::plot(wagesResult)
```

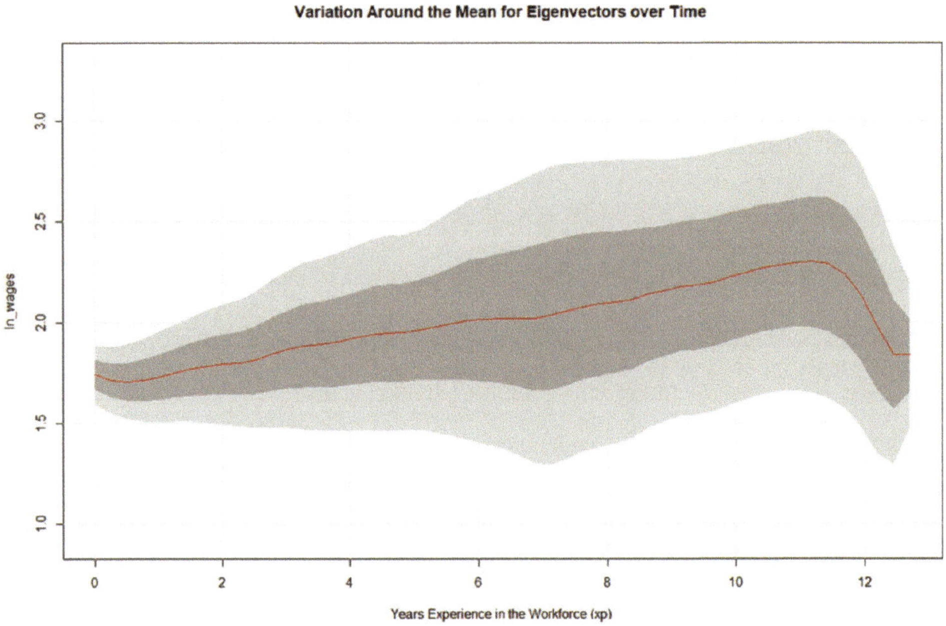

Figure 15.3 Variation of eigenvectors over time, for wagesResult.

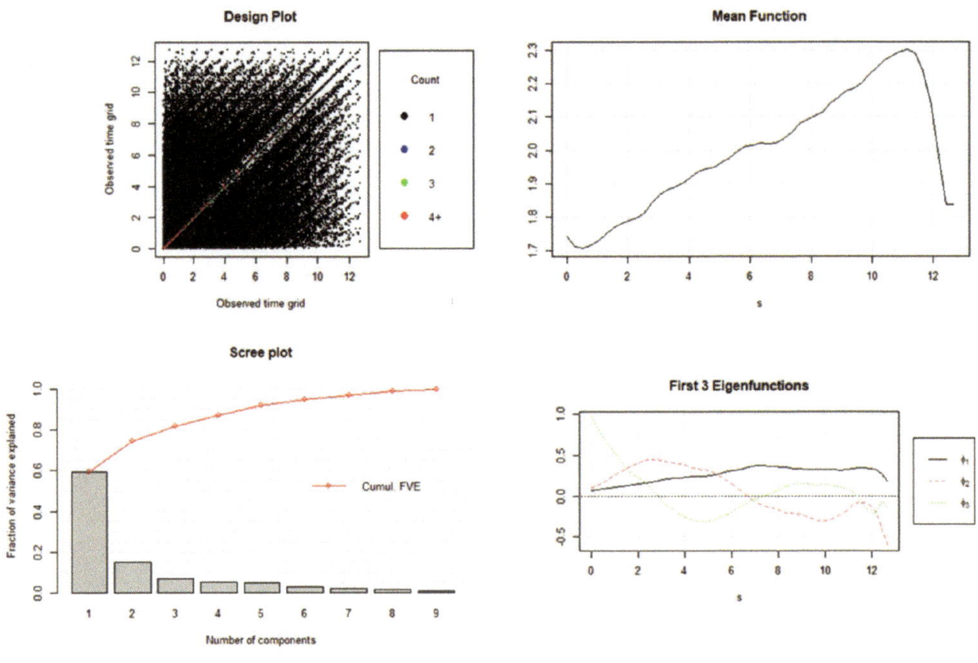

Figure 15.4 Diagnostic plots for wagesResult.

- "Design Plot": This shows how dense or sparse the data are across the support (the time variable, here xp). For the example, data are dense through 11 years of experience in the workforce but are sparse for year 12.
- "Mean Function" plot: This plot shows trends in the target variable (here, the natural log of wages) across the time variable (here, xp). The x axis is labeled "s" for "support. This is the default label for the time variable, which in this example is xp (experience in the workforce in years). For the wages data, mean wages (measured on a natural log basis) rise through Experience (xp) of about 11 but the drop off sharply for Experience at the high end, around 12.
- "Scree plot": This plot shows the percent of variance explained by each eigenfunction. The red line is cumulative FVE (fraction of variance explained). The first two components explain about three quarters of the variance. The first three explain over four-fifths. It takes five components to explain 90%. All nine components explain over 99%. The actual percentages are in the "$cumFVE" element of the result object.

```
base::round(wagesResult$cumFVE, 3)
[Output:]
    [1] 0.594 0.744 0.814 0.868 0.917 0.946
    [7] 0.967 0.985 0.994
```

- irst 3 Eigenfunctions" plot. The black, red, and green curves in the stand for the first, second, and third eigenfunctions (components), respectively. The x axis is labeled "s" for "support. This is the default label for the time variable, which is xp in this example. Since larger eigenvalues explain more of the variance, the plot shows the relative importance of the first three eigenfunctions (components) in explaining the variance in wages at any given time point on the x axis. At first, component 3 is most important, for experience in the 2 to 4 years range, component 2 is most important, then from around 6 to the end (12 years), component 1 is most important. While the first component explains almost 60% of the variance overall (shown in the scree plot), this is an average across the time variable.

Outlier detection

The outlier detection plot shown in Figure 15.5 displays the first component (FPC1) on the x axis and the second component (FPC2) on the y axis. The percentages in the axis labels are the percent of variance explained by each of the two components. The plot is created using the function `CreateOutliersPlot()`.

```
# The "ks" library is needed for the KDE variant.
base::library(ks)

# Create the outliers plot and outliers object. This plot is Figure
15.5.
base::set.seed(123)
outliers <- fdapace::CreateOutliersPlot(wagesResult,
    main= "Outlier Plot for wagesResult",
    sub = "99% outliers are shown in gold, 95% outliers are shown in
    red",
    optns = base::list(variant = "KDE"))
```

Explanation

- The first argument (wagesResult) in the command above is a results object from `FPCA()`.
- In the options, variant = "KDE" specifies using a kernel density estimator as the ranking method. The "ks" library must have been loaded.

To get outliers at or beyond the 99% level, access the "$p99plus" element from the outliers object created above in the CreateOutliersPlot object. This returns three data row numbers in the df3 data frame being used. Note these are row numbers, not the id numbers. Note also that besides "$p99plus" there are elements for "$p95to99", "$p50to95", and "$p0to50".

```
outliers$p99plus
[Output:]
    [1] 384 793 808
```

To get the corresponding id numbers for outliers:

```
outrows <- outliers$p99plus
outlierID <- df3$uid[outrows]
outlierID
[Output:]
    [1] 5990 10432 10545
```

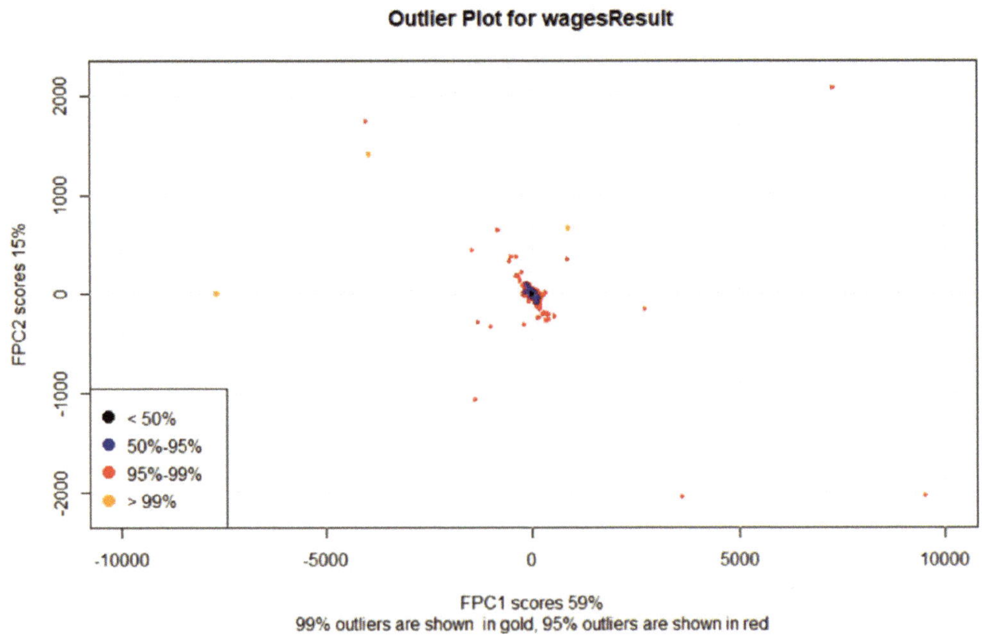

Figure 15.5 Outlier plot for wagesResult.

Scores

Subject scores on each of the components are contained in the "$xiEst" element of the result object. We use the scores to compute the joint mean, then display it as a green asterisk in Figure 15.6, which otherwise is the same as Figure 15.5.

```
wagesResult$xiEst
[Partial output for first three subjects on the first two
components:]
                  [,1]          [,2]
  [1,]  3.689485e-01 -7.130421e-01
  [2,] -1.697994e+00  2.340346e+00
  [3,] -3.270000e+02  1.596875e+

# Put scores for first two components into the numeric vector vari-
ables FPC1 and FPC2:
FPC1 <- wagesResult$xiEst[,1]
FPC2 <- wagesResult$xiEst[,2]

# Add the scores variables to the dataset.
df3$FPC1 <- FPC1
df3$FPC2 <- FPC2

# Get the means of the two FPC variables
x <- base::mean(df3$FPC1)
y <- base::mean(df3$FPC2)
```

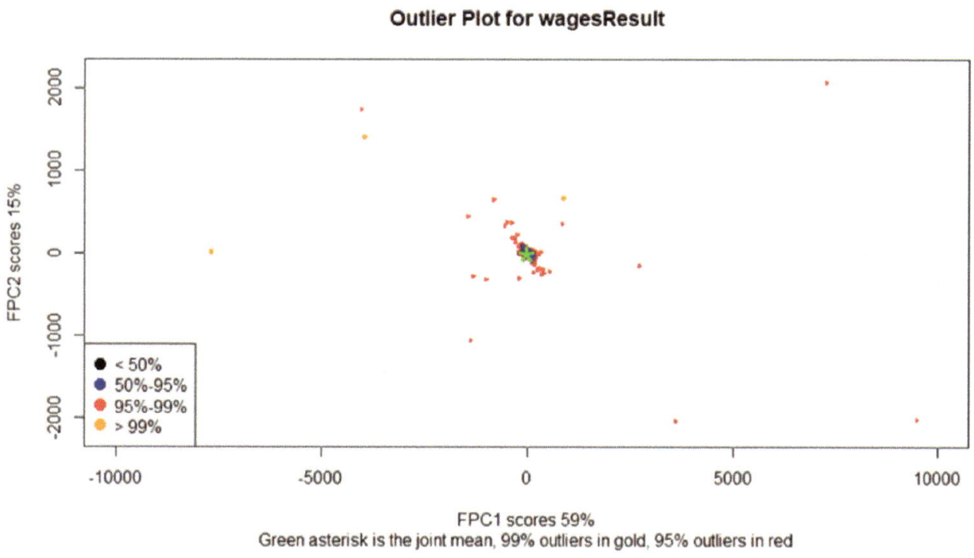

Figure 15.6 Outlier plot with joint mean.

```
# Re-create the outlier plot
outliers2 <- fdapace::CreateOutliersPlot(wagesResult,
   main= "Outlier Plot for wagesResult",
   sub = "Green asterisk is the joint mean, 99% outliers in gold,
   95% outliers in red",
   optns = base::list(variant = "KDE"))

# Add a green asterisk representing the joint means. This is Figure
15.6.
graphics::points(x,y, pch="*", col="green", cex=3)
```

Other R packages for functional PCA

In addition to the "fdapace" package described above, there are other R packages which implement functional principal component analysis, including these:

- *fpca*: Restricted MLE for Functional Principal component analysis. This is a geometric approach to MLE for functional principal components.
- *FPCA2D*: Two-Dimensional Functional Principal Component Analysis. This package computes the two dimension functional principal component scores for a series of two-dimensional images.
- *FPCA3D*: Three-Dimensional Functional Component Analysis. This package runs three-dimensional functional principal component analysis and returns the three-dimensional functional principal component scores.
- *fpcb*: Predictive Confidence Bands for Functional Time Series Forecasting. This package contains functions to represent functional objects under a Reproducing Kernel Hilbert Space (RKHS) framework, with implementation of the Autoregressive Hilbertian Model for functional time series using RKHS and predictive confidence bands construction.
- freqdom.fda: Functional Time Series: Dynamic Functional Principal Components. This package implements functional dynamic principal component analysis.
- KFPCA: Kendall Functional Principal Component Analysis. This package implements Kendall functional principal component analysis. Kendall functional principal component analysis is a robust functional principal component analysis technique for non-Gaussian functional/longitudinal data.
- MFPCA: Multivariate Functional Principal Component Analysis for Data Observed on Different Dimensional Domains. This package calculates a multivariate functional principal component analysis for data observed on different dimensional domains.

Note

1 Also available as a text file at https://stats.idre.ucla.edu/stat/r/examples/alda/data/wages_pp.txt.

Appendix I
Datasets used in this volume

Examples in this volume use the following files, listed alphabetically. Additional files created in the text as part of examples are not listed.

ability.cov

This dataset is one of the datasets in R's "datasets" package, which is part of the System library. The data are from a psychology study concerning six ability and intelligence tests which were administered to 112 individuals. Further description is in the "Example data" subsection of the "Common factor analysis (PFA)" section.

Adult Self-Transcendence Inventory (ASTI)

This data frame is supplied as part of the MPsychoR package. This has 25 categorical variables measuring various personality traits.

cobb_survey.csv

This is a survey dataset used in Appendix 2 to illustrate handling missing values.

fragilestates.csv

This dataset contains 17 variables on 179 countries. The fragility variables are columns 6–17. Col. 1 is Country ID; col. 2 is Country (country name); col. 3 is Year (2021 for all rows); col 4 is Rank (1st through 179th); and column 5 is Total (the sum of all fragility variables for that row). Description is in this endnote.[1] It comes as a .csv text file (see below).

gironde

The "PCAmixdata" package contains the "gironde" data. This is a list object containing four datasets pertaining to French cities: housing, employment, services, and environment.

GSS93subset.csv

This dataset contains 103 variables on 1,500 observations from the 1993 General Social Survey of US opinion data. It comes as a .csv text file (see below). It is not used per se but is the source for musicvars.csv.

Holzinger.9 dataset

This dataset is used in the section on oblique rotation. It contains nine variables which were ability tests measuring verbal, mental speed, and spatial dimensions. Type `help(Holzinger.9)` for more information on this dataset. See Holzinger & Swineford (1937).

house_votes84

This dataset is supplied with the "logisticPCA" package. There are 435 observations (one for each U. S. Representative) and 16 columns (the 16 roll-call votes). All data are binary, coded 0 or 1 for voted against or voted for respectively.

iris

This dataset is one of R's built-in example datasets. Widely used for educational purposes, it contains five variables: Species (observations are setosa, virginica, or versicolor species) and four predictor variables (sepal width and length, petal width and length). Note that it is lowercase.

Iris

This is a version of the "iris" dataset compatible with the dimRed package as presented in Part II of this book. Note that it is upper case.

musicvars.csv

This dataset contains 18 variables on 1,500 observations, taken from the GSS93subset.csv. It comes as a .csv text file.

mydata.csv, mydata.sav, mydata.dta

These are comma-separated-values, SPSS, and Stata versions of a tiny data file used only in Appendix 2.

mytest.csv

This is a variant of mydata.csv with missing values, used only in Appendix 2.

USArrests

The USArrests data frame is part of R's built-in "datasets" package. It contains four variables: "Murder", "Assault", "UrbanPop", and "Rape".

USJudgeRatings

The USJudgeRatings data frame is part of R's built-in "datasets" package. It contains 43 observations on 12 numeric variables. For information, type `help(USJudgeRatings)`.

wages

This dataset is supplied with the "brolgar" package.[2] It consists of wages data from the National Longitudinal Survey of Youth (NLSY). The sample covered male high-school dropouts, ages 14–17 inclusive.

Datasets which are bundled with R packages are invoked using the data() command after installing and loading the corresponding package with the base::library() command. Additional information may usually be obtained by typing `help(name_of_data_frame)`. For instance:

```
utils::install.packages("brolgar")
base::library(brolgar)
```

```
utils::data(wages)
View(wages)
utils::help(wages)
```

Three of the csv files are linked in the url's below and will be available on the publisher's Companion Website:

Go to http://www.statisticalassociates.com/data/GSS93subset.csv
to download GSS93subset.csv.

Go to http://www.statisticalassociates.com/data/musicvars.csv
To download musicvars.csv.

Go to http://www.statisticalassociates.com/data/fragilestates.csv
To download fragilestates.csv.

R commands to read in csv files once downloaded to a local directory such as "C:/Data" (your folder location may well differ) are shown below.

```
base::setwd("C:/Data")
gss <- utils::read.csv("GSS93subset.csv", header=TRUE, sep =
", ", stringsAsFactors=TRUE)
music <- utils::read.csv("musicvars.csv", header=TRUE, sep =
", ", stringsAsFactors=TRUE)
fs <- utils::read.csv("fragilestates.csv", header=TRUE, sep =
", ", stringsAsFactors=TRUE)
```

Notes

1 Fragile States Index, from The Fund for Peace
https://fragilestatesindex.org/excel/

Security	C1: Security Apparatus
Factions	C2: Factionalized Elites
Grievance	C3: Group Grievance
Economy	E1: Economy
Inequality	E2: Economic Inequality
Flight	E3: Human Flight and Brain Drain
Legitimacy	P1: State Legitimacy
Services	P2: Public Services
Rights	P3: Human Rights
Demog	S1: Demographic Pressures
Refugees	S2: Refugees and IDPs
Intervention	X1: External Intervention

Variable groupings as conceived by
https://fragilestatesindex.org/indicators/

Cohesion
- C1: Security Apparatus
- C2: Factionalized Elites
- C3: Group Grievance

Economic
E1: Economic Decline
E2: Uneven Economic Development
E3: Human Flight and Brain Drain

Political
P1: State Legitimacy
P2: Public Services
P3: Human Rights and Rule of Law

Social and cross-cutting indicators
S1: Demographic Pressures
S2: Refugees and IDPs
X1: External Intervention

Indicators
The Fragile States Index is based on a conflict assessment framework – known as "CAST" – that was developed by FFP nearly a quarter-century ago for assessing the vulnerability of states to collapse. The CAST framework was originally designed to measure this vulnerability and assess how it might affect projects in the field, and continues to be used widely by policy makers, field practitioners, and local community networks. The methodology uses both qualitative and quantitative indicators, relies on public source data, and produces quantifiable results.
Twelve conflict risk indicators are used to measure the condition of a state at any given moment. The indicators provide a snapshot in time that can be measured against other snapshots in a time series to determine whether conditions are improving or worsening. Below is the list of indicators used both in the CAST framework and also in the Fragile States Index.
Higher is more fragile.

2 Also available as a text file at https://stats.idre.ucla.edu/stat/r/examples/alda/data/wages_pp.txt.

Appendix 2
Introduction to R and RStudio*

Why R?

R is a language which is used for statistics, data analysis, text analysis, and machine learning. While it may be employed using simple commands to implement common procedures such as linear regression, it is also a programming language which allows the user to create original functions, programs, and packages. This has made R arguably the fastest-growing and leading statistical tool for researchers. Even if you do not wish to program yourself, you can take advantage of thousands of cutting-edge programs, including ones for agent-based modeling, Bayesian modeling, cluster analysis, correlation, correspondence analysis, data management, decision trees, descriptive statistics, economics, factor analysis, forecasting, generalized linear modeling, instrumental variables regression, logistic regression, longitudinal and time series analysis, machine learning models, mapping and spatial analysis, mediation and moderation analysis, multiple linear regression, multilevel modeling, network analysis, neural network analysis, panel data regression, path analysis, partial least squares modeling, power analysis, reliability analysis, significance testing, structural equation modeling, survey research, text analytics, visualization of data – and many more. New state-of-the-art R packages are added daily in an ever-expanding universe of research tools, many created by leading scholars in their fields.

R is free and thus liberates the researcher from dependency on the willingness of his or her purchasing department to acquire new software. Moreover, it is platform-independent and may be used with any operating system. Also, R is open-source, with all source code available to those inclined to look "under the hood". Statistical algorithms are not locked in proprietary "black boxes". One benefit of being open source is that bug fixes and added features may be put forward from the R user community, members of which do not need to wait upon a vendor for new releases which may be a year or more away.

R packages can import from and export to a variety of data formats, such as those for SPSS, SAS, Stata, and Excel, to name a few. Although R is quite full-featured in its own right, it also may be integrated with other programming environments, such as Python, Java, and C/C++. Starting with version 1.4, RStudio now offers access to Python tools and packages through its Python interpreter, the "reticulate" package, and as well as by way of other avenues. All of this is supported by a very large user community with a full array of mailing lists (through which help questions may be posed and answered), blogs, conferences, journals, training opportunities, and archives.

* At this writing it is anticipated that RStudio will rename itself to Posit but Posit should be backward compatible to Appendix 2.

Installing R and RStudio

While it is perfectly possible to install R without using a graphical user interface, in this tutorial we recommend the most popular approach, which is using the free RStudio package. Installation of RStudio will also install the latest version of R. RStudio's functionality is summarized at https://www.rstudio.com/products/rstudio/features/. To install R Studio, follow the steps below.

To install RStudio, go to https://www.rstudio.com/products/rstudio/download/.

Click the "Download" button for the "Free" version of "RStudio Desktop".

Under "All Installers", click the RStudio link for your operating system

(Usually the Windows 10/11 choice, but macOS and various flavors of Linux are also available. Note that RStudio requires a 64-bit system. If yours is 32-bit, you will have to use an older version of RStudio.)

After clicking to download, an executable file (.exe) will be downloaded to your computer. Depending on your browser and operating system, this file may be launched automatically or you may have to choose to run the .exe file manually.

When the .exe file is run, a setup wizard starts. Accept the defaults and click the "Finish" button at the end.

RStudio will now be in the Windows start menu. You can right-click its icon and choose to pin it to the taskbar if you wish.

It is a good idea to check from time to time to see if you have the current version of R and RStudio. In the RStudio menu system, select Help > About RStudio to see the RStudio version. Issue the command `version` to see the loaded version of R.

Installing RStudio will not automatically install RTools. However, installing some R packages requires RTools and so we recommend installing it. Go to https://cran.rstudio.com/bin/windows/Rtools/ and follow the download directions to install the 64-bit version.

While it is possible to install R without RStudio, we recommend the installation procedure above. RStudio will automatically use the most recent version of R installed on the user's computer. Typically this will be the version installed when RStudio is installed.

Example data

Further below we use a very small dataset called mydata.csv (comma-separated values format), mydata.sav (SPSS format), or mydata.dta (Stata format). These are supplied in the online Companion Website for this book. However, this dataset is small enough to enter into R manually, as shown below.

The dataset contains ten fictional cases, each with two variables: gender and happy. The "happy" variable is an ordinal item with values coded from 1 = very unhappy to 5 = very happy. Gender is a binary item coded male = 0 and female = 1. Actual data are listed below and may be pasted into any statistical package or spreadsheet, from which a comma-separated values (.csv) or other format dataset may be created.

The data may be read into an object called "mydata" from the supplied comma-separated values (.csv) file by running the R commands below. Note that lines beginning with the pound sign (#) are comments not acted upon by R. Note also that all three commands, including comments, can be cut and pasted as a batch into RStudio and run.

happy	*gender*
4	0
1	0
3	0
5	0
3	0
3	1
2	1
4	1
5	1
1	1

```
# Set the working directory (yours may differ)
base::setwd("C:/Data")

# Read the .csv file into the object "mydata".
mydata <- utils::read.table("mydata.csv", header = TRUE, sep = ",")

# Verify that mydata is of data class "data.frame", the most common
R format for storing data.
base::class(mydata)
Output:]
  [1] "data.frame"
```

Alternatively, the mydata data frame may be created manually using the commands below. The assignment operator (<-) puts whatever is on its right-hand side into the object named on its left-hand side. The `c()` combine function combines values into a vector or list. Below it combines the ten values listed for each variable into numeric vectors called happy and gender.

```
#  Create numeric vector objects for happy and gender
happy <- base::c(4,1,3,5,3,3,2,4,5,1)
gender <- base::c(0,0,0,0,0,1,1,1,1,1)

# Create a data frame called"(mydata" from gender and happy.
mydata <- base::data.frame(happy,gender)

# View the contents of mydata. Note that "View" must be capitalized,
unlike most R commands.
# The display of mydata will appear in RStudio in the upper left.
View (mydata)
```

In addition to viewing a data frame like mydata, we may summarize it. The `summary()` command provides basic descriptive statistics about variables in the data set.

```
base::summary(mydata)
[Output:]
```

```
     happy            gender
 Min.    :1.00    Min.    :0.0
 1st Qu.:2.25    1st Qu.:0.0
 Median :3.00    Median :0.5
 Mean    :3.10    Mean    :0.5
 3rd Qu.:4.00    3rd Qu.:1.0
 Max.    :5.00    Max.    :1.0
```

In this introduction to R, we also use a larger dataset, gssdata. The file "gssdata.csv" is supplied on the Companion Website to this text. Other .csv files used in this tutorial are cobb_survey.csv and world.csv. For illustrating importing SPSS data, we use mydata.sav. All are found in the Data section of this text's Companion Website.

Quick start: computing a correlation

As a quick start example, we will use RStudio to import an SPSS data file, then we will compute the correlation coefficient between "gender" and "happy". We assume the "mydata" object has already been created as in the previous section.

After starting RStudio, its basic user interface appears as shown in Figure Appendix 2.1.

In Figure Appendix 2.1, the R user interface is shown. In the upper left the results of the `View()` command are displayed, revealing spreadsheet-style the contents of a data frame or other object. The upper right area displays the Environment, History, Connections, and Tutorial, depending on which tab is selected. In the figure, the Packages tab is selected, showing packages currently loaded or available for loading into the R environment. The lower left shows the area where commands may be entered and text

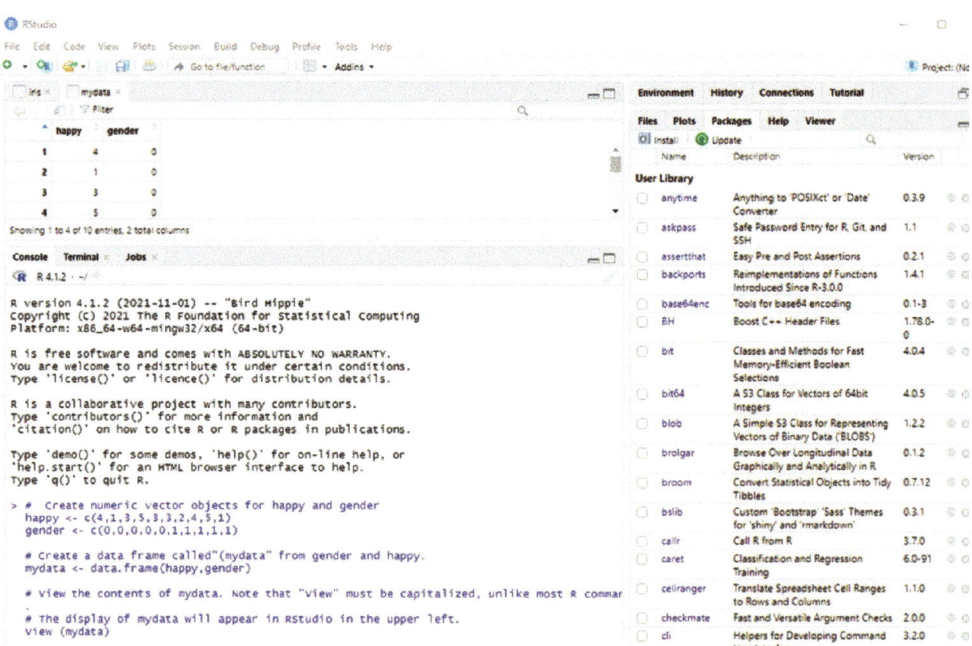

Figure Appendix 2.1 The RStudio interface.

output from commands, as well as error messages, are displayed. The lower right area displays Files, Plots, Packages, or the Viewer, depending on which tab is selected.

It is a good idea to do three things right at the start and one thing at the end of an RStudio session:

Select Session > New Session from the RStudio menu. This creates a new instance of RStudio with a cleared environment.

Choose Session > Set Working Directory. Many researchers make a practice of creating a separate file folder for each R project. Set the working directory to point to your project folder. This is the same as issuing the `setwd()` command used above.

At the end, select Session > Quit Session. Respond "Yes" when asked if you want to save your workspace image in an .RDATA file in your working directory.

To continue with the "quick start" example of computing the correlation of the variable gender and happy, make sure the example mydata object was created as described earlier. The data frame object "mydata" should appear in the Environment tab of RStudio.

The correlation command in R is `cor()`. Because it is in the "stats" package, which is part of R's "System library", it may be used directly without having to install a package and without having to use the `base::library()` command to activate it. As this is in the System library, it is not necessary to use the package prefix "stats::" but doing so may help clarify which packages are being used.

```
myoutput <- stats::cor(mydata$happy,mydata$gender)

myoutput
[Output:]
   [1] -0.0727393
```

This simple example provides the following lessons:

- We are performing an operation on a data frame (mydata), which is the most common format in R for storing data.
- This is an example of using the assignment operator (<-) to assign the results of the right-hand side of the operator to the object named on the left-hand side.
- We may list results by typing the name of the output object, here myoutput. However, had we entered only the `cor()` expression on the right-hand side without any assignment, the results would also have been printed.
- R is case sensitive. Be careful to enter terms in the proper case. Most commands in R are lower case, but not all.
- Variable names are entered in long form, which is the name of the data frame (mydata), a dollar sign, then the name of the column (happy). If a command had a `data=mydata` option (not the case here and not supported by the `cor()` command), then the short variable name (e.g., happy) could be used. It is also possible to issue the command `attach(mydata)` before the correlation command, then issue the command `detach(mydata)` afterward. This allows the use of short variable names. However, attaching and detaching is derogated for reasons given in the FAQ at the end of this document. Many recommend using long form variable names in all circumstances, even if just for clarification of which variable comes from which data frame.

- When a data frame is the source, variables correspond to column names in the data frame.
- The `cor()` command defaults to Pearsonian correlation. However, `cor()`, like most R commands, supports options. Type `help(cor)` to see the options. We may similarly type `help(name or command)` for almost any command. It is a good idea to check the options for any command if only to see what defaults are being used.
- Among the options for `cor()` are the `method=` option to generate Kendall's correlation tau (used for ordinal variables) instead of Pearson's correlation r (used for continuous or binary variables).

```
myoutputKendall <- stats::cor(mydata$happy,mydata$gender,
method="kendall")

myoutputKendall
[Output:]
   [1] -0.06405126
```

Another option for the `cor()` command is to omit mentioning of a pair of variables to correlate and instead entering just the name of the data frame. If variables in the data frame are numeric, we then get the correlation matrix of all variables. If saved to an object like "mycormatrix" below, the correlation matrix may be used as input to some other statistical command. Note that while we do not do so here, we could have added the option `method="kendall"` to get a matrix of Kendall's tau rather than Pearson's r coefficients

```
# Compute a correlation matrix, putting it in the object
"mycormatrix".
mycormatrix <- stats::cor(mydata)

# List the matrix.
mycormatrix
[Output:]
           happy     gender
    happy  1.0000000 -0.0727393
    gender -0.0727393  1.0000000
```

Importing data

Acquiring data from some source is often one of the first tasks to be undertaken in a research project. Fortunately, R has the ability to import data from a very wide range of formats. R Studio provides menu options for common types of data importation. Thus, one may select File > Import Dataset > From SPSS to import an SPSS .sav data file. Other choices are From SAS, From Stata, From Excel, or From Text (for .csv files, for instance). RStudio will create the necessary code for you, displaying it in the lower right of the "Import Statistical Data" window. Browse to the file you wish to import, give it an output name, and click the "Import" button. Note that loading one data file does not erase others from memory, provided it is not of the same name. There are multiple other ways to create or import data, listed below.

Manual data entry

Manual data entry was illustrated earlier in the creation of the "mydata" data frame in the "Example Data" section.

Excel data and .csv files

One of the most common data storage strategies is to keep data in an Excel or similar spreadsheet, then share the data by saving it in comma-separated values (.csv) format. The .csv format is a type of "generic" data medium. A very large number of statistics and data management packages, including R, can read and write .csv data.

If one chooses to archive data in Excel or some other non-R software, the researcher should keep in mind these general R data format requirements:

- One variable per column.
- All columns have the same number of rows.
- It is common for variable names to be in row 1.
- No interior spaces in variable names.
- Names are case-sensitive.

Do not use reserved words in names like "integer" or "break". For a list of reserved words in R, type `help(reserved)`. Use of reserved terms may lead to unpredictable results.

For missing data values, enter "NA" without the quote marks. If the original data contain some other missing code, such as -99, these cases must be recoded inside R. For example, the command below recodes −99 to NA missing for the variable happy:

```
mydata$happy[mydata$happy==-99] <- NA
```

For continuous data entered as vectors, enter an integer (e.g., 76) or a decimal value (e.g., 76.832). Note that the expression `newvar = 76L` will assure newvar is of type integer, not type numeric as would be for the expression `newvar2 = 76`.

Categorical data may be entered as integer values such as 1L, 2L, and 3L. Factors also may be entered as text values (e.g., female, male) or logical values (e.g., T, F).

Assuming that a .csv file has been moved to one's working directory, the `read.csv()` command may be used to read in the data as shown below. The `read.table()` command functions in the same way. The `stringsAsFactors= option` is recommended (assuming this is consistent with the research purpose) but is not required. If set to TRUE (or T) then strings in the data will be of factor data class. If set to FALSE (or F) then strings in the data will be of character data class. This can make a difference for certain statistical procedures. The header = TRUE option specifies that row 1 contains the variable names, which is usual.

```
base::setwd("C:/Data")

mydata <- utils::read.csv("mydata.csv", header=TRUE, sep = ",",
stringsAsFactors = TRUE)
or:
mydata  <- utils::read.table("mydata.csv", header = TRUE, sep =
",", stringsAsFactors = TRUE)
```

Alternatively, it is possible to use a Windows-style file selection window to browse the text file that is wanted (e.g., .txt, .csv) using the `file.choose()` command. The command below will enter the selected file into the object "mydata". Of course, what is chosen must be mydata.csv to have the mydata content discussed above. This method gives the option of browsing to the correct file when it is not in the working directory.

```
mydata <- utils::read.csv(file.choose(),header=TRUE, sep = ",",
stringsAsFactors = TRUE)
```

Type `help(read.table)`, with or without the quotation marks, to view more options for this command, which is not restricted to comma-separated values format.

Importing Stata .dta files

A commonly-used utility package in R is "haven", which is used for importing data in various formats, including Stata's .dta format. We must, of course, install and load the haven package before importing Stata files. This can be done by the RStudio menu method or by the command method.

The menu method is to select the "Packages" tab in RStudio, selecting "Install", as shown in Figure Appendix 2.2. Most packages are installed from the CRAN repository, also shown in the figure, but there are other archives. Note that "install dependencies" is the default. This means that other packages needed by the haven package will also be installed.

Alternatively, use the command method of installation below. Note that unlike the `base::library()` command, for `install.packages()` the package name must be within quotation marks.

```
utils::install.packages("haven")
```

Like all packages not in the R System library, haven must be loaded after it is installed. This is done with the `base::library()` command from the "base" package.

```
base::library(haven)
```

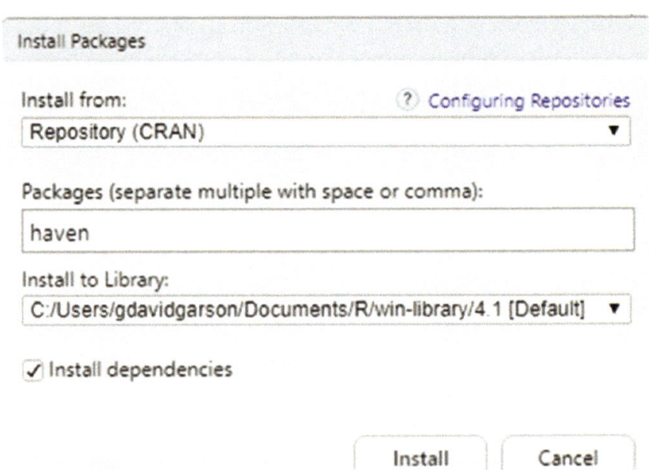

Figure Appendix 2.2 RStudio install packages window.

The command below reads in a Stata .dta file which had been placed in the named folder.

```
mydataStata <- haven::read_dta("C:/Data/mydata.dta")
```

Then inspect the data:

```
View(mydataStata)
```

In addition to `read_dta()` and `write_dta()`, the haven package has these functions:

`read_sas(), write_sas()`	Read and write SAS files.
`read_xpt(), write_xpt()`	Read and write SAS transport files
`read_sav(), write_sav()`	Read SPSS (.sav, .zsav, .por) files. Write .sav and .zsav files.

Importing SPSS .sav files

The haven package discussed above for Stata also imports SPSS .sav data files. The commands are identical, except the `read_sav()` command is used instead of `read_dta()`.

```
base::library(haven)
mydataSPSS <- haven::read_sav("C:/Data/mydata.sav")
View(mydataSPSS)
```

Alternatively, if the "foreign" package is used instead of "haven" to importing SPSS files, then variable names and value labels will be supported in commands such as `table()`.

```
utils::install.packages("foreign")
base::library(foreign)

mydataSPSS2 <- foreign::read.spss("C:/Data/mydata.sav",
use.value.labels=TRUE, to.data.frame=TRUE, max.value.labels=Inf,
trim.factor.names=FALSE)
```

Now when the data are viewed and put in a table, value labels will appear.

```
View(mydataSPSS2)
base::table(mydataSPSS2$happy,mydataSPSS2$gender)
[Output:]
                            male female
  Very unhappy                 1      1
  unhappy                      0      1
  Neither happy nor unhappy    2      1
  happy                        1      1
  very happy                   1      1
```

Warning: Before importing SPSS data, read the section further below on "Checking and handling missing data" as SPSS poses special problems when there are missing data.

Specifically, R wants missing data to be coded NA, but SPSS allows a variety of other codings. This does not apply to the sample mydata.sav file, however.

In addition to `read.spss()` and `write.spss()`, the "foreign" package has these functions:

`read.dbf(),write.dbf()`	Read and write a DBF File
`read.dta(), write.dta()`	Read and write files in Stata format
`write.foreign()`	Write text files and code
`read.mtp()`	Read MiniTab files
`read.systat()`	Read Systat files
`read.xport()`	Read SAS export files

Importing SAS .sas7bdat files

Importing SAS files may also be done by an almost identical haven method, substituting the command read_sas(). We assume that the haven package has already been installed as described above.

```
base::library(haven)
mydataSAS <- haven::read_sas("C:/Data/mydata.sas7bdat")
View(mydataSAS)
```

SAS export files (.xpt format) may also be loaded with the "Hmisc" package. This is not illustrated here but information may be obtained using the following commands:

```
utils::install.packages("Hmisc")
base::library(Hmisc)
utils::help(sasxport.get)
```

Saving data

The commands for saving data largely parallel the commands discussed above for reading and importing data. In this section, we assume that the data frame object "mydata" is still in memory.

Remember to set the working directory to which you wish to save. For example:

```
base::setwd("C:/Data")
```

Saving to .csv format

For R users, the comma-separated values (csv) format is the most widely used, even more so than the native R format discussed below. The command below is used to write the mydata data frame object to a .csv file. The col.names= option determines if column (variable) names are to be written to row 1 (TRUE is the default). The row.names= option determines if row names are to be written to column 1 (TRUE is the default).

```
utils::write.csv(mydata, file = "mydata2.csv", col.names = TRUE,
row.names = FALSE)
```

Saving to .txt format

To write to plain text format (.txt), as for purposes of writing a web page to a raw, uncleaned text file, one may use the `write.table()` command. The `sep=` option specifies what separator is to be used between values in any row of mydata. The example below specifies a space as the separator.

```
utils::write.table(mydata,"mydata2.txt", sep=" ",col.names =
FALSE, row.names = FALSE)
```

However, ordinarily, a data frame would have been written to a comma-separate values file rather than a text file. A better example would be saving a web page to a .txt file. This can be done easily using the "htm2txt" package, used here to capture the web page of the governor of California.

```
utils::install.packages("htm2txt")
base::library(htm2txt)
CA <- htm2txt::gettxt("https://www.gov.ca.gov/")
utils::write.table(CA, file = "myCA.txt", quote = FALSE, sep =
"\n", col.names = FALSE)
```

Saving and reading in R format (.rds)

As in other statistical packages, data in memory in R will not be saved to disk unless the researcher explicitly asks to do so. R's native file format extension is .rds, as in mydata.rds. Ironically, this data format is eclipsed by the .csv format even among R users. The `saveRDS()` and `readRDS()` functions are in the "base" package in the R System library and can be accessed directly without installing any package.

```
# Save to .rds format
base::saveRDS(mydata, file = "mydata.rds")

# Read from .rds format
mydata <- base::readRDS("mydata.rds")
```

Adding value labels to data

Many data objects (datasets) in R are kept in a rectangular rows-and-columns format called a "data frame". The newer "tibble" format can handle more types of data. A data frame is a subset of a tibble, handling most numeric or character in common use. The "mydata" example data contain the numeric vectors "happy" and "gender". Such a data frame may be displayed as a table but unless the original data had a value label and was imported data using a package like "foreign" which supports value labels, by default initially a data frame is apt to lack value labels for binary and categorical variables. The lack of value labels makes it difficult to read as shown in the table output below.

```
base::table(mydata)
[Output:]
  gender
happy 0 1
    1 1 1
    2 0 1
    3 2 1
    4 1 1
    5 1 1
```

The table above has no value labels. The `factor()` function may be used to assign value labels, for example, for the gender variable. The command below assigns gender to the data class factor, with the value labels "male" and "female".

Warning: Note that we start by making a copy of mydata. The copied data frame is "mydataLabeled". This is an important step because adding value labels will put the labels, which are character strings, in place of the numeric codes. If the labeled data are then saved to, say, mydata.csv, the original numeric codes will be lost.

```
mydataLabeled <- mydata

mydataLabeled$gender <- base::factor(mydataLabeled$gender,levels
= base::c(0,1), labels = base::c("male", "female"))
```

For ordinal variables, one may use the `ordered()` command to add value labels. The "happy" variable is a Likert item with five levels. The command below assigns value labels for the five levels coded from 1 through 5, substituting the five listed labels.

```
mydataLabeled$happy <- base::ordered(mydataLabeled$happy,
   levels = base::c(1, 2, 3, 4, 5),
   labels = base::c("very unhappy", "unhappy",
      "neither happy nor unhappy", "happy", "very happy"))
```

Now we create a table of happy by gender. Output now has value labels.

```
base::table(mydataLabeled$happy,mydataLabeled$gender)
[Output:]
                           male female
  very unhappy                1      1
  unhappy                     0      1
  neither happy nor unhappy   2      1
  happy                       1      1
  very happy                  1      1
```

Note that if one inspects mydataLabeled using the View() command or if one saves this data frame to disk using the command below, the actual cell entries will be of character type, with values such as "very unhappy" and "male". The 0's, 1's, and 2's in the table output above are simply counts of these character strings. The original numeric coding (1-5 for happy, 1-1 for gender) is lost. The command to save mydataLabeled to disk is:

```
utils::write.csv(mydataLabeled, file = "mydataLabeled.csv",
row.names = FALSE)
```

In summary, the method of adding value labels described above calls for maintaining two datasets: (1) the original one containing numeric codes, and (2) the value-labeled version containing character labels. Either the labeled data frame is saved under a different name (e.g., "mydataLabeled.csv") or the labeling commands are saved as a text file to cut and paste as a batch of commands when needed. This achieves its purpose in producing output with value labels, such as the table above, but there is an alternative to this dual data frames approach.

Most functions in R do not use variable labels and even drop them. Though not discussed here, the package "expss" supports SPSS-style variable names and value labels. If variable names and value labels will be used more than briefly in the user's research, use of this package is recommended. The expss package integrates value labels support into base R functions and into functions from other R packages

Inspecting data

As a research project unfolds, there are apt to be many R objects created. These are listed under the "Environment" tab in RStudio. For data frames and objects which can be coerced into a data frame (e.g., by `as.data.frame(mymatrix)` for the object "mymatrix"), data may be inspected by the `View()` command, which opens spreadsheet-like window in the upper left of the RStudio interface. Certain objects which are not a data frame, such as a text Corpus, must be viewed with the `inspect()` command, not `View()`.

```
View(mydata)
```

A way to see a list of all objects in the environment is to issue the objects() command.

```
objects()
[Output:]
   [1] "CA"                "mydata"            "mydataLabeled"
```

Use the `class()` function to reveal the data class of any object. Here mydata is a data. frame while gender is an integer vector.

```
base::class(mydata)
[Output:]
   [1] "data.frame"

base::class(mydata$gender)
[Output:]
   [1] "integer"
```

We can also use the `sapply()` command to get the data class of all columns (variables) in a data frame. While mydata has only two columns, this method is even more useful when there are numerous variables. It is important to know the data class of each variable because some R functions will want a certain class of data as input. Also, this helps debug data class errors, as when a function gives an error message because it wanted integers such as 0 or 1, but a given variable was actually of character data class with values of "0" and "1". Another common problem occurs when a procedure wanted a variable of the factor class but the actual variable was of the character class. Bottom line: it is good to know what the data classes of your variables are. When you type `help(insert_name_of_command)`, the resulting help page on RStudio will tell you what data class was wanted as input by a given command. Here, both happy and gender are integer vectors.

```
base::sapply(mydata, class)
[Output:]
    happy    gender
"integer" "integer"
```

We can make sure all variables are of the desired data class at the time they are read into R by using the `colClasses=` option within the `read.csv()` command. While data classes may be assigned to columns after data are read in, it is more convenient to do so at the time of read-in, using the `colClasses =` option. For instance, if all variables were factors, with none meant to be of character type or numeric (even if coded with numerals), then we could read the data in this way:

```
gssdata  <-  utils::read.csv("C:/Data/gssdata.csv", header=TRUE,
sep = ",", stringsAsFactors = TRUE, colClasses="factor")
```

Often, however, data are mixed in data type. This is the case for the example file "gssdata.csv", which is supplied on the Companion Website for this text. Below we read in the data, assigning columns their appropriate data classes. In this example, the first 6 columns are numeric, then 1 is factor, then 7 are numeric, then 5 are factors, then 4 are numeric, then 1 is factor, for a total of 24 columns. This is specified using the `rep()` replicate command nested within the `c()` combine command.

```
gssdata <- utils::read.csv("C:/Data/gssdata.csv", header=TRUE,
sep = ",", stringsAsFactors = TRUE, colClasses =
base::c(base::rep("numeric",6), "factor",base::rep("numeric",7),
base::rep("factor",5), base::rep("numeric",4), "factor"))
```

To verify the data classes of all columns we reissue the sapply() command, asking that it apply the class() command repeatedly to columns of the gssdata data frame.

```
base::sapply(gssdata, class)
[Output:]
     year         id     adults        age     attend
"numeric" "numeric" "numeric" "numeric" "numeric"
   babies      bible       born     childs      class
"numeric"  "factor" "numeric" "numeric" "numeric"
   degree       fund    gender1      happy     hhrace
"numeric" "numeric" "numeric" "numeric"  "factor"
 hispanic    marital    partyid      relig    respnum
 "factor"   "factor"   "factor"   "factor" "numeric"
   satfin        sex   vetyears    wrkstat
"numeric" "numeric" "numeric"   "factor"
```

Each row in a data frame is an observation. The sample size is therefore the number of rows:

```
base::nrow(gssdata)
[Output:]
   [1] 1785
```

Each column in a data frame is a variable. The `length()` command returns the number of variables and the `names()` command returns their names.

```
base::base::length(gssdata)
[Output:]
   [1] 24
```

```
base::names(gssdata)
[Output:]
 [1] "year"     "id"       "adults"   "age"      "attend"
 [6] "babies"   "bible"    "born"     "childs"   "class"
[11] "degree"   "fund"     "gender1"  "happy"    "hhrace"
[16] "hispanic" "marital"  "partyid"  "relig"    "respnum"
[21] "satfin"   "sex"      "vetyears" "wrkstat"
```

The `summary()` command is one of the simplest ways to inspect an object. The partial output below shows us that all data in gssdata are for the year 2012. For the numeric variable adults (in family), the minimum code is 1, the maximum is 6, and the median is 2. Mean age is 48.4 years. For the factor variable "bible", the frequency distribution is returned. Some 802 respondents are of code 2 on this variable. Above, we saw that gssdata has 1,785 observations but below we see its maximum is 1,973, meaning that some id numbers are skipped.

```
base::summary(gssdata)
[Partial output:]
      year            id               adults
 Min.   :2012   Min.   :   1.0   Min.   :1.000
 1st Qu.:2012   1st Qu.: 504.0   1st Qu.:1.000
 Median :2012   Median : 989.0   Median :2.000
 Mean   :2012   Mean   : 990.7   Mean   :1.885
 3rd Qu.:2012   3rd Qu.:1481.0   3rd Qu.:2.000
 Max.   :2012   Max.   :1973.0   Max.   :6.000

      age           attend           babies        bible
 Min.   :18.0   Min.   :0.000   Min.   :0.0000   1:590
 1st Qu.:34.0   1st Qu.:1.000   1st Qu.:0.0000   2:802
 Median :48.0   Median :3.000   Median :0.0000   3:367
 Mean   :48.4   Mean   :3.525   Mean   :0.1894   4: 26
 3rd Qu.:62.0   3rd Qu.:7.000   3rd Qu.:0.0000
 Max.   :89.0   Max.   :8.000   Max.   :4.0000
. . .
```

More descriptive statistics may be obtained by using the `describe()` command found in the "psych" package. After installing and loading this package, type `help(describe)` for an explanation of the different statistical coefficients listed as column headers. Note that factor variables like bible are flagged with an asterisk.

```
utils::install.packages("psych")
base::library(psych)
```

```
psych::describe(out.table <- gssdata, IQR = TRUE)
[Output:]
        vars    n    mean     sd median trimmed    mad
year       1 1785 2012.00   0.00   2012 2012.00   0.00
id         2 1785  990.66 566.22    989  990.84 724.99
adults     3 1785    1.88   0.83      2    1.78   0.00
age        4 1785   48.40  17.70     48   47.72  20.76
attend     5 1785    3.53   2.82      3    3.44   4.45
babies     6 1785    0.19   0.52      0    0.05   0.00
bible*     7 1785    1.90   0.77      2    1.86   1.48
. . .
         min  max range  skew kurtosis    se IQR
year    2012 2012     0   NaN      NaN  0.00   0
id         1 1973  1972  0.00    -1.20 13.40 977
adults     1    6     5  1.28     2.80  0.02   1
age       18   89    71  0.27    -0.82  0.42  28
attend     0    8     8  0.15    -1.43  0.07   6
babies     0    4     4  3.06     9.71  0.01   0
bible*     1    4     3  0.36    -0.67  0.02   1
. . .
```

We may be interested in the highest and lowest observations on some variable, such as age. This requires us to sort the data frame with the `order()` command, then view the top and bottom observations with the `head()` and `tail()` commands respectively. Of course, this assumes we sort descending, otherwise top and bottom are flipped. We see below that the top five cases are all age 89 and the bottom five are all age 18. The sorting variable was gssdata$age. The minus sign before it asked for a descending sort.

```
# Sort gssdata on age, descending, putting results in the data frame
"gssdataSorted".
gssdataSorted <- gssdata[base::order(-gssdata$age),]

# List the top five observations.
utils::head(gssdataSorted,5)
[Partial output:]
    year  id adults age attend babies bible born childs
200 2012 228      1  89      3      0     3    1      2
226 2012 256      1  89      7      0     1    1      8
231 2012 261      1  89      7      0     2    2      1
```

```
294 2012 334      1  89      2      0     3    1      1
369 2012 415      1  89      2      0     2    1      2
. . .

# List the bottom five observations
utils::tail(gssdataSorted,5)
[Partial output:]
     year   id adults age attend babies bible born childs
628  2012  702      3  18      2      0     3    1      0
828  2012  921      2  18      5      0     1    1      0
1251 2012 1388      1  18      0      0     2    2      0
1468 2012 1619      2  18      7      0     1    1      0
1522 2012 1679      3  18      3      0     3    1      0
```

Finally, we may also be interested in the value of a particular observation on a particular variable. Above, one of the youngest observations was observation 1522. This is the person with id = 1679. That person is in row 1522 of gssdata, the unsorted data frame (not in gssdataSorted). If we wanted to know this person's code on church attendance (the "attend" variable), we would ask for it as below. Note that the sequence in R is always row first, then column (the variable). We may ask for the "attend" column by its column number or by its name. The format is data frame name, begin bracket, row, column, end bracket.

```
gssdata[1522,5]
[Output:]
[1] 3

gssdata[1522,"attend"]
[Output:]
[1] 3
```

By way of summary, some useful data inspection commands in R are listed:

```
class(mydata)         # Shows data class (data frame, matrix, numeric,
                      etc.)
dim(mydata)           # Shows dimension of mydata, here 10 rows,2
                      columns
head(mydata,10)       # Print the first 10 rows of mydata (if 10
exist)
ls()                  # Lists objects in the environment, like mydata
mode(mydata)          # Shows the data type of mydata
names(mydata)         # Shows the variable names in mydata
utils::str(mydata)    # Shows structure of mydata in brief format
summary(mydata)       # Displays descriptive statistics for mydata as
                      shown earlier
tail(mydata, n=10)    # Print the last 10 rows of mydata
View(mydata)          # Displays mydata as a spreadsheet-style table;
                      capitalize "View"
```

R data structures

R supports many complex data structures, which is one reason for its popularity. A common novice problem in using R for statistics is error messages arising from using the wrong class of data object for the selected statistical command. For purposes here, however, we discuss only the very basics. In general, it is a good idea to type `class(my-object)` to be sure you know the data `class of "myobject"`. Data frames, tibbles, vectors, factors, and other data structures are all "objects" in R parlance.

Data frames and tibbles

A "data frame" is what social scientists usually think of as a dataset. Each row is an observation. Each column is a variable, which may be a numeric vector, a factor (a categorical variable), a date, or other data types. Whatever the type, each has the same number of observations, thus forming a rectangular dataset. The data frame is the most common structure for storing data in R, which calls it a "data.frame". Note that a data frame is a type of matrix. Cells in the matrix may be retrieved by adding "[rownumber, columnnumber]" to the data frame object name, as discussed in Section 1.8. Thus:

```
# List the contents of a cell at row 1, column 1
mydata[1,1]
[Output:]
   [1] 4

# List the contents of row 1
mydata[1,]
[Output:]
  happy gender
1     4      0

# List the contents of column 1
mydata[,1]
[Output:]
   [1] 4 1 3 5 3 3 2 4 5 1
```

A "tibble" is a newer type of to a data structure, providing enhanced functionality compared to traditional data frames. A traditional data frame is one type of tibble. Most operations one can do on data frames can be done on tibbles, but not necessarily vice versa. Tibbles may be conceived as a superset of data frames. Sometimes they are called "tibble data frames". When the `class()` command is used on a tibble object, one may something like that below. This object is of three types simultaneously: tibble data frame, tibble, and data.frame. The mytibble object may be used with programs that require any of these three types of input. For further discussion of tibbles, see https://cran.r-project.org/web/packages/tibble/vignettes/tibble.html

```
install.packages("tibble")
library(tibble)
mytibble <- tibble::as_tibble(mydata)
base::class(mytibble)
```

```
[Output:]
   [1] "tbl_df"     "tbl"        "data.frame"
```

Vectors

A "vector" is what social scientists usually think of as a variable, corresponding to a single column in a dataset. Usually "vector" means that it is a numeric vector, containing numeric data. However, a vector may also be a character vector. This is illustrated by the simple examples below

```
x <- base::c(1,2,3)
x
[Output:]
   [1] 1 2 3
base::class(x)

[Output:]
   [1] "numeric"

x <- base::c("1","b","3")
base::class(x)
[Output:]
   [1] "character"
```

Occasionally one needs to know to be sure that a variable is classed as an integer. An integer type has only whole numbers, in contrast to the default numeric vector type, which is double precision. To force a numeral to be read as an integer, add "L":

```
x <- base::c(1L, 2L, 3L)
# x is of data class integer
base::class(x)
[Output:]
   [1] "integer"

# Therefore the first element in x is an integer.
base::is.integer(x[1])
[Output:]
   [1] TRUE
```

Factors

A "factor" is what social scientists usually think of as a categorical variable, corresponding to a column in a dataset, containing alphanumeric data. It is different from a character vector. A character vector is simply a listing of words or characters (some or all of which may be numerals). Sometimes character vector elements are called "strings". In contrast, a factor has categories, called levels. The R program you are using may give an error message if you try to use a character vector instead of a factor, even though the contents "look the same" to you. Fortunately, a character vector like "party" below can be coerced into being a factor like "partyFactor".

```
party <- base::c("Republicans", "Democrats", "Independents")
base::class(party)
[Output:]
   [1] "character"

partyFactor <- base::as.factor(party)
base::class(partyFactor)
[Output:]
   [1] "factor"
```

The command below converts all character variables in a data frame called "df" to factors.

```
df <- gssdata

df[base::sapply(df, is.character)] <- base::lapply(df[sapply(df,
is.character)], as.factor)
```

It is also possible to separate a data frame into its atomic vectors, which may be numeric, character, or logical (illustrated further below). The `as.integer()` and `as.factor()` functions are examples. Once created in this way, note that simple names are used in commands. For instance, there is no mydata$happyInteger, only happyInteger.

```
happyInteger <- base::as.integer(mydata$happy)
genderFactor <- base::as.factor(mydata$gender)
```

Many arithmetic operations can be performed on atomic vectors.

```
base::sum(happyInteger)
[Output:]
   [1] 31
```

However, what works for a numeric vector does not necessarily work for a factor:

```
base::sum(genderFactor)
[Output:]
    Error in Summary.factor(c(1L, 1L, 1L, 1L, 1L, 2L, 2L, 2L, 2L,
    2L), na.rm = FALSE) : 'sum' not meaningful for factors
```

The `is.vector()` function determines if an object is a numeric vector. The `is.factor()` function does the same for character vectors. The commands and output below assume happy and gender are numeric and character atomic vectors respectively and are also embedded in the data frame mydata. The embedded versions return "FALSE" because they are not atomic vectors. The `is.data.frame()` function checks if an object is a data frame.

```
# These three commands return TRUE
base::is.vector(happyInteger)
base::is.integer(happyInteger)
base::is.factor(genderFactor)
```

```
# These three commands return FALSE
base::is.factor(happyInteger)
base::is.factor(mydata$happy)
base::is.factor(mydata$gender)
```

We see that although data frames (e.g., mydata) may contain numeric or factor data, they are not vectors or factors in data type until put on an atomic basis by a command such as this:

```
genderFactor2 <- mydata$gender
base::class(genderFactor2)
[Output: The mydata$gender variable is treated as integer because it
was coded 0, 1]
   [1] "integer"

genderFactor3 <- base::as.factor(mydata$gender)
base::class(genderFactor3)
[Output: genderFactor3 is a factor because we used the as.factor()
function to coerce mydata$gender into status as a factor]
   [1] "factor"
```

It is also possible to go the other way, creating a data frame from atomic vectors or factors. This was done earlier when using the `data.frame()` command as one way of creating mydata.

Lists

The list data format allows mixing of data types such as vectors, factors, and logical vectors. However, due to this mixing, where one may use `View()` with data frames, `View()` does not work with lists. Below we create a list called "mylist" and demonstrate some of its properties.

```
# Create a numeric atomic vector called mpg:
mpg <- base::c(28, 25, 30, 34)

# Create a character atomic vector (factor) called brand:
brand <- base::c("Ford","Chevrolet","Honda", "Mazda")

# Create a logical atomic vector called owned:
owned = base::c(TRUE, FALSE, FALSE, TRUE)

# Combine the atomic vectors in a list called mylist
mylist <-  base::list(mpg, brand, owned)

# Optionally, remove the original atomic vectors from the R
environment:
base::rm(brand,mpg,owned)
```

```
base::summary(mylist)
[Output: ]
        Length Class  Mode
   [1,] 4      -none- numeric
   [2,] 4      -none- character
   [3,] 4      -none- logical

# To better read the summary output, give names to list components.
# Make sure to use the same order as in the list() command above.
base::names(mylist) <- base::c("mpg", "brand", "owned")

# Display named contents of mylist:
base::summary(mylist)
[Output: ]
         Length Class  Mode
   mpg   4      -none- numeric
   brand 4      -none- character
   owned 4      -none- logical

# We can print the contents of individual list elements like brand.
base::print(mylist$brand)
[Output: ]
   [1] "Ford"      "Chevrolet" "Honda"     "Mazda"
```

The `lapply()` function in base R provides certain statistics repetitively. Below, we get the mean of all variables. However, as brand is of character data type, we cannot compute a mean and an error message is returned. For a logical variable like owned, a mean is returned using the numeric equivalents of 0 = FALSE and 1 = TRUE. It is 0.5 because half the cases were FALSE and half were TRUE.

```
base::lapply(mylist, mean)
[Output: ]
    $mpg
       [1] 29.25

    $brand
       [1] NA

    $owned
       [1] 0.5

    Warning message:
    In mean.default(X[[i]], ...) :
      argument is not numeric or logical: returning NA
```

Most statistical commands will want the data to be in a data frame, not a list. We may convert the mylist list to a data frame called myframe. This assumes all list components are atomic vectors, which are numeric (synonym: double), character, logical, integer, complex, or raw in data type.

```
myframe <- base::as.data.frame(mylist)
utils::View(myframe)
base::class(myframe)
[Output: ]
   [1] "data.frame"
```

Changing from one data type to another

As we have seen above, in many cases it is possible to coerce data of one class to be of another class. Below, for instance, we take the tibble data frame "mydata" and create the matrix "mydata_matrix" object.

```
base::class(mydata)
[Output: ]
    [1] "tbl_df"     "tbl"        "data.frame"

mydata_matrix <- base::as.matrix(mydata)
base::class(mydata_matrix)
[Output: ]
   [1] "matrix" "array"
```

Selected conversion functions in the "base" package in the R system library are listed below. Other packages such as the "data.table" package have other conversion functions as well. For more information on these conversion functions, type `help(name of function)`, as in `help(as.Date)`.

Function	*Convert to*
`as.character()`	Character vectors
`as.character.Date()`	Date conversion to and from character
`as.data.frame()`	Coerce to a data frame
`as.data.frame.Date()`	Date class
`as.data.frame.table()`	Cross-tabulation and table creation
`as.Date()`	Date conversion functions to and from character
`as.difftime()`	Time intervals/differences
`as.double()`	Double-precision vectors
`as.factor()`	Factors
`as.integer()`	Integer vectors
`as.list()`	Lists—generic and dotted pairs
`as.logical()`	Logical vectors
`as.matrix()`	Matrices
`as.matrix.noquote()`	Class for "no quote" printing of character strings
`as.numeric()`	Numeric vectors
`as.ordered()`	Factors
`as.table()`	Cross-tabulation and table creation
`as.vector()`	Vectors

Handling missing values

Most real-world data have missing values. R has many packages for treating missing values. Handling missing values is treated in Garson (2022) and many other texts. Though the topic is very large we can discuss some highlights here. To illustrate, we use the file "cobb_survey.csv", also available on this book's website. Using this file, we create the data frame "cobb". The cobb data frame contains survey responses of 338 individuals (rows) on 88 items (columns).

```
cobb <- utils::read.csv("C:/Data/cobb_survey.csv", header=TRUE,
sep = ",", stringsAsFactors = TRUE)
```

Missing values in R are coded NA. If data have been imported from SPSS or some other package using different missing value codes, the researcher must convert to NA coding. Once missing values are properly coded NA, one may check for rows with missing data.

The `summary()` command displays descriptive statistics for all columns. If any column has missing values, such as "rvpos" below, the NA count is given.

```
base::summary(cobb)
[Partial output:]
       id               debwatch             rvpos
 Min.   :  1.00    Min.    :1.000    Min.    :1.000
 1st Qu.: 85.25    1st Qu.:1.000    1st Qu.:1.000
 Median :169.50    Median :1.000    Median :2.000
 Mean   :169.50    Mean    :1.429    Mean    :1.592
 3rd Qu.:253.75    3rd Qu.:2.000    3rd Qu.:2.000
 Max.   :338.00    Max.    :2.000    Max.    :2.000
                                     NA's    :76
```

The more complex command below gives a simple count of missing values by variable.

```
na_count  <- base::sapply(cobb, function(y)
base::sum(base::length(base::which(base::is.na(y)))))


na_count
[Partial output:]
          id      debwatch         rvpos       rposvst         gsupv
           0             0            76            64           168
       bsupv      rposuhc      rposuhcs        rposdp      rstposdp
         144            32            34            43            33
. . .
```

It is also possible to get a long listing of all rows (observations) with at least one missing value. For the Cobb data, this is most of them. In syntax below, the exclamation point in R means "not", so we are checking for rows which are not complete cases, meaning rows which have missing data.

```
cobb[!complete.cases(cobb),]
[Partial output:]
   id debwatch rvpos rposvst gsupv bsupv rposuhc rposuhcs
1   1        2     2       1     1     2       1        1
2   2        1     1       1    NA     2       1        1
3   3        2     2      NA     2     1       1       NA
4   4        2     2       1    NA    NA       1        1
5   5        1    NA       2     2    NA       1        1
6   6        1     1       1     2     1       2        2
. . .
```

If missing data are found, one may create a new dataset in which rows with missing data have been dropped listwise. However, imputation of missing values is usually preferred. Data imputation is discussed in Garson (2022) and many other texts. For the cobb data, we see below that 330 of the 338 observations are dropped, leaving only 8! This is clearly unacceptable and might prompt the researcher to go back and retain only variables with few missing values.

```
cobbComplete <- stats::na.omit(cobb)

base::nrow(cobbComplete)
[Output:]
   [1] 8
```

We now turn to the question of what to do if imported data have missing values but do not use the "NA" code for them. SPSS, for instance, allows user-created missing values such as −99. If data are exported from SPSS or another package, R may not recognize the ways these other packages specify missing values. In R, missing data must be coded NA, without quotes.

We illustrate how to handle this problem for SPSS as an example. In SPSS, we change all system-missing or user-missing values to some numeric value, such as minus 99. To do this, in SPSS select Transform > Recode into the same variables. In the Transform dialog, move all numeric variables into the variable list. Click Old and New Values. Check "System-missing or user-missing". Let the "New Value" be −99. Click Add. Click Continue, then OK. Only then export to .csv, which SPSS can read: File > Export > CSV Data. Save under some name such as "mytest.csv".

Then in R, import mytest.csv as exported by SPSS. If exported from Excel, the .csv file will not have these problems. With export from Excel, blanks are converted to NA's and there is no naming glitch as described below.

Note: mytest.csv is supplied with this text. It can be created by the user in SPSS. Syntax below is illustrative but your data will differ. The mytest object is like mydata, but with a couple -99 codes in the variable "happy".

Steps in R:

```
# 1. Import mytest.csv into R in the usual way.
     base::setwd("C:/Data")

     mytest <- utils::read.csv("C:/Data/mytest.csv", header=TRUE,
     sep = ",", stringsAsFactors = TRUE)

     # 2. Change -99s to NAs

     # Before: No missing are reported because -99 not recognized as
     a code for missing.
     base::is.na(mytest)
              happy  gender
       [1,]   FALSE  FALSE
       [2,]   FALSE  FALSE
       [3,]   FALSE  FALSE
       [4,]   FALSE  FALSE
       [5,]   FALSE  FALSE
       [6,]   FALSE  FALSE
       [7,]   FALSE  FALSE
```

```
 [8,]    FALSE  FALSE
 [9,]    FALSE  FALSE
[10,]    FALSE  FALSE

# Replace -99's with NA. Note changes are not saved to file un-
til you do so explicitly.
mytest[mytest == -99] <- NA

# After: Missings are now recognized.
base::is.na(mytest)
         happy  gender
 [1,]    FALSE  FALSE
 [2,]    FALSE  FALSE
 [3,]    FALSE  FALSE
 [4,]    FALSE  FALSE
 [5,]     TRUE  FALSE
 [6,]    FALSE  FALSE
 [7,]    FALSE  FALSE
 [8,]     TRUE  FALSE
 [9,]    FALSE  FALSE
[10,]    FALSE  FALSE
```

Replacement may also be done with the "naniar" package. This command replaces a missing value code for data from an external source with NA, which R wants. Below we replace −99 values with NA values. (The xdf data does not have any −99 values, so here this just checks). The piping operator (|>), available since R version 4.1, is used in the code below. (Alternatively, the older "%>%" the piping operator from the "dplyr" package may be used.)

```
mytest2 <- utils::read.csv("C:/Data/mytest.csv",
header=TRUE, sep = ",", stringsAsFactors = TRUE)

utils::install.packages("naniar")
base::library(naniar)
mytest2 <- mytest2 |>
naniar::replace_with_na_all(condition = ~.x == -99)

# Check result, which now has NA coding.
mytest2
[Output:]
    # A tibble: 10 x 2
       ï..happy gender
          <int>  <int>
    1         4      0
    2         1      0
    3         3      0
    4         5      0
    5        NA      0
    6         3      1
```

```
      7          2       1
      8         NA       1
      9          5       1
     10          1       1

# 3. A known glitch in SPSS csv export will/may mislabel the first
variable
#The code below illustrates and corrects the name of the first
column.
base::names(mytest)
[Output:]
    [1] "ï..happy" "gender"

base::names(mytest)[1]<-"happy"

base::names(mytest)
[Output:]
   [1] "happy"   "gender"

# 4. Then write the corrected data frame to a .csv file, possibly
with a new name
utils::write.csv(mytest, file = "mytest2.csv", row.names = FALSE)
```

The `freq.na()` command from the "questionr" package will print a frequency table of missing values, such as those for the cobb data frame.

```
utils::install.packages("questionr")
base::library(questionr)
questionr::freq.na(cobb)
[Partial output:]
             missing %
. . .
defense          105   31
rposdef          104   31
pid2pt            92   27
rstdef            81   24
vouchers          77   23
. . .
```

Finding useful packages to install

Packages in the R System library do not need to be installed but others do. One may browse the R System library by clicking on the "Packages" tab in RStudio. At the top of the listing will be the "User Base::library", which is the packages you have installed thus far. Scroll below this and you will see the "System library", which is a list of R's "built-in" packages such as base, utils, foreign, and graphics. Though no package in the System library requires installation, only packages with a check in the check box are loaded. Checking an empty check box issues the `base::library()` command for that package, loading it.

How do people find appropriate R packages to use? All kinds of ways. Here are a few ideas:

- *Browser search.* Because the R community is so large, the probability of finding a useful package in your area of interest is high. For example, simply type "cluster analysis in R" (with quotes to keep these terms together) to find listings for this subject. At this writing, there were over a quarter million hits for this popular topic! Many hits link to articles which not only identify useful packages but also provide R code for using them.
- *Article search.* Research articles that you encounter in conducting your literature review for your research project may cite R packages which have proven to be useful to other researchers in your field. R syntax (code) for using the package may be available as a supplement to the article or by writing to the authors.
- *Archive search.* While CRAN is not the only archive of R packages, it is the largest. One can find a listing of packages by name or date at https://cran.r-project.org/web/packages/. At this writing, there were over 19,000 packages available on CRAN, listed alphabetically by name or by publication date. A second major website is that of the R-Forge project at https://r-forge.r-project.org/.
- *Blog search.* While there are a great many R-oriented blogs, a notable one is r-Bloggers at https://www.r-bloggers.com/. This blog republishes blog posts of others and publicize popular packages, including code to use them. Every month it lists the "top 40" new R packages. RPubs (https://rpubs.com/) lists recent interesting examples. There is also a search box on the home page to search by keyword.
- *Ask on a list.* A popular list which may be used to ask about new packages is stackoverflow (https://stackoverflow.com/), which also maintains an archive of past questions and answers. Register for free membership or just click "search content" under the "Join the community" button on the home page. Another list is R-Help (https://stat.ethz.ch/mailman/listinfo/r-help), for which a free subscription is required.
- *Use the "sos" package* described below to search by keyword, such as "reliability".

Unfortunately, R commands (programs) may be found on websites or in articles without mention of the name or names of the required packages. More confusingly, sometimes the same command term may be available in more than one package, not always with the same functionality. For instance, there are 11 packages which use the command `reliability()` for differing purposes. In some cases, it may be necessary to add a package prefix when invoking the command (e.g., `psych::reliability(…)`). Using the `help()` command may suffice in identifying the package containing a command of interest (e.g., `help("reliability")`). If one is aware of a command name one wants to use but is unsure which package contains it, or if one simply wants to search for functions containing a keyword, there is an "sos" package to help. This package contains the `findFn()` function. For instance, to find all packages referring to "reliability":

```
utils::install.packages("sos")
base::library(sos)
findFn("reliability",maxPages=1000,sortby="Function")
```

Results appear in one's browser. The "findFn Results" page contains a table listing hundreds of reliability-related links. Look under the "Function" column to find the name of a command, then look across in the "Packages" column to see the package it comes from. For instance, the `reliability.plot()` command is in the "verification" package. Click on the link in the last column (the Description and Link column) of this row to see more information about this command.

Installing packages

Before trying to install a package, one may want to see if it is already installed. Let's say one is considering installing the "lme4" package, which is for linear mixed modeling (multilevel modeling). Simply use the `base::library()` command and if there is no error message, it has been installed.

```
# The package does not exist:
base::library(lme8)
[Output:]
    Error in base::library(lme8) : there is no package called 'lme8'

# When the package exists but is not installed:
base::library(lme4)
[Output:]
    Error in library(lme4) : there is no package called 'lme4'

# When the package is exists and may be installed:
utils::install.packages("lme4")
base::library(lme4)
[Output - only the prompt sign:]
>
```

Alternatively, one may use the "Packages" tab in the RStudio interface:

1 In the "Packages" tab search box, enter the name of the desired package.
2 If the package does not appear, the user has never installed it. Click the "Install" tab, enter the name of the package, then click the "Install" button to install it.
3 If it appears but the check box is blank, it has been installed but is not loaded. Enter a check in the checkbox and the base::library() command will be issued automatically.

One may use the `getoption()` command to see packages which are loaded automatically when R starts up. These are all in the R system library.

```
base::getOption("defaultPackages")
[Output:]
   [1] "datasets" "utils" "grDevices"
   [4] "graphics" "stats" "methods"
```

There are other packages in the R System library which do not need to be installed but are not loaded automatically. They are listed at the bottom of the listing from the "Packages" tab in RStudio in the "System library" section. One, for example, is the "cluster" package, which has the built-in dataset "animals".

```
# Before loading the cluster package, we cannot access animals,
# even though cluster is in the R System library.
utils::head(animals)
[Output:]
   Error in head(animals) : object 'animals' not found

# We load cluster and then can access animals.
# Alternatively, in the "Packages" tab enter a check in the checkbox
for the cluster program.
base::library(cluster)
utils::head(animals)
[Output:]
    war fly ver end gro hai
ant   1   1   1   1   2   1
bee   1   2   1   1   2   2
cat   2   1   2   1   1   2
cpl   1   1   1   1   1   2
chi   2   1   2   2   2   2
cow   2   1   2   1   2   2
```

Optionally, one may view all currently installed packages, including their version numbers, dependencies, and other information using the `installed.packages()` function, also part of the "utils" package in the R System library.

```
utils::installed.packages()
[Output is too long to list here]
```

As another installation consideration, in the rare event that there are conflicts, the `base::library()` command will give an "Error in unloadNamespace" error. This error message means that the one package is trying to unload and install a given package but it can't because one or more other packages already have the given package as a dependency. The solution is to close R and reload with the package of primary interest loaded first.

Note that sometimes the unloadNamespace error cites a package you did not recall installing. The cited package is a dependency of a package which you did load. For instance, gower is a dependency of recipes which is a dependency of caret. The gower package may be cited in the error but you only recall loading caret. The remedy is the same as above: restart and issue the `base::library()` commands again. You may have to load gower manually by the command `install.packages("gower")`, then restart and issue the `base::library()` commands again.

When loading a package, you may get conflict warnings such as that below.

```
base::library(caret)
[Output:]
   Loading required package: lattice
```

```
Attaching package: 'caret'
The following object is masked from 'package:purrr':
   lift
```

This means that both lattice and purr have the command `lift()`. If lattice was loaded most recently, then its version of `lift()` will be used. However, this can be overridden by using package prefixes. The command `purrr::lift()` assures that the purr version is used. Using package prefixes routinely may be easier than keeping track of such conflicts.

In summary:

- If a package is not in the R System library, it must be installed. For packages in the CRAN archive (most packages), this is done with the `install.packages()` command. The package name within the parentheses must be within quote marks. Documentation for non-CRAN packages will contain different installation instructions.
- Packages not on the R System library default list must then be loaded using the `base::library()` command.
- On loading a package, take note of error messages and warnings described above.
- Installing a specified package may install automatically other packages on which it depends ("dependencies").
- Once a package is installed and activated, one may type `help(name-of-command)` to get details about any of the package's commands and options.

Alternative to command-line installation as above, it is also possible in RStudio to click on the "Install" button for the "Packages" tab and an "Install Packages" window will appear, as shown in Figure Appendix 2.3. In this figure, "rel" is the name of the package to be installed. Note that the default for the drop-down "Install from" text box is

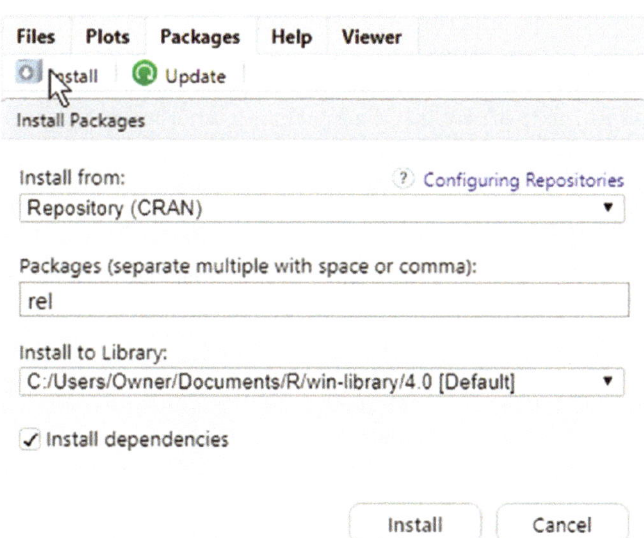

Figure Appendix 2.3 Install packages window in RStudio.

"Repository (CRAN)". While unusual, the alternative is "Package Archive File", used to install packages downloaded from the Internet in archived (.zip, .tar, .gz) formats. If this is the case, in the same dialog screen browse to and select the file address when you saved the archive (usually your download directory) and enter it in the "Package Archive" text box of RStudio.

Updating packages

Checking R and package version numbers

For update purposes and because of possible incompatibilities, occasionally the researcher may need to know the version number of a package and/or the version number of R itself. To get the version number of R, simply type `version`. At the time this command was issued, we were running R Version 4.1.3 on a Windows computer with 64-bit architecture:

```
version
[Output:]                    _
    platform       x86_64-w64-mingw32
    arch           x86_64
    os             mingw32
    system         x86_64, mingw32
    status
    major          4
    minor          1.3
    year           2022
    month          03
    day            10
    svn rev        81868
    language       R
    version.string R version 4.1.3 (2022-03-10)
    nickname       One Push-Up
```

To see the version number of any installed package, whether active or not, click on the "Packages" tab in RStudio and look in the "version" column.

Updating R and RStudio

When R commands are run, a warning may appear that the package does not support one's version of R. When a new version of R appears, the user is not notified. Instead, it is necessary to check frequently.

Installing RStudio will also install the latest version of R. Note that RStudio does not automatically update. To see the version of RStudio currently in use, from the RStudio menus, select Help > About Studio. To see the newest version of RStudio available, this is listed at and may be downloaded from https://rstudio.com/products/rstudio/download/.

Alternatively, R itself may be updated using the "installr" package. While there are other methods, installr makes the process easy. Issue the commands below and follow the prompts. Afterwards, close RStudio and relaunch it, and it will list the new version of R being in use.

```
utils::install.packages("installr")
base::library(installr)
installr::updateR()
```

Updating installed packages

The code for a task typically starts with a series of `base::library()` commands which load the packages needed for the task. Before undertaking the task, it is a good idea to see if any of these packages need updating. This can be done using the "Update" button as shown in Figure Appendix 2.3. An alphabetical list will pop up of installed packages needing updates. Check the ones that your task will be using, then click "Install Updates". The same "Update Packages" window will come up if in the RStudio menu, you select Tools > Check for Package Updates. It is good practice to do this at the start of any project.

Sometimes you will want to update specific packages. One reason would be getting an error message such as "Error: package or namespace load failed for 'caret' .... namespace 'rlang' 0.4.5 is already loaded, but >= 0.4.7 is required." That is, the rlang package needs updating. The most reliable update method is to close and restart R and then before any other command is issued, in RStudio click the "Update" button which is located under the "Packages" tab. A list will appear: select the ones wanted or click "Select All" (could be time-consuming), then click "Install Updates". This function is also found under Tools > Check for Package Updates in the RStudio menu.

To remove a package altogether (will require re-installation if needed later), use the `remove.packages()` command.

```
utils::remove.packages("installr")
```

Using, saving, and loading packages and sessions

Using R keyboard shortcuts: From the RStudio menu, select Help > Keyboard Shortcuts Help (or type Alt-Shift-K) to view a listing of keystrokes used for scrolling and editing (e.g., the Esc key aborts the R interpreter and returns to the R prompt (to ">"). Keep in mind that R is case-sensitive (e.g., Ctrl + L is not the same as Ctrl + l).

Setting the working directory: An early step is to set the working directory. This declares where files are to be loaded or saved. For example, `setwd("C:/Data")` sets the "Data" folder as the location from which to load or to which to save files. Note also that a forward slash is used in the path name. The `getwd()` command lists the current working directory.

Clearing RStudio: The work environment may become cluttered in the course of a project. The researcher can start fresh by selecting Session > New Session from the RStudio menu. However, re-running RStudio for the same session will not accomplish this as the saved environment is reinstated. Instead, the researcher may consider using the following commands:

Clear the console: Ctrl-L – This will clear the console window in the lower left of RStudio, which is where commands are entered and where output appears. However, command history is not cleared. The up/down arrows may still be used to scroll through past commands. This is equivalent to selecting Edit > Clear Console from the RStudio menu.

Clear all datasets from the environment: Select the "Environment" tab in RStudio and click the broom icon. Alternatively, select Session > Clear Workspace. As a third alternative, issue the command `rm(list=ls())`. All objects are removed from the "Global Environment" in the upper right of RStudio.

Clear history: Select the "History" tab and click the broom icon. This will clear the history of all commands. The up/down arrow keys will no longer have effect until you enter more commands.

Clear datasets shown in upper left "View" window: Click the "x" option for each dataset to close it. When RStudio is closed, answer "yes" when queried whether you wish to save the current environment. When RStudio is run again, it will not list the datasets that were closed. Datasets are not removed from saved files on disk. To clear a specific dataset use the `rm()` command. For example: `rm(mydata_df)`. To clear all datasets, use `rm(list=ls())`

Clear plots: Select the "Plots" tab and click the broom icon.

Clear workspace: Select Session > Clear Workspace, which removes all objects from the R environment.

Saving and loading the workspace: The workspace is the current R working environment, including any objects the researcher has created, such as data frames and functions. At the end of any R session, the researcher is to be prompted to save the workspace. Then the next time R is started, the saved workspace will be reloaded automatically. The workspace is saved in the current working directly with the default name ".RData" but the researcher can assign it a unique filename if desired.

Saving and loading the command history: R keeps a log of commands that have been issued. Use the up and down arrow keys to recall them, then click the Enter key to re-issue the selected previous command. When an R session is quit, the command history is lost unless saved. Below, the working directory is set, two commands used elsewhere in this tutorial are given, the `savehistory()` command is given, and the R session is terminated. History will be saved to "C:/Data/.RHistory".

```
base::setwd("C:/Data")
utils::savehistory()
base::quit()
```

When R is restarted, history will be empty so the up/down arrow keys will not scroll through previously used commands until the `loadhistory()` command loads the saved history of commands. The `history()` command displays these commands in the "History" tab of RStudio.

```
base::setwd("C:/Data")
utils::loadhistory()
history()            # Switches to display the history tab in RStudio.
```

It is also possible to save or load under specific filenames. Files go to the current working directory.

```
utils::savehistory(file="Tutorial2.Rhistory")
utils::loadhistory(file="Tutorial2.Rhistory")
```

Saving the RStudio session: The researcher may wish to save and even share an RStudio session. Any of the techniques listed below may be used.

Cut and paste commands and output. The researcher can drag the mouse over a section of output, which will include commands too. Then type "Ctrl-C" to copy, go to Word or some other program, type "Ctrl-V" to paste.

To save figures, go to the Plots tab and click the down arrow for "Export", and select one of the choices there. Choices are Copying to Clipboard, Save as Image (any of seven graphics formats), or Save as PDF.

To save commands without output, go to the "History" tab in RStudio and click on the save icon (a computer disk). The list of all commands will be saved to a directory and filename set by the researcher, with the file extension ".RHistory". To save just a portion of history, go to the "History" tab and block out what is wanted. Click on the first line, scroll down to the last line, "Shift-click" to multi-select down to the last line, then cut and paste. The researcher can also save history using the `history()` command and can retrieve it with the `loadhistory()` command. Files must have a .Rhistory extension.

Save output with `sink()`. The `sink()` command from R's "base" package can redirect output (and only output, not prompts, commands, comments, warnings, or plots) to a file. Output is not sent to the screen until the researcher reverts output back as shown below. This makes the `sink()` method useful if one wants output only and if one has a prepared batch of commands to enter in the syntax section illustrated below. Note: the `capture.output()` command uses `sink()` and is very similar. The `file.show()` command displays the sunk output, which is a text file in the working directory. For example: `file.show("myoutput_sink.txt")`. Type `help(sink)` for more information and examples.

Saving everything with RMarkdown. RMarkdown is the most versatile way to save R work. It can save code, output, graphics, and more. It can save to html, Word, pdf, and other formats. What is saved can be "live" such that viewers on the web could run the researcher's program with the same or other inputs. Scholars have written entire books in RMarkdown. This impressive versatility comes at the price of a nontrivial learning curve, however.

There are many available tutorials on using RMarkdown. A few are:

- "RMarkdown from RStudio: An Introduction", from RStudio, at https://rmarkdown.rstudio.com/lesson-1.html. This is the "Getting Started" button on the RStudio RMarkdown page at https://rmarkdown.rstudio.com/.
- "Getting Started with RMarkdown", from the Coding Club at https://ourcodingclub.github.io/tutorials/rmarkdown/
- "RMarkdown: The Definitive Guide", by Yihui Xie, J. J. Allaire, & Garrett Grolemund, at https://bookdown.org/yihui/rmarkdown/
- "RMarkdown Cheat Sheet", from RStudio, at https://www.rstudio.com/blog/the-r-markdown-cheat-sheet/

Visualization and graphics in R

Data visualization is one reason researchers may choose R as their preferred statistical environment. This is an enormous area outside the scope of this tutorial. However, a great starting point is the website http://r-graph-gallery.com/. It presents hundreds of different graph types with examples and illustrative working code. Even mapping and animated graphs are treated. It illustrates visualization with many packages, including base (has `plot()`), graph, graphics (has `barplot()`), igraph, leaflet, and ggplot2.

The most widely used package for data visualization is "ggplot2", which was used for many figures in this textbook. However, the versatility of ggplot2 may seem daunting to beginners. Some free tutorials on ggplot2 are found at https://www.tutorialspoint.com/ggplot2/index.htm and http://r-statistics.co/Complete-Ggplot2-Tutorial-Part1-With-R-Code.html to name just two. There are also many video tutorials on ggplot2, found by searching YouTube, Bing Videos, or similar websites.

The "esquisse" package supports creation of attractive plots automatically through drag and drop methods in R. In a browser, type "r esquisse tutorial" (without the quotes) for a variety of tutorials.

```
utils::install.packages("esquisse")
base::library(esquisse)
# Sample data
utils::data(USArrests)
# Launch the program in a browser, select graph options, and save
the R code
esquisse::esquisser()
```

Data management basics

This section deals with selected data tasks common in social science research. For example purposes, we use the "mydata" object created earlier:

```
base::setwd("C:/Data")
mydata <- utils::read.table("mydata.csv", header = TRUE, sep = ",")
```

We start by making a copy called "xdf" and viewing it. Making a copy safefguards the original file.

```
xdf <- mydata
View(xdf)
```

Data frame, delete a variable (column)

The syntax below deletes the first column in a data frame. In most cases, columns correspond to variables. Find column numbers by looking at the data in View or by using the `names(xdf)` command. The desired column may be referenced by column number or by variable name.

```
xdf[1] <- NULL
```

or

```
xdf["happy"] <- NULL
```

Restore xdf:

```
xdf <- mydata
```

Data frame, rename a variable in

This command renames column 2 (gender) to become sex. Then the data frame is restored to its original.

```
base::colnames(xdf)[2] <- "sex"
View(xdf)
```

Restore xdf:

```
xdf <- mydata
```

Dummy variables, creating

Sometimes the researcher wishes to convert all categorical variables to sets of dummy variables. Below, "world" is the original data frame. The "worldDummy" object is the new data frame with dummy variables. The last level will be the omitted reference level. For example, infdeaths in world is character-coded as Low or High. In worldDummy, there is only infdeaths.Low, with the High category being the reference category. Type names(worldDummy) to see all variable names.

```
# Set the working directory
base::setwd("C:/Data")

# Read in the text file called world.csv into a data object called
"world"
world <- utils::read.csv("C://Data/world.csv", header=TRUE, sep
= ",")

# Make a copy of world called xdf2
xdf2 <- world

# Then create dummy variables. Object worldTypcDummyvars is of data
class"dummyvars".
# The "~." specification asks for all variables in xdf2.
utils::install.packages("caret")
base::library(caret)
worldTypeDummyvars <- caret::dummyVars(" ~ .", data =
xdf2,fullRank = T)

# worldDummyDf is a data frame (like world, but with dummy
variables)
worldDummyDf <-
base::data.frame(stats::predict(worldTypeDummyvars, newdata =
xdf2))
```

Where the world data frame had one column for the variable labeled "country", the worldDummy data frame has separate dummy variable columns for each country except the last (reference) country. The United States is now the dummy variable countryUSA, for instance. It has also converted to dummy variables the former region, infdeaths, and litgtmean variables. The infdeaths variable is now the dummy variable infdeathsLow, with the left-out High category being the reference.

```
base::names(worldDummyDf)
[Output not shown]
```

We can use the `table()` command on the original data (world) to see the levels of a categorical variable like infdeaths, and also the number of observations in each level. This also shows the order of the levels, here High = 0 and Low = 1.

```
base::table(world$infdeaths)
[Output:]
High  Low
  71  141
```

The sum of the corresponding variable in the dummy version of the data is the same count as the "1" category – here 141 observations are "Low" on infdeaths and this is saved in the dummy variable infdeathsLow.

```
base::sum(worldDummyDf$infdeathsLow)
[Output:]
   [1] 141
```

Objects, testing if the same

The identical(object1,object2)command below returns TRUE if two R objects are the same. We use two data frame objects from the discussion above on dummy variables.

```
base::identical(world$deathrate, worldDummyDf$deathrate)
[Output:]
   [1] TRUE
```

Variables, binning

The "cut" command creates a categorical variable split on given cutpoints. For illustration, let the bins be 0–2, 3–7, and >7. First, we get the descriptives for the un-binned variable "literacy".

```
base::summary(world$literacy)
[Output:]
   Min. 1st Qu.  Median    Mean 3rd Qu.    Max.
17.60   75.55   92.50   83.28   98.00  100.00

world$literacy_categories <- base::cut(world$literacy, breaks =
base::c(-1, 75.56, 83.29, 98.01, 100.1), right=TRUE, labels =
base::c("Q1","Q2","Q3", "Q4"))
```

Below is the distribution of observations by bin.

```
base::summary(world$literacy_categories)
[Output:]
Q1 Q2 Q3 Q4
53 19 93 47
```

Dealing with error messages

Wrong data class error

Consult `help(name of command)` to see what data type the command needs, then use `class(name of object)` to see if your object is the right class for the command you are using.

Wrong data length error

Some programs need all vector variables to have the same number of elements. `Use length(name of vector variable)` on each variable to check.

Complete cases error

Check for missing values as some operations want complete cases data. Reduce to complete cases or impute missing data.

Common quotation mark error

The user may encounter "Unexpected input" error messages when using quote marks. R wants straight quote marks, such as around "text". Unfortunately, Word defaults to smart quotes: "text". To change this Word default, do this in Word:

1 On the File tab, click Options.
2 Click Proofing, and then click AutoCorrect Options.
3 In the AutoCorrect dialog box, do the following: Click the AutoFormat As You Type tab, and under Replace as you type, uncheck the option labeled "Straight quotes with smart quotes".
4 Click OK.

Common minus sign error

A similar error may occur when trying to type the minus sign. The symbol in Word is not the same as what R needs even though they look the same. Avoid cutting-and-pasting minus signs from Word and instead, type the minus sign directly in RStudio.

Pasted:

```
4–1
     Error: unexpected input in "4 –"
```

Typed in RStudio:

```
4-1
     [1] 3
```

Common capitalization error

Most command words in R are lower case and R is case-sensitive. However, by default Word "auto-corrects", making words at the beginning of a line or after a period upper case. Be careful that Word has not done this if you are cutting-and-pasting from Word.

Plot margins error

If your RStudio plot panel is too small for the plot you are creating, you may get the error message, "Error in plot.new(): figure margins too large". Correct this by making your plot panel window larger and re-running the command which generated the plot. You can avoid this error by checking your margins with the par("mar") command. You should get: `"[1] 5.1 4.1 4.1 2.1"`. If needed, reset with `par(mar=c(5.1,4.1,4.1,2.1))`. For maximum margins enter `par(mar=c(1,1,1,1))`.

Installing packages, "not available" error message

If you get an error message that a given package does not work with your version of R, in RStudio try selecting Tools > Install packages and install that way rather than with the `install.packages()` command. Most packages load automatically but sometimes it is necessary to install Rtools using the commands below. After installing Rtools, try the `install.packages()` command again. At this writing, the current version of RTools was 4.2 (for R >= 4.2.0). The installation page at https://cran.r-project.org/bin/windows/Rtools/ has links for earlier versions of R.

RStudio logs for error messages

RStudio keeps logs of errors, warnings, and exceptions. This information may be useful for debugging purposes. The researcher can access the names of the log files in RStudio by selecting Help > Diagnostics > Show Log Files. Log files are text files that can be loaded into a word processor, but only after the session is terminated and RStudio has closed the file. Right-click on a log name and select Properties to see where it is located on your computer. For more detail, select Help > Diagnostics > Write Diagnostics Report. Or issue the command rstudioDiagnosticsReport(). The file diagnostics-report.txt will be written to your Documents folder (e.g., C:/Users/Owner/Documents/rstudio-diagnostics/diagnostics-report.txt). It contains listings of attached base and other packages, system and platform information, R version and home path, environmental variables, loaded packages, installed packages, site profile with options settings, user preferences, the error log file, and more.

Obtaining data

Getting data for R projects is no different from getting data for any project. However, for practice purposes, R comes with sample data and demonstrations.

Many datasets are listed in the "datasets" package, which is in the R System library. To view some of the available datasets, simply type:

```
utils::data()
```

To view datasets available in any package installed in the User or System Libraries:

```
utils::data(package = .packages(all.available = TRUE))
```

One of the available datasets is USJudgeRatings. To load it into to a data frame called "judges":

```
judges <- base::data.frame(USJudgeRatings)
```

View the loaded dataset in the ordinary manner:

```
View(judges)
```

Get documentation on the dataset:

```
utils::help(USJudgeRatings)
```

Next. use the dataset in a command. Below we create a bare-bones plot of judges' decision promptness by judges' worthiness of retention with a lowess smoothing line through the points:

```
base::plot(judges$DECI,judges$RTEN)
graphics::lines(stats::lowess(judges$DECI, judges$RTEN))
```

Some packages have datasets not from the "datasets" package in the R System library. For instance, the package "carData" has the "Arrests" data on marijuana possession arrests. You may need to install the package first: `install.packages(carData)` first.

```
base::library(carData)
base::class(Arrests)
      [1] "data.frame"
View(Arrests)
```

Getting help

Vignettes

Many packages have "vignettes" which describe how to use the package. The `vignette()` command is part of the "utils" package in the R System library.

```
utils::vignette(package="psych")
[Output:]
    Vignettes in package 'psych':

    intro            Introduction to the psych package
                     (source, pdf)
    overview         of the psych package for
                     psychometrics (source, pdf)
```

To access any vignette, click on the "Packages" tab in RStudio, browse to the package of interest, then click on the link "User guides, package vignettes and other documentation".

Demos

In the same way, R packages also may come with built-in demonstrations. Only some statistical programs have demos. The `demo()` command is part of the "utils" package in the R System library. For information, type `help(demo)`. To bring up a listing of the many demos in packages in the system library type:

```
utils::demo()
```

Packages may have multiple demos. To list all demos in a particular package:

```
utils::demo(package = "lattice")
```

To run a specific demo:

```
utils::demo("intervals", package = "lattice")
```

To run an available demo in the `demo()` listing, such as that for one of the linear and generalized linear modeling demos from 'An Introduction to Statistical Modelling' by Annette Dobson (available in the stats package), type:

```
utils::demo(lm.glm)
```

Use the escape key to terminate any demo.

Cheatsheets

Cheatsheets are concise summaries of R commands in a given area. Some come with RStudio under the menu sequence Help > Cheatsheets. These pdf documents concisely summarize many widely used R packages and commands, with very brief illustrations of R code. For example, one is "Data Transformation Cheatsheet" and another is "Python with R and Reticulate Cheatsheet." A large number of cheat sheets are listed at https://www.rstudio.com/resources/cheatsheets/ . From this page, one may subscribe to cheatsheet updates.

With the "cheatsheet" package the researcher may download many more cheatsheets in pdf format:

```
utils::install.packages("cheatsheet")
base::library(cheatsheet)

# Save to a specified existing directory such as C:/References/R, but adding the folder "/cheats"
# Note: Use forward slashes.
# If the local path is only "cheats", it will be created in the Documents folder of your User area.
cheatsheet::get_all_cheatsheets(local_path = "C:/References/R/cheats", tidyverse_only = FALSE)
```

View the cheatsheets, such as "base-r.pdf", as you would any other pdf document.

Search

A searchable listing of thousands of R packages for statistical analysis is found at https://advanceddataanalytics.net/r-packages/.

More

From the R prompt, the following commands illustrate other ways to get help:

- Get basic information about file manipulation functions in R: `utils::help(files)`
- Get basic information about R functions for base stats: `utils::help(stats)`
- Get documentation listing commands available for the stats package: `base::library(help = "stats")`
- Get help for the cor (correlation) command, one of the commands in the stats base::library: `utils::help(cor)`
- Use the search facility at https://rdrr.io/search

Similarly, you can also insert other package names in the `help()` command. For instance, `help(graphics)` gives help on the "graphics" package and `help(data)` gives help on the data function of the "utils" package, which includes help on loading data, formats supported, etc. To list all locally installed packages issue the command `installed.packages()`. Not all packages have help pages, however.

Sometimes `help()` fails, in which case the double-question mark method may be used, causing the R console to search more broadly, as illustrated below.

```
utils::help(lme4)
> No documentation for 'lme4' in specified packages and libraries:
you could try '??lme4'

??lme4
> Search Results
[Search results appear web-style in RStudio's lower right, with
hypertext links.]
```

As a web search will show, there are many free tutorials available for learning R. One is the swirl package. It contains a variety of lessons, some of which are video-based, some are self-review tests, and some are hands-on exercises. User responses are tested for correctness and hints are given if appropriate. Progress is automatically saved so that a user may quit at any time and later resume, without losing work, by typing `swirl()`.

A note on using the `attach()` command

Note that an object in the View window is not necessarily in R's search path. The `attach()` function puts an object in the search path. When in the search path, variables like "happy" can be referenced with their simple name instead of their full name (that is by "happy" rather than "mydata$happy").

Why the use of `attach()` is derogated: The `attach()` method does not update changes made to variables in the data frame, requiring the researcher to detach and then re-attach. Also, in more complex projects the `attach()` method may find more than one object of the same name, requiring reverting to the more conservative practice of using full variable names. For these reasons, and for variable-naming clarity, the `attach()` method often is derogated. If it is used, remember to issue the `detach()` command at the end.

In summary:

- `attach()` is read-only and will always refer to the values in your data frame at the time you issued this command. It will not recognize subsequent changes.
- Do not use `attach()` if you have multiple objects with the same variable name such as income. The attach() command will only reference names in the most recently attached object.
- You must remember to `detach()` at the end.
- If you have an R script and come back to it later, it will be a lot easier to read if only long names have been used. Short names are bad programming practice. The `attach()` command is mainly for short exploratory or instructional purposes.
- Consider using `with` instead of `attach()`. The format is with(data frame name, command(you may use simple variable names). Because the `with()` command names the data frame, short names may be used. For example:

```
base::setwd("C:/Data")
mydata <- utils::read.table("mydata.csv", header = TRUE, sep =
",")
base::with(mydata, base::class(happy))
[Output:]
   [1] "integer"
```

Appendix 3
Frequently asked questions

How to report factor analysis

The American Psychological Association (APA) is for social science the authoritative source for writing up statistical research. In a web browser, search for "APA guidelines 'factor analysis'" (without the double quote marks). One will find a number of useful guides and examples. A link specifically to formatting factor analysis tables is found at https://apastyle.apa.org/style-grammar-guidelines/tables-figures/sample-tables#factor.

Watkins (2020: 104–105) suggests that factor analytic reports should contain nine elements:

- Describe the indicator variables and give the rationale for their use in the model.
- Describe the unit of analysis (cases, participants) and justify sample size and sample representativeness.
- Descriptive statistics should address the assumptions of normality, outliers, and linearity; provide a correlation matrix; discuss missing values analysis.
- Provide KMO and Bartlett's test of sphericity.
- Give the rationale for choosing between PCA and PFA (or other dimension reduction procedures).
- Give the rationale for the choice of extraction method.
- Give the rationale for the type of correlation matrix used (Pearson, polychoric, Spearman, Kendall, mixed, etc.).
- Report the method of estimating communalities based on correlations.
- Give researchers prior criteria for selecting how many factors to retain. Discuss multiple criteria for this decision.
- Give the rationale for selection of orthogonal or oblique rotation, and for selection of a particular type (e.g., varimax, promax).
- Discuss findings in terms of loadings, communalities, reliability, and model fit. For oblique rotation discuss both pattern and structure coefficients and discuss inter-factor correlations.

What are "data modes" in factor analysis?

Though the term "data modes" comes from a vocabulary rooted in the early days of factor analysis. Though terms such as "R-mode" and "Q-mode" are little used today, they are still worth noting because they point up different ways in which factor analysis may be used, whether EFA or CFA. These are enumerated below.

- "R-mode": In the most common mode, on the input side, rows are units of observation (e.g., subjects, cities, organizations, or countries) and columns are variables. On the output side, rows are variables and columns are the dimensions (factors). Thus. the factors represent clusters of indicator variables which have been measured for a set of people (or other entities) at a given point of time.
- "Q-mode": On the input side, rows are variables and columns are the observations (units of analysis). On the output side, the rows are the observations and columns are the dimensions. In this mode, the factors represent clusters of observations for a set of variables at a given point in time. Q-mode may be used, for example, to establish the factional composition of a group such as the U. S. Senate on a set of environmental issues at a given point in time. Partly because of the problem of negative loadings in Q mode,[1] cluster analysis is now more common for grouping observations.
- "O-mode": On the input side, data are collected on a single entity (e.g., one U. S. Senator), the columns are years, and the rows are measures (e.g., environmental votes). In this mode, factors show which years cluster together on a set of measures for a single observation. Based on this, one can compare individual cases or one can differentiate periods for purposes of explanation of behavior of the given individual.
- "T-Mode": On the input side, rows are observation (typically people) and columns are years (or other time unit). Data are gathered for a single variable. In T-mode, the factors show which years cluster together on that variable for a set of people or other entities. One might investigate, for instance, if Senators' positions on military spending are differentiated between war years and peacetime years.
- "S-mode": On the input side, observations are columns (e.g., Senators) and time units (e.g., years) are rows. Cell entries measure a single variable. In S-mode, factors would show which Senators cluster together over a period of years on a single variable. S-mode would be used, for instance, to establish the underlying factional composition of a group on an issue over a long period of time.

What is KMO? What is it used for?

When access to computers was not widespread, there was a need to assess beforehand whether to engage in the computation-intensive effort required by factor analysis. Kaiser (1970) and his colleagues developed the Kaiser-Meier-Olkin (KMO) procedure with its statistic of overall MSA (measure of sampling adequacy, from KMO). MSA of 0.60 or higher indicates data will factor well. MSA coefficients are available both overall and for each predictor variable.

KMO today has two uses:

Predicting if data will factor well and thus will be likely to be interpretable. If overall MSA is equal to or higher than some cut-off (usually 0.60), the criterion is met for judging the data sample to be adequate to proceed (hence MSA is sometimes called a "test of sample adequacy"). A level of 0.6 or higher is considered adequate and 0.8 or higher is considered good factorability. That is, KMO is a predictor of whether data will factor well enough to be meaningfully interpretable.

Dropping variables from the analysis. An MSA statistic is generated for each predictor. Predictors whose MSA does not rise to some criterion level (e.g., 0.6 or higher) may be dropped from the analysis. Doing so is dropping predictors based on low partial correlation. The more prevalent criterion for dropping predictors is low communality, based on factor analysis. As the two dropping criteria (MSA and communality) may differ, the latter is generally preferred though both may be considered when evaluating a given input variable. Low MSA indicates the variable in question may be too multicollinear with others in the model. Low communality indicates that the variable is not well explained by the factor model. If MSA is used as the dropping criterion, one strategy is to drop the indicator variables with the lowest individual MSA one at a time until overall MSA rises above 0.60.

Computation: To compute MSA overall, the numerator is the sum of squared correlations of all variables in the analysis (except the 1.0 self-correlations of variables with themselves, of course). The denominator is this same sum plus the sum of squared partial correlations of each variable with each other variable j, controlling for others in the analysis. The concept is that the partial correlations should not be very large if one is to expect distinct factors to emerge from factor analysis. See Hutcheson and Sofroniou (1999: 224). KMO was treated in the earlier section on "PCA for R with `prcomp()`".

Is it necessary to standardize one's variables before applying factor analysis?

No. Results of factor analysis are not affected by standardization, which is built into the procedure. Note, however, that standardization (subtracting the mean, dividing by the standard deviation) scales data in a sample-specific way. If the research purpose is to compare factor structures between two or more samples, then one should use the covariance matrix rather than the correlation matrix as input to factor analysis (Kim and Mueller, 1978b: 76). However, the covariance method has problems when the variables are measured on widely different scales (ex., income measure to $100,000 and education measured to 22 years). Kim and Mueller recommend multi-sample standardization for this case (subtracting the grand mean of all samples, and dividing by the standard deviation of all cases) prior to computing the sample covariance matrix (p. 76). However, in practice factor comparison between samples often is based on ordinary factor analysis of correlation matrices and correlation is standardized covariance.

Can you pool data from two samples together in factor analysis?

Yes, but only after you have shown both samples have the same factor structure through factor comparison, discussed in the next topic. The danger of pooling is that pooling may obscure sample-specific factors unique to a given sample as compared to a different sample or even of a sample at one time point with the same sample at a different time point. For warnings about the complexity of pooling, see Comrey and Lee (1992) and Tabachnick and Fidell (2019).

How does factor comparison of the factor structure of two samples work?

The *pooled data method* has the researcher pool the data for two samples, adding a dummy variable whose coding represents sample/group membership. The factor loadings of this dummy variable indicate the factors for which the groups' mean factor scores would be most different.

The *comparison measures method* requires computation of various measures which compare factor attributes of the two samples. Factor comparison is discussed by Levine (1977: 37–54), who describes these factor comparison measures:

1 *RMS, root mean square.* RMS is the root mean square of the average squared difference of the loadings of the variables on each of two factors. RMS varies from 0 to 2, reaching 0 in the case of a perfect match between samples of both the pattern and the magnitude of factors in the two samples. An RMS of 2 indicates all loadings are at unity but differ in sign between the two samples. Intermediate values are hard to interpret.
2 *CC, coefficient of congruence.* The coefficient of congruence is the sum of the products of the paired loadings divided by the square root of the product of the two sums of squared loadings. Like RMS, CC measures both pattern and magnitude similarities between samples. There is a tendency to get a high CC whenever two factors have many variables with the same sign.
3 *S, salient variable similarity index.* The salient variable similarity index is based on classifying factor loadings into positive salient ones (over +0.1), hyperplane ones (from −0.1 to +0.1), and negative salient ones ((below −0.1). Hyperplane loadings, which approach 0, indicate variables having only a near-chance relationship to the factor. The S index will be 0 when there are no salient loadings, indicating no factor congruence between the two samples. An S of 1 indicates perfect congruence, and −1 indicates perfect negative (reflected) congruence. Note that calculating S for all possible pairs of factors between two samples risks coming to conclusions on the basis of chance.

Note

1 A Q-mode issue has to do with negative factor loadings. In conventional R-mode factor analysis, loadings are those of variables on factors and a negative loading indicates a negative relation of the variable to the factor. In Q-mode factor analysis, loadings are loadings of observations on factors and a negative loading indicates that the observation/individual displays responses opposite to those observations which load positively on the factor. In conventional factor analysis, a loading approaching zero indicates the given variable is unrelated to the factor. In Q-mode factor analysis, a loading approaching zero indicates the given case is near the mean for the factor. Cluster analysis is now more common than Q-mode factor analysis when the research objective is grouping observations. Note, however, that correlations in factor analysis are treated in a general linear model which takes control variables into account, whereas cluster analysis uses correlations simply as similarity measures. For this reason, some researchers still prefer Q-mode factor analysis for clustering analysis.

References

Abdi, H. & Williams, L. J. (2010). Principal component analysis. *Wiley Interdisciplinary Reviews: Computational Statistics* 2: 433–459.

Auerswald, M. & Moshagen, M. (2019). How to determine the number of factors to retain in exploratory factor analysis: A comparison of extraction methods under realistic conditions. *Psychological Methods* 24(4): 468–491.

Aulakh, P. S. & Gencturk, E. F. (2000). International principal–agent relationships—control, governance and performance. *Industrial Marketing Management* 29: 521–538.

Bandalos, D. L. (2018). *Measurement theory and applications in the social sciences.* New York: Guilford Press.

Bandalos, D. L. & Finney, S. J. (2019). Factor analysis: Exploratory and confirmatory. Pp. 98–122 in Hancock, G. R.; Stapleton, L. M.; & Mueller, R. O., eds., *The reviewer's guide to quantitative methods in the social sciences.* Second ed. Milton Park, Abingdon-on-Thames, Oxfordshire: Routledge.

Bartholomew, D. J. (1987). *Latent variable analysis and factor analysis.* London: Charles Griffin and Company.

Bartholomew, D. J. & Knott, M. (1990). *Latent variable analysis and factor analysis.* Second edition. London: Arnold.

Bartholomew, D. J., Knott, M., & Moustaki, I. (2011). *Latent variable analysis and factor analysis.* Third edition. New York: Wiley.

Bartlett, M. S. (1951). The effect of standardization on a χ^2 approximation in factor analysis. *Biometrika* 38(3/4): 337–344.

Beauducel, A. (2005). How to describe the difference between factors and corresponding factor-score estimates. *Methodology: European Journal of Research Methods for the Behavioral and Social Sciences* 1(4): 143–158.

Belkin, M. & Niyogi, P. (2003). Laplacian eigenmaps for dimensionality reduction and data representation. *Neural Computation* 15: 1373–1396.

Benson, J. & Nasser, F. (1998). On the use of factor analysis as a research tool. *Journal of Vocational Education Research* 23(1): 13–33.

Bentler, P. M. (2005). *EQS 6 structural equations program manual.* Encino, CA: Multivariate Software.

Boehmke, B. & Greenwell, B. M. G. (2019). *Hands-on machine learning with R.* Boca Raton, FL: Chapman & Hall/CRC.

Boker, S. et al. (2011). OpenMX: an open source extended structural equation modeling framework. *Psychometrica* 76(2): 306–317.

Briggs, N. E. & MacCallum, R. C. (2003). Recovery of weak common factors by maximum likelihood and ordinary least squares estimation. *Multivariate Behavioral Research* 38(1): 25–56.

Broc, G. & Gana, K. (2019). *Structural equation modeling with lavaan.* New York: ISTE/Wiley.

Bryant, F. B. & Yarnold, P. R. (1995). Principal component analysis and exploratory and confirmatory factor analysis. Chapter 4 in Grimm, L. G.; & Yarnold, eds., *Reading and understanding multivariate analysis.* American Psychological Association Books.

Carroll, J. B. (1978). How shall we study individual differences in cognitive abilities? - Methodological and theoretical perspectives. *Intelligence* 2(2): 87–115.

Carroll, J. B. (1985). Exploratory factor analysis: A tutorial. Pp. 25–58 in Detterman, D. K., ed., *Current topics in human intelligence.* New York: Ablex Publishing Company.

Carroll, J. B. (1989). Factor analysis since Spearman: Where do we stand? What do we know? Pp. 43–70 in Kanfer, R.; Ackerman, P. L.; & Cudeck, R.; eds., *Abilities, motivation, and methodology: The Minnesota Symposium on Learning and Individual Differences*, Volume 10. Milton Park, Abingdon-on-Thames, Oxfordshire: Routledge.

Carroll, J. B. (1993). *Human cognitive abilities: A survey of factor-analytic studies.* New York: Cambridge University Press.

Cattell, R. B. (1952). *Factor analysis: An introduction and manual for the psychologist and social scientist.* Oxford: Harper.

Cattell, R. B. (1966). The scree test for the number of factors. *Multivariate Behavioral Research* 1(2): 245–276.

Cattell, R. B. (1978). *The use of factor analysis in behavioral and life sciences.* New York: Plenum Press.

Chalmers, R. P. & Flora, D. B. (2015). faoutlier: An R package for detecting influential cases in exploratory and confirmatory factor analysis. *Applied Psychological Measurement* 39(7): 573–574.

Chavent, M., Kuentz-Simonet, V., Labenne, A., & Saracco, J. (2017). Multivariate analysis of mixed data: The R package PCAmixdata. https://arxiv.org/pdf/1411.4911.pdf.

Cheung, M. W. (2015). metaSEM: An R package for meta-analysis using structural equation modeling. *Frontiers in Psychology* 5: Article 1521.

Child, D. (2006). *The essentials of factor analysis.* Third edition. London: Continuum International Publishing Group.

Cho, S.-J., Li, F., & Bandalos, D. (2009). Accuracy of the parallel analysis procedure with polychoric correlations. *Educational and Psychological Measurement* 69(5): 748–759.

Coifman, R. R. & Lafon, S. (2006). Diffusion maps. *Applied and Computational Harmonic Analysis* 21: 5–30.

Collins, M., Dasgupta, S., & Schapire, R. E. (2001). A generalization of principal component analysis to the exponential family. Pp. 617–624 in Dietterich, T.; Becker, S.; & Ghahramani, Z., eds., *Advances in neural information processing systems*, Volume 14. Cambridge, MA: MIT Press.

Comrey, A. L. & Lee, H. B. (1992). *A first course in factor analysis.* Second edition. Mahwah, NJ: Lawrence Erlbaum Associates, Inc.

Conti, G., Frühwirth-Schnatter, S., Heckman, J. J., & Piatek, R. (2014). Bayesian exploratory factor analysis. *Journal of Econometrics* 183(1): 31–57.

Crawley, M. (2007). *The R book.* West Sussex: John Wiley & Sons Ltd.

Cronbach, L. J. (1951). Coefficient alpha and the internal structure of tests. *Psychometrika* 16(3): 297–334.

Cureton, E. E. & Mulaik, S. A. (1975). The weighted varimax rotation and the promax rotation. *Psychometrika* 40: 183–195.

Curran, P. J., West, S. G., & Finch, J. F. (1996). The robustness of test statistics to nonnormality and specification error in confirmatory factor analysis. *Psychological Methods* 1(1): 16–29.

Dai, X., Hadjipantelis, P. Z., Han, K., & Ji, H. (2021). fdapace: Functional data analysis and empirical dynamics. R package (Version 0.5.7). Available online: https://cran.r-project.org/package=fdapace (accessed on 19 July 2021).

de Leeuw, J. (2006). Principal component analysis of binary data by iterated singular value decomposition. *Computational Statistics and Data Analysis* 50(1): 21–39.

De Silva, V. & Tenenbaum, J. B. (2004). Sparse multidimensional scaling using landmark points. Technical report, Department of Mathematics, Stanford University. http://www.cmap.polytechnique.fr/~peyre/cours/x2005signal/dimreduc_landmarks.pdf.

DeSimone, J. A., Harms, P. D., & DeSimone, A. J. (2015). Best practice recommendations for data screening. *Journal of Organizational Behavior* 36(2): 171–181.

DeVellis, R. F. (2017). *Scale development: Theory and applications.* Fourth edition. Thousand Oaks, CA: Sage Publications.

DiStefano, C. (2002). The impact of categorization with confirmatory factor analysis. *Structural Equation Modeling* 9(3): 327–346.

DiStefano, C. (2016). Examining fit with structural equation models. Pp. 166–193 in Schweizer, K.; & DiStefano, C.; eds., *Principles and Methods of Test Construction: Standards and Recent Advances.* Newburyport, MA: Hogrefe Publishing Corp.

Dombrowski, S. C., Beaujean, A. A., McGill, R. J., Benson, N. F., & Schneider, W. J. (2019). Using exploratory bifactor analysis to understand the latent structure of multidimensional psychological measures: An example featuring the WISC-V. *Structural Equation Modeling* 26(6): 847–860.

Donoho, D. L. & Grimes, C. (2003). Hessian eigenmaps: Locally linear embedding techniques for high-dimensional data. *Proceedings of the National Academy of Science* 100(10): 5591–5596.

Drasgow, F. (1986) Polychoric and polyserial correlations. Pp. 68–74 in Kotz, S.; & Johnson, N., eds., *The encyclopedia of statistics*, Volume 7. New York: Wiley.

Dwyer, P. S. (1937). The determination of the factor loadings of a given test from the known factor loadings of other tests. *Psychometrika* 3: 173–178.

Escofier, B. & Pagès, J. (1994). Multiple factor analysis (AFMULT package). *Computational Statistics & Data Analysis* 18(1): 121–140.

Fabrigar, L. R. & Wegener, D. T. (2012). *Exploratory factor analysis.* Oxford: Oxford University Press.

Fabrigar, L. R., Wegener, D. T., MacCallum, R. C., & Strahan, E. J. (1999). Evaluating the use of exploratory factor analysis in psychological research. *Psychological Methods* 4(3): 272–299.

Fanaee-T, H. & Thoreson, M. (2019). Performance evaluation of methods for integrative dimensionality reduction. *Information Sciences* 493: 105–119.

Fava, J. L. & Velicer, W. F. (1992). The effects of overextraction on factor and component analysis. *Multivariate Behavioral Research* 27: 387–415.

Field, A. (2013). *Discovering statistics using IBM SPSS Statistics.* Thousand Oaks, CA: Sage Publications.

Finch, W. H. (2013). Exploratory factor analysis. Pp. 167–186 in Teo, T., ed., *Handbook of quantitative methods for educational research.* Rotterdam: Sense Publishers.

Finch, W. H. (2020). *Exploratory factor analysis.* Thousand Oaks, CA: Sage Publications.

Flora, D. B. (2018). *Statistical methods for the social and behavioural sciences: A model-based approach.* Thousand Oaks, CA: Sage.

Flora, D. B. & Curran, P. J. (2004). An empirical evaluation of alternative methods of estimation for confirmatory factor analysis with ordinal data. *Psychological Methods* 9(4): 466–491.

Fox, J. (2006). Structural equation modeling with the sem package in R. *Structural Equation Modeling* 13: 465–486.

Fruchterman, T. M. J. & Reingold, E. M. (1991). Graph drawing by force-directed placement. *Software: Practice and Experience* 21: 1129–1164.

Garrido, L. E., Abad, F. J., & Ponsoda, V. (2013). A new look at Horn's parallel analysis with ordinal variables. *Psychological Methods* 18(4): 454–474.

Garson, G. D. (2014). *The Delphi method in quantitative research.* Asheboro, NC: Statistical Associates.

Garson, G. D. (2015). *Cluster analysis.* Asheboro, NC: Statistical Associates.

Garson, G. D. (2018). *Structural equation modeling.* Asheboro, NC: Statistical Associates.

Garson, G. D. (2019). *Missing values analysis and data imputation.* Asheboro, NC: Statistical Associates.

Garson, G. D. (2022). *Data analytics for the social sciences: Applications in R.* Milton Park, Abingdon-on-Thames, Oxfordshire: Routledge, 2022.

Gifi, A. (1990). *Nonlinear multivariate analysis.* New York: Wiley.

Gilley, W. F. & Uhlig, G. E. (1993). Factor analysis and ordinal data. *Education* 114(2): 258–265.

Giordano, C. & Waller, N. G. (2020). Recovering bifactor models: A comparison of seven methods. *Psychological Methods* 25(2): 143–156.

Golino, H. F. & Epskamp, S. (2017). Exploratory graph analysis: A new approach for estimating the number of dimensions in psychological research. *PLoS One* 12: e0174035. https://doi.org/10.1371/journal.pone.0174035.

Gorsuch, R. L. (1974). *Factor analysis*. Philadelphia, PA: W.B. Saunders.

Gorsuch, R. L. (1983). *Factor analysis*. Hillsdale, NJ: Lawrence Erlbaum. Orig. ed. 1974.

Gorsuch, R. L. (1988). Exploratory factor analysis. Pp. 231–258 in Nesselroade, J. R.; & Cattell, R. B., eds., *Handbook of multivariate experimental psychology*. Second edition. Boston, MA: Springer.

Gorsuch, R. L. (1997) New procedure for extension analysis in exploratory factor analysis. *Educational and Psychological Measurement* 57: 725–740.

Gower, J. C. (1966). Some distance properties of latent root and vector methods used in multivariate analysis. *Biometrika* 53: 325–328.

Graham, J. M., Guthrie, A. C., & Thompson, B. (2003). Consequences of failure to meet assumptions underlying the fixed effects analysis of variance and covariance. *Review of Educational Research* 10(1): 142–153.

Greene, C. N. & Organ, D. W. (1973). An evaluation of causal models linking the received role with job satisfaction. *Administrative Science Quarterly* 18: 95–103.

Greenwald, A. G., Pratkanis, A. R., Leippe, M. R., & Baumgardner, M. H. (1986). Under what conditions does theory obstruct research progress? *Psychological Review* 93(2): 216–229.

Grice, J. W. (2001). Computing and evaluating factor scores. *Psychological Methods* 6(4): 430–450.

Guadagnoli, E. & Velicer, W. F. (1988). Relation of sample size to the stability of component patterns. *Psychological Bulletin* 103(2): 265–275.

Gunn Haig, B. D. (2018). *Method matters in psychology: Essays in applied philosophy of science*. Berlin, Heidelberg: Springer.

Hair, J.F., Jr., Anderson, R. E., Tatham, R. L., & Black, W. C. (1998). *Multivariate data analysis with readings*. 5th edition. Englewood Cliffs, NJ: Prentice-Hall.

Hair, J.F., Jr., Anderson, R. E., Tatham, R. L., & Black, W. C. (2019). *Multivariate data analysis*. 8th edition. Boston, MA: Cengage Learning.

Han, K., Hadjipantelis, P. Z., Wang, J.-L., Kramer, M. S., Yang, S., Martin, R. M., et al. (2018) Functional principal component analysis for identifying multivariate patterns and archetypes of growth, and their association with long-term cognitive development. *PLoS One* 13(11). https://doi.org/10.1371/journal.pone.020707.

Harman, H. H. (1976). *Modern factor analysis*. Third edition. Chicago, IL: University of Chicago Press.

Harris, C. W. (1962). Some Rao-Guttman relationships. *Psychometrika* 27: 247–263.

Hendrickson, A. E. & White, P. O. (1964). Promax: A quick method for rotation to oblique simple structure. *British Journal of Mathematical Psychology* 17(1): 65–70.

Hatcher, L. (1994). *A step-by-step approach to using the SAS system for factor analysis and structural equation modeling*. Cary, NC: SAS Institute. Focus on the CALIS procedure.

Hinton, G. E. & Roweis, S. T. (2003). Stochastic neighbor embedding. Pp. 857–864 in Becker, S.; Thrun, S.; & Obermayer, K., eds., *Advances in neural information processing systems* 15: Cambridge, MA, USA: MIT Press.

Hoelzle, J. B, & Meyer, G. J. (2013). Exploratory factor analysis: Basics and beyond. Pp. 164–188 in Weiner, I. B.; Schinka, J. A.; & Velicer, W. F., eds., *Handbook of psychology: Research methods in psychology*, Volume 2. Second edition. New York: Wiley.

Hofmann, R. J. (1978). Complexity and simplicity as objective indices descriptive of factor solutions. *Multivariate Behavioral Research* 13: 247–250.

Holzinger, K. & Swineford, F. (1937). The bi-factor method. *Psychometrika* 2(1): 41–54.

Holzinger, K. & Swineford, F. (1939). A study in factor analysis: The stability of a bifactor solution. *Supplementary Educational Monograph, no. 48*. Chicago, IL: University of Chicago Press.

Horst, P. (1941). The prediction of personnel adjustment. *Social Science Research and Council Bulletin* 48: 431–436.

Howard, M. C. (2016). A review of exploratory factor analysis decisions and overview of current practices: What we are doing and how can we improve? *International Journal of Human-Computer Interaction* 32(1): 51–62.

Hu, L.-t. & Bentler, P. M. (1999). Cutoff criteria for fit indexes in covariance structure analysis: Conventional criteria versus new alternatives. *Structural Equation Modeling* 6(1): 1–55.

Hutcheson, G. & Sofroniou, N. (1999). *The multivariate social scientist: Introductory statistics using generalized linear models.* Thousand Oaks, CA: Sage Publications.

Hyvarinen, A. (1983). Independent component analysis: recent advances. *Philosophical Transactions: Mathematical, Physical and Engineering Sciences.* 2012 Dec 31;371(1984): 20110534. doi: 10.1098/rsta.2011.0534.

Hyvarinen, A. (1999). Fast and robust fixed-point algorithms for independent component analysis. *IEEE Transactions on Neural Networks* 10: 626–634.

Iverson, R. D. & Maguire, C. (2000). The relationship between job and life satisfaction: Evidence from a remote mining community. *Human Relations* 53: 807–839.

Jennrich, R. I. & Sampson, P. F. (1966). Rotation for simple loading. *Psychometrika* 31(3): 313–323.

Jensen, A. R. (1998). *The g factor: The science of mental ability.* Westport, CT: Praeger.

Johnson, S. L., Swerdlow, B., Tharp, J. A., Chen, S., Tackett, J., & Zeitzer, J. (2021). Social dominance and multiple dimensions of psychopathology: An experimental test of reactivity to leadership and subordinate roles. *PLoS One* 16(4). https://doi.org/10.1371/journal.pone.0250099.

Jones, C. S. (2001). Extracting factors from heteroskedastic asset returns. *Journal of Financial Economics* 62(2): 293–325.

Jordan, P. J. & Troth, A. C. (2019). Common method bias in applied settings: The dilemma of researching in organizations. *Australian Journal of Management* 45(1): 3–14.

Joreskog, K. G. & Sorbom, D. (1986). *PRELIS: A preprocessor for Lisrel.* Mooresville, IN: Scientific Software, Inc.

Jung, S. & Takane, Y. (2008). Regularized common factor analysis. *New Trends in Psychometrics* 141–149. Retrieved 12/11/2021 from https://www.researchgate.net/profile/Yoshio-Takane/publication/266333420_Regularized_Common_Factor_Analysis/links/564b62c208ae020ae9f80979/Regularized-Common-Factor-Analysis.pdf.

Kaiser, H. F. (1958). The varimax criterion for analytic rotation in factor analysis. *Psychometrika* 23: 187–200.

Kaiser, H. F. (1960). The application of electronic computers to factor analysis. *Educational and Psychological Measurement* 20: 141–151.

Kaiser, H. F. (1970) A second generation little jiffy. *Psychometrika* 35(4):401–415.

Kamada, T. & Kawai, S. (1989). An algorithm for drawing general undirected graphs. *Information Processing Letters* 31: 7–15.

Kelley, T. L. (1927). *Interpretation of educational measurements.* Oxford: World Book Company.

Kim, J.-O. & Mueller, C. W. (1978a). *Introduction to factor analysis: What it is and how to do it.* Thousand Oaks, CA: Sage Publications, Quantitative Applications in the Social Sciences Series, No. 13.

Kim, J.-O. and Mueller, C. W. (1978b). *Factor analysis: Statistical methods and practical issues.* Thousand Oaks, CA: Sage Publications, Quantitative Applications in the Social Sciences Series, No. 14.

Kim, J. & Park, H. (2008). Sparse nonnegative matrix factorization for clustering. Technical report. GT-CSE-08-01, Computational Science and Engineering, Georgia Institute of Technology.

Kline, R. B. (1998). *Principles and practice of structural equation modeling.* New York: Guilford Press.

Kline, R. B. (2012). Assumptions in structural equation modeling. Pp. 111–125 in Hoyle, R., ed., *Handbook of structural equation modeling.* New York: Guilford.

Kline, R. B. (2013). Exploratory and confirmatory factor analysis. Pp. 171–207 in Petscher, Y.; Schatschneider, C.; & Compton, D. L., eds., *Applied quantitative analysis in education and the social sciences.* Milton Park, Abingdon-on-Thames, Oxfordshire: Routledge.

Kolenikov, S. & Angeles, G. (2004). The use of discrete data in principal component analysis with applications to socio-economic indices. *CPC/MEASURE Working Paper No. WP-04-85*. Available at http://www.cpc.unc.edu/measure/publications/wp-04-85.

Kowarik, A. & Templ, M. (2016). Imputation with the R package VIM. *Journal of Statistical Software* 74(7): 1–16.

Kraemer, G., Reichstein, M., & Mahecha, M. D. (2018). dimRed and coRanking—Unifying dimensionality reduction in R. *The R Journal* 10(1): 342–358. https://journal.r-project.org/archive/2018/RJ-2018-039/index.html.

Kraemer, G., Reichstein, M., & Mahecha, M. D. (2019). dimRed and coRanking—Unifying dimensionality reduction in R. Part of online documentation for the dimRed package in R. (This is a version of Kraemer et al. *2018) modified by the authors to be published as an online vignette,)

Kruskal, J. B. (1964). Nonmetric multidimensional scaling: A numerical method. *Psychometrika* 29: 115–129.

Kuhn, M. (2008). Building predictive models in R using the caret package. *Journal of Statistical Software* 28(5): 1–26.

Kurka, M. (2019). GLRM model. Available at https://github.com/h2oai/h2o-tutorials/blob/master/best-practices/glrm/README.md.

Lancaster, B. P. (1999). Defining and interpreting suppressor effects: Advantages and limitations. Paper presented at the Annual Meeting of the Southwest Educational Research Association (San Antonio, TX, January 21–23, 1999). Available at https://files.eric.ed.gov/fulltext/ED426097.pdf.

Lance, C. E, Butts, M. M., & Michels, L. C. (2006). The sources of four commonly reported cut-off criteria: What did they really say? *Organizational Research Methods* 9(2): 202–220. Discusses Kaiser and other criteria for selecting number of factors.

Landgraf, A. (2016). An introduction to the logisticPCA R package. Published as a vignette on CRAN as part of package documentation, at https://cran.r-project.org/web/packages/logisticPCA/vignettes/logisticPCA.html.

Landgraf, A. J. & Lee, Y. (2015). Dimensionality reduction for binary data through the projection of natural parameters. Technical Report No. 890. Columbus, OH: Department of Statistics, The Ohio University.

Laparra, V., Malo, J., & Camps-Valls, G. (2015). Dimensionality reduction via regression in hyperspectral imagery. *IEEE Journal of Selected Topics in Signal Processing* 9: 1026–1036.

Larsen, K. R. & Bong, C. H. (2016). A tool for addressing construct identity in literature reviews and meta-analyses. *MIS Quarterly* 40(3): 529–551.

Lawley, D. N. and A. E. Maxwell (1971). *Factor analysis as a statistical method.* London: Butterworth and Co.

Lawrence, E. J., Shaw, P., Baker, D., Baron-Cohen, S., & David, A. S. (2004). Measuring empathy: Reliability and validity of the empathy quotient. *Psychological Medicine* 34(5): 911–919.

Le, H., Schmidt, F. L., Harter, J. K., & Lauver, K. J. (2010). The problem of empirical redundancy of constructs in organizational research: An empirical investigation. *Organizational Behavior and Human Decision Processes* 112(2): 112–125.

Lee, D. D. & Seung, H. S. (1999). Learning the parts of objects by non-negative matrix factorization. *Nature* 401: 788–791.

Lee, J, A. & Verleysen, M. (2009). Quality assessment of dimensionality reduction: Rank-based criteria. *Neurocomputing* 72(7): 1431–1443.

Lee, J. A. & Verleysen, M. (2010). Scale-independent quality criteria for dimensionality reduction. *Pattern Recognition Letters* 31(14): 2248–2257.

Lee, J. A., Renard, E., Bernard, G., Dupont, P., & Verleysen, M. (2013). Type 1 and 2 mixtures of Kullback-Leibler divergences as cost functions in dimensionality reduction based on similarity preservation. *Neurocomputing* 112: 92–108.

Lee, J. A., Peluffo-Ordóñez, D., & Verleysen, M. (2014). Multiscale stochastic neighbor embedding: Towards parameter-free dimensionality reduction. Conference paper, presented at

"2014 European Symposium on Artificial Neural Networks, Computational Intelligence and Machine Learning (ESANN 2014)", Bruges (Belgium) (April 23–25, 2014). Published in 2015 in *Proceedings of ESANN 2014* 169: 177–182.

Levenson, M. R., Jennings, P. A., Aldwin, C. M., & Shiraishi, R. W. (2005). Self-transcendence: Conceptualization and measurement. *The International Journal of Aging and Human Development* 60: 127–143.

Levine, M. S. (1977). *Canonical analysis and factor comparison*. Thousand Oaks, CA: Sage Publications, Quantitative Applications in the Social Sciences Series, No. 6.

Lewin-Koh, S.-C. & Amemiya, Y. (2003). Heteroscedastic factor analysis. *Biometrika* 90(1): 85–97.

Lim, S., & Jahng, S. (2019). Determining the number of factors using parallel analysis and its recent variants. *Psychological Methods* 24(4): 452–467.

Lozano, L. M., Garcia-Cueto, E., & Muniz, J. (2008). Effect of the number of response categories on the reliability and validity of rating scales. *Methodology* 4(2): 73–79.

Mair, P. (2018). *Modern psychometrics with R*. Berlin, Heidelberg: Springer.

Mansolf, M. & Reise, S. P. (2016). Exploratory bifactor analysis: The Schmid-Leiman orthogonalization and Jennrich-Bentler analytic rotations. *Multivariate Behavioral Research* 51(5): 698–717.

Mardia, K. V. (1970). Measures of multivariate skewness and kurtosis with applications. *Biometrika* 57(3): 519–530.

Martin, S., Brown, W. M., & Wylie, B. N. (2007). *Distributed recursive (graph) layout*. Albuquerque, NM: Sandia National Laboratories, Report No. dRl; 002182MLTPL00.

Matthews, G., Deary, I. J., & Whiteman, M. C. (2003). *Personality traits*. Second edition. Cambridge: Cambridge University Press.

McDonald, R. P. (1999). *Test theory: A unified approach*. Hillside, NJ: Erlbaum.

McInnes, L., Healy, H., Saul, N., & Großberger, L. (2018).UMAP: Uniform manifold approximation and projection. *The Journal of Open Source Software* 3(29): 861. https://doi.org/10.21105/joss.0086.

Morin, A. J. S., Myers, N. D., & Lee, S. (2020). Modern factor analytic techniques. Pp. 1044–1073 in Tenenbaum, G.; & Eklund, R. D., eds., *Handbook of sport psychology*. Fourth edition. New York: Wiley.

Mueller, R. O. & Hancock, G. R. (2019). Structural equation modeling. Pp. 445–456 in Hancock, G. R.; Stapleton, L. M.; & Mueller R. O., eds., *The reviewer's guide to quantitative methods in the social sciences*. Second edition. Milton Park, Abingdon-on-Thames, Oxfordshire: Routledge.

Mulaik, S. A. (1972). *The foundation of factor analysis*. New York: McGraw-Hill.

Mulaik, S. A. (2018). *Fundamentals of common factor analysis*. Pp. 211–251 in Irwing, P.; Booth, T.; & Hughes, D. J., eds., *The Wiley handbook of psychometric testing: A multidisciplinary reference on survey, scale and test development*. New York: Wiley.

Mundfrom, D. J., Shaw, D. G., & Ke, T. L. (2009). Minimum sample size recommendations for conducting factor analyses. *International Journal of Testing* 5(2): 159–168.

Muthen, B. O. & Kaplan, D. (1985). A comparison of some methodologies for the factor analysis of non-normal Likert variables. *British Journal of Mathematical and Statistical Psychology*38: 171–189.

Newman, D. A. (2014). Missing data: Five practical guidelines. *Organizational Research Methods* 17(4): 372–411.

Norman, G. R. & D. L. Streiner (1994). *Biostatistics: The bare essentials*. St. Louis, MO: Mosby.

Norušis, M. J. (2011). *IBM SPSS statistics 19 statistical procedures companion*. Upper Saddle River, NJ: Prentice-Hall.

Nunnally, J. C.& Bernstein, I. H. (1994) The assessment of reliability. *Psychometric Theory* 3: 248–292.

O'Connor, B. P. (2000). SPSS and SAS programs for determining the number of components using parallel analysis and Velicer's MAP test. *Behavior Research Methods, Instrumentation, and Computers* 32: 396–402.

Olsson, U. (1979) Maximum likelihood estimation of the polychoric correlation coefficient. *Psychometrika* 44: 443–460.

Osborne, J. W. & Banjanovic, E. S. (2016). *Exploratory factor analysis with SAS*. Cary, NC: SAS Institute.

Osborne, J. W., Costello, A. B., & Kellow, J. T. (2008). Best practices in exploratory factor analysis. Pp. 205–213 in Osborne, J. W., ed., *Best practices in quantitative methods*. Thousand Oaks, CA: Sage Publishing.

Pearson, K. (1901). On lines and planes of closest fit to systems of points in space. *Philosophical Magazine* 2(6): 559–572.

Pearson, R. H. & Mundfrom, D. J. (2010). Recommended sample size for conducting exploratory factor analysis on dichotomous data. *Journal of Modern Applied Statistical Methods* 9(2): 359–368.

Pettersson, E. & Turkheimer, E. (2010). Item selection, evaluation, and simple structure in personality data. *Journal of Research in Personality* 44(4): 407–420.

Podsakoff, P. M., MacKenzie, S. B., & Lee, J.-Y. (2003). Common method biases in behavioral research: A critical review of the literature and recommended remedies. *Journal of Applied Psychology* 88(5): 879–903.

Puth, M.-T., Neuhäuser, M., & Ruxton, G. D. (2015). Effective use of Spearman's and Kendall's correlation coefficients for association between two measured traits. *Animal Behaviour* 102: 77–84.

Raubenheimer, J. E. (2004). An item selection procedure to maximize scale reliability and validity. *South African Journal of Industrial Psychology* 30(4): 59–64.

Reise, S. P. (2012). The rediscovery of bifactor measurement models. *Multivariate Behavioral Research* 47(5): 667–696.

Reise, S. P., Scheines, R., Widaman, K. F., & Haviland, M. G. (2013). Multidimensionality and structural coefficient bias in structural equation modeling: A bifactor perspective. *Educational and Psychological Measurement* 73(1): 5–26.

Reise, S. P., Waller, N. G., & Comrey, A. L. (2000). Factor analysis and scale revision. *Psychological Assessment* 12(3): 287–297.

Revelle, W. (2019). *An introduction to the psych package*. Evanston, IL: Northwestern University.

Revelle, W. (2021). *How to: Use the psych package for factor analysis and data reduction*. Evanston, IL: Psychology Department, Northwestern University. Published online at http://personality-project.org/r/psych/HowTo/factor.pdf.

Revelle, W. & Rocklin, T. (1979). Very simple structure: An alternative procedure for estimating the optimal number of interpretable factors. *Multivariate Behavioral Research* 14: 403–414. Available at https://personality-project.org/revelle/publications/vss.pdf.

Rickert, J. (2021). Exploratory functional PCA with sparse data. *R Views*. 2021–07–08. Available online at https://rviews.rstudio.com/2021/07/08/exploratory-fda-with-sparse-data/.

Ripberger, J. T., Gupta, K., Silva, C. L., & Jenkins-Smith, H. C. (2014). Cultural theory and the measurement of deep core beliefs within the advocacy coalition framework. *Policy Studies Journal* 42(4): 509–527.

Ripley, B. D. (1996). *Pattern recognition and neural networks*. Cambridge: Cambridge University Press.

Roberson, R. B., Elliott, T. R., Chang, J. E., & Hill, J. N. (2014). *Exploratory factor analysis in rehabilitation psychology: A content analysis. Rehabilitation Psychology* 59(4): 429–438.

Robert, C. (2007). *The Bayesian choice*. 2nd edition. Paris: Springer.

Rosseel, Y. (2012). lavaan: An R package for structural equation modeling. *Journal of Statistical Software* 48(2): 1–36. http://www.jstatsoft.org/v48/i02/.

Rouquette, A. & Falissard, B. (2011). Sample size requirements for the validation of psychiatric scales. *International Journal of Methods in Psychiatric Research* 20(4): 235–249.

Roweis, S. T. & Saul, L. K. (2000). Nonlinear dimensionality reduction by locally linear embedding. *Science* 290: 2323–2326.

Rummel, R. J. (1970). *Applied factor analysis*. Evanston, IL: Northwestern University Press.

Saunders, C., Gammerman, A., & Vovk, V. (1998). Ridge regression learning algorithm in dual variables. In *Proceedings of the Fifteenth International Conference on Machine Learning*, ICML '98, pp. 515–521, San Francisco, CA, USA. San Francisco, CA: Morgan Kaufmann Publishers.

Scharf, F. & Nestler, S. (2019). Should regularization replace simple structure rotation in exploratory factor analysis? *Structural Equation Modeling: A Multidisciplinary Journal* 26(4): 576–590.

Schmid, J. & Leiman, J. M. (1957). The development of hierarchical factor solutions. *Psychometrika* 22(1): 53–61.

Schmitt, T. A. (2011). Current methodological considerations in exploratory and confirmatory factor analysis. *Journal of Psychoeducational Assessment* 29(4): 304–321.

Schmitt, T. A., Sass, D. A., Chappelle, W., & Thompson, W. (2018). Selecting the "best" factor structure and moving measurement validation forward: An illustration. *Journal of Personality Assessment* 100(4): 345–362.

Schölkopf, B. (1998). Nonlinear component analysis as a kernel eigenvalue problem. *Neural Computation* 10(5): 1299–1319.

Schölkopf, B., Smola, A., & Mūller, K.-R. (1998). Nonlinear component analysis as a kernel eigenvalue problem. *Neural Computation* 10(5): 1299–1319.

Singer, J. D. & Willett, J. B. (2003). *Applied longitudinal data analysis*. Oxford: Oxford University Press.

Shapiro, S. E., Lasarev, M. R., & McCauley, L. (2002). Factor analysis of Gulf War illness: What does it add to our understanding of possible health effects of deployment? *American Journal of Epidemiology* 156: 578–585. Retrieved 3/7/08 from http://aje.oxfordjournals.org/cgi/content/full/156/6/578.

Sokal, R. R. & Rohlf, F. J. (1962). The comparison of dendrograms by objective methods. *Taxon* 11(2): 33–40.

Spearman, C. (1904). "General intelligence", objectively determined and measured. *American Journal of Psychology* 15(2): 201–293.

Stevens, J. R. (2002). *Applied multivariate statistics for the social sciences*. Fourth ed. Hillside, NJ: Erlbaum.

Streiner, D. L. (2018). Commentary no. 26: Dealing with outliers. *Journal of Clinical Psychopharmacology* 38(3): 170–171.

Tabachnick, B. G., & Fidell, L. S. (2019). *Using multivariate statistics*. Seventh ed. London: Pearson. Sixth ed., 2013.

Tenenbaum, J. B., de Silva, V., & Langford, J. C. (2000). A global geometric framework for nonlinear dimensionality reduction. *Science* 290(5500): 2319–2323.

Thompson, B. (2004). *Exploratory and confirmatory factor analysis: Understanding concepts and applications*. Washington, DC: American Psychological Association.

Thorndike, E. (1904). *An introduction to the theory of mental and social measurements*. New York: Science Press.

Thurstone, L. L. (1931). The measurement of change in social attitude. *The Journal of Social Psychology*, 2: 230–235.

Thurstone, L. L. (1935). *The vectors of mind: Multiple-factor analysis for the isolation of primary traits*. Chicago: University of Chicago Press.

Thurstone, L. L. (1937). Current misuse of the factorial methods. *Psychometrika*, 2(2): 73–76.

Thurstone, L. L. (1940). Current issues in factor analysis. *Psychological Bulletin*, 37(4): 189–236. 02.

Thurstone, L. L. (1947). *Multiple factor analysis*. Chicago: University of Chicago Press.

Udell, M., Horn, C., Zadeh, R., & Boyd, S. (2016). Generalized low rank models. *Foundations and Trends: R in Machine Learning* 9(1): 1–118. Retrieved from https://web.stanford.edu/~boyd/papers/pdf/glrm.pdf.

van der Maaten, L. J. P. & Hinton, G. E. (2008). Visualizing high-dimensional data using t-SNE. *Journal of Machine Learning Research* 9: 2579–2605.

Velicer, W. F. (1976). Determining the number of components from the matrix of partial correlations. *Psychometrika* 41: 321–327.

Velicer, W. F., Eaton, C. A., & Fava, J. L. (2000). Construct explication through factor or component analysis: A review and evaluation of alternative procedures for determining the number of factors or components. Pp. 41–71 in R. D. Goffin & E. Helmes, eds., *Problems and solutions in human assessment.* Boston, MA: Kluwer.

Velicer, W. F. & Jackson, D. N. (1990). Component analysis versus common factor analysis: Some issues in selecting an appropriate procedure. *Multivariate Behavioral Research* 25(1): 1–28.

Venna, J. & Kaski, S. (2001). Neighborhood preservation in nonlinear projection methods: An experimental study. In Dorffner, G.; Bischof, H.; & Hornik, K., eds., *Artificial neural networks — ICANN 2001.* ICANN 2001. Lecture Notes in Computer Science, Volume 2130. Berlin, Heidelberg: Springer. https://doi.org/10.1007/3-540-44668-0_68.

Venna, J. & Kaski, S. (2006). Local multidimensional scaling. *Neural Networks* 19: 889–899.

Voeten, E. (2013). Data and analyses of voting in the UN General Assembly. Pp. 54–66 in Reinalda, B., ed., *Routledge handbook of international organization.* Milton Park, Abingdon-on-Thames, Oxfordshire: Routledge, 2013.

Waller, N. G. (2020). *fungible: Psychometric functions from the Waller Lab.* R package, version 1.95.4.8.

Walsh, B. D. (1996). A note on factors that attenuate the correlation coefficient and its analogs. Pp. 21–31 in Thompson, B., ed., *Advances in social science methodology,* Volume 4. Greenwich, CT: JAI Press.

Warner, R. M. (2007). *Applied statistics: From bivariate through multivariate techniques.* Thousand Oaks, CA: Sage.

Watkins, M. W. (2020). *A step by step guide to exploratory factor analysis with R and RStudio.* Milton Park, Oxfordshire: Taylor & Francis.

Wheaton, B., Muthén, B., Alwin, D. F., & Summers, G. F. (1977). Assessing reliability and stability in panel models. Pp. 84–136 in Heise, D. R., ed., *Sociological methodology 1977.* San Francisco, CA: Jossey-Bass.

Widaman, Keith F. (2018). On common factor and principal component representations of data: Implications for theory and for confirmatory replications. *Structural Equation Modeling: A Multidisciplinary Journal* 25(6): 829–847.

Widaman, K. F. & Herringer, L. G. (1985). Iterative least squares estimates of communality: Initial estimate need not affect stabilized value. *Psychometrika* 50(4): 469–477.

Wilkinson, L., Blank, G., & Gruber, C. (1996). *Desktop data analysis with SYSTAT.* Upper Saddle River, NJ: Prentice-Hall.

Yao, F., Müller, H.-G., & Wang, J.-L. (2005). Functional data analysis for sparse longitudinal data. *Journal of the American Statistical Association* 100(470): 577–590.

Zoski, K. & Jurs, S. (1996). An objective counterpart to the visual scree test for factor analysis: The standard error scree. *Educational and Psychological Measurement* 56(3): 443–451.

Zwick, W. R. & Velicer, W. F. (1986). Comparison of five rules for determining the number of components to retain. *Psychological Bulletin* 99: 432–442.

Index

Note: *Italic* page numbers refer to figures and page numbers followed by "n" denote endnotes.